The Plant Explorer's Guide to New England

The diversity of New England's vegetation manifests itself even in small spaces. At Notchview Reservation, high in the Massachusetts Berkshires, flowering bunchberry plants (center) grow amid ferns, conifers, mosses, sphagnum mosses, and lichens. —The Trustees of Reservations

The Plant Explorer's Guide to New England

Raymond Wiggers

Mountain Press Publishing Company
Missoula, Montana 1994

All photographs by Raymond Wiggers
unless otherwise noted.

Front cover photograph © 1994 by David Muench

Back cover photograph © 1994 by Steve Mulligan

Maps by Bob Perrier

Short quotation (page v) and poem (page 371) from *The Narrow Road to the
Deep North and Other Travel Sketches* by Matsuo Basho, translated by
Nobuyuki Yuasa (Penguin Classics, 1966) copyright © Nobuyuki Yuasa, 1966.
Reproduced by permission of Penguin Books Ltd.

Library of Congress Cataloging-in-Publication Data

Wiggers, Ray.
    The plant explorer's guide to New England / Raymond Wiggers.
        p. cm.
    Includes bibliographical references (p.   ) and index.
    ISBN 0-87842-306-0 : $18.00
    1. Botany–New England–Guidebooks. 2. Natural history–New England–
Guidebooks. 3. New England–Guidebooks. 4. New England–Tours. I. Title.
QK121.W49 1994   94-23184
508.74–dc20   CIP

Printed in U.S.A. on recycled paper.

Mountain Press Publishing Company
P.O. Box 2399 • Missoula, MT 59801
406-728-1900 • Fax 406-728-1635

For my father,
Raymond P. Wiggers, Sr.,
who taught me the love of the open road

We all did our best to help one another, but since none of us were experienced travellers, we felt uneasy and made mistakes, doing the wrong things at the wrong times. These mistakes, however, provoked frequent laughter and gave us the courage to push on.

—Matsuo Basho, A Visit to Sarashina Village

Index Map of Tours

Numbers mark the starting point of each tour; black lines follow the tour routes.

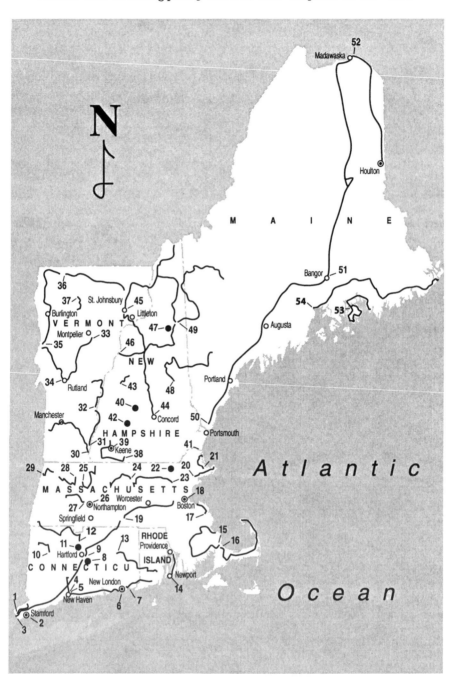

Contents

The Tours

Acknowledgments

The idea for this book first took form in the spring of 1990, when my decision to stop pursuing two simultaneous careers freed enough time for me to travel in New England more comprehensively than before. These far-ranging trips from my home in southwestern New Hampshire did more than teach me something of the region's plant life. They also demonstrated that there are many knowledgeable persons, in many walks of life, in this region, and that they are willing to share their expertise and insights with a stranger. So much for Yankee aloofness. Those mentioned below have done much to improve this book, and certainly nothing to hurt it. Any errors are mine alone.

In my own area of the Granite State, several individuals were especially helpful. My good friend and fellow gardening enthusiast Patricia Clark assisted me on some of my research travels, rallied my flagging hopes, read my tour texts, and kept a close and informed eye on the young street trees of Keene as they grew or succumbed over the course of several seasons. My Alstead neighbors Dale and Lorraine Dustin better acquainted me with a whole range of insights, from the rigors of gardening near the bank of the well-named Cold River to tips on maple sugaring and the uses of various hardwood and softwood trees. Additional woodland and woodlot lore was provided by Rodney Kiblin, of East Deering, and Carroll Warren, of Greenfield. Marge Graves, of the *Keene Sentinel* staff, introduced me to the tasty subject of fern-fiddlehead cuisine; I can recommend her recipes highly. My thanks also go out to several other people at the *Sentinel*, especially Angie Ruest, who helped launch me on my horticultural career in New Hampshire, and to the paper's editor and president, James Rousmaniere, who published my articles on plant care and botany long before my first plant book hit print. Also, my grateful remembrance of the late David Loveland, also of Keene. Dave, originally from northern Massachusetts, worked in the tobacco fields of the Connecticut River valley as a young man; his vivid accounts of that unique form of Yankee agriculture helped me describe the world of Pioneer Valley. And Susan Follansbee, garden director of North Hampton's Fuller Gardens, took the time to chat with me long-distance, when I was in the Sonoran Desert country of southern Arizona, far away from her lovely seaside setting.

In Massachusetts, several officials of The Trustees of Reservations took special pains to provide me with the information I needed. Foremost among them was Lisa McFadden, deputy director for public information. Lisa agreed to meet with me, supplied me with crucial photos, and read part of my text. Her suggestions, those of a talented editor with one eye on the reader's needs, were always pertinent and incisive. By the same token, I thank Larry Simpson, Long Hill Reservation superintendent, for commenting on the tour that includes his domain. At Harvard University's Arnold Arboretum, I was assisted by Richard Schulhof, assistant director for external relations, who kindly read my tour of that facility and alerted me to impending changes at the arboretum. I also profited much from conversations with several staff members at the New England Wild Flower Society's Garden in the Woods. And a very large debt of gratitude is owed to Richard Little, professor of geology at Greenfield Community College and author of two highly recommended general texts on Bay State earth science. Dick agreed to check several of my sections for their geologic veracity. He also made sure that I gave Edward Hitchcock, a renowned nineteenth-century Massachusetts paleontologist and geologist, his proper due.

In Connecticut, Marge Kuhlmann, the public relations and membership official at Yale's Peabody Museum, went to great effort to see that the pertinent tour section was reviewed and approved by science staff. My very special thanks to her. Glenn Dreyer, the Connecticut Arboretum director, promptly reviewed and corrected parts of my text. Gaye Mote, administrative assistant at the Bartlett Arboretum of the University of Connecticut, checked my tour text and, in a very hectic time for her, confirmed my suspicions about the true identity of one blueberry shrub along the swamp boardwalk. And my thanks to the officials in the Cornwall town office who related to me one gloomy spring day the sad details of the storm that had ravaged the lofty Cathedral Pines. In Rhode Island, Blithewold Arboretum horticulturist Julie Morris took the time to give me informative material and the benefit of her own perspective, even though she was busy dealing with the aftermath of a recent hurricane.

One of the most helpful people I met face-to-face was Ray Allen, owner of the Vermont Wildflower Farm in Charlotte, Vermont. I had an extended conversation with him late one summer's afternoon, when his property was at the peak of its glory; later, he supplied me with photos, supporting material, and delightful annotations that clearly revealed the wide range of his knowledge of Vermont ecology and history. Far to the eastward, Barbara Cole, of Bar Harbor, Maine, helped me with the Mount Desert Island–Schoodic Point tour and brought it to the attention of her fellow Wild Gardens of Acadia committee members. One clarification of hers was especially helpful; had she not made it, tour takers on that route would have spent more time scratching their heads in confusion than in seeing the natural wonders.

I've also received substantial assistance from people currently residing in other parts of the country. First, I give my warmest thanks to the Mountain Press staff, who encouraged my efforts and my book's new format, even when it became apparent that my double-spaced manuscript would have to be moved from office to office in a wheelbarrow. I count myself particularly fortunate to have worked with Kathleen Ort, the natural history editor at Mountain Press; without her patience, professionalism, and excellent sense of humor, I would have been hard pressed to complete the project. By the same token, copy editor Jane Taylor did a wonderful job of reviewing the completed manuscript, trimming excess verbiage, and keeping the reader's needs foremost. More than once she kept the wrong things from floating to the surface. My thanks to Montana graphic artist Bob Perrier, who skillfully deciphered my barely legible southpaw descriptions to produce the tour maps, and to Kim Ericsson, of Mountain Press, who artfully designed and laid out this text.

Separately, it was my good fortune during my stint in the National Park Service to have Leila Raim and Catherine Wisniewski as my friends and colleagues. Both had recently been staffers at Connecticut's Dinosaur State Park. Leila made a special effort to garner material from her contacts at the park. On top of that, she gave me the full story of the park's almost miraculous origin, and read and emended my tour copy. Catherine, subsequently assigned to Acadia National Park, reviewed my tour text for that lovely locale. My mother, Irene Mary Wiggers, whose career accomplishments include those of editor and research librarian, unearthed elusive literary references for me. From across the Atlantic, my German cousin Rainer Wiggers provided me with important documents relating to the eighteenth-century botanist Friedrich Heinrich Wiggers. And more recently, several of my new colleagues at the Illinois State Museum have given me a better understanding of various aspects of the plant and earth sciences. My gratitude also to Springfield, Illinois, artist Karen Lommler, who provided one illustration for this book as well as helpful background material.

Perhaps just as important, I should thank all the helpful people whose names I never learned. They ranged from patient and personable state government librarians and archivists to park naturalists, and from gas station attendants who helped me get back on the right track after I'd missed a turn, to one particular federal enforcement officer in a seacoast wildlife refuge. He began by trying to apprehend me for suspicious behavior on the edge of a salt pan. When he realized I was not trying to kidnap a nesting pair of piping plovers, but was instead meditating on the growth habit of cord grass, an hour-long conversation about plant ecology ensued—all while the flashing lights of his cruiser created a gapers' block of gawking birders. These individuals have given me many lifetimes' worth of memories. It has been a privilege to learn from them.

Introduction

On the beautiful summer evening of August 5, 1851, a little-known American writer opened his journal and began to write his daily entry. After noting that the moon was half full, he described the fading colors of the western horizon, then turned his attention to things closer at hand:

> It is 8 o'clock. The farmer has driven in his cows, and is cutting an armful of green corn fodder for them. Another is still patching the roof of his barn, making his hammer heard afar in the twilight, as if he took a satisfaction in his elevated work,—sitting astride the ridge,—which he wished to prolong. The robin utters a sort of cackling note, as if he had learned the ways of man. The air is still. I hear the voices of loud-talking boys in the early twilight, it must be a mile off. The swallows go over with a watery twittering.

The writer was Henry David Thoreau, and his vantage point was his native Concord, Massachusetts. Here he portrayed a vision of the New England landscape that still appeals powerfully to the American sense of nostalgia. His words paint a reassuring picture of human industry and order, set against a backdrop of singing birds and nature reduced to pastoralism.

Thoreau's vision of New England and nature was not confined to dreamy sunsets or pleasant vignettes. As one who traveled extensively in the region, he was quick to point out its staggering diversity. On the upper reaches of Mount Katahdin, on the windswept beaches of Cape Cod, and especially within the more modest bounds of his hometown, Thoreau encountered numberless natural dramas, some edifying, some unsettling, which he duly recorded in thousands of separate observations in his journal. And nothing fascinated him more than the abundant vegetative life around him. He understood its predominant role in defining the New England landscape, from the great white pines that towered over inland Maine to the least obvious lichens and liverworts that added their subtle tints and textures to the rock-strewn surface.

———————— ❧ ————————

It is true that in many ways the New England of today bears little resemblance to the New England of Thoreau's time. Freeways and industrial parks replace the old dirt roads and family farms, and blur the outlines of traditional village settings. Tangles of second-growth trees have claimed fields once painstakingly cleared for crops or grazing.

Diversity persists, however, even if its components have changed and shifted places. Few other sections of the United States offer so many different things to see over such a short distance. What is true about New England in terms of its geology, history, and architecture is just as true of its plant life—the place abundantly rewards anyone who seeks to learn. All that's needed to unlock this treasure chest is the willingness to do what Thoreau did—to pause, consider, and use the naturalist's eye to find the wonders that lie just below the level of the obvious.

———————— ❧ ————————

1

This book, therefore, is an invitation to explore New England's rich legacy of botany, ecology, and horticulture. It is intended for beachcombers and bog-trotters, gardeners, native-plant fanciers, professional and amateur naturalists, and indeed any inquisitive person. But please understand what the book is not: it is not a comprehensive checklist of New England vegetation. It does not attempt to duplicate current field guides to the region's plants, lichens, and marine algae. This book is a set of itineraries designed to give you a sampling of what the region has to offer. It can be used to good advantage in conjunction with field guides.

Further, this book was never intended to be a thoroughly objective treatise. I've yet to see anyone with a horticultural background who refrains from the dangerous duty of opinion making. I am no exception. In addition, I've taken the liberty of including information from other disciplines—geology, history, architecture—to help place the vegetative world in a broader perspective. I hope this approach enlivens what might otherwise have been a dreary litany of plant names and travel directions. I have leaned heavily on the observations of earlier generations, made by persons ranging from Japanese poets and English mystics to American Indian power brokers, Swedish explorers, and Yankee jacks-of-all-trades. My intention is that the human element will provide the reader with a familiar point of departure. Any account of a tree or a mountain range is helped by the knowledge that some earlier, significant soul marveled at it, too.

While I trust the book has much solid information to impart, I've tried my best to avoid unilluminated technical language wherever possible. Of all the sciences, botany has what is often considered the most daunting vocabulary (though I'd personally give the crown to organic chemistry). Whenever I think such terminology is justified or interesting in itself, I use it and define it. Otherwise it's kept to a minimum. However, if a particular term in the text perplexes you, check the Glossary at the back.

Also, I have used the common English-language names of plants, algae, and lichens whenever possible, despite the inherent bother of doing so. The official taxonomic names are vastly superior for shop talk among the initiated. But it's too much to ask of most people, however bright or well educated, to retain *Metasequoia glyptostroboides* or *Dryopteris spinulosa* as anything more than jangling strings of nonsense syllables. If you'd like to know the scientific names that correspond to the common ones, refer to the list of taxonomic equivalents at the back of the book.

———————————— ∾ ————————————

One last word about the tours. They were selected to provide you with a broad cross section of sights, subject matter, and terrains. A few can be completed in the space of an hour or two; others, if done thoroughly from start to finish, require a full day's drive or even longer. The scenery has always been an important factor in plotting the exact routes, so even casual nature-loving souls not fired with botanic zeal will find the trips worthwhile. Leaf-peepers please note that certain sections are especially good for fall foliage sightseeing. These are indicated at the beginning of the "highlights" blurb shown at the head of each tour.

Don't be disappointed if you fail to find all the plants detailed in each description; and don't be surprised if I've overlooked some fascinating aspect that's apparent to you at once. It is a task outside the bounds of a single human life to see all the sites described here, in every season, even over the course of several years. A book is always an exercise in selective consciousness. The most important purpose of these tours is to provide a framework, which you can modify as your needs and interests dictate. Once again, don't forget to take along good field guides to assist you in identifying the flora. If you need help finding the right ones, consult the Bibliography. In addition, read the Mounting Your Expedition section, which precedes Tour 1.

Have a terrific time exploring. As Thoreau noted in his journal, "Our life should be so active and progressive as to be a journey." And as he demonstrated so well, New England is the perfect place to make that journey.

The New England of today: forest, meadow, and wetland along the Housatonic River, at Bartholomew's Cobble, Massachusetts. The earliest ancestors of these plants first colonized this region almost a half billion years ago. –The Trustees of Reservations

Stone fences, a familiar New England sight, are testament to the massive clearing and farming of the New England landscape in the colonial and Federal eras. –The Trustees of Reservations

A Brief Botanical History of New England

∾

T he classic Yankee attitude toward anything unknown is a watchful reserve: don't commit yourself until you know what you're dealing with. This isn't the worst approach toward New England's natural world. Appearances definitely can be deceiving. When you take a stroll along a woodland path in Connecticut, or wander down a Cape Cod beach, or drive through a New Hampshire mountain notch, there is an almost overwhelming sense of the primeval. The world seems older here than it does out west, where earthquakes still rumble and geysers and volcanoes erupt. The perception is aided and abetted by New England's human environment, too. In each of the region's six states, the earlier centuries live on, in chaste upland towns and preserved or restored seaport settlements. By American standards at least, New England glows with the essence of venerableness.

For these reasons, many people are surprised to discover that New England's modern appearance is vastly different from the way things were in the colonial, Revolutionary, or Civil War eras. The truth is that if there's one constant factor in the natural world here, it's change. And not all the change has been due to human intervention. Hundreds of millions of years ago, long before our species arrived on the scene, the ebb and flow of habitats and plant types was well under way. However dramatic the human impact has been here, the greatest transformations to date have been the work of the earth itself.

ERA	PERIOD	EVOLUTION OF THE EARTH	EVOLUTION OF PLANTS	MILLIONS OF YEARS AGO
Cenozoic "The Age of Mammals"	Neogene	Ice Age, beginning ca. 2 million years ago; part of worldwide cooling	Spread of grasses & other herbaceous plants; conifers dominate cooler regions	24
	Paleogene	New England climate similar to that of modern Florida	Flowering plants continue to diversify	65
Mesozoic "The Age of Reptiles"	Cretaceous	New England is a low, gently rolling landscape	The rise of flowering plants	144
	Jurassic	New England and the rest of North America break away from Africa and Europe	Conifers, ginkgos, cycads, cycadeoids, & other naked-seed plants dominant	208
	Triassic	Pangaea still intact; earliest dinosaurs		245
Paleozoic	Permian	The supercontinent of Pangaea; Alleghenian mountains rise	Naked-seed plants become predominant	285
	Carboniferous	Extensive tropical coal swamps in North America and Europe	Both free-sporing & early naked-seed plants	360
	Devonian	The minicontinent of Avalonia collides with New England; Acadian mountains rise	The first trees; free-sporing club moss & horsetail ancestors	408
	Silurian	Taconian mountains eroded away	The rise of vascular plants (early free-sporing forms)	438
	Ordovician	Bronson Hill island arc collides with western New England; Taconian mountains rise	First indication of plants on land	505
	Cambrian	A shallow sea covers New England	First evidence of red algae	570
Precambrian Time		Ancient North America (including Scotland) breaks away from the main part of Europe	Multicellular organisms appear	700
			Primitive one-celled green algae	1000 (1 billion)
			First evidence of organisms with advanced cell structure	1400 (1.4 billion)
			First evidence of stromatolites and other primitive organisms	3500 (3.5 billion)
		Formation of earth's crust		4600 (4.6 billion)

Much of the region is underlain by tough igneous or metamorphic rocks, which bear testament to New England's tumultuous past. This crystal-studded rock, called Kinsman Quartz Monzonite, dates from the formation of the Acadian mountains, roughly 400 million years ago.

A Paleobotanical Background

In the many decades that have passed since their founding, the sciences of geology and paleobotany have proposed intriguing theories and provided fascinating facts about New England's prehistory. Paleobotanists, the specialists who study the evolution of plants through geologic time, have less than ideal conditions in New England, because much of the region's bedrock is composed of metamorphic or igneous units that do not contain ancient plant remains. To some extent, however, their investigations in neighboring fossil-rich regions have helped. By the close of the twentieth century, it is possible to piece together a general picture of New England's plant life through earth history.

Precambrian Time (4.6 billion to 570 million years ago)

Our planet's earliest organisms appeared very far back in earth history, not later than 3.5 billion years ago. In other words, they were on the scene approximately a billion years after the formation of the earth's crust. These rudimentary, bacteria-like creatures were prokaryotic, which means they lacked a high level of cellular organization. Their kind included photosynthetic forms, the earliest precursors of plants. Curiously, evolution seems to have been on the back burner for a lengthy stretch afterward. Not until about 1.4 billion years ago—when almost half of all earth history had passed—did eukaryotic organisms, the more advanced type, enter the record.

At about the same time that one-celled eukaryotic creatures were coming into their own, another important phenomenon was occurring. Plate tectonics, the process whereby sections of the earth's crust move relative to one another, had begun. The first known mountain-building event in the greater Northeast region took place when an ancestral part of northern Europe collided with what then passed for North America. The dance of the continents had begun. Some time

7

thereafter, perhaps by 700 million years before present, multicellular life evolved and began its explosive expansion through the oceans' myriad habitats. As yet, no organisms had permanently established themselves on the dry land. Seawater provided an excellent medium for all sorts of life forms, but the bare ground—rocky, soilless, and bathed in deadly ultraviolet light—remained uncolonized. In the sea and coastal shallows, there no doubt were many photosynthetic organisms, including the algae that built the impressive stromatolites preserved as fossils in the strata of upstate New York. Still, there were no true plants. Theirs would be the last of the five great kingdoms of life to appear.

The Paleozoic Era (570 to 245 million years ago)

At some point close to the dividing line between the Precambrian and the Paleozoic, Laurentia (the early version of North America that apparently also contained Scotland) split free from Baltica, the forerunner of northern Europe. In those days, a shallow sea covered the western New England area, as the Cambrian period limestones there testify. The next period, the Ordovician, has provided the first microscopic evidence of real plant life, found in such nearby locations as Nova Scotia and Pennsylvania. It is likely that this first plant community was composed of humble prostrate forms, perhaps resembling modern liverworts. At the same time, a new chain of mountains arose where the Green Mountains and Berkshires stand today. This was the Taconian mountain-building phase. It owed its origin to the approach and collision of an arc of volcanic islands now known as the Bronson Hill Zone.

The region then received a respite. It was in the next, calmer period, the Silurian, that the first plant-fossil type visible without a microscope, dubbed *Cooksonia*, appeared on the scene some 420 million years ago, in sites ranging from New York to Russia. In one sense, *Cooksonia* was quite advanced. It exhibited a branching pattern and was a vascular plant; that is, it had well-developed internal water-conducting tissue. Still, it lacked true leaves and roots and stood only about 2 inches tall. Paleobotanists have suggested that this and the even earlier Ordovician plant forms arose from one or more relatively complex forms of the green algae. Experts have also speculated that the Ordovician and Silurian periods marked the first time the ultraviolet-blocking ozone layer was sufficiently thick to permit the plants' development in the open air.

To survive in their new environment, these novel organisms had to deal with several unique challenges that did not affect their aquatic precursors. They had to resist desiccation, they had to develop a method of anchoring themselves to the substrate, and they had to devise a way of obtaining carbon dioxide directly

A generalized panorama of plant and animal evolution, beginning on the right with the middle of the Paleozoic era, and progressing leftward (and forward in time) to the end of the Mesozoic era. —The Age of Reptiles, a mural by Rudolph F. Zallinger. Copyright 1966, 1975, 1985, and 1989 by the Peabody Museum of Natural History, Yale University, New Haven, Connecticut

Humble but ancient survivors, these club mosses keep a low profile among the forest's leaf litter on Quaggy Joe Mountain, in northern Maine. They are the descendants of a line of plants that dominated the great Carboniferous period coal swamps.

from the atmosphere and photosynthesizing effectively. Before long, plants had come up with some important adaptations: a protective cuticle layer that retarded water loss; stomata, or minute breathing pores for gas exchange; the ability to synthesize large inert molecules; and lignin for structural support and cutin for the cuticle.

The succeeding period, the Devonian, was a boom time for the plant kingdom. Fossil-bearing rocks reveal that this was when the world's first trees developed. In addition, two important plant groups, the lycopsids and the sphenopsids, were present. Today, their ancient family lines are carried on in New England woods and wetlands by the club mosses and the horsetails. Another important Devonian development was the creation of mycorrhizae, the symbiotic relationship between plant roots and fungi that allows vegetation to extract nutrients, even in poor soils. And geologically speaking, things were getting complicated again. Yet another mountain-building episode, the Acadian, was under way—the result of New England's collision with the minicontinent Avalonia.

In time the lofty heights of the Acadian mountains gave way to the low-lying tropical coal swamps of the Carboniferous period. Because of continued continental drift, New England was at or very near the equator. Impressive lycopsid and sphenopsid trees dominated the flora (community of plants) of this steamy, heavily vegetated world; but there were some important newcomers as well. Among these were true ferns, and gymnosperms, the first seed plants. Before

9

Giant female cones on a prickly cycad in the Lyman Plant Houses of Smith College. Now extant in relatively few and scattered populations across the globe, cycads were nevertheless an important component of the plant life of the Mesozoic era.

the advent of the gymnosperms, all plants reproduced by the cumbersome method of spore dispersal. While most fossils of the period are found farther to the west, there are some Carboniferous formations in Rhode Island.

The earth suffered a dramatic change during the next period, the Permian. Lush conditions were replaced by drier, colder regimes, and glaciers appeared in some locations. One by one, the wandering land masses had combined into Pangaea, a supercontinent that extended from roughly one pole to the other. As Europe and Africa plowed into maritime Canada and New England, the last great eastern mountain range, the Alleghenian, was formed. In these more trying times, the lycopsids and sphenopsids largely gave way to the gymnosperms, just as the amphibians were yielding the dance floor to the reptiles. The Permian ended with what apparently was the largest mass extinction in the whole history of life. According to some accounts, 96 percent of all animal species disappeared from the record. In actuality, this is the theoretical upper limit. (Nothing makes a paleobotanist smile more smugly than animal-centered scientists who use such extinction events as the benchmarks of earth history. For whatever reason, the plant kingdom is largely immune to such dramatic turnovers in species. New plant types definitely do overtake the old, but the process is usually a gradual affair. After receiving such a pummeling from climatic change, asteroids, or whatever else, perhaps the animals should ask their green friends to teach them some survival skills.)

The Mesozoic Era (245 to 65 million years ago)

Woe unto the third-grader who doesn't know that the Mesozoic was the Age of Dinosaurs. During the first period, the Triassic, New England was snugly locked in the heart of Pangaea. But by the onset of the second period, the Jurassic (about 200 million years ago), the supercontinent was ripping apart. Our region once more fronted an ocean, this time the fledgling Atlantic. Throughout this span of time, gymnosperms continued to dominate. At their forefront were four main plant types: conifers (in this case, the cone-bearing ancestors of modern pines, spruces, and redwoods); ginkgophytes (one species survives as the famous ginkgo "living fossil"); cycads (a few are still with us as the sago palm and related species); and the now extinct cycadeoids. Remains of some of these plants, not to mention hundreds of dinosaur footprints, have been preserved in the sedimentary rocks of the Central Lowland in Massachusetts and Connecticut.

If the saga of plant evolution contains little in the way of mass extinctions, it makes up for it in mass arrivals. In the final Mesozoic period, the Cretaceous, a new major group known as the angiosperms burst upon the scene. More

An ornamental horse chestnut tree in full bloom in New Haven's Lighthouse Point Park. Flowering plants are now the dominant forms of plant life in the temperate and tropical portions of the earth.

commonly referred to as the flowering plants, these quickly rose to prominence in two different forms, dicots and monocots. Fanciful Hollywood reconstructions notwithstanding, the world of *Tyrannosaurus* and *Triceratops* was dominated not by outlandish, primitive backdrops but by laurels, dogwoods, and true palms. Perhaps the great reptiles even paused from time to time to sniff the sweet fragrance of the magnolias. While New England was no doubt sharing in this floral revolution, its geologic setting was rather humdrum. Now relegated to the trailing edge of a westward-moving continent, its terrain was a low, eroded surface without much in the way of substantial hills.

The Cenozoic Era (65 million years ago to the present)

In the first part of the Cenozoic, the Paleogene period, New England looked much the way it had in the Cretaceous. Whatever it really was that did the dinosaurs in at the end of the Mesozoic, it had little lasting effect on the angiosperms. But that success didn't mean that their evolutionary story had come full circle. Many other developments were under way. Up to the latter part of the Paleogene, most plants had been woody—hence they were trees, shrubs, or scrambling vines. At this point, however, the herbaceous or nonwoody types

Snow mantles the boughs of fir and spruce trees at Quoddy Head State Park, Maine, a gentle aftertaste of the icy conditions that beset the region in the recent geologic past.

expanded greatly. Chief among them was the monocot grass family. It invaded many preexisting habitats and created several more, including the savanna and the prairie.

The second of the two Cenozoic periods, the Neogene, is our own. It has been marked by an overall cooling of the earth's climate and, in recent times, by repeated Ice Age glacial advances. The most successfully enduring gymnosperms, the conifers, have adapted well to this situation, and have dominated vast tracts of the more frigid parts of the Northern Hemisphere. In New England, meanwhile, the surface of the land was lifted up several thousand feet about twenty to twenty-five million years ago. As a result, the rivers cut more deeply than they had previously, and new mountains—the Whites, the Greens, the Berkshires, and the Taconics—emerged. Unlike their greater Taconian and Acadian predecessors, they were not the direct product of plate tectonics, but of the much simpler process of uplift and subsequent erosion.

When the glaciers descended on the New England area a little over one and a half million years ago, no plants could stand in their path. Grinding sheets of frozen water formed a vast ice desert similar to those of Greenland and Antarctica today. At their deepest, the glaciers even overtopped the summit of mighty Mount Washington, well over 6,000 feet above the nearby seacoast. Whenever the ice retreated for a while, plants began to recolonize the barren ground. The latest interglacial time is the one we know the best, because it contains the entire span of human civilization (though our species has its origin a bit further back). When the latest mass of continental ice departed New England, a succession of plant communities followed in its rock-strewn wake. First came the tough tundra plants—prostrate shrubs and herbaceous species adapted to the coldest locales. Next, and perhaps sharing the landscape with the smaller plants in a patchwork pattern, were the forests of white spruce. They in turn were eventually followed by pines and such broad-leaved trees as the birches, maples, and beeches. In southern New England, conditions moderated enough to allow another group of plants, including oaks, chestnuts, and hickories, to predominate. This was the setting the first New Englanders, the Indians, knew so well.

THE AGE OF *HOMO SAPIENS*

The entire span of English-speaking culture in New England is contained in four hundred years. American Indian peoples have known the region for a period about twenty-five times as long. When they first arrived, in the early postglacial phase, they were nomadic hunters. Over the course of the centuries, however, these people, now referred to as the Algonquins, developed a distinctive form of agriculture, especially in southern New England. This involved the use of a wide array of Western Hemisphere plants, ranging from the all-important maize to beans, pumpkins, watermelons, squashes, and tobacco. In direct contrast to European custom, men did not have a major role in crop raising; instead, they concentrated on hunting. It was the women who were the professional farmers. According to one estimate, female know-how and muscle power provided three-quarters of a community's nourishment.

A reconstruction of a traditional Wampanoag structure at Hobbamock's Homesite, Plimoth Plantation, Massachusetts. The region's first human inhabitants were particularly adept at using plant materials. Here, note the framework made of young tree stems, and the cottonwood-bark siding.

Various stereotypes persist about precolonial Indian life. One of the most enduring portrays the native as an untainted child of the wilderness who wisely refrained from altering the finely tuned ecological balance of the forest primeval. This particular image appealed to Europeans and white Americans brought up on the Old Testament account of the Garden of Eden. More recently, it has been given a new lease on life by New Age devotees, who have the rather dogmatic notion that low-tech cultures are intrinsically more astute, environmentally and otherwise.

The Indians of the Northeast effected significant change on their environment, within the limits of their own technology. When the first Europeans arrived on these shores, they found not untouched forest but a carefully tended parklike setting, with tall overstory trees and little undergrowth. This is not a natural state of affairs. From the coast of southwestern Connecticut to the lower shoreline of Maine, the Indians had used controlled burning extensively. Their reasons for doing so were several: the fire removed the dense brush that impeded crop plantings, secluded game animals, and harbored noxious insects and other vermin. In some cases, the burning got out of hand. For example, the forests of the

Narragansett Bay and Boston areas had been quite thoroughly destroyed by the time the first whites arrived. One can only speculate what changes the Indians would have wrought had they possessed cattle, horses, plows, the wheel, and iron implements. The story of that technology is reserved for the European invasion.

The first century of sustained English settlement, from 1600 to 1700, is a chronicle of some initial privation followed by widespread success. This is not so much an indication of native docility or superior Old World firepower; rather, it is the legacy of largely unintended genocide. Before the earliest days of the Plymouth Colony, visiting European voyagers had unwittingly introduced the Indians to diseases for which they had little or no immunity. Epidemics reduced the native population by as much as 90 percent. The famous Pilgrim-aiding Squanto, for instance, was the sole survivor of his entire village. As Daniel Denton wrote in the latter half of the seventeenth century, "It hath been generally observed that where the English come to settle, a divine hand makes way for them by removing or cutting off the Indians, either by wars one with another or by some raging mortal disease." And, as Denton here points out, illness was not the only factor favoring white expansion. The local Indians were also suffering from numerous internecine conflicts, due at least in part to the recent arrival of hostile tribes from the west.

A Trustees of Reservations crew member conducts a controlled burning at the Albert F. Norris Reservation, in Norwell, Massachusetts. This time-honored technique—used by pre-contact Indians as well as by modern parks personnel—helps keep woodland undergrowth in check.

The European colonization of New England resulted in the introduction of many Old World plant species. Here a tall ornamental Norway spruce graces the town green of Woodstock, Vermont.

One dramatic result of the swift disappearance of Algonquin culture was the return of thicket growth throughout the region. The English settlers took advantage of as much of the previously fire-cleared land as they could; but some inland sections soon became overgrown and almost impenetrable. In the long run, though, this wild vegetation gave way to the dictates of European-based agriculture. English colonists were willing to adopt selected Indian planting practices at first, but the Old World techniques of total woodland removal and row tillage became more thoroughly entrenched with each passing generation. Before long, New England had become the colonial breadbasket, the center of production for those imported cereals, wheat and rye.

New England's native vegetation reeled from this onslaught of Yankee agriculture. Estimates vary widely, but at the peak of the land-clearing mania, in the early or middle 1800s, roughly 70 percent of the region was deforested, with southern sections above that average, and northern areas somewhat below it. As amazing as it seems, practically no first-growth timber was left standing. There is little doubt that in this era the individual farmer accounted for more woodland destruction than did loggers. In fact, some attempts were made to limit the loss of woodlands, but the needs of the domestic fireplace, coupled with the economic incentives of many lumber-based industries, encouraged the landowner to cut down as much as possible. And another substantial transformation took place as well. The Europeans, willingly or not, brought with them aggressive weed species—plantains, the dandelion, chicory, Queen Anne's lace, red and white clovers, and a host of others. On top of that, one of the most invasive trees in modern New England, the black locust, was introduced from the Appalachians to the south.

One especially important eyewitness to the culmination of the era of deforestation was Henry David Thoreau. In the popular imagination, he was a philosophic hermit who wandered endlessly through vast forests. In actuality, the New England of his time was not much different from a hilly version of modern Iowa. In his native Concord, more than 80 percent of the land was cleared for cultivation and pasturage—and that is no freak statistic for the region. In his day, one had to go and find selected pieces of insurgent woodland. Today, one has to try fairly hard to escape it.

The latest change in New England's vegetation patterns concerns regional agriculture. Taken in overall terms, it began to decline dramatically beginning in the late nineteenth century—the result of overwhelming competition from Midwestern farms, and from the economic effects of the industrial revolution. Ironically, while the regional population continued to increase, more and more people abandoned the countryside. New tree communities proliferated. Distinctly inferior to the original stands of timber, these second-growth forests have nevertheless been cut repeatedly—not by farmers, who would have kept the land clear afterward, but by logging concerns both large and small. The result is a woodland community that occupies approximately 70 percent of New England's land surface, even though it is locked into a state of perpetual adolescence. As you'll see in the following tours, few wild trees match the size of older ornamental specimens in cities and towns.

Helping to feed an insatiable demand for wood products: a lumber truck in northern Maine. In modern times, New England's forests are kept ever immature by widespread logging.

New England's Living Landscapes

∾

To a botanist fresh from the Amazon jungle, even the most rampant New England thicket or meadow seems at least a little boring. It lacks the rain forest's seemingly endless diversity of species. On the other hand, a specialist in arctic tundra plant life would be impressed by the luxuriant growth and the various overlapping habitats of those same New England locales. These two extremes highlight the fundamental character of the region: it is a place between, a land of transitions. A single day's drive can take you from a forest with close family ties to Pennsylvania and Virginia to a treeless zone resembling northern Labrador. Of the 250,000 extant plant species on the globe, New England can claim about 2,000—not counting the additional roster of introduced plants. According to the New England Wild Flower Society, some 480 of these are rare or endangered.

The grand determinant controlling this and every other plant environment is climate. The same forces that make New England white and icy in January and gray and muddy in April are responsible for the vegetation growing everywhere it possibly can. As Mark Twain noted:

> There is a sumptuous variety about the New England weather that compels the stranger's admiration—and regret. The weather is always doing something there; always attending strictly to business; always getting up new designs and trying them on people to see how they will go. . . . [There is] weather to spare; weather to hire out, weather to sell; weather to deposit; weather to invest; weather to give to the poor.

Obviously, the plants of the region are capable of enduring a prolonged winter with frequent bouts of freezing temperatures. What they need, however, is abun-

Tundra plant life. White-flowered mountain sandwort shares the wind-blasted summit of Vermont's Mount Mansfield with clumps of grasslike Bigelow's sedge.

dant water during the growing season, when they must photosynthesize, create food reserves, grow new tissues, contend with pests and diseases, and reproduce. Fortunately, New England's geographic position and other factors conspire to provide water seemingly without end. Throughout the region, annual rainfall averages 30 or even 40 inches a year—a *yard* of precipitation across the span of twelve months.

If climate is the paramount consideration, then certainly the region's underlying geological character is next in line. In this area, the dominant rock types do not contain lime; the only significant exceptions are the eastern section of Maine's Aroostook County and the lowlands of far western New England. Without the sweetening influence of lime, water percolates through the topsoil and leaches away any basic (i.e., alkaline) compounds present. As a result, the soil becomes acidic, especially where the leaf litter is composed mainly of conifer needles. Acidity can interfere with a plant's ability to take up moisture and nutrients, even if those components are in plentiful supply. For that reason, many New England species have developed symbiotic relationships with special fungi or bacteria that help overcome the problem. Some botanists now suggest that these partnerships have played a crucial role in plant evolution.

NEW ENGLAND'S PLANT REGIMES

As I suggested in the Introduction, New England contains many different floral habitats, or plant regimes, in a relatively compact setting. As a matter of fact, the entire six-state region takes up only a little more than half the land area of a single western state, Arizona. Over the years, geographers and ecologists have presented many different ways of describing its formidable ecological diversity. The following list of plant regimes is my own interpretation, though it's based on many others. It definitely does not represent any hard-and-fast blueprint of nature's true will. Rather, each regime is an idealized concept, and is merely the first step in understanding the real vegetation of any locale. Bear in mind that these categories usually overlap. If they didn't, the system of life would lack the complexity that is the hallmark of its amazing resiliency.

We begin in New England's highest reaches, and then work our way downward:

1 *Alpine Tundra.* In terms of area, this is by far the smallest regime. It is found only on the region's highest or most exposed peaks, where the weather is severe enough to simulate the conditions of northern Canada. This is a frigid, windswept zone without trees. Its plants are prostrate, dwarf shrubs and highly adapted herbaceous species. Several forms of lichen are also characteristic. These isolated groups are thought to be remnants of a much more widespread population that once lived at the edge of the Ice Age glaciers. Such rare and extremely fragile "island communities" are nothing less than windows onto New England's pre-history. This book describes visits to three alpine tundra communities: Vermont's Mount Mansfield (Tour 37), and New Hampshire's Mount Cardigan (Tour 43) and Mount Washington (Tour 47). Selected alpine species are also found in other coastal and northern tours.

2 *Taiga.* Of all the New England plant regimes, this one has the largest number of alternative names. They include the Spruce-Fir zone, the North Woods zone,

Spruce species and balsam fir dominate this taiga setting near the bank of the Magalloway River, near Errol, New Hampshire. The relative youth of modern taiga trees in northernmost New England is one sign of relentless logging.

the Boreal zone, and the Northern Evergreen zone. I prefer the Russian term *taiga* because it's short, evocative, and as accurate as any other. In North America, the best example of the taiga is the great belt of spruce and fir trees that lies just below the tundra in Canada. This belt does not extend contiguously southward into lowland areas of New England in anything like a pure form. However, it does mix with the next zone, the Northern Hardwoods, in Vermont's Northeast Kingdom, in northernmost New Hampshire, and in parts of inland Maine. Still, it does exist in a rather dominant fashion along Maine's wet and cold Down East coast, on higher mountainsides, and in a few other locations, such as cool lakesides and stream banks. The fir tree of the taiga regime in our region is always the balsam fir, but the spruce species vary. In northern Canada and on the northeastern Maine seacoast, the white spruce is predominant; elsewhere in New England, red spruce is king. Most of the taiga areas covered in the book are in red spruce territory, but in Tours 53 and 54, you should spot the more elusive white spruce, too. The montane version of the taiga regime, often referred to in this book, consists of red spruce and balsam fir that live below treeline on the mountain heights. The latter species is somewhat more cold-hardy, and it often grows at higher altitudes than the former.

3 *Northern Hardwoods.* A glorious zone indeed, this is the northernmost regime in which broad-leaved trees predominate. It is defined by the presence of sugar maple, American beech, and yellow birch. Much of northern New England and the Massachusetts Berkshires belong to this community. Generally, it is optimum territory for fall foliage color, thanks to the flame-orange tones of the sugar maple, which often mix with the scarlets and burgundies of the associated red maple. Unlike the evergreen firs and spruces of the taiga, the main tree species of the northern hardwoods regime are deciduous—they drop their leaves in the autumn. For that reason, the forests in this zone look much barer in winter and provide

less shelter for wildlife. While taiga soil is generally nutrient starved, the Northern Hardwoods provide a much better growing medium for a host of understory shrubs and herbaceous plants. Some of the nonwoody species are spring ephemerals. They unfurl their leaves and flowers in early spring, to take advantage of the higher light level that exists before the foliage of the overstory trees fills in. The ephemerals usually die back by midsummer, their short life cycle already complete. In contrast, other plants such as the ferns and club mosses persist in the twilit gloom of the forest floor throughout the warm season, in a state of semidormancy. They have developed the capability to conduct most of their important life processes before and after the trees are mantled in foliage. Two important conifers often encountered in both this regime and the next are the mighty white pine and the Canadian hemlock. In some locations, they are as numerous as the hardwoods.

4 *Central Hardwoods.* This regime is also typified by deciduous trees—in this case, by oaks (especially the red and white species) and hickories (including the shagbark, mockernut, bitternut, and pignut). Before the advent of a devastating blight disease, the magnificent American chestnut tree was another indicator. Now, it is virtually obliterated, except for its persistent but largely foredoomed stump sprouts. The Central Hardwoods zone is common throughout the southern half of New England and the seacoast sections of New Hampshire and southern

Many wildflowers of the northern and central hardwoods regimes grow and flower in spring, before light levels on the forest floor begin to drop. One of the loveliest of all spring ephemerals is the showy orchis. —The Trustees of Reservations, S. Waldo Bailey

Freshwater wetlands are a common sight in New England. Here, at Vermont's Missisquoi National Wildlife Refuge, red maples and other swamp tress grow luxuriantly in the waterlogged soil.

Maine. It too is blessed with a relative wealth of wildflowers, pteridophytes (free-sporing vascular plants such as ferns, club mosses, and horsetails), and understory shrubs (including the magnificent mountain laurel and rosebay rhododendron). The Canadian hemlock and white pine are often two more important players in this zone.

5 *Pine Barrens.* This is a composite classification including both inland and seacoast locations. The main determinant of this zone is its sandy soil. In coastal sites,

A small but powerful competitor, sphagnum moss is a dominant plant type in the peatland along the Atlantic White Cedar Trail of Cape Cod National Seashore. Note the highbush blueberry leaf resting on the sphagnum mat.

the sand is largely deposited by ocean currents and wind; farther inland, the deposits are usually the work of river systems or glacial meltwater streams. New England's two best inland sites, the Concord and Ossipee Barrens, both lie in New Hampshire. By far the best seacoast locale is Cape Cod. Barrens communities are defined by pitch pine, scrub oak, black oak, and sometimes also by a variety of rare shrubs and wildflowers having affinities with the South.

6 *Freshwater Wetlands.* This zone contains several distinct and important habitats. In general, wetlands are areas where the water table is at or near the surface throughout the year. River bottomlands are fairly self-explanatory; they are often populated by flood-tolerant trees and shrubs such as silver maple, box elder, willows, and the American elm. In effect, swamps are forests with wet feet. Their indicator species include red maple, black ash, tupelo, buttonbush, clethra, poison sumac, speckled alder, and skunk cabbage. Marshes, on the other hand, are largely open and free of substantial woody plant growth. Their characteristic herbaceous plants are tussock sedge, pickerel weed, arrow arum, common cattail, and one aggressive pest from Europe, purple loosestrife. Peatlands, a wonderful world unto themselves, are wetlands with organic soils. In many places in New England, the organic content is mainly due to the presence of a mat-forming bryophyte, or nonvascular plant, called sphagnum moss. Ecologists have split peatlands into two basic types: fens, where there is at least some oxygenating surface water flow, and bogs, which have rainfall as their sole source of water. While fens are sufficiently acidic to cause them to be colonized by specially adapted species, bogs are the extreme case, and pose the greatest challenge to

The Albert F. Norris Reservation, in Norwell, Massachusetts, is one of the tour sites in this book that features tidal wetland. Here, salt-tolerant herbaceous plants hold sway. –The Trustees of Reservations

25

A sculptor's garden. Residential plantings, such as this lovely setting at the Saint-Gaudens National Historic Site in Cornish, New Hampshire, have been an important part of the region's landscape since colonial times.

plant survival. Among the peatland indicator plants are black spruce, tamarack, Atlantic white cedar, swamp loosestrife, large and small cranberry, selected orchids, cotton grasses, and such carnivorous plants as the sundews, the bladderworts, and the pitcher plant. Also present are a number of broad-leaved shrubs, some of which may also be found near or at the alpine tundra zone.

7 *Tidal Wetlands*. Also known as salt marshes. In New England, they reach from the coast of southwestern Connecticut all the way to the estuaries of Down East Maine. These extremely abundant environments have long been appreciated by Yankee farmers for the salt hay they provide, but in recent times they have been under siege by real estate expansion. Salt marshes are breeding grounds for many organisms, and they produce a greater mass of vegetation than any other natural regime. Their characteristic species include the salt meadow and saltwater cord grasses, marsh elder, salt marsh aster, sea lavender, the glassworts, orache, seaside plantain, and other plants tolerant of saline conditions.

8 *Coastal Dunelands*. This is a fascinating regime that contains several subenvironments existing on or near seaside sand dunes. Plants here are experts in water retention and survival in nutrient-poor conditions. They include bay-berry, beach plum, the introduced rugosa rose, woolly hudsonia, beach grass, beach pea, wild dusty miller, seaside goldenrod, and various lichens. This regime is visited in Tour 5 (Harkness State Park, Connecticut), Tour 16 (Cape Cod), and Tour 21 (Parker River National Wildlife Refuge, Massachusetts).

9 *Intertidal Regime*. The name is accurate, but make sure you distinguish this zone from salt marshes, which also lie between the ocean's high and low tide boundaries. This is the only regime mentioned in the book that is not a plant community.

It is occupied by marine algae, or, in common parlance, the seaweeds. Originally grouped within the plant kingdom, seaweeds are now placed in the separate kingdom Protista. New England seaweeds include members of three different divisions: the red, brown, and green algae. Most often, they are found on rocky coastlines, where they can best attach themselves and resist the powerful action of the surf. One of the seashore's most enchanting subenvironments, the tidal pool, is included in this regime. At low tide, its still water and graceful residents are a sharp contrast to the nearby world of crashing waves.

To the preceding natural regimes I add two created by our own species.

10 *Agricultural and Horticultural Lands*. Though no longer as all-encompassing as it was in colonial and Federal times, this regime still has a powerful effect on New England's character and scenic appeal. It includes the most uplifting of all human works, the orchard, as well as tillage (land devoted to row crops), hayfields, and dairying and other forms of pasturage. While their positive role in the living landscape is often overlooked, commercial plant nurseries, rural garden centers, and even roadside farm stands should be included here. When not encumbered with too many whirligigs, reflecting globes, and ceramic gnome ensembles, they add significantly to the pastoral effect.

11 *Cityscape*. If you have the opportunity to observe the urban areas of the Connecticut River valley from a nearby hilltop, you will see one of the most heartening aspects of modernity: our downtowns are often more treesy than the surrounding countryside. This regime includes neighborhood parks, arboreta, and residential and commercial plantings. It also contains a traditional subenvironment that has been getting renewed emphasis in recent decades—the street tree habitat. For more on that particular aspect of urban greenery, see Tour 39.

The New England cityscape is noted for its parks and commons. Here, residents of Brattleboro, Vermont, enjoy their town green on a sunny day in late April.

Lake Willoughby and the glacially scoured flank of Mount Hor, in Vermont's Northeast Kingdom. Tour trips to remote areas like this require planning, careful attention to the weather forecast, and proper hiking attire—including ooze shoes.

Mounting Your Expedition

Plan ahead. Before you actually launch out on a particular route, check the tour section that describes it. If the blurb at the beginning of the tour sounds appealing, read through the full trip text ahead of time and see if there are any special considerations worth noting. When you do set out, have someone who is not behind the steering wheel keep track of the mileage and directions. In the interest of good morale, it often helps to give this person an impressive title—something like Lord (or Lady) High Navigator, or Right Honorable Narrator of the Roadside Botanical Wonders. If you're alone, don't attempt to read the text while you're driving. Botany should not be overtly terminal. One trick I've learned on solo trips is to record the text of a tour on cassette tape ahead of time, and then play it back as I go.

Call ahead. Pick up the phone and check whether the main stops on your tour are actually open the day you want to visit them. While it's difficult to telephone the windswept summit of New Hampshire's Mount Kearsarge or many of the other wild spots in the book, your itinerary may take you to state, federal, or commercial facilities that have information numbers or offices. If you're venturing out in the cold season, when roads may be slick, or if you're doing one of the Roadside Vegetation Tours along an interstate, you may wish to call the local state police headquarters and ask if there are any dangerous spots or heavy construction you should avoid.

A note on the driving directions in this book. In giving distances to turns and points of interest, I have tried to be accurate to the nearest tenth of a mile, as

A painted trillium in full flower at Rhodendendron State Park, New Hampshire. Consult a good field guide to catch the blooming periods of plants you particularly want to see.

measured by the odometer in my pickup, or by measuring topographic and atlas maps. Nonetheless, my experience is that odometers vary more than they're probably supposed to. If you find the proper turn is located 2.4 miles down the road, instead of the listed 2.7 miles, don't panic; follow your common sense. As a backup, I have also referred to distinctive landmarks whenever possible.

There ain't no such thing as a free park visit (anymore). Many of the sites described in these tours require an entry fee or a suggested donation. Be prepared. It is difficult to generalize, and fruitless to particularize, about the costs incurred on any one tour, since prices seem to vary considerably from year to year. If you want to find out ahead of time, call and check. Incidentally, if you plan to do more than one or two Massachusetts tours involving Trustees of Reservations sites, you may find it in your interest to become a member of their laudable nonprofit organization. In this way, you can benefit a noble cause while you forgo the bother of feeling obligated to pay repeated entry fees. To join The Trustees, write them at 572 Essex Street, Beverly, MA 01915, or call (508) 921-1944.

Never underestimate the weather. This is especially true if you're heading to higher elevations. Call the local weather number, or better yet, get an inexpensive weather-band radio at an electronics specialty store. Not only will you receive the forecast straight from the National Weather Service, you'll also get to hear the announcers' colorful Yankee pronunciations. Resist the temptation to take mountain summit roads on days when clouds are low or thunderstorms are likely. The first time I ascended Mount Washington, it was socked in. I didn't learn much botany or geology or see any nice views, but I did find out what the inside of a Ping-Pong ball looks like.

Note that state and federal parks may have short operating seasons. In recent times, many governmental facilities are on reduced operating schedules due to budget problems and short staffing. With the exception of fall foliage trips and early spring wildflower expeditions, the best time to travel to see all the sights is between Memorial Day and Labor Day. In some cases, you can get into officially closed state parks after their season is over. In other cases, you may find that the walk from the locked entrance gate just isn't worth it. If all else fails, call the parks and recreation headquarters in the appropriate state capital.

If you go hiking during hunting season, do not dress like a whitetail deer. Unfortunately, the imaginative skills of hunters are sometimes heightened by their hip flasks. Many hunters, especially local folks, are skilled and careful outdoorsmen and outdoorswomen, with a discerning eye and a keen love of nature. A few others, though, have mistaken holstein cows, logging trucks, and cruise-altitude Boeing 747s for the buck of their dreams. Avoid clothing yourself in basic tans, browns, whites, or military camouflage. Do not wear headgear that resembles antlers. Do not snort or audibly paw at the ground. Go with bright colors—yellow, red, or better yet, Day-Glo orange. I'm serious.

Take along a pair of ooze shoes. When you hike in the woods, even during a dry spell, the chances are good that you'll come upon soggy or downright submerged ground. Waterproofed hiking boots or rubberized shoes, including the low-cut kind, are vastly preferable to deck shoes or porous sneakers.

Be prepared for coordinated airborne attack. As seventeenth-century naturalist John Josselyn wrote of New England, "The Countrey is strangely incommodated with flyes, which the English call Musketaes. . . . They will sting so fiercely in summer as to make the faces of the English swell'd and scabby, as if the small pox. . . . Likewise there is a small black fly no bigger than a flea, so numerous up in the Countrey, that a man cannot draw his breath, but he will suck of them in: they continue about Thirty dayes say some, but I say three moneths, and are not only a pesterment but a plague to the Countrey." Amen. Few are the northern New England gardeners and orchardists who haven't suffered the springtime martyrdom resulting from black flies. The mosquitoes in their turn are a formidable kind of distraction. On hikes through wet and windless low areas, be prepared for the worst. The insider's insect repellent is a particular brand of skin lotion that will make you smell like the world's oldest business establishment for an indefinite period afterward. I prefer one of the more mainline products, though some people would shy away from it for health or political reasons. I'm no friend of the chemical industry, either, but this is war.

Take special precautions to prevent Lyme disease. This potentially serious affliction is caused by a bacterium carried by the deer tick and, less frequently, by the American dog tick and the lone star tick. When you plan to walk through heavily vegetated areas, such as woods, meadows, or beaches, wear long-sleeved shirts and trousers. Also, tuck the trouser legs into socks, if possible, and don't worry about the resulting fashion statement. At the risk of sounding heretical, I suggest one's health is more important than one's Total Look. Use repellent if you can, preferably the kind formulated specifically for ticks, and follow the application instructions explicitly. When you return home, examine yourself *thoroughly.* With children, pay special attention to their neck and scalp. If you find a tick has attached itself to you, it does not necessarily mean you'll get Lyme disease. Remove the pest carefully (petroleum jelly should smother it and make it pull out of the flesh), then save the tick and wait to see if symptoms develop. These may include a bull's-eye rash around the bite, as well as headache, chills, and pains in the joints or muscles. When in doubt, check with a physician.

Do NOT—I repeat, NOT—try the herbal remedies described in this book. New England's plants, lichens, and algae have a long record of herbal use. I have described some of their applications as an interesting sidelight, but I make no claims whatsoever about their actual efficacy. Please do not attempt to use this book as a guide to self-healing or home medicine. The danger of accidental poisoning is substantial.

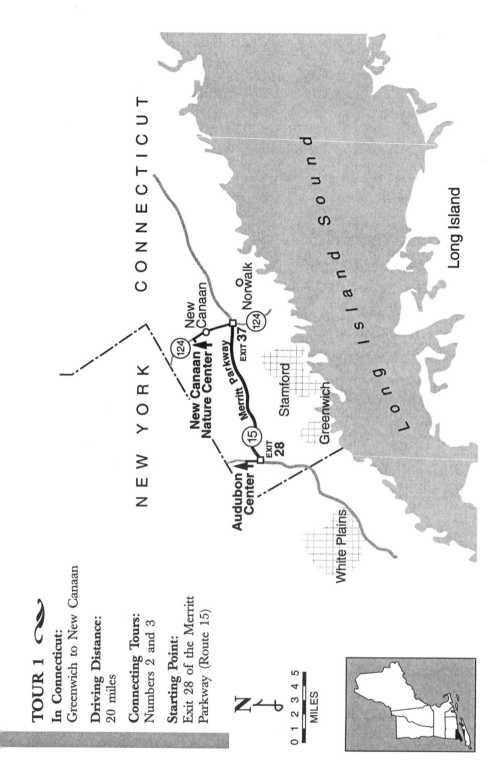

TOUR 1 ❧

In Connecticut:
Greenwich to New Canaan

Driving Distance:
20 miles

Connecting Tours:
Numbers 2 and 3

Starting Point:
Exit 28 of the Merritt
Parkway (Route 15)

New
Canaan

Norwalk

New Canaan
Nature Center

Merritt Parkway

EXIT 37

124

124

Stamford

Greenwich

Audubon
Center

EXIT
28

15

White Plains

Long Island Sound

Long Island

CONNECTICUT

NEW YORK

N

0 1 2 3 4 5
MILES

Tour 1
A Suburban Plant Tour

HIGHLIGHTS

The Audubon Center of Greenwich ∾ Why Flowers Come in
Colors ∾ Aliens Among Us ∾ The World's Smallest Flowering Plant ∾
A Deadly Tree Disease ∾ A Shrub to Ward Off the Devil ∾ A Showplace of
Spring Ephemerals ∾ Dandelion Sandwiches ∾ The New Canaan Nature
Center ∾ A Solar-Powered Greenhouse ∾ Great Trees, Foreign and Domestic

*T*he southwesternmost corner of Connecticut, a region of great historical
interest, is better known for premium suburban real estate and the New
York City–oriented culture than for the traditional New England way
of life. The emphasis on outdoor horticulture is obvious in the fastidiously
maintained lawns and garden beds that adorn the homes of the well-to-do. On
any summer's day, landscaping contractors' trucks almost outnumber the Mercedes
and BMWs. Not all the greenery here serves private ends, though. This lovely
area boasts two public nature centers that demonstrate intriguing aspects of the
world of plants.

From the starting point, leave the Merritt Parkway and proceed to our first
stop, the Audubon Center of Greenwich. Head north from Exit 28 on Round
Hill Road. After 1.4 miles, turn left onto John Street and wind along it for another
1.4 miles to the intersection with Riversville Road. You'll see the Audubon Center
on your right. Make the sharp turn at the entrance road and park in the lot
just across from the building housing the office and gift shop. Don't forget to
get a map of the Nature Center grounds in the gift shop before you set out on
the trails. (If you wish, you can purchase a trail guide booklet that provides
additional information.)

The walk to the Audubon Center's main body of water, Mead Lake, is an
easy and rewarding one. The botanical delights begin almost immediately. Just
down the roadway from the parking lot, past the Interpretive Building, jewelweed,
a wild representative of the well-known genus *Impatiens,* mixes with wild grape

A lone black cherry tree blooms before leaf break, on the shore of the Audubon Center's Mead Lake.

species and Virginia creeper. Another aggressive vine found here, trumpet creeper, produces tubular orange-red flowers that are irresistible to hummingbirds. The characteristic color of these blossoms is not just a random occurrence. Flower coloration is often connected with the type of animal pollinator each species has. Birds, especially hummingbirds and orioles, are attracted to reds and oranges; insects are attracted most often to blues, yellows, and whites. But plants that depend on bats—Arizona's saguaro cactus is one—rely not so much on the blossoms' color as on their strong scent and the fact that they open at night, when bats are active.

Farther along, in the vicinity of Indian Spring Pond, you may chance upon an alien—not a visitor from planet Neptune, for better or worse, but a plant species that originated in another part of the world. In most cases, alien plants were intended only for agricultural applications or for use in the garden. They made good their escape because they found a habitat where they could compete successfully with native wild species. Many aliens gravitate to disturbed ground, where the natives are not already overwhelmingly established. As a consequence, they're most often encountered on roadsides, along paths, in abandoned lots, and at construction sites. The alien of note here is the alluringly named wineberry, a cane-forming raspberry shrub from the Far East. It has prickly stems and leaves divided into three leaflets, dark green above and white below.

If you choose the most direct trail to Mead Lake, pause for a moment at small Vernal Pool. If it isn't dry, check to see if there's a light green film on the water surface. If there is, you'll find that what appears to be an unattractive algal scum is actually duckweed. The tiny members of this floating community belong to a genus that includes the world's smallest flowering plants—extremely simplified yet extremely specialized relatives of the familiar arum family, which includes everything from woodland wildflowers to popular houseplants. As its common name suggests, duckweed is a favorite food for waterfowl; in fact, it multiplies most effectively by being carried on their feet, from one body of water to the next. Research has shown that duckweed also has the remarkable ability to adjust its photosynthetic apparatus to varying levels of light. On cloudy days, the chlorophyll-containing parts of its cells orient themselves horizontally, to increase their efficiency; on sunny days, they move into a vertical plane to avoid getting burned.

Next, the trail winds through woodland drained by the Byram River. One of the most prevalent and eye-catching trees here is a northern hardwood, the American beech. Note its typically smooth, gunmetal gray bark. Several big and especially impressive specimens are within sight of the path. As magnificent a tree as it is, the American beech has been afflicted in some other New England locales with the sinister beech bark disease. First noticed in Nova Scotia in 1920, this malady has slowly spread across most of its host's natural range. It is caused by the joint action of a scale insect and a *Nectria* fungus. Cankers form on infected trunks, and eventually the trees die back to stump sprouts.

Another northern hardwood you'll probably notice is New England's most famous plant, the sugar maple. See if you can also locate white ash, tulip tree, Canadian hemlock, black cherry, and black birch. Black birch is common in the southern half of the region, though its true identity is often not surmised, since it lacks the showy, light-toned bark of its close relatives, the paper and yellow birches.

The shrubs of the forest zone definitely deserve some attention as well. Keep a lookout for two tall-growing natives, witch hazel and sassafras. The latter is a plant with the unusual penchant of producing leaves with three different shapes, featuring two, one, or no lobes. It was highly prized by early English explorers, who believed it had remarkable curative powers. As a matter of fact, sassafras holds the distinction of having been New England's first exported cash crop. So highly were its powers touted that it soon reached into the realm of the metaphysical. It was taken as a treatment to arrest the aging process; it was made into boxes to protect Bibles from the minions of Satan; and its wood was even incorporated into the hulls of seagoing vessels to preserve them from shipwreck.

At one point, Sir Walter Raleigh held a monopoly on the plant's importation into England. Interestingly, the name sassafras was given to this species by the great Spanish botanist Nicolas Monardes, apparently because it reminded him of saxifrage (which, as it turns out, is in another family altogether).

In this area you should also find a native member of the great *Viburnum* genus. It goes by the rather misleading common name of American cranberry bush, which the shrub earned by having fruit vaguely similar to that of the true cranberry. Here and there it is joined by the highly invasive (not to mention downright ratty) Japanese barberry. This thorny species is another example of a woody alien.

Perhaps the best part of the botanical show awaits you along the edge of Mead Lake. Near the dam, the native shrub leatherwood puts in an appearance, as does that ubiquitous water lover, speckled alder. Indians of this region employed leatherwood's strong but supple bark in making ropes and baskets; the colonists used it as a binding material and, as the explorer Pehr Kalm observed, for switches to discipline their children. If you visit in midsummer, you may locate the pale purple, ankle-high blossoms of the sharp-winged monkey flower. Without doubt you'll get a good look at the chaste white blooms of the fragrant water lilies in the lake's shallower reaches. All these species, woody or herbaceous, are angiosperms, or representatives of the gigantic division of flowering plants. Among the angiosperms' many notable traits is one you'll never directly witness on a nature hike. It's called double fertilization. After a flower is pollinated, two male gametes, not just one, survive. One of the gametes fuses with the egg in the accustomed way, while the other helps create the endosperm tissue that provides nourishment for the embryo. If you come across a beechnut or cherry pit as you walk down the trail, take a good look at it. It is an angiosperm seed, containing the tiny plant-to-be, surrounded by the crucial endosperm food supply that keeps the embryo going until the first leaves unfurl and start drinking in life-giving sunlight.

While the lake's summertime flower display is impressive, an earlier peak in the blooming takes place from March through the beginning of May. Then the fringing woods become a virtual museum of spring ephemerals, with skunk cabbage, trilliums, dwarf ginseng, Dutchman's breeches, bloodroot, wild leek, Solomon's seal, May apple, cut-leaved toothwort, and other wildflowers all on display. Practically every species listed here has had its herbal applications. Trilliums, for instance, were used by both Native Americans and English-speaking settlers to check bleeding and to disinfect wounds. The traditional use of the toothwort might seem pretty obvious—it must be a toothache remedy. Alas, not so; the name refers to the toothlike projections on the plant's underground stems.

The second and final destination of this tour is the New Canaan Nature Center. To reach it, return to the starting point at Exit 28 of the Merritt Parkway. Get on the parkway heading east and proceed 10.3 miles to Exit 37. Take that turnoff and head north on Route 124 approximately 3.0 miles, until you see the

Nature Center entrance on your left. The final leg takes you through downtown New Canaan, so follow the Route 124 signs carefully.

On your way, scan the lawns you pass for the one most reviled suburban plant in modern America, the dandelion. This persistent colonizer of turf and disturbed ground was introduced to this continent in early colonial times. John Josselyn noted its presence in New England, in a book written around 1672. One of the best-known species of the world's largest plant family, the composites, the dandelion has managed to survive the onslaught of hand weeders and herbicides for decades. In the nineteenth century, it was actually sought after, both as a salad green and as a good flower plant for honeybees. The indefatigable Mrs. Grieve, author of *A Modern Herbal,* gives this recipe: "Young Dandelion leaves make delicious sandwiches, the tender leaves being laid between slices of bread and sprinkled with salt. The addition of a little lemon-juice and pepper varies the flavour. The leaves should always be torn to pieces, rather than cut, in order to keep the flavour." If for some reason you decide to try this yourself, make sure you choose plants not covered with weed killer.

The moment you enter the grounds of the New Canaan Nature Center, you'll understand that it isn't a mere repetition of what you've already visited in Greenwich. On the access road to the parking lot, you pass an arboretum (an outdoor collection of specimen trees) well worth exploring on foot. And

Cut-leaved toothwort (center), *one of many spring ephemerals that bloom in the woods of the Audubon Center of Greenwich.*

adjacent to the parking lot stands an impressive solar-powered greenhouse. Currently, the greenhouse is used primarily as a teaching facility, and though it doesn't have labeled plant displays, it too is worth investigating, since it contains good examples of indoor foliage plants used in New England homes and offices. While you're at it, swing by the Activities Building across the parking lot, where you can get a map of the grounds and find out more about this important facility's special programs.

To me at least, the high point of this stop is a tree located just beyond (to the west of) the greenhouse. With a little hunting in the vicinity of the brook and the center's wildflower garden, you'll find it. The plant is an excellent example of one of the handsomest ornamentals used in the Northeast. This is the Amur cork tree, a native of eastern Asia with spreading branches and striking tan, furrowed bark. Its species belongs to the rue family, and therefore is a relative of citrus trees and the prickly ash shrub, which grows wild in western Vermont (see Tour 34). Placing it in a larger context, the amur cork is a member of the dicots, one of the two main angiosperm groups. In common with the unrelated pines and other conifers, woody dicots exhibit two different modes of growth.

Both modes are centered on the meristem, the "growth zone" of their tissues. Primary growth is caused by the meristem located at the tips of shoots or roots. Its activity results in the lengthening of those stems. Secondary growth is generated by a separate kind of meristem that causes trunks, branches, and roots to thicken. This beautiful tree illustrates the joint effect of primary and secondary growth: the plant stems will continue to gain both length and girth until death overtakes them. This is not generally the case with the other great angiosperm group, the monocots. A good example of a monocot tree is the palm. It experiences only primary growth; thus, the trunk stays narrow while it keeps getting taller and taller. This gives the palm its characteristically slender, graceful aspect.

Not far from the Amur cork tree, on the edge of the woods by the wildflower section, you can find good examples of a Connecticut native, the noble tulip tree. Note its strange leaves, lobed yet flat tipped. They resemble those of no other North American species. Unlike the Amur cork, this lofty magnolia relative produces a straight, uninterrupted, columnar trunk that soars skyward as the plant reaches maturity. Consider for a moment how this massive organism can manage to bear such an immense weight vertically. Plants do not have the internal supporting skeletal structure that vertebrate animals do. Instead, they have developed lignin, a polymer that strengthens their cell walls. This stiffening agent allows woody plants to grow tall and better deploy their leaves in the sunlight. It's almost impossible to overrate the evolutionary importance of lignin; had it not appeared, all plants might well have been relegated to the humble size and habit of mosses.

Unfortunately, the tulip tree's natural range in New England is restricted to the region's southernmost reaches. Here and in warmer states the species has been prized for its wood, which reputedly has more uses than any other type. Cabinetmakers have relied on it as a prime material for drawers and other furniture parts; Indians made canoes from its straight trunks; and early European settlers derived both a malaria cure and a rich golden dye from its bark.

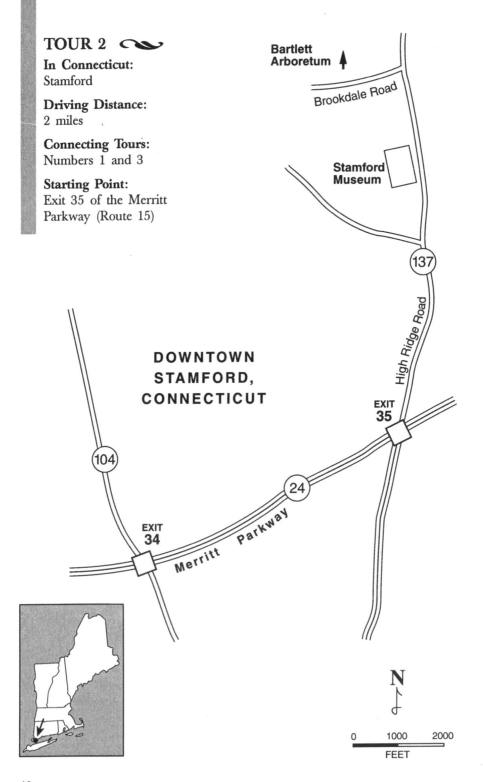

TOUR 2 ∾

In Connecticut:
Stamford

Driving Distance:
2 miles

Connecting Tours:
Numbers 1 and 3

Starting Point:
Exit 35 of the Merritt Parkway (Route 15)

Bartlett Arboretum ↑

Brookdale Road

Stamford Museum

137

High Ridge Road

DOWNTOWN STAMFORD, CONNECTICUT

EXIT 35

104

24

EXIT 34

Merritt Parkway

N

0   1000   2000
FEET

Tour 2
The Bartlett Arboretum

HIGHLIGHTS

Dwarf Conifers and Showy Ericads ❧ An Outdoor Gallery of New England's
Trees and Shrubs ❧ Blueberries and a Scarlet Beauty ❧ The Tussock Sedge ❧
Surviving in a World of Water ❧ Poison Ivy's Next of Kin ❧ How Leaves
Breathe ❧ A Marigold That Isn't ❧ The Bald Cypress and Its
Knobby Knees ❧ Imported Lovelies and a Tree-Sized Grass

*T*o the plant lover, the state of Connecticut has a botanical and horticultural importance that far outstrips its modest size. Among the chief
attractions are two living museums, showplaces of the plant kingdom:
the Bartlett Arboretum of the University of Connecticut and the Connecticut
College Arboretum in New London (see Tour 6). The Bartlett facility is located
in a fine residential section of Stamford. To reach it, head north on High Ridge
Road (also known as Route 137) from the Exit 35 starting point. Go 1.5 miles—
you'll pass the Stamford Museum on the way—and turn left onto Brookdale Road.
The entrance to the Arboretum is just 0.3 mile down and to the right. Take the
access road past the visitor center and education building to the lower, main
parking lot. You'll probably want to begin by walking back up to the visitor center,
to get a map. I suggest you also purchase the inexpensive Ecology Walk trail guide.

Two horticultural highlights of the arboretum are the dwarf conifer plantings
and the collection of ericads (plants belonging to the heath family). These include
heathers, azaleas, rhododendrons, and mountain laurels. One especially impressive ericaceous specimen, a spreading Japanese andromeda shrub, stands in
front of the visitor center, facing the road entrance. In late April 1991, I had
the good fortune to catch it at the height of its bloom, when it produced a billowing
cloud of tiny white flowers. It was a splendid example of what even a commonly
used ornamental shrub can do when it likes the treatment it's getting.

Once you've seen the various ornamental woody plants on the grounds,
return to the parking lot and enter the Ecology Walk trail at the parking lot's

In spring, the Japanese andromeda shrub by the Bartlett Arboretum visitor ccenter produces masses of small, bell-shaped flowers.

southeast corner. Follow the triangular yellow blazes. At first the trail heads through woods containing an excellent mixture of northern hardwoods, especially American beech and yellow birch, and such central hardwoods as hickories, white oak, and tulip tree. A good blend of understory shrubs grows here, too, with splendid late fall-flowering witch hazel, as well as spicebush, arrowwood, and American cranberry bush, the last two of which are viburnums. Arrowwood's common name derives from the fact that Indians used its straight stems for arrow shafts. As noted in Tour 1, the cranberry bush is not truly a cranberry plant at all, though its red fruit bears a superficial resemblance to the genuine article.

There is a dramatic change of environment as the trail descends into a secluded freshwater wetland. The boardwalk gives you a frog's-eye view of swamp and marsh species that have successfully adapted to life where the water level remains above the ground, or at least close to it, throughout the year. Red maple is the dominant tree here; shrubs include buttonbush, speckled alder, the powerfully fragrant clethra, and assorted dogwoods and blueberries. (As a point of personal pedantry, I'd like to postulate that what the trail guide suggests is *blue* highbush blueberry is actually *black* highbush blueberry, a closely related species that is in full bloom just as its leaves emerge in the spring. After checking, the arboretum staff agrees with this assessment. It may be of little importance to anyone else, perhaps even to the plants themselves, but botanists will have their harmless fun.)

The herbaceous plants are the real showstoppers here. In midsummer, the spectacular cardinal flower, one of our native lobelias, lights up the undergrowth with spikes of scarlet blooms. And this is a good time to spot its avian namesake hopping about in the boughs above. Other water-loving wonders here include the arrow arum, the tiny film-forming duckweed, and tussock sedge. As its name implies, this important monocot grows in raised, clumped communities that constitute one of the most familiar sights in New England marshes. Distant relatives of the grasses, sedges are a group of highly evolved but often overlooked plants that manage to thrive in a multitude of different environments without the trappings of colorful flowers.

The survival techniques of these various wetland plants may differ considerably, but all must deal with a superabundance of water and a lack of oxygenated soil. One trick that some species use is to pump atmospheric oxygen from their stems or leaves into the roots, where it's most crucially needed. Woody swamp plants often have bumpy lenticels in the bark; these function as gas-exchange devices. In turn, herbaceous plants frequently develop large internal air chambers, known as aerenchyma.

Arrow arum, a classic marsh plant, grows amid the duckweed in a wetland section of the Bartlett Arboretum.

On the far end of the boardwalk, the trail reenters the dry forest. Among the understory plants stands dwarf or fragrant sumac, a close but harmless relative of poison ivy. It too has leaves composed of three shiny leaflets, but they are distinctly smaller than those of its rash-producing counterpart. Also present are sassafras, bigtooth poplar trees, and a familiar invader of the southern New England woods, Japanese barberry.

At this point, ask yourself a basic question: What do all the plants you've seen so far have in common? One obvious answer is that they all have leaves, at least in the warm season. As you can easily discern in this densely vegetated setting, leaves come in all sizes and shapes. Sometimes there is a correlation between foliage size and the amount of water at hand. But regardless of their design, all leaves share a crucial, though usually microscopic, feature. This is the stoma, or breathing pore (plural, stomata). Each leaf blade contains a huge number of stomata; most often they're located on the underside. An average tobacco leaf, for example, has about twelve thousand per square centimeter. In most species, the stomata open during the day and close at night, but in some plants in arid environments, the pores do just the opposite. Stomata are an essential part of a plant's makeup, and they have been for a long time—even the earliest plant fossils contain them. They enable leaves to partially regulate transpiration, the loss of water from their tissues. Several other factors influence the rate of transpiration; for instance, as the wind or air temperature rises, water loss rapidly increases. In contrast, as the humidity rises, the loss is somewhat slowed.

Farther down the trail lies a second wetland, which is at its fullest glory in early spring, when the marsh marigolds put forth cheerful yellow blooms. This species is not a true marigold, thank goodness, but a member of the buttercup family. All parts of it are allegedly irritating to the skin, though I personally haven't suffered any discomfort. Notwithstanding its reputation, or perhaps because of it, the plant has served as a traditional cure for warts.

As you cross the small stream at the edge of the pond, note the striking pair of bald cypresses thriving in the open water. This species of the redwood family is the most distinctive wetland tree of the southeastern United States; it is also found in swamps up the Mississippi River valley as far as southern Illinois. It's an unusual conifer because it is deciduous—it drops its needle foliage each fall (most other cone-bearing plants are evergreen). When bald cypress specimens grow in standing water, they commonly develop bumps or knobs, called knees, on the exposed roots. For many years, botanists thought the knees were pneumatophores, or "air roots," that helped the plants obtain badly needed oxygen from the air. But subsequent research has suggested this is not their main function. While this tree has been known and revered for centuries, a close relative and look-alike, the dawn redwood, has only recently been thrust into the limelight. (For the amazing story of dawn redwood, see Tour 4.)

On your way back to the parking lot, you will find a number of handsome imported specimen plants, including a tree-sized European alder and a Chinese cork tree, representing a species closely allied to the Amur cork tree seen near the end of Tour 1. Before you reach these, however, you will pass an impressive stand of bamboo. If you inspect the bamboo stems, you'll discover they are giant versions of the grass culms that wave in New England meadows. Indeed, the various species of bamboo include the world's largest grasses. These often attain tree size. When they are well established, their rate of growth can be astonishingly high. Some plants will elongate several feet in the span of a single day.

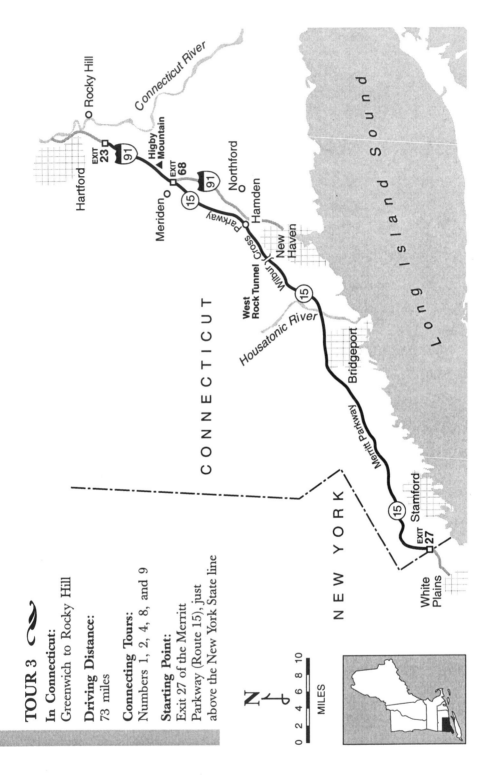

TOUR 3 ❧

In Connecticut:
Greenwich to Rocky Hill

Driving Distance:
73 miles

Connecting Tours:
Numbers 1, 2, 4, 8, and 9

Starting Point:
Exit 27 of the Merritt
Parkway (Route 15), just
above the New York State line

48

Tour 3
A Roadside Vegetation Tour: Greenwich to Rocky Hill

*N*ormally, botany is a science of intense scrutiny. To gain more than a superficial acquaintance with the world of plants, you must be willing to learn the art of careful, sustained observation of details. For that reason, botanizing on a superhighway, at 55 or 65 miles an hour, might seem a frustrating and almost impossible task. Many individual plants do pass by in a blur, but it is possible, nonetheless, to get a good feel for the overall scheme of the local vegetation. Accordingly, in this tour and several others like it, the emphasis more often than not will be on the macroscopic—on plant types and communities, rather than on individual specimens. If you want a primer in the plant ecology of southern New England, this trip, especially when coupled with Tour 9, will fit the bill nicely.

The journey's Exit 27 starting point is the turnoff for Port Chester and Armonk, via King Street, and the final exit before New York. Head east (north) on the Merritt Parkway, toward Stamford and New Haven. This leafy thoroughfare, completed in 1940, holds the distinction of being Connecticut's first expressway. It may lack some of the design innovations found on more modern roads, but it is inestimably more pleasant a drive than the dreary stretch of Interstate 95 just to the south.

From the very beginning of this tour, trees form the most significant portion of the plant community, both on the roadsides and on the median strip. Such classic central hardwoods as white and red oaks are numerous; red cedars, characteristic of more open areas, show up periodically. The red oak had a number of old-time commercial uses. It provided furniture makers, for example, with a good stain. Barbados merchants used it for sugar and molasses barrels. Such was not the case in the wine business, however; unlike its white equivalent, red oak is too porous, and barrel staves made from it soon began to leak.

In this area you may spot the distinctive foliage of southern New England's native member of the magnolia family, the tulip tree. Its common name derives from the large green- and orange-toned flowers that superficially resemble real tulips. A little more than 1.0 mile from the starting point, the parkway passes through several road-cuts—rock outcrops exposed by highway construction. Many geologists believe that these crumpled rocks, almost half a billion years old, originally formed a section of the floor of the Iapetos Ocean, an older version of the Atlantic that lay between ancestral North America and other lands to the east. Another 0.3 mile beyond, the quintessential urban weed tree, the ailanthus, or tree of heaven, is a common sight. In late summer, female ailanthus plants are easy to spot because they bear large pink or yellow clusters of seeds. In its native China, the species is used as a traditional remedy for intestinal complaints.

At this mention of ailanthus, a slight digression is in order. It might seem odd to describe some plants as being one sex or the other. Indeed, individuals of many flowering types possess both male and female reproductive structures. In some species, however—and the ailanthus is a good example—the individual plants are either one gender or the other, and never both. Botanists say such plants are dioecious; other species that have individuals with both male and female parts borne on separate flowers are called monoecious. Plants with both genders wholly represented in each flower are termed "perfect," or bisexual. These different modes of procreation suggest that the plant kingdom has a more varied and exciting sex life than anything dreamed of by prime-time TV producers. (Some specialists frown on the practice of ascribing maleness and femaleness to plants, because they believe the reproductive methods and structures of plants are sufficiently distinct from anything the animals enjoy. Even if that's true, the reference to female and male flower parts makes a good working metaphor.)

Other familiar tree types can be seen as you approach Exit 28, including white ash, American elm, and red maple. Before the onset of the deadly Dutch elm disease, the stately American elm was used as a town and city street tree across much of the nation (see Tour 34). The wood has an interlocking grain pattern and can be fiendishly hard to split, but it has been employed nonetheless in the manufacture of yokes, ship blocks, wheel hubs, and feeding troughs. Additionally, its thick, plated bark has seen duty as chair-bottom material and cordage.

On this leg of the trip, keep an eye on the median strip. You should notice trees with drooping lower branches and telltale flaring trunk bases. These are examples of the pin oak, a native swamp-loving species that has seen extensive horticultural use as a lawn and boulevard planting.

In summer, another kind of basic plant type, the vine, is in evidence almost everywhere. If unchecked, vine growth can choke out or significantly diminish the vigor of trees and shrubs. In this area, poison ivy and Virginia creeper are joined by bittersweet and wild grape species, and all do their best to clamber over the other woody plants in a never-ending battle for sunlight. The much maligned poison ivy, incidentally, owes its reputation to a clear or yellowish oil called urushiol (pronounced "you-*roo*-she-all"). As everyone knows, this produces a severe skin rash on some people, but has little or no effect on others. The plant has even been used by homeopathic doctors to treat ringworm—perhaps by distracting the amazed patient from the original problem. The sap of poison ivy was once sought after as an excellent linen-marking ink.

Identifying herbaceous plants from a moving car can be a difficult proposition, but in midsummer you probably will recognize some familiar roadside species, such as chicory and black-eyed Susan, until they are replaced by their late summer and fall equivalents, common mullein, wild lettuce, Queen Anne's lace, and assorted goldenrods. The striking blue-blossomed chicory, a taprooted perennial alien, is something of a pariah to native plant fanciers. In earlier times, though, it was a popular salad green, and to this day it remains a fairly common coffee additive, especially in southern states. Chicory is said to be a relaxant and to counteract coffee's caffeine hype somewhat. Another highly successful alien colonizer of roadsides is Queen Anne's lace. Few people realize it, but Queen Anne's lace is the parent species of the cultivated carrot. Besides providing a herbal cure for dropsy and kidney diseases in general, it once lent its services to the dictates of high fashion at the court of James I, where women adorned their dresses with its fernlike foliage.

At 2.6 miles past Exit 28, another important (though imported) New England tree shows up in the thickets. Black locust, prized by generations of farmers for rot-resistant wood that is ideal for fence posts and many other uses, is a member of the large and diverse legume order. It originally hails from more southerly parts of the Appalachians. As with its far-flung relatives, the garden pea, the mesquite trees of the American Southwest, and clovers found on various continents, the black locust produces distinctive seed pods. Many legumes have compound foliage, and this species is no exception. Each of its leaves is composed of smaller, elliptic leaflets.

The stretch of road between Exits 28 and 37 is an excellent place to observe different types of tree growth. Competition for room and light is intense along the parkway, and individual trees rarely develop spreading, symmetrical crowns. Each plant tends to be narrow, and growth is concentrated at the top or at openings, where leaves can truly earn their keep. On the median strip, the situation is different. The relatively few trees that grow there are protected by a pavement barrier. They have more shoulder room and more light. Their response to this opportunity is to produce the maximum number of leaf-bearing branches, so that their species' characteristic habit, or overall shape or silhouette, is most fully revealed.

In the midst of all this dense vegetation, it's difficult to believe one well-documented fact: in the early 1800s, at its agricultural peak, southern New England was about 75 percent deforested. There were many reasons for this, including wholesale clear-cutting by farmers and loggers. Many of the felled trees fed the immense demand for domestic and industrial fuel wood. The best heat-producing tree types are the oaks, hickories, sugar maple, American beech, and hop hornbeam; the worst are pines and the other softwoods. A typical preindustrial Yankee household used a staggering thirty to forty cords of firewood in a year's time. (A cord is a volume equal to a pile of 4-foot logs stacked 4 feet high and 8 feet long.) A single family consumed roughly an acre of forest annually for that purpose alone. Farm folk liked to have their woodlots located on a hillside near their dwellings whenever possible, since it was easier to sled or drag logs downhill. But the supply would be exhausted within a few years, and in the case of newly established towns, merely a decade or so would pass before the nearby woods had receded an unacceptable distance. Even strict laws against overcutting had little effect. With the coming of the Industrial Revolution, other voracious firewood consumers appeared. Regional railroads burned tens of thousands, perhaps hundreds of thousands of cords each year; and the potash and charcoal industries also took a large toll.

As the forest passed away, more terrain was opened up to pasturage and cultivation, but this was achieved at the cost of dramatically increased soil erosion, compaction, wildlife habitat destruction, and serious floods. The woodlands had been a first-rate conservation and recycling system; cleared farmland was just the opposite. Henry David Thoreau, who saw the problem at its worst, summarized the situation: "If a man walks in the woods for love of them half of each day, he is in danger of being regarded as a loafer; but if he spends his whole day as a speculator, shearing off those woods and making earth bald before its time, he is esteemed an industrious and enterprising citizen. As if a town had no interest in its forests but to cut them down!"

At a point about 1.3 miles north of Exit 37, the region's most prevalent large conifer, the white pine, comes into view. Another 6.8 miles beyond that, you pass more large road-cuts that expose metamorphic rocks. This is an ancient faults zone, one legacy of the Taconian mountain-building event that occurred in western New England some 450 million years ago. Had you been making this trip in those days, you would have to negotiate a range of peaks perhaps 10,000 feet tall—but at least you wouldn't have to worry about getting lost in the woods. The earth's surface was as yet devoid of any but the humblest forms of plant life.

The road crosses the Housatonic River 14.8 miles farther on, just before you reach Exit 54. This great waterway is the heart of southwestern New England's drainage system, and for what it's worth, it was also a favored subject for Connecticut's best-known classical composer, Charles Ives. It stretches from its mouth at Long Island Sound to headwaters in the Massachusetts Berkshires. Not long after the Housatonic bridge, the road becomes the Wilbur Cross Parkway. Past the river another 8.4 miles, you will see the imposing ridge of West Rock looming up ahead. You are now entering the extensive Central

Approaching the West Rock Tunnel.

Lowland, which runs northward from New Haven up the middle of Connecticut and Massachusetts. Without knowing it, you've jumped ahead some 250 million years, from the ancient rocks we saw earlier to these, formed in the Age of Dinosaurs. By then, plants had been established on the land for a considerable time. Dominant shrubs and trees included cycads (kin of the so-called sago palms now found in florists' shops and indoor botanical collections) and other cone-bearing types. True flowering plants, such as grasses, wildflowers, and broad-leaved trees, had yet to arrive on the scene.

After passing a big sedimentary rock outcrop on the left, the road plunges into the West Rock Tunnel. Somewhere above you, believe it or not, grows New England's one indigenous cactus, a species of *Opuntia*, or prickly pear. The plant survives on some of the least hospitable portions of the ridge. On the tunnel's far side, meanwhile, wild grape vines spill over everything in sight; but a sharp eye can detect old naturalized apple trees, catalpas, staghorn sumacs, cotton-woods, and, in midsummer, possibly even the pink blossoms of the herbaceous everlasting pea. Just before Exit 61 is a low area on the right where *Phragmites*, or giant reed, has established itself. New England's largest native grass, *Phragmites* spreads rapidly in damp, disturbed areas. It will be a common sight during the rest of this tour, and in many others.

As you drive through the Hamden area, you are not far west of Northford, a town that was once one of Connecticut's silk-production centers. Silk implies the existence of cultivated mulberry trees, and Northford had thousands of them. This industry, so impractical in retrospect, was strongly encouraged throughout New England in the late eighteenth and early nineteenth centuries. Challenging climatic and economic forces conspired to make this most unlikely of all Yankee agricultural enterprises a failure. For more on the fascinating mulberry mania, see Tour 18.

———————————— ∾ ————————————

When you approach Meriden, keep a lookout for Exit 68. When you reach it, bear right onto Interstate 91 North. Just to the east is Higby Mountain, another traprock ridge. It too is formed of resistant igneous rock that towers over the softer, more easily eroded sedimentary formations of the valley floor. Sedimentary rocks are a common sight in the Central Lowland, just as they are in parts of New York, the Midwest, and the Far West. Still, they make up only about 5 percent of the earth's crust.

The plant community on the median strip is the proverbial mixed bag: cottonwood, red cedar, and staghorn sumac, as seen before, but also poplar and gray birch. In the final 10 miles of the trip, weeping willows—definitely not native, but planted here for ornamental effect—seem to spring up everywhere. These flamboyant trees have had a long history of use in eastern North America; apparently, they were introduced as early as 1730.

Our journey ends at Exit 23 of I-91, the best turnoff for Dinosaur State Park. Refer to Tour 8 if you wish to visit that wonderful facility (and if you have children along, I dare you not to). Also, you may wish to continue straight past Exit 23 to begin Tour 9, the connecting Roadside Vegetation Tour to Sturbridge, Massachusetts.

Red cedars are a common sight along the Interstate 91 median strip.

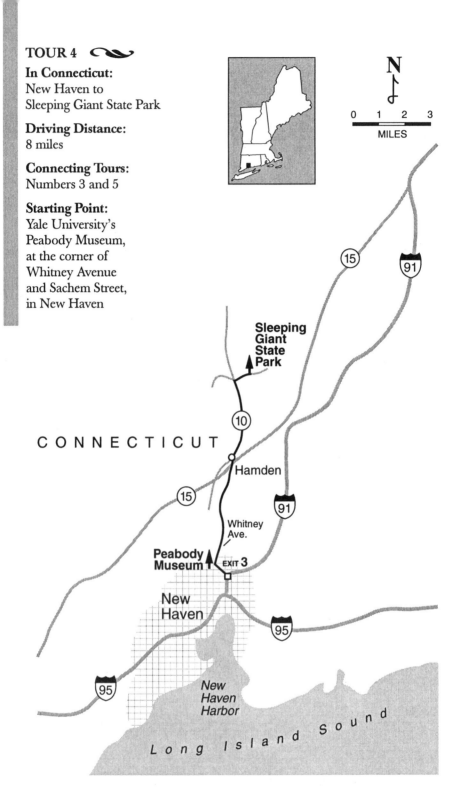

TOUR 4 ◌◟◠◞

In Connecticut:
New Haven to
Sleeping Giant State Park

Driving Distance:
8 miles

Connecting Tours:
Numbers 3 and 5

Starting Point:
Yale University's
Peabody Museum,
at the corner of
Whitney Avenue
and Sachem Street,
in New Haven

N

0   1   2   3
MILES

Sleeping
Giant
State
Park

CONNECTICUT

Hamden

Whitney
Ave.

Peabody
Museum   EXIT 3

New
Haven

New
Haven
Harbor

Long Island Sound

Tour 4
A Long Look at Prehistory
and a Hike up the Sleeping Giant

HIGHLIGHTS

A Living Symbol of Ancient Times ∿ A Gallery of Priceless Fossil Plants ∿
Seed Ferns and Ancestral Oaks ∿ The 300-Million-Year Mural ∿
Coal Swamps and Cordaites ∿ A Revolution of Flowers ∿ The Street
Trees of New Haven ∿ Ascending a Dolerite Ridge ∿ Desire Lines and
Parasitic Plants ∿ Woodland Wildflowers on Display ∿ A Sundered
Supercontinent ∿ A High Perch for Chestnut Oaks and
Mountain Hollies ∿ The View from a Gothic Romance

*T*his tour is a hybrid, combining a visit to a museum on an urban
university campus with a hike and a hillside climb in a forested state
park. If you wish, you may do just the indoor half of the trip, or just
the outdoor portion, but if you do both, you'll find the two complement each
other in unexpected ways.

The starting point is Yale's Peabody Museum. If you're coming from out
of town, your best bet is to get on Interstate 91 and take it to Exit 3, located
in downtown New Haven. Once you're on the exit ramp, continue straight (west)
onto Trumbull Street. After 0.2 mile, bear right (north) at the second stoplight
onto Whitney Avenue. The museum is located another 0.2 mile up and on the
left. Park in the first available metered parking space.

Before you head into the museum, pause to admire the Christmas tree look-
alike planted in front of the main entrance. If you visit in winter, you may wonder
if the poor plant is dead, since its branches are bare. Actually, you're looking
at a deciduous conifer—a cone-bearing plant that has absolutely no intention of
being evergreen in solidarity with the vast majority of its relatives. This unusual
species, known as the dawn redwood, or *Metasequoia,* does the unexpected and
drops its needle foliage each fall, in the same way most of New England's broad-
leaved trees do.

But that isn't the only intriguing thing about the dawn redwood. It shares
a rare distinction with the more famous ginkgo tree, by being what is commonly
termed a "living fossil." From about 90 to 15 million years ago, through the last

part of the Age of Dinosaurs and well into the Age of Mammals, this tree type was one of the most prevalent conifers in the Northern Hemisphere. It even grew in the Arctic, when the climate was less harsh. Apparently, it had little trouble adapting to the long nights and days of the polar light cycle. In recent times, the earth's cooling climate and a change in the rainy seasons caused the dawn redwood's range to shrink tremendously, until it became restricted in modern times to one small area in China. There it was discovered in the 1940s. While ancient fossil remains of the dawn redwood had been known to science for decades, they had been mistakenly assigned to another genus. Ironically, the true identity of both the living trees and the fossils was uncovered in the same year.

More recently, this once almost extinct tree has become a sensation. Harvard's Arnold Arboretum, always a leader in the introduction of promising woody plants from the Orient, has been instrumental in returning the species to its ancient place of honor in North America. In this country, it can now be found thriving in parks, botanic collections, and even in private yards. The dawn redwood shares many traits (including its atypical deciduousness) with its closest extant kin, the bald cypress (see Tour 2). Horticulturists often praise the tree for its remarkably fast growth. One young dawn redwood of my acquaintance (a street planting located next to the Brooklyn Botanic Garden) gained so much height between my visits that I kept thinking it had died and been replaced with a much older tree.

--------------------- ∾ ---------------------

The dawn redwood standing in front of the museum is the perfect prelude to what lies inside. The building contains much of interest in the earth sciences, zoology, and anthropology, but the main objective is the Great Hall, located to the right of the Museum Shop as you enter. At first glance, the fascinating dinosaur exhibits seem to rule this roost, but if you look at the displays located under the 110-foot mural, *The Age of Reptiles,* you'll find an oft-overlooked treasure. A precious collection of plant fossils is on display, including everything from the

Vegetation of the Paleozoic era. On the right, lycopsids and other early plants occupy the warm, wet Carboniferous environment; on the left, gymnosperms grow in the drier conditions of the Permian. –The Age of Reptiles, a mural by Rudolph F. Zallinger. Copyright 1966, 1975, 1985, and 1989 by the Peabody Museum of Natural History, Yale University, New Haven, Connecticut

strange pteridosperms, or seed ferns of the Carboniferous coal swamps, to the more recent ancestors of our modern oak, pine, laurel, and sassafras species. And, not so coincidentally, it won't take long to find some ancient dawn redwood remains. Bear in mind that these remains predate our friend outside by tens of millions of years.

After you've had a chance to scrutinize the fossil plants, take a good look at the painting above them. The striding monsters and exotic landscapes have tremendous entertainment value, but they also reveal some important ideas and discoveries from the annals of paleobotany. The monumental mural is the work of artist Rudolph Zallinger, who completed it in 1947. It depicts a vast sweep of geologic time, roughly 300 million years in extent, ranging from the middle of the Paleozoic (Early Life) era, through the end of the Mesozoic (Middle Life) era, the heyday of the dinosaurs. It's interesting to note that Zallinger presented this epic of prehistory in southpaw fashion, from right to left, so the scenes would line up chronologically with some of the exhibits below. The process he used for this mural was painstaking. First, he applied coats of plaster to the bare brick wall; when that surface dried, he applied the visual framework in the form of line drawings. Next came an extensive monochrome underpainting; only then did Zallinger bring the wonderful creation to life by adding the full colors and dimensional effects. He used this *fresco secco* (dry fresco) technique, rather than painting in wet plaster, because it better revealed the many intricate details that make this masterpiece both stirring and plausible.

Our understanding of paleobotany has moved forward since the World War II era, and we now suspect that some of the renderings aren't quite correct. For example, look for two probable misunderstandings at the far left-hand end of the painting. In the foreground, between those two perennial dinosaur favorites, *Triceratops* and *Tyrannosaurus,* stands a magnolia shrub, blooming before its leaves emerge, just as some modern garden magnolias do. In fact, this plant most likely would have been evergreen. Also, there is the curiously horticultural form of a weeping willow, skulking behind assorted palms and conifers at the base of the nearest volcano. It looks as though it just escaped from a nearby industrial park and is hiding as best it can. When Zallinger consulted with the experts, they told him they'd found fossil willow leaves contemporaneous with the later dinosaurs. It appears that the fossils weren't actually willow leaves after all. Indeed, the willows seem to be a more advanced line that didn't even appear until the next era was well under way. (If you'd like to test your perceptive skills further, you can look for one other extremely subtle misinterpretation. Roughly in the middle of the mural, just over the line from the Triassic period into the Jurassic, there is a small, odd-looking plant, *Cycadeoidea,* just below and to the right of the big *Allosaurus* with the bad table manners. If you look closely at the plant's beehive-shaped stem, you'll notice star-shaped cones that have opened up the way flower petals spread open. It now appears that these cones did not split apart at all; they were either self-pollinated, or insects bored into the sealed interiors.)

Such relatively minor objections do not diminish the drama or the immediacy of the artist's vision. To the person with a special interest in plants, the right-hand side of the mural is nothing less than the lost Eden. Strange primitive trees, *Lepidodendron* and *Sigillaria,* tower above a golden age of greenery, when the coal swamps of the Carboniferous period spread across thousands of square miles. Like their humble New England club moss descendants, these plants reproduced the old-fashioned way, by dispersing spores rather than seeds. Immediately to their left, notice the lone example of a later, Permian period tree called *Cordaites.* It was one of the early gymnosperms, or naked-seed plants. Its broad, strap-shaped leaves do not remind you of modern pines, hemlocks, and spruces; but these later plants are gymnosperms, too.

There is one last thing to consider before you leave Zallinger's convincing dreamscape. Take one more careful look at the whole length of the mural. All the fearsome animals excepted, what makes the world of the far left-hand side so different from the rest of the scene? The answer is flowers. The plants that produce these highly advanced reproductive structures are the angiosperms. They did not appear in the fossil record until the Cretaceous period, the final era of the Age of Dinosaurs. Notice how familiar *Tyrannosaurus'* home is to our human eyes; in contrast, the landscapes of earlier ages, devoid of all flowering plants, strike us as almost utterly alien. It could be a different planet, orbiting a different sun. Flowering plants, from oaks and palms to petunias and crabgrass, are an integral part of our world, yet they are the newest of newcomers. Their arrival marked one of the greatest revolutions in the history of life. And that revolution isn't over yet.

When you've seen as much as you wish indoors and are ready to forge on to nearby Sleeping Giant State Park, depart the museum and drive north along Whitney Avenue for 4.8 miles, to the junction of Route 10. Along the way, note the mixture of weed trees and ornamental plantings on and near the roadside. These include ailanthus, red oak, black locust, American elm, and that ever popular import, the Norway maple. A native of the forests of Europe, this species closely resembles North America's sugar maple in leaf shape and habit. The best way to tell the two types apart at a little distance is by their bark. At maturity, the sugar maple's trunk surface is somewhat scaly, patchy, or peeling, and often has a faint pink or orange tinge to it. The Norway maple has more regularly furrowed, vertically ridged bark. At closer range, there's an even better test–a leaf stem removed from the Norway maple bleeds milky sap, while the sugar maple's sap is clear. In this country, the Norway maple's role is restricted to horticultural applications, but Old World cultures have used its hard wood for furniture and the bark for a reddish brown dye.

On reaching the Route 10 junction, continue straight ahead; you are now on 10 North. Proceed another 2.6 miles and then turn right at the intersection. You should see a state park sign here. Go down the road 0.3 mile and bear left into the entrance and parking lot. The hike up to the Sleeping Giant summit takes an hour or a little less, along a fairly gentle slope the whole way.

Walk uphill into the park woods until you come to the intersecting Tower Trail (it's the only one that looks like a one-lane road rather than a narrow footpath). Take this track. As you'll notice, the trail has a roundabout approach, and as a result, numerous footworn shortcuts have been blazed along the way. Park planners, who can be rather poetic when they try, call these unplanned paths "desire lines." Resist the temptation to use them, since they hasten erosion, cause soil compaction, and threaten plant communities.

On the lower slope, you'll see a combination of the basic plant types shown in the Peabody Museum mural. Underfoot are the relatively primitive, spore-producing plants such as Christmas and hay-scented ferns. They are well adapted to the environments of the forest floor and the sides of the path. Next are the gymnosperms, woody plants that form seeds but not flowers. They're represented here by tall-growing white pines and somber, dark green Canadian hemlocks. Last but most prominent of all are the angiosperms, the flowering plants, in their widely varying forms. The broad-leaved crowns of red oak, red maple, and black birch stretch overhead; at the understory level are native viburnum shrubs, the invasive Japanese barberry, and large communities of mountain laurel. Lower to the ground are the herbaceous species—grasses, sedges, and wildflowers. In the spring, look for Solomon's seal, pussytoes, and plantain-leaved sedge; near

Trees growing amid the talus at Sleeping Giant State Park.

A trout lily in flower on the Sleeping Giant hillside.

summer's end, you should find woodland sunflower and the leafless, yellowish dodder that twines itself around joe-pye-weed and other victims. As its pale color suggests, dodder has no food-producing chlorophyll. Instead, it develops rootlike structures called haustoria that dig into a victim's tissues and rob them of nourishment. Apparently, dodder is initially attracted by telltale chemicals given off by the host plant. Given its predatory habits and otherworldly aspect, it's little wonder that dodder has earned such alternative common names as hellweed and devil's guts.

You've reached the halfway point, approximately, when you come upon the large slope of boulders fallen from the cliff face above. Geologists call this kind of rock rubble "talus" or "scree." In this case, the rock is an igneous variety known as dolerite. At the beginning of the Mesozoic era, the major land masses of the world were joined into the supercontinent Pangaea (pronounced "pan-*jee*-uh"). About 200 million years ago, the supercontinent began to break up; Africa and Europe, previously joined to the eastern seaboard of the United States and Canada, started to pull away. The Central Lowland of Connecticut and Massachusetts suffered from this rifting, and in the midst of the upheaval, masses of molten rock, including the dolerite, welled up from deep in the earth's interior. Here, it formed a structure known as a stock. Earth scientists have established that this rock, unlike some of the igneous units in other parts of the valley, solidified before it reached the surface; subsequent erosion exposed it. Since it was more resistant than the surrounding sedimentary rocks, it formed a high, humped ridge that reminded earlier generations of a reclining giant.

Along this section of the trail, sugar maple, dogwood species, and witch hazel are quite common. In early spring, the slopes abound with classic New England wildflowers, including bloodroot, wild ginger, Dutchman's breeches, blue violets, Canada mayflower, and jack-in-the-pulpit. Later on, in middle to late summer, the stage is reset for horse balm, Virginia knotweed, and the pretty but poisonous white snakeroot. Wild ginger contains an oil once used for perfumes, and its roots were the basis of a herbal cure for dropsy and chest problems. Jack-in-the-pulpit was a favorite American Indian cough remedy.

As you make the final approach to the summit, you'll notice that another plant community, better suited to dry and rocky soil, takes over. Two good indicators of this regime are chestnut oak and red cedar. They are joined by basswood, hickories, sugar maple, and other oaks, and even by one naturalized apple tree. Mountain holly and lowbush blueberry help form the understory. Take a good look at the chestnut oak in particular, since it is rarely found in the lowlands. While this species has never been as common as the red and white oaks, it was occasionally used as the source of a light brown dye for wool clothing.

At the very summit of Sleeping Giant stands a masonry tower that was completed in 1937. As you'll see, it makes the perfect setting for a really bad gothic romance. The view from the top is well worth one last climb—up the winding stairs. It may be too late to see Africa pulling away in the distance, but there in the background is the glittering surface of Long Island Sound, framed by masses of treetops below you.

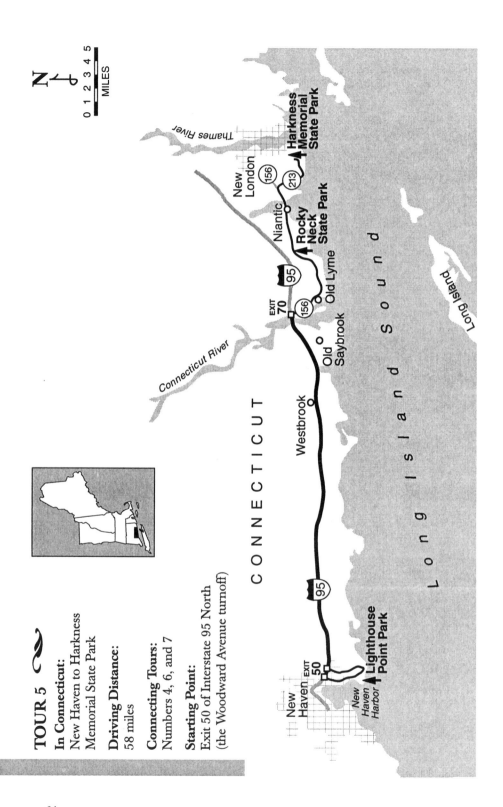

TOUR 5 ∽

In Connecticut:
New Haven to Harkness
Memorial State Park

Driving Distance:
58 miles

Connecting Tours:
Numbers 4, 6, and 7

Starting Point:
Exit 50 of Interstate 95 North
(the Woodward Avenue turnoff)

Tour 5
Long Island Sound

When we think of the New England seacoast, images of Maine's craggy Down East shore or the dunelands of Cape Cod quickly spring to mind. But the shorter waterfront of Connecticut, especially the stretch eastward from New Haven, offers its own enchanted world. This tour features three lovely parks along the southern fringe of New England, from the eastern edge of the state's highly urbanized waterfront corridor through resort and residential towns that manage to retain their quaintness, even amid traffic. As you might expect, this is an excellent opportunity to learn some beachcomber botany; but a variety of other treats awaits you, ranging from horticulture on a grand scale to the subtle mysteries of forest lichens.

The first goal is New Haven's Lighthouse Point Park. Depart Interstate 95 at the Exit 50 starting point and turn right at the stoplight at the base of the exit ramp. Make sure you keep track of your mileage. Follow the signs for Fort Nathan Hale, and when you reach it, continue past it. At 1.7 miles from the start, you'll reach a T-intersection; turn right onto Townsend Avenue and continue south another 0.8 mile until you reach the park entrance. Inside, you have your choice of several big parking lots. Any one will do.

If you're starting this tour early in the morning, so much the better, because Lighthouse Point is then at its peaceful and luminous best. This is true even at the height of summer, when heavy use from the day before might make the place look a trifle unkempt. One of the first things to catch the plant lover's eye is the smattering of trees in the open area near the shore. Among these stand horse

A proud but lonely survivor: the persimmon tree at Lighthouse Point Park.

chestnuts, members of a species native to the Balkans, but long used for boulevard and lawn plantings in the cities of northern Europe and America. Horse chestnut leaves are distinctively shaped, with usually seven large leaflets radiating from a central point. Unfortunately, the tips of the leaflets often turn an ugly brown by midsummer. This is the result of sun scald, a condition caused not by pests or disease organisms but by the environment itself. It occurs when a plant's roots cannot absorb as much water as the leaves lose through transpiration (see Tour 2). Despite this all-too-common problem, horse chestnuts are still park favorites because they're excellent shade trees and put on a springtime flower display few other trees can match. If you've ever seen them in full bloom, when they are bedecked with upright "candle" clusters of blossoms, you'll understand their appeal. The tree's herbal applications can't match its ornamental value, but the bark has been used in a treatment to allay fevers.

The horse chestnut isn't a North American original, but several other trees nearby, including hackberry, pin oak, and red maple, are. (For more on the interesting hackberry, see Tours 32 and 39.) Walk down to the long beach front and head toward the carousel building. Nearby is the last plant of an old stand of persimmon trees, a lonely and vulnerable sentinel. Its bark is characteristic of the species, composed of square and rectangular projections that form an attractive tessellated pattern. While it is unclear whether the persimmon grove that once stood here was wild or planted by the human hand, it is very possible that this enduring tree is now the northernmost of its kind on the eastern seaboard.

In the woodland behind and to the north of the lawn area, white and red oaks compete with hickories and black locusts. The understory contains chokecherry, sassafras, and arrowwood, all interlaced with poison ivy and Japanese honeysuckle. See if you can spot at least one wolf tree—one that has managed to grow tall enough to spread its crown over shorter neighbors. Wolf trees are the bane of woodlot managers because they steal an inordinate share of sunlight; thus, fewer trees in a given area can produce healthy, sustained growth. Still, nature as a whole has grander priorities than human timber production, and wolf trees are common in wild settings. In the end, they redress their lifetime of conspicuous consumption by dying and suddenly opening a large swath of growing space, eagerly exploited by a host of plants beneath them. Even after their death, the big trees play a vital role by providing homes for numerous kinds of wildlife. Incidentally, red oaks are among the species most likely to be wolf trees. Is that the case here? This species fills its wolfish role so well for a very good reason. Many trees, including maples and birches, grow fairly rapidly at first, but after twenty years or so, they slow down. Red oak continues to grow at the same steady rate, and so overtops its competitors.

Lighthouse Point's main attraction, to the majority of its human visitors at least, is the broad beach that stretches beyond the lone persimmon. More botanically interesting is the short stretch of shoreline to the right of the beach, facing westward toward New Haven harbor. In that vicinity, up and a little bit away from the water, is a thicket fronted by wildflowers. Look for stargrass, a small but ravishing member of the lily family, and wild geranium. Herbalists have used wild geranium leaves and rhizomes to control both external and internal bleeding.

Closer to the water's edge, you'll find a separate wildflower community specially adapted to saline environments. Its members include sea rocket, an annual of the mustard family; the leathery-leaved seaside goldenrod; and another tough annual, orache, with drab green flowers and distinctive triangular foliage. Just as intriguing, if somewhat remote from our normal concept of what plants should be, are the marine algae, or seaweeds, often seen here among the rocks. According to modern experts, seaweeds aren't plants at all.

In the traditional scheme of things, all forms of life were grouped into two great kingdoms, officially known as Plantae and Animalia. However, many organisms—fungi, for instance—couldn't be shoehorned into either group without ignoring some of their essential attributes. As a result, a new scheme, incorporating no fewer than five kingdoms, has been suggested and largely accepted by biologists. Seaweeds, including members of the red, brown, and green algae well known to New England beachcombers, are now assigned to the kingdom Protista. Protists may be one-celled or many-celled; some, including the seaweeds, contain chlorophyll and use sunlight to manufacture food, in what certainly seems to be a very plantlike way. Many paleobotanists believe that the first real plants arose from protists of the green algae group.

And speaking of green algae, you may come across a familiar example of it near the high tide mark, in the form of sea lettuce. This alga has a wide, bright green blade, somewhat reminiscent of its salad-making namesake. Such red algae as *Gracilaria* and so-called Irish moss also put in an appearance. The crinkily edged Irish moss is one of several seaweeds with considerable cultural and commercial value. In Canada, for instance, it is collected and processed for various industrial uses. In some places it is part of the human diet and has even been valued as a cure for pulmonary problems. In addition to their chlorophyll, Irish moss and other red algae contain special pigments, phycobilins, that help them utilize the relatively poor light of deep or murky water.

The next destination is Rocky Neck State Park, farther east on the sound. From the Lighthouse Point Park exit, go north on Concord Street to the junction of Townsend Avenue. Turn right onto Townsend, continue on it through the Main Street intersection, and then bear right onto Interstate 95 North. In the first few miles, you'll be seeing an assortment of Connecticut roadside regulars: moisture-loving cottonwoods, weeping willows, staghorn sumacs, red oaks, American elms, red cedars, and the usual manic assortment of wild grape vines. At a point about 13.5 miles from your turn onto the interstate, white pines appear with the red maples. We've left the soft rock (sedimentary) geology of the Central Lowland behind, and are now proceeding into the hard-rock metamorphic landscape. Once part of an ocean bottom, this vast assemblage of rocks has been dubbed the Iapetos Terrane. (*Terrane* is not misspelled; it is postmodern earth-science lingo signifying a particular region and the sum of its underground rock units, seen as a three-dimensional structure and not just as a surface. The older term, *terrain,* still refers to the surface of the land, all by itself. Got it?)

Another 7.0 miles down the road is a colony of ailanthus, the much maligned Asian import that has become the champion urban weed tree of the Northeast. Ailanthus sprouts practically anywhere, even in tiny cracks in the pavement, and its tolerance for heat, bad soil, and air pollution is nothing less than transcendent. It sounds as though this species would be the answer to a city planner's dreams, since it could survive in even the most stressful curbside and traffic island locations. As usual, though, there's a catch: ailanthus is one of the undisputed rabbits of the plant kingdom. It grows quickly, produces the next generation in startling numbers, then cheerfully expires in the blink of an eye. This kamikaze lifestyle is a superb adaptation for a species that must occupy as much bare ground as possible while competitors are few and far between. So it does not exactly endear itself to planners or landscape architects who want a cost-effective species that will not need replacement for decades to come.

The mouth of the Connecticut River, one of the greatest sights along this or any of the other tours, comes into view about 8.0 miles farther east. This is New England's greatest inland waterway and transportation corridor. At its source, in northernmost New Hampshire, it is a rock-strewn little trout stream; here, at its culmination, it forms a sweeping tidal estuary roughly half a mile wide where the interstate crosses it. In a moment, you'll be leaving the interstate at Exit 70, but as you cross the bridge, think of the mass of water rolling under you. Also think about how it was brought here, to its meeting with the ocean.

This is a vivid example of how our planet drains its surface of the liquid compound that has been called the universal solvent. In our first feeble attempts to explore other worlds, we've learned that this chemical, water, is not particularly common, at least not in its liquid state. Here in New England, its superabundance fools us, and most of the time we don't even recognize its crucial role. That attitude would not be shared by African farmers who've lately seen their fields swallowed up by encroaching sand dunes.

When you turn off at Exit 70 on the river's far bank, bear left onto Route 156 South. This is a lovely area. The surrounding towns are some of New England's most historic locales. (It was in nearby Saybrook, by the way, that one of the region's earliest ornamental gardens was established in the mid-1600s, by one Lady Fenwick.) The road passes a mixed seafront landscape of residential lots, salt marshes, and abandoned ground. *Phragmites*, or giant reed, marks the upper edge of the wetland zones, while smaller, more salt tolerant cord grasses of the genus *Spartina* grow nearer the open water. The entrance to Rocky Neck State Park is on the right, approximately 7.1 miles from the interstate turnoff.

Drive down the main access road toward the beach parking lot (in Park-speak, this is known as the "Day Use Area"). Along the way, note the tamaracks and white oaks. These trees, like many others along the coast, sustained considerable damage from Hurricane Bob in August 1991. Before reaching the parking lot, the road passes over a salt-marsh inlet. Marsh mallow, a tall-growing wildflower relative of hollyhocks and the greenhouse hibiscus, puts on a late-summer floral show with large pink blossoms that stand out handsomely from the other wetland vegetation.

A trail leads into the marsh farther down, at the picnic area across from the beach parking lot; recently, it has been closed off because of the presence of nesting ospreys. Nonetheless, a stroll around the marsh periphery is rewarding. Besides good examples of sassafras, there are tupelo trees—they produce splendid red leaf color in autumn—and native rose azaleas. These deciduous members of the rhododendron genus are a delight to behold when they bloom in late spring. Make sure you also note the specimens of white oak. This species is the Connecticut state tree. For centuries it has provided the preferred wood for fencing, firewood, and shipbuilding. In modern times, its timber has been exported to Europe in large quantities.

You might wish to stop at the beach office and pick up a nature trail guide before you proceed up the paved road to the Pavilion parking lot. Whether you take the short self-guiding trail or not, don't miss the Pavilion and the seaward view from the promenade. In front of the building, you may spot the languid flower tassels of downy chess, an alien grass of the brome genus. Farmers once believed that wheat could turn into the unwanted chess, in a sort of spontaneous metamorphosis. In actuality, chess seed had been inadvertently mixed with their wheat seed.

If you decide to venture into the woods nearby, you'll find one of the highlights of the nature trail—excellent specimens of lichens growing on boulders. Some of the lichens are foliose, or shaped vaguely like a leaf; others are umbilicate,

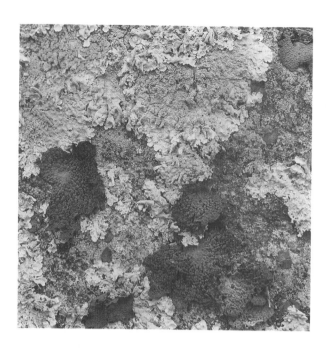

Foliose and umbilicate lichens compete for lebensraum *on a boulder at Rocky Neck State Park.*

or attached to the rock by short, narrow strands at their centers. These slow-growing organisms are not plants, despite what some older texts say, but rather symbiotic associations of fungi and photosynthetic green algae (or, sometimes, blue-green bacteria). The lichens' composite structure presents quite a reproductive challenge, since each member of the team has its own separate way of propagating; it's thought that most lichens rely on their isidia or soredia, small fragments from the main organism's body. Each fragment contains a tiny algal component surrounded by some fungus tissue. When it finds a suitable new home, it begins to develop as its parent lichen did.

You might be surprised to learn that the basic symbiotic nature of the lichen was first identified by Beatrix Potter (1866–1943), one of the most promising English botanists and illustrators of her day. Later, after being stonewalled by her male colleagues, she turned to another field entirely and created some of the most famous children's books of all time. Why no one has honored her by naming a lichen species after her (perhaps *Peterrabbitana potterensis?*) I do not know. All students of lichens are indebted to her.

———————————— ❧ ————————————

To reach the third and final stop on this tour, return to Route 156 and continue east through the beautiful bay settlement of Niantic. The junction of Route 213 is on the far side of town. Turn right onto 213 and go 3.0 miles to the entrance of Harkness Memorial State Park. Formerly a private estate, this magnificent site combines many horticultural high points with botany and seacoast ecology. The mature specimen trees are especially notable. Near the main house stands a huge purple-leaved cultivar of the European beech, undoubtedly the king of the lot. Other Old World species present include fine specimens of sycamore maple and our previous acquaintance, the horse chestnut. Be sure to make a circuit of the parking area, for it too contains interesting plantings, including two Asian ornamentals often used as urban street trees, the fluffy-blossomed Kwanzan cherry and the ginkgo. (For more on the latter, see Tour 16.) Also, walk to the far (eastern) side of the grounds, where there is a well-maintained collection of Japanese maples and perennial beds featuring irises and peonies. Other horticultural delights appear elsewhere, ranging from wisteria pergolas and hedge plantings to annual displays and Japanese cedars.

For one last look at the ocean, take the walkway over the shore dunes, noting as you go the plants whose roots help stabilize these great barriers of windblown sand. Most crucial of these, without a doubt, is beach grass, the true hero and point man of sandy shores from the Atlantic Coast to the westernmost Great Lakes. Here this best of dune fixers is joined by beach pea, dusty miller, poison ivy, dwarfed red maple, and the exquisite beach plum shrub. Despite their well-developed vegetational cover, the dunes—in common with some of the park trees inshore—were hard hit in 1991 by Hurricane Bob. (Farther down the dunes, to the east, you may notice a colony of large, coarse herbaceous plants with huge leaves and stems that resemble bamboo. Unless I am mistaken, this is a fairly uncommon ornamental introduction, Sakhalin knotweed. It seems to have successfully escaped whatever nearby garden bed it once terrorized.)

At the ocean's edge, you have a second opportunity to get to know the seaweeds of Connecticut; in fact, Harkness Park is reputedly one of the best places in the state to see them. Among the species you're likely to spot is knotted rockweed, a most prolific marine alga of the New England littoral. It is distinguished by rubbery olive stems and by small bubbles—air bladders—on its branches. Rockweed and the other brown algae derive their characteristic color from the pigment fucoxanthin, which masks the brighter green chlorophyll.

Wisteria and grape vines scramble over a pergola at Harkness Memorial State Park. Note the wisteria's long, dangling flower clusters.

TOUR 6 ～

In Connecticut:
New London

Driving Distance:
1 mile

Connecting Tours:
Numbers 5 and 7

Starting Point:
The intersection of
Interstate 95 and Route 32

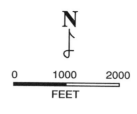

N

| 0 | 1000 | 2000 |

FEET

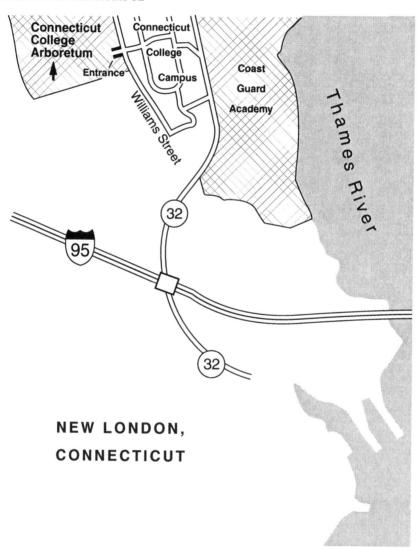

Connecticut College Arboretum

Connecticut College

Entrance

Campus

Coast Guard Academy

Williams Street

Thames River

32

95

32

NEW LONDON, CONNECTICUT

Tour 6
The Connecticut College Arboretum and Campus

*N*ew Englanders may not have much use for self-aggrandizing pride, but
they certainly have the right to be boastful about the quality and
quantity of their institutions of higher learning. Among these are several
with particularly fine campuses that balance good architecture with excellent
landscaping. One of the most fortunate sites is the one enjoyed by Connecticut
College in New London. Situated within sight of Long Island Sound, and just
across the road from the more famous but stark-looking U.S. Coast Guard
Academy, the college grounds include a delightful arboretum and a fine assortment of well-labeled campus trees. Few other liberal arts schools have such an
obvious penchant for living botanical collections and the study of plant science.
(As you can see in Tour 27, Smith College is no slouch in that department, either.)

To reach the college from the starting point, take Route 32 North for 0.6
mile and then turn left at the second traffic light into the main campus entrance.
Follow the road north and then west across the grounds to Williams Street, where
the Connecticut College Arboretum is situated. Park your car along Williams.
All the sights of the tour are within walking distance of this spot.

The arboretum, created in 1931, combines several natural areas with systematic collections organized by plant families. The facility has definitely had its
share of hard knocks. Some of these have been due to New London's location,
in what might be dubbed Hurricane Alley. In 1985, for example, the effects of
Hurricane Gloria required the removal of 250 trees; then in 1991, Gloria's
younger brother, Bob, swept through the area, leaving more wreckage in his wake.

73

A section of the dwarf conifer collection at the Connecticut College Arboretum.

In addition to the onslaught of the elements, two insects now bedeviling Connecticut woodlands have been doing their deadly business here. These are the red pine scale (see Tour 10) and the hemlock woolly adelgid (described in Tour 9). The first of these pests has caused the removal of a grove of red pines. Over the years, however, this beautiful setting has triumphed more or less intact, and has even grown to include several additional properties in the New London area. (It is the main part, here, that has the most to offer the general visitor.)

Begin your stroll at the arboretum gateway, where self-guiding trail brochures are usually available. Directly in front of you lies what is undoubtedly the best-loved feature of the whole place, the Laurel Walk. This long, descending promenade is bordered with mature mountain laurel specimens. To my way of thinking, the mountain laurel is the quintessential Yankee shrub, often assuming ornery attitudes in old age, but always full of hidden virtues—hidden, that is, until flowering time in June. Taken individually, the plant's blooms are less showy than those of its rhododendron and azalea relatives, but each blossom is a miniature miracle, ranging from pure white to white with pink markings or even a lustier pinkish rose.

The next sight is the pond, originally an unbroken expanse of water that has in recent years become silted in and marshy. In 1992, the north end was dredged and a portion of the open-water habitat was restored. The pond's shallow reaches and periphery are home to a variety of wildflower species, most notably jewelweed, pickerel weed, common cattail, fragrant water lily, and spotted joe-pye-weed. In middle and late summer, there is also the fascinating Virginia meadow beauty. It gets my vote for Most Interesting Flower Design—numerous long anthers, resembling nothing so much as tiny elephant trunks, are framed by four hot-pink petals. The anthers are of special interest because they shed pollen in an unusual way, through tiny pores at their tips. The plant's family, none too well represented in this region, is related to the much larger myrtle family.

Water-loving shrubs, especially buttonbush and clethra, are also very much in evidence in the pond area. At about the same time the Virginia meadow beauty blooms, clethra bursts forth with narrow clusters of powerfully fragrant white flowers. The scent is very pleasant, at least at first; after a while, you might think you've been caught in an explosion at the perfume factory. Clethra is an excellent wetland indicator and is one of New England's best contributions to the often overlooked world of shrubs.

Past the pond, check out the bog and forested wetland sections. The small but intriguing bog is an area of peaty soils, formed by sphagnum moss that has spread over a wet depression. (Here, especially, be sure to stay on the trails. The bog is particularly susceptible to human disturbance.) Even when you maintain a respectful distance, you should be able to spot the bizarrely beautiful pitcher plants as they trap insect prey and so extract nutrients lacking in the substrate. (For more on the pitcher plant, see Tour 42.) And not far away, you'll discern acid-loving highbush blueberry shrubs on the bog's edge. In this leafy setting, pause a moment, not to ponder the types of plants or their habitat but to consider the photosynthetic process that directly or indirectly drives all life on our planet. The green color of the leaves is due to chlorophyll pigmentation. When light of the proper wavelength hits a chlorophyll molecule, that molecule emits an electron. This reaction, repeated many times over, produces the power output that ultimately turns atmospheric carbon dioxide into carbohydrates, and simple nitrogen compounds into life-building proteins.

In the forested wetland, located to the east of the bog and also to the south, two incomparable wildflower natives, the cardinal flower and the great blue lobelia, hold sway in July and August. The cardinal flower is the Hester Prynne of the plant world, with breathtaking scarlet blossoms that appear like a bolt of fire in the woodland gloom. In contrast, the great blue lobelia blooms are a deep royal blue, with pale, longitudinal stripes. While the one plant seems the essence of brashness, the other is the very soul of dignity. On closer inspection, you'll see how similar the two plants really are, in terms of floral shape, habit, and habitat. As it turns out, both are species in the genus *Lobelia*. The generic epithet honors Matthias de Lobel, a sixteenth-century Flemish botanist.

When you return to the vicinity of the arboretum entrance, take a good look at the cultivated plant displays. In addition to a worthy holly collection, you'll see some excellent tree specimens. One of these is a large, handsome river birch, a species native to the eastern United States but for some reason more common as an ornamental in the Midwest. Probably more difficult to locate, since it stands away from the path near the Williams Street fence, is a young example of

Franklinia, or the Franklin tree. Named for clever old Benjamin himself, this species is the apparent victim of what paleobiologists call an extinction event—albeit a very small one. This rare tree was last seen in its native Georgia locale shortly after the American Revolution. Since its disappearance from the wild, it has survived only in gardens and botanical collections. Its demise as a true woodland plant will never kindle the kind of interest generated by the sudden fall of the dinosaurs, but it makes a fitting local symbol for the process obliterating untold species in tropical rain forests worldwide.

On leaving the arboretum grounds, cross Williams Street and head past the chapel. Here, on the central campus, you can see more than a score of impressive specimen trees from the continents of the Northern Hemisphere. A guide to these plants is available outside the botany department office, which is Room 206 in New London Hall. If you can, visit the Japanese stewartia just to the south of the library. It has one of the handsomest bark patterns imaginable. Next, find the well-named giant fir, a massive far western relative of New England's own balsam fir that fronts the north side of New London Hall. It is the largest plant of its species in New England. Another conifer worth hunting down is the Japanese umbrella pine, close by Unity House. As is often the case with common names, "umbrella pine" is a misnomer: this species, still a matter of some taxonomic controversy, is probably best placed in or near the redwood family. In any case, its whorls of large needle leaves make it a striking plant indeed.

If you're willing to venture a little farther afield, visit the pretty 4-acre Caroline Black Garden located across Route 32 from the college's main entrance. The plantings there emphasize cultivated varieties of conifers, as well as woody flowering plants.

The leaves and fruit of a hybrid chestnut in the Connecticut College Arboretum nut tree grove.

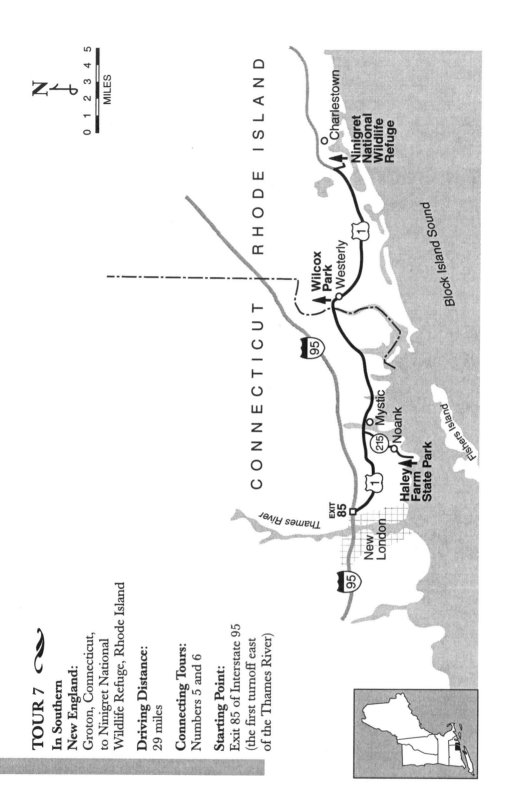

TOUR 7

In Southern New England:
Groton, Connecticut, to Ninigret National Wildlife Refuge, Rhode Island

Driving Distance:
29 miles

Connecting Tours:
Numbers 5 and 6

Starting Point:
Exit 85 of Interstate 95 (the first turnoff east of the Thames River)

N

0 1 2 3 4 5
MILES

CONNECTICUT RHODE ISLAND

Charlestown

Ninigret
National
Wildlife
Refuge

Wilcox
Park
Westerly

1

Block Island Sound

95

Mystic
Noank

215

Haley
Farm
State Park

Fishers Island

Thames River

EXIT
85

1

New
London

95

Tour 7
East to Little Rhody

Highlights

Haley Farm State Park ᔆ The Elaeagnus Invasion ᔆ Knobby Knapweeds
and a Native Legume ᔆ A Not-So-Dwarf Sumac ᔆ Hardhack and Hybridizing
Oaks ᔆ A Railroad Peatland ᔆ Wildflowers of the Sphagnum Mat ᔆ Mystic
Demystified ᔆ A Juniper by Any Other Name ᔆWelcome to the Big RI ᔆ
Westerly and Wonderful Wilcox Park ᔆ A Drive on the Charlestown
Moraine ᔆ Ninigret Wildlife Refuge ᔆ Life on the Asphalt ᔆ Stalking the
Wild Ninebark ᔆ Bayberry's Soul-Restoring Scent ᔆ
An Uneasy Walk Through Vegetative Mayhem

*T*his second excursion along the lovely eastern half of the Connecticut
coast begins just across the Thames River inlet from the point where
the first one, Tour 5, ended. But this is a distinctly different sort of
trip. Not only does it extend into neighboring Rhode Island, it also features new
kinds of plants, an artificial peatland, an alluring city park, and even a glimpse
at greenery run absolutely amok. What plant explorer could resist it?

Our first destination is Haley Farm State Park, located east of Groton, on
Fishers Island Sound. When you reach the tour's starting point, depart the
interstate by turning off at Exit 85. Bear right and follow the signs for Route
1 North, and by all means ignore the fact that you're actually heading south.
After 4.2 miles of stoplights and modern American clonescape, you'll reach the
junction of Route 215, where a sign directs you toward Groton Long Point and
Noank. Turn right onto 215 and go 1.2 miles to the *second* entrance onto Brook
Street; then bear right onto Brook and head down it 0.3 mile. Finally, turn left
onto Haley Farm Lane. The parking lot is just ahead, on the right.

The park entrance and trailhead are visible close by. Walk down the trail
approximately 200 yards, until you come to a fork. To the right is a thicket of
elaeagnus (pronounced "ell-ee-*ag*-nuss"), a high-growing shrub well established
in these parts. In fact, there are three different elaeagnus species, all native to
either Europe or Asia, that are now widespread—many would say out of control—
in North America. Originally praised and enthusiastically planted because of its
attractive silvery foliage and its tolerance for dry and salty soil, elaeagnus soon

proved that it didn't really need human advocacy in the least. In this area, it quickly became the mafioso of shrubs, staking out its territory in poor soils and abandoned fields, to the virtual exclusion of less aggressive native plants. This setting shows how it can take over an entire parcel of land.

Notwithstanding the local dominance of the elaeagnus, there are several wildflowers that manage to make a go of it along the sunny paths. One of these is the indigenous showy tick trefoil. It is a member of the legume order, and its pink summertime flower spikes eventually develop small, seed-bearing pods. Another herbaceous plant frequently seen in this location is the Tyrol knapweed. As its common name indicates, it's an Old World import, as are the other knapweeds now found in the Northeast. (There is one species native to this continent, but its range is centered in Texas and the Southwest.) The plant's common name is a corruption of the older "knobweed," which referred to the knobby flower heads characteristic of the genus. In the 1300s, medical practitioners in Europe used a knapweed ointment to treat symptoms of the plague.

Now take the left-hand fork in the trail. As you proceed down the path, all sorts of things merit inspection. Look for a shrub that is one of several New England members of the cashew family. It's known variously as dwarf, shining, or flameleaf sumac. The first moniker is a bit suspect, since it here grows 7 or 8 feet tall, and has the potential to reach two or even three times that height. The compound leaves are distinctive, even before they turn bright red in autumn, because their leaf stems are winged between the leaflets. Other woody plants present include bayberry (the seacoast shrub par excellence), red cedar, arrow-wood, and some oak specimens. The oaks most closely resemble the pin oak species, because their leaf sinuses tend to be U-shaped rather than C-shaped, as is the case with scarlet oak. What makes the issue less than certain is this location's dry, sandy habitat, which normally is prime scarlet oak country. The pin oak prefers swampy settings. As every experienced oak aficionado ruefully knows, the genus has the supreme bad taste to hybridize freely and blur its characters. It shows a fundamental lack of consideration for the feelings of budding naturalists, but at least it gives them a sampling of the kind of humility the art of plant identification requires.

One other plant found along the way presents no such dilemma. Hardhack, also known by the prettier name of steeplebush, is a wild spiraea that puts up narrow, soaring clusters of deep pink or rose-tinted flowers. The undersides of its leaves are woolly and rust colored. Although the plant appears to be herbaceous, its stems are at least partly woody, and so it is often classified as a subshrub.

If you take the trail all the way to its end, you'll reach a railroad embankment. A little before that, to your right, is a sizable pond of particular interest, in that it's fringed by a sphagnum moss mat containing rare plants of the bog and fen. The pond can be reached by paths leading off the main trail. The story of this body of water is a lesson in how easily coastal wetland ecosystems can be disturbed and transformed by human activity. Originally, this was an arm of the nearby salt marsh. When the train line was laid across it, this section was cut off from tidal access. The environment gradually turned less saline, and species dependent on fresh water replaced those tolerant of brackish conditions. When the sphagnum arrived, it began to grow outward onto the pond's surface. For more on sphagnum's role in peatlands, see Tours 10 and 22.

If you venture down toward the water, you'll be well rewarded for your curiosity—but be very careful not to harm the unusual plants. In addition, don't mistake the sphagnum mat that rings the true shore for dry land. In most cases it will support your weight, but if you stay put too long, you'll sink down to your ankles. An assortment of peat-loving herbaceous plants, such as buckbean, sundew, marsh St. Johnswort, and yellow-eyed grass, all blossom in the summer months. On higher ground a little distance from the mat grow shadblow, tupelo, speckled alder, and bigtooth poplar.

——————————— ❧ ———————————

When you've bidden farewell to Haley Farm Park, return to Route 215 and retrace your way north to the Route 1 junction. There turn right onto 1 North. The road winds through residential areas with typical New England lawn trees: oaks, Norway spruces, weeping willows, and purple-leaved Norway maples. Before long you enter downtown Mystic, a nineteenth-century center of clipper ship construction and whaling. One can only wonder, darkly, what the stern Yankee sea captains and their rum-swilling crews would have made of the art galleries, Oriental rug emporia, and gourmet fudge boutiques that grace the main thoroughfare in these more enlightened times.

Ornamental lotus plants in shallow pond water at Wilcox Park.

The road east of Mystic passes through an area with numerous tidal inlets. True to its rough-and-tumble nature, our old friend elaeagnus is a common "escape" (this verb-become-a-noun is insider trade talk for "escaped plant"). Red cedars, too, are a frequent sight. In southern New England, they are often the first trees to sprout on abandoned or disturbed ground. Once again we have an example of a misleading common name: these hardy conifers are not true cedars but upright junipers, as evidenced by their fleshy, berrylike cones.

Within a few miles, you reach the town of Westerly, just over the line into Rhode Island. Welcome to the nation's smallest state. Geographically diminutive, Little Rhody is by no means the smallest in other ways. According to a 1986 estimate, its population is almost twice that of Vermont. And it boasts more residents than Delaware, titanic Alaska, and five other western states. Downtown Westerly is the site of our next stop, Wilcox Park. This is surely one of the nicest and most botanically interesting urban open spaces in the region.

Route 1 goes right through town, so find a parking place along it, preferably in the vicinity of the court building. The park is just across from that noble edifice. When you enter the park's leafy oasis and head toward its center, you'll notice that the trees are largely planted on the fringes, to block out much of the flurry and bustle of the streets beyond. People sit singly or in small groups on the lawn, or stroll down the paths to the lotus pool. There is a tendency for everyone to speak softly. If your luck holds and the ambulance sirens and portable radios don't intrude, you may see the park at its tranquil best, when it resembles nothing so much as a Seurat landscape stirred slightly into gentle motion.

Beyond this general impression are the trees themselves. Development of the park into a full-fledged arboretum began in the 1960s, and today many specimens are clearly labeled. On the courthouse side of the grounds is a young but beautiful cedar of Lebanon—and it's a true cedar at that. Elsewhere, within an easy stroll, are two native and relatively primitive flowering plants, tulip tree and sweet gum, as well as striking Japanese maples, a big sycamore maple, and a copper-leaved European beech. The alluringly named Garden of the Senses is in truth a thoughtfully designed, raised-bed display with Braille interpretive signs; nearby, you'll find a rock garden setting devoted to dwarf conifers. With absolutely no disparagement intended, horticulture is no more impervious to trendiness than any other department of human conduct. Dwarf conifers are now all the rage, as are multimillion-dollar Japanese Zenscapes full of artificially grotesque pine trees. A few years ago, it was massive plantings of ornamental grasses. If global warming really does come, maybe the next bandwagon for the botanically chic will be loaded with Mexican fan palms and organ-pipe cacti.

———————— ∽ ————————

When you're ready to press on, take Route 1 North (east) out of Westerly. Once you're past the local airport, it's difficult to believe Rhode Island is in fact the most densely populated state in the union. Here the road goes over the Charlestown Moraine, a hummocky expanse formed when the last wave of Ice Age glaciers deposited their load of accumulated sediments. To be a little more precise, it is what geologists call an end moraine. The term denotes a ridge of glacial debris that forms at the leading edge of an ice sheet when it stops advancing and begins to retreat. As you would expect here, the road-cuts reveal sand instead of rock, and boulders stick out of the hillsides. The tree community is formed mostly of oaks, red cedar, and red maple.

Bear right onto Route 1A at the turnoff 9.4 miles from downtown Westerly— look for the Ninigret Park sign. After a short distance on 1A, you'll come to the park entrance on the right. Take it, and drive past the sports area into Ninigret National Wildlife Refuge. Park in the lot on the edge of an old air strip, at the end of the road.

The wildlife refuge, formerly a U.S. Naval Air Station, comprises more than 400 acres and contains 2.5 miles of footpaths. The main trailhead is at the edge of the runway; look for the large map sign. Before marching off onto one of the paths, though, take a moment to study the plant life along, and even on, the expanse of the old pavement. If you visit in early August, for instance, you will probably see lance-leaved coreopsis, or tickseed, producing attractive golden yellow flower heads amid the alien weeds. This cheerful little sweetheart of a plant actually hails from the North American prairie, but over the years, it has been used extensively in New England gardens, for obvious reasons. The plants here, and those found growing in other transitional places, are escapes from those gardens. Quite unwittingly, humankind has increased its distribution in the wild.

The trail that bears off to the left of the map sign is a fairly short loop, and it presents you with a good chance to bone up on shrub identification. One species found here, the shaggy-stemmed ninebark, is a representative of the rose family often more difficult to find in natural settings than in ornamental plantings. If you'd like a good view of the coastal wetland, take the path to the right of the sign; it leads to an observation platform. Along the way, in spots where the path comes close to the water's edge, there are colonies of bayberry, beach pea, and seaside goldenrod. Bayberry fruit, a favorite source of candle wax, gives off one of the two most pleasing olfactory experiences New England provides. (The other is trailing arbutus, described in Tour 12 and elsewhere.)

The most lasting impression garnered along this path is not so reassuring. I refer to the expanse of rampant vine growth. In some directions, the landscape takes on an overtly nightmarish aspect, as Oriental bittersweet, Japanese honeysuckle, poison ivy, Virginia creeper, and wild grape species swarm over every surface in sight. It would be the ideal set for a good, low-budget killer-plant movie.

Rampant vine growth takes advantage of unobstructed sunlight at Ninigret National Wildlife Refuge.

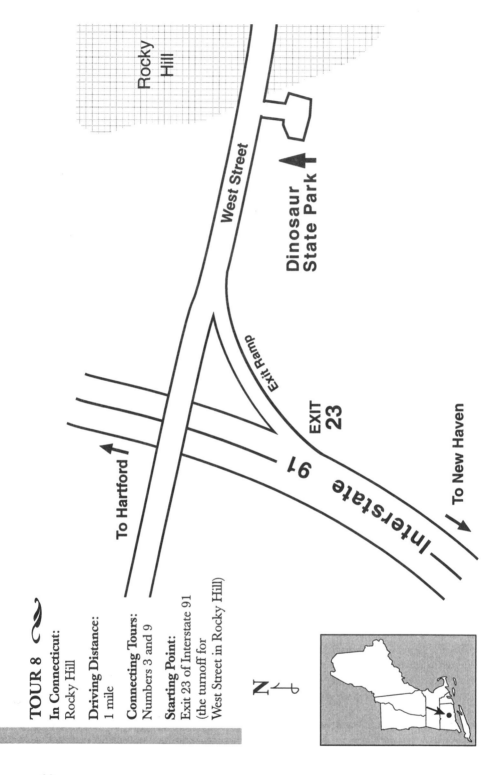

TOUR 8 〜

In Connecticut:
Rocky Hill

Driving Distance:
1 mile

Connecting Tours:
Numbers 3 and 9

Starting Point:
Exit 23 of Interstate 91
(the turnoff for
West Street in Rocky Hill)

N

Rocky
Hill

West Street

Dinosaur
State Park

Exit Ramp

EXIT
23

Interstate 91

To Hartford

To New Haven

Tour 8
Dinosaur State Park

HIGHLIGHTS

Homage to Ed McCarthy ∾ The Earth's Awesome Calendar ∾
Highlights of 420 Million Years of Plant Evolution ∾ Trees with Ancient
Pedigrees ∾ Fossils Make Bad Firewood ∾ A Mesozoic Traffic Jam? ∾
A Window on the Connecticut of 200 Million Years Ago ∾ Triassic and
Jurassic Flora ∾ Gymnosperms and Angiosperms ∾ A Walk Through a
Modern Swamp and Marsh ∾ The Reclusive Black Ash

The year 1966 was a high point for dinosaur fans. That was the year bulldozer operator Ed McCarthy flipped over a section of rock while he was clearing the way for a new state highway department facility in the town of Rocky Hill, Connecticut. The stone slab had suspicious markings, and McCarthy, curious about them, asked his foreman to check with authorities to see if it was anything worth saving. This simple act lead to a most remarkable discovery. Police were dispatched to stand guard over the precious find, and ultimately, scientists unearthed an expanse of bedrock containing almost 1,500 three-toed footprints made by dinosaurs some 200 million years ago.

Many other prehistoric tracks have been found in Connecticut and Massachusetts before and since. What makes the Rocky Hill site unique is how it was handled after it was discovered. Eternal praise is due to McCarthy, his foreman, and the various state officials who acted quickly to stop the construction project and turn the property into an educational park. In one sense, the full assemblage of fossils, still intact, is the best possible monument to the many government workers, in Connecticut and elsewhere, who transcend the uncaring-bureaucrat stereotype and do good work to fulfill the public trust.

Most people who enjoy the unique exhibit at Dinosaur State Park are probably unaware that the first lot of tracks has actually been covered up again, ever so carefully, to protect its soft rock matrix from exposure to the elements. A second, smaller set of five hundred footprints, found in 1967, is what visitors see today. It is the core of a dramatic display sheltered by a geodesic dome. Two other aspects

A young Japanese umbrella pine at Dinosaur State Park.

of the park make it an eminently suitable stop for this plant-centered book. One involves an indoors investigation of the vegetation of the ancient past; the other, a look outdoors at unusual examples of modern plant life inhabiting an intriguing environment.

To reach Dinosaur State Park, get off the interstate at Exit 23. You should spot Dinosaur State Park signs to help guide you. At the head of the ramp, turn east and proceed 0.7 mile; the entrance and parking lot are on the right. Begin the tour with a visit to the dome, officially known as the Exhibit Center. The sidewalk leading up to the building may be one of the most humbling paths you'll ever take: it contains a geologic time scale that shows, among many other things, the span of human existence in its true proportion to other events of earth history.

The issue of human insignificance notwithstanding, the time scale offers an appropriate cue to review some of the highlights in the epic story of plant evolution. First, find the Paleozoic era. This is the major subdivision that precedes the dinosaurs' own era, the Mesozoic. The probable starting point of land plants comes during the Ordovician period of the Paleozoic, about 450 million years ago. According to a prevailing theory, this was when the earliest plants, perhaps resembling today's flat-lying liverworts, began to colonize coastal areas. In the hundreds of millions of years that followed, plants developed many ingenious ways of coping with life in the open air; for instance, they invented leaves, seeds, flowers, fruit, and true roots. But here in the Ordovician, the most rudimentary ancestors of today's plants were able to make a go of it, however tenuously, without those advanced structures.

Another point of interest, later in the Paleozoic, is the Carboniferous period (which in the United States is usually split into two sections, the Mississippian and Pennsylvanian). This span, from 360 to 285 million years ago, was the heyday of the great coal swamp forests of North America. Then, approximately 245 million years ago, the Mesozoic era began. In the great sweep of time, it lies surprisingly close to the present; geologically speaking, the great reptiles both appeared and left the stage quite recently. In the final period of the Mesozoic, the Cretaceous, flowering plants suddenly exploded into prominence in the fossil record.

Finally, there is our own short era, the Cenozoic, starting some 65 million years ago. One of its most significant developments has been the rise of the grasses and other nonwoody plants. Before their advent, global vegetation had been predominantly trees, shrubs, and woody vines. From this point on, though, many parts of the globe came to be dominated by the seemingly humble, but extremely successful, herbaceous plants. See if you can locate the place occupied by the Ice Age. It stands near the very end of the time scale, within the last two million years or so. Even on the large scale of this representation, it appears to be a part of modern history.

As you continue on toward the Exhibit Center, you'll see some plantings that make use of trees with ancient pedigrees. One of these is the splendid katsura tree, whose genus apparently grew in North America while *Tyrannosaurus, Triceratops,* and duckbill dinosaurs roamed the landscape. Also present are two young Japanese umbrella pines. As noted in Tour 6, these handsome plants are actually not pines; they are probably more closely allied to the redwood family. Fossil finds suggest that relatives of this species were growing in the last days of the dinosaurs or, at any rate, soon after their disappearance. There are examples of another, more famous, member of the redwood clan on the other side of the building, facing the road. They are dawn redwoods, the curious deciduous conifers once thought to be extinct. (For a fuller description of their significance, see the beginning of Tour 4.)

The main attraction inside the Exhibit Center is, naturally, the exposed dinosaur tracks. Their density in the display area is so high that you are almost forced to conclude that rush-hour traffic predates human evolution. However long it may have taken to make all these imprints, this must have been a popular spot. While geologists are quite sure this was a lakeshore environment, a good watering or fishing hole, I can't resist the alternative theory, here published for the first time, that this was actually a prehistoric Lotto or offtrack betting outlet.

In fact, speculation on the nature of this region's dinosaur prints and the other Mesozoic remains has a long, and now entertaining, history. In the early 1800s, a man of questionable mental acuity found a petrified conifer trunk near Southbury, to the southwest. His hypothesis was that the fossil was just more firewood, but when he tried to chop it up, he ruined his axe. In the 1830s, Edward Hitchcock, the renowned Massachusetts geologist and Amherst College president,

Dinosaur State Park is more than fascinating footprints. This wetland boardwalk is part of the network of outdoor nature trails.

collected fossil track specimens from the Connecticut Valley and studied them extensively. He published his findings in a beautiful volume entitled *Ichnology of New England.* He reasoned that the prints must have been made by large, extinct birds. A decade later, the great British geologist Sir Charles Lyell visited the region and saw track specimens. Having previously observed modern bird footprints on the tidal flats of Canada's Bay of Fundy, he too surmised that these fossil imprints had been made by avian feet. In the years that followed, however, the true identity of the printmakers was ascertained.

The other exhibits here under the dome may be sideshows, so to speak, but they're fascinating nonetheless. Be sure to scrutinize the wonderful Triassic-Jurassic diorama, which includes an audio narrative by naturalist Leila Raim. It shows the plant and animal life of this region at the time the tracks were made, about 200 million years ago. In the background, you'll see the towering Eastern Highlands, the product of faulting associated with the breakup of the supercontinent Pangaea. In front stands the lowland vegetation: the tall-growing conifers *Elatocladus, Araucarites,* and *Brachyphyllum* soar over an understory composed of several species, including the cycad *Leptocycas.* Meanwhile, a relative of the modern horsetails, known as *Neocalamites,* stakes its claim in the shallow water. It has been proposed that some of the bigger trees grew 180 feet tall, which would have put them in the same league with the largest surviving specimens of the extant white pine. Paleobotanists further believe that some of the conifers shown here were related to the modern Norfolk Island pine, a Southern Hemisphere evergreen sold widely in this country as a popular (if temperamental) houseplant.

At this point in the Mesozoic era, the angiosperms—the flowering plants—apparently had not yet arrived on the scene. Instead, the dominant species were the more primitive gymnosperms, the so-called naked-seed plants that reproduce without aid of flowers or fruit. The conifers—pines, spruces, cedars, junipers, hemlock trees—are the most prevalent gymnosperms in modern times, but in the Age of Dinosaurs other types, including the cycads and their look-alikes the cycadeoids, were common as well. Unlike the conifers, though, these other gymnosperm types passed their zenith long ago. The cycadeoids are now totally extinct, and the cycads are reduced to a smattering of species found in far-flung, isolated populations in the warmer portions of the planet. In New England, the only place you'll see a cycad is indoors—in a greenhouse display, a florist's shop, or here, where during my most recent visit several live specimens faithfully enhanced the prehistoric ambiance. If you see one of them, take a good look at it. Its subdivided, leathery foliage suggests a palm (indeed, some cycads are known as sago palms), but the resemblance is thoroughly superficial. Palms are card-carrying angiosperms, complete with flowers and fruit, while cycads form cones—some of which, incidentally, are truly gigantic. (For an example of this, see Tour 27.)

On leaving the Exhibit Center you return to our own age of flowering plants. To your right as you depart the dome is the entrance to the park's network of three nature trails. One trail, the Red, features a short path and boardwalk through a typical Connecticut freshwater wetland. The first sights along the boardwalk are the classic woody angiosperms of the swamp: red maple, clethra with its sweet-smelling blossoms, highbush blueberry, speckled alder (also known as smooth alder), and buttonbush. Other shrubs, including leatherwood, swamp rose, and sweet gale, reside in a transitional area. The plant that really defines the true marsh section is the herbaceous tussock sedge. This grasslike species grows in distinct clumps poised just above the water.

The best treat of all comes a few steps past the far end of the boardwalk. Keep a sharp lookout for an ash tree, most easily identified in summer by its characteristic compound leaves. If you do locate the tree, you may think you've found the ubiquitous white ash, one of New England's most frequently encountered forest trees. In fact, you've discovered the rarer black ash, a distinct species that can sometimes be distinguished by its bark, which is scaly, rather than vertically ridged. A more reliable clue, however, is the foliage; individual leaflets lack stems at their bases and are attached directly to the main axis of the leaf. Very tolerant of waterlogged conditions, black ash is an excellent swamp indicator species—it's a plant that generally sticks to just one type of environment. By doing so, it helps to define it.

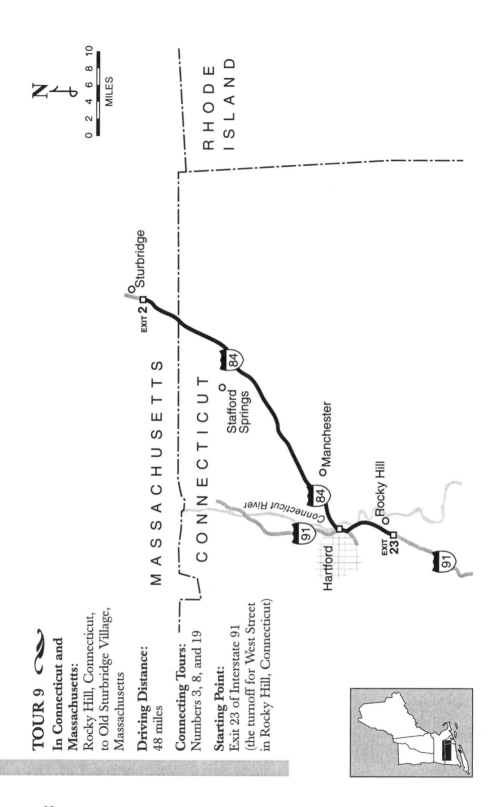

TOUR 9 ∾

In Connecticut and Massachusetts:
Rocky Hill, Connecticut, to Old Sturbridge Village, Massachusetts

Driving Distance:
48 miles

Connecting Tours:
Numbers 3, 8, and 19

Starting Point:
Exit 23 of Interstate 91 (the turnoff for West Street in Rocky Hill, Connecticut)

Tour 9
A Roadside Vegetation Tour: Rocky Hill to Sturbridge

HIGHLIGHTS

Rocky Hill's Natural Dam ❧ When a River Becomes a Glacial Lake ❧ Tundra Plants, White Spruce, and Mastodons ❧ The Modern Temperate Flora ❧ Connecticut's Capital ❧ Crossing the Great River of New England ❧ Lovely, Crumbly Brownstone ❧ Up to the Eastern Highlands ❧ An Appreciation of the Poverty Birch ❧ Land of the Central Hardwoods ❧ A Noble but Susceptible Conifer ❧ The Fine Art of Birch Identification ❧ The Bay State's Southern Reaches ❧ A Darling of the Highway Department

*T*o the naturalist interested in the geologic story of New England, the town of Rocky Hill is an important focal point. As noted in the preceding tour, it is the site of hundreds of fascinating dinosaur footprints, now carefully preserved in a popular state park. But the significance of the place definitely doesn't end there. Millions of years after the last dinosaur made an impression on the local Jurassic lakefront, this area suffered a climatic upheaval that was the general fate of New England and much of the Northern Hemisphere. In a number of distinct pulses, huge sheets of ice, some well over a mile thick, descended from a central point in Labrador and buried the region. The weight of the glaciers was so great that the continental crust actually sank somewhat into the underlying mantle.

At the conclusion of the most recent glacial advance, when a lobe of the decaying ice began its slow retreat up the Connecticut River valley, sediments borne by meltwater formed a large barrier deposit at Rocky Hill. This deposit was the foundation for a natural dam, which retarded the flow of the reemerging main river. As the glaciers continued their northward withdrawal, a long, narrow lake flooded this lowland area, all the way from here to St. Johnsbury, in Vermont's Northeast Kingdom. This impressive body of water, which geologists have named Lake Hitchcock, lasted between one and three thousand years, until the Rocky Hill dam was finally breached. Remnants of the dam can still be seen on the lower end of town. Experts disagree on just when the lake drained, but it was

In New England's wet climate, trees grow practically everywhere–including in the Portland Arkose outcrops along Interstate 84.

somewhere in the neighborhood of 10,500 B.C. The earliest human beings in this region may have hunted on its shores.

It is at the southern tip of old Lake Hitchcock that we begin this journey. Head north on I-91 toward Hartford. Where cold, gray water laden with icebergs once flowed sluggishly, there are now a busy expressway, office parks, and plenty of vegetation enjoying the fertile soil the Connecticut River has provided in more recent times. A plant tour in Lake Hitchcock days wouldn't have been as overtly rewarding. By then, only a few tundra shrubs and wildflowers, and perhaps the first forests of white spruce, had begun the plant kingdom's reconquest of the otherwise barren and ice-sheared land. It certainly wouldn't have compared with the diversity of today. When the temperate-climate plant species returned, most did so by slowly extending their range from warmer sanctuary areas to the south. While the presence of early human hunters is open to debate, chances are that the ancient spruce woods fringing the lake had another important mammalian visitor, the mastodon. This cold-climate pachyderm preferred woodland habitats, while the woolly mammoth stuck to the open tundra far to the northwest.

The modern trees on this first leg of our excursion include three softwoods, or gymnosperms: red cedar, red pine, and white pine. They are outnumbered by typical bottomland hardwoods, or angiosperms: black willow, cottonwood, black locust, and American elm. Also present are the native staghorn sumac and the alien ailanthus, two woody species often mistaken for one another because of their similarity in habit and compound foliage.

About 6.4 miles from the starting point, the skyline of Hartford, the state capital, is clearly visible. World famous as a major insurance center, Hartford can lay claim to many other interesting facts. For instance, it is one of the two oldest chartered cities in the United States (the other is New Haven, which for many years was the co-capital in an unusual arrangement engineered to satisfy the state's seacoast and inland constituencies). Also well known as a magnet for immigrants, Hartford has long had the blessings and challenges associated with racial and ethnic diversity. One of the earliest non-English peoples to arrive, the Irish, reportedly preferred the Hartford area because it reminded them of the open, green countryside of their island homeland.

Looking at the city's larger geographic context, the lower Connecticut River valley as a whole, you find a long history of agricultural endeavor. Up and down its length in this state and Massachusetts, the broad lowland served as America's first breadbasket, where wheat and rye predominated, and as a production center for everything from flax, onions, and orchard fruit to horses, tobacco, and garden seeds. And it comes as no surprise that maize, or corn, the greatest grass-family crop produced here, has often brought a yield-per-acre twice that of the surrounding uplands.

Barring construction or rush-hour delays, you'll soon reach the downtown area. After an additional 2.1 miles, bear right at Exit 30 onto Interstate 84. If traffic permits, take a good look at the mighty Connecticut River as you cross it. Since it is a tidal waterway throughout most of the state, hydroelectric plants, a common sight upstream in Massachusetts and New Hampshire, are not as feasible here. At 5.5 miles from the turn onto I-84, clumps of staghorn sumac try their best to keep a toehold in the road-cuts. The sedimentary rock exposed here is the famous Portland Arkose, often better known simply as brownstone. The handsome, sandy-textured arkose has been quarried and used extensively for building construction in Northeastern cities—hence the famous brownstone residences on Manhattan's Upper West Side. As a matter of fact, the rock has proven to be a rather poor architectural material, for all of its appeal. It disintegrates more quickly, for instance, than granite, an igneous rock.

At a point about 7.4 miles from the arkose outcrops, you'll note how the geographical setting has changed. You have risen out of the river valley and into the Eastern Highlands, a much older terrane composed of resistant metamorphic rocks. The roadside vegetation is a mishmash of cottonwoods, introduced forsythia shrubs, black locusts, and red pines. One significant if often overlooked character on the scene is the little gray birch, a hardy pioneer that likes to settle in the most unlikely places. See if you can spot it sprouting from crevices in the outcrops. Old-timers often referred to this white-barked species as poverty birch, presumably because it frequently colonized the unmowed pastures abandoned by bankrupt farmers.

The landscape in this vicinity also speaks of New England's shifting human settlement patterns. Even when the population of the region as a whole has grown, the countryside has had a tendency to become emptier. In the half century from 200 to 150 years ago, as much as 95 percent of all New Englanders lived on farms; but then the Industrial Revolution and western agricultural competition triggered a flight to urban centers. Ironically, a countervailing, back-to-the-country

movement is now under way, though it is largely confined to yuppies and other affluent people who rarely alter either the population statistics or the look of their land in any startling way.

Consider another community here. This country is a showplace of the central hardwoods regime that typifies much of southern and middle New England. Oaks and ash trees are especially evident. Keep an eye out for the rest area on the right; if you stop there for a few moments, you'll be able to inspect the area's characteristic plant life at closer range.

Another 1.7 miles past the rest area, a new conifer, resembling a chubby, dark green Christmas tree, comes into view. This is the Canadian hemlock, an important member of the New England upland forest community. Vulnerable to fire and insect attacks but tolerant of deep shade, the hemlock is often a sign of moist soil conditions. Don't confuse this plant with the poison hemlock that the Athenians served to Socrates. The latter plant is an extremely toxic herbaceous member of the carrot family, and it closely resembles Queen Anne's lace. Originally an Old World native, it is now found throughout much of the eastern United States, especially on roadsides and in other disturbed habitats.

The Canadian hemlock is a truly beautiful tree, and horticulturists have used it in many settings, including formal hedges (see Tour 32). Sadly, the species is now under siege in many parts of southern New England. Its formidable foe is an insect pest called the hemlock woolly adelgid (pronounced "ah-*dell*-jid"). Although it has been present in other parts of the country for a long time, this aphid relative was first reported in this region in 1985—apparently having been blown in, quite literally, by Hurricane Gloria. It harms hemlocks by feeding on the sap in their young stems; an infestation can prove fatal in as little as a year or less. Currently, there seems to be no effective way to protect wild trees. Interestingly, however, this is not the first onslaught the hemlock has had to endure. Pollen studies suggest that the species was laid low five thousand years ago by some pest or disease that almost swept it away entirely. It took about five centuries for the hemlock population to recover.

Just before you reach Connecticut's boundary with the Commonwealth of Massachusetts, you'll spot more white-barked trees on the hillsides to both left and right. Given the earlier presence of gray birches, you might conclude, reasonably enough, that these are larger versions of the same species. See if you can confirm or debunk that hypothesis. The only other native plant that distinctly resembles gray birch is its close relative, paper birch. The latter's bark naturally peels away in curved or coiled strips; the bark of gray birch remains smooth and intact. Further, the branches of the gray birch are usually noticeably thin and wispy; those of the paper birch are stouter. When you're up close, there's no problem telling them apart, at least in the warm season. Gray birch leaves are small and triangular with long, tapered tips, while paper birch foliage is larger and broader in outline.

Just 0.6 mile after you cross the Bay State line, you will come to another pull-off, this time a picnic area with a pleasant grove of white pines. Along the median strip stands a mixture of native black cherry trees and introduced elaeagnus plants. These silver-leaved shrubs are a favorite selection of state highway departments because they are unusually tolerant of the rock salt used to de-ice roadways in winter. But in some areas, the plants have become a serious nuisance, as can be observed in Tour 7.

The journey ends at the interstate's Exit 2, the turnoff for Route 131 and Old Sturbridge Village via its southern access road. If you'd like to continue on to that delightful attraction, turn to Tour 19.

TOUR 10 ∽

In Connecticut:
Litchfield to Cornwall

Driving Distance:
19 miles

Starting Point:
The White Memorial
Foundation, on
Route 202,
2 miles west of
Litchfield town center

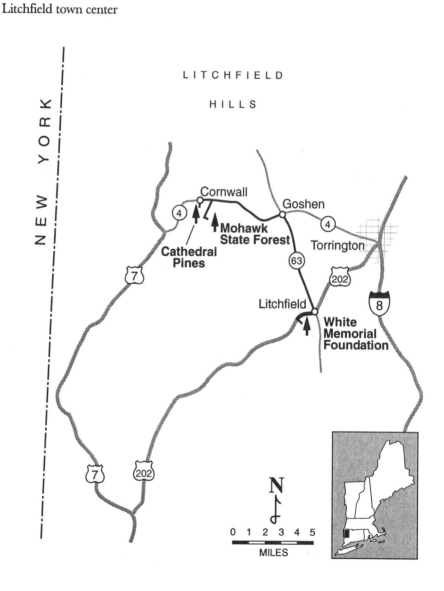

MASSACHUSETTS

CONNECTICUT

Tour 10
The Woodlands of
Northwestern Connecticut

HIGHLIGHTS

The White Foundation ᕙ What's in a Wetland? ᕙ The Herbal Virtues
of Skunk Cabbage ᕙ An Ancient Family Finds Modernity to Its Liking ᕙ
A Beginner's Guide to Pine Cone Aerodynamics ᕙ A Roman Philosopher and
the Catlin Woods ᕙ Wanderings in the Land of Goshen ᕙ New England's Worst
Shrub ᕙ Mohawk State Forest ᕙ A Forest Bog and Boardwalk ᕙ What It Really
Means to Be Nonvascular ᕙ The Carnivorous Sundew: Look Who's Coming
for Dinner ᕙ The Cathedral Pines, or Lo How the Mighty Have Fallen

I f you had to describe the territory covered in this tour in a single phrase,
the only real option would be "country charm." But no Yankee landscape can easily be reduced to a catch phrase. It doesn't take long to
recognize that the arts of civilized life and the works of the natural world are
dovetailed together in this hilly and decentralized setting. There is a sense of
peaceful and productive seclusion. Still, as we'll see at tour's end, northwestern
Connecticut is no more immune to unpredictable forces than any more turbulent
environment.

We begin with two of the state's best woodland preserves. When you've
reached the Route 202 entrance to the White Memorial Foundation, turn in and
take the gravel access road about 0.5 mile to the main parking area. Before you
set out on foot, stop by the museum. I highly recommend purchasing the modestly
priced, detailed map of the grounds, which is available at the adjoining gift shop.
It will keep you from much aimless wandering in an area almost five times as
large as Manhattan's Central Park.

Established just before the onset of World War I, the privately administered
White Foundation facility is more than 4,000 acres in extent, and it is the largest
wildlife refuge in the state. Once you've left the museum and its taxidermy behind,
head due east toward the Bantam River. One of the first sights is characteristic
of New England–a row of grand old sugar maples lining the road. Note their
shaggy, faintly orange-tinged bark. This distinguishes them from the gray-barked,
vertically ridged Norway maple. According to the American Forestry Association,

the largest known sugar maple is a Connecticut tree, located to the south in Norwalk. It stands 91 feet tall and is a massive 269 inches in trunk circumference. (For a description of this species' contribution to the American sweet tooth, see Tour 45.)

At the intersection of the main road-sized trails, and before you reach the gateway ahead of you, bear left toward the wetland area. The Bantam River is close at hand to your right; soon, on your left, you'll see a low, open area dotted with clumps of tussock sedge. This is an excellent point at which to pause and consider what sort of ecosystem this is. In common usage, terms such as marsh, swamp, and bog are often used interchangeably. Science just wouldn't be science, though, if it didn't have its precise definitions; so, as you might expect, each of these words has a specific meaning to ecologists and botanists. The first step in clarifying the terminology is to realize that bogs, swamps, and marshes are all included in the overall concept of the wetland. A wetland is any habitat that has water above or near the surface throughout the year. Wetlands make up about 6 percent of the earth's surface. That might not seem to be a particularly significant amount, but in fact it comprises an area three times that of the United States. In our region, this type of environment has a crucial role to play. Besides providing natural flood control, it helps filter out water pollution and neutralize agricultural fertilizer runoff. The abundant vegetation in wetlands is an excellent source of atmospheric oxygen required by animal life; in addition, it's instrumental in preserving the diversity of species.

A classic New England marsh: clumps of tussock sedge in the standing water of a White Memorial Foundation wetland.

One special kind of wetland is called peatland, an ecological unit character-ized by the presence of soils containing mostly organic matter, or peat. Peatlands are often classified either as bogs (stagnant habitats where all the water is supplied by rainfall) or as fens (which have some surface water flow). In many New England peatlands, the most important plant type is the bryophyte called sphagnum moss. It has the impressive ability to form extensive mats of growth over damp surfaces and even open water. Peatlands can be extremely challenging places for other plants, especially when the conditions are highly acidic and nutrient starved.

Two other important freshwater wetlands do not have the peatlands' high concentration of organic soils. A swamp is specifically a wetland containing plenty of tree or shrub growth. In New England, swamps may look much like upland forests—except that their trees are growing in muck or standing water. In contrast, a marsh is a more open wetland that lacks woody plants. Its dominant vegetation is herbaceous. With those definitions in mind, it's clear that the wetland to your immediate left is a marsh. In fact, the tussock sedge here is a prime indicator of this habitat.

As you continue down the track a little farther, you will come to the In-terpretive Nature Trail bearing off to the left. Take it and continue across the boardwalk into the woods a bit; then retrace your steps to the main trail by the river. While you're on the boardwalk, you'll probably spot other marsh plants: turtlehead, skunk cabbage, bulrush species, forget-me-not, and the highly invasive alien, purple loosestrife. Skunk cabbage is by far the most striking. In early spring, its strange green-and-purple flower structures poke out of the snow, and its huge, coarse foliage persists until midsummer or so. As rank smelling as it undeniably is, this species has been sought out by herbalists, who have used its roots as the base for a soothing ointment. When certain parts of the skunk cabbage are taken internally, however, they supposedly produce nausea, headaches, and vertigo.

Also take a good look at the marsh fringe, where woody plants wait in the wings. See if you can locate the buttonbush shrub and the vining fox grape. In the same general locale, look for the splendid examples of pteridophytes, or free-sporing plants. Chief among these are the bracken, sensitive, hay-scented, royal, and cinnamon ferns. The last two are members of the especially venerable osmunda family. This group of ferns apparently had its origin in the final part of the Paleozoic era, some 250 million years ago. As ancient as these seedless plants are, it has been suggested that they are now more prevalent and successful than at any other time in their lengthy history. If so, it's a rather heartening thought.

Now return along the river trail to the main intersection by the gateway. Turn left (east) through the gate and continue over the river, along the winding Mattatuck Trail, until you reach a paved two-lane road. This is variously called Whites Woods Road and Alain White Road, depending on which side of the foundation map you look at. Carefully cross the thoroughfare; the trail continues after a small jog to the right (south). Here the woodland contains American hazel shrubs, a wide-spreading white oak wolf tree, black cherry, Canadian hemlock, and a number of hefty white pines, which in the early 1990s were heavily infested with adelgids, insect pests that leave a cottony white residue on the bark. In addition, there are sprouts of the blight-blasted American chestnut, as well as healthy specimens of jewelweed, tall meadow rue, and other trailside wildflowers.

Despite their current problems with pests, the big white pines here are ample proof that the gymnosperms have by no means been swept off the scene by the more advanced angiosperm trees. Instead of producing flowers, conifers such as pines form pollen and seeds in two types of cones—the microsporangiate kind, analogous to the male form, and the megasporangiate type, analogous to the female. For many years, the accepted thinking was that the male cones shed their windborne pollen in a willy-nilly fashion, with the vast majority of the pollen coming nowhere near its target, the female cone. This supposedly demonstrated the primitive nature of gymnosperms—they were wasteful and clumsy, inefficient. But recent wind tunnel experiments have shown that pine pollination is not entirely a hit-or-miss affair. Air patterns around the female cone actually influence pollen passing nearby. If a pollen grain is successfully wafted toward the cone so that it strikes it, the grain is sucked up between the cone scales and sticks to a drop of liquid on the ovule inside. Once that tricky part is over, it might seem that fertilization would proceed without delay. However, it's a long time between pollination and fertilization—usually between twelve to fourteen months. When the second process finally does occur, four to eight embryos form, even though there was only one fertilized egg. In the end, all but one of the competing embryos loses out. This type of polyembryony is almost unique to the conifers.

Before long, the path crosses Miry Brook and reveals another aquatic scene, featuring joe-pye-weed and more purple loosestrife and tussock sedge. Can you identify the type of wetland here? If not, check the definitions given earlier in this tour. A little farther in, you enter the profound and splendid gloom of Catlin Woods. The red oaks and especially the Canadian hemlocks here are nothing short of magisterial. This sort of scene was best described by the first-century Roman philosopher Lucius Annaeus Seneca: "When you find yourself within a grove of exceptionally tall, old trees, whose interlocking boughs mysteriously shut out the view of the sky, the great height of the forest and the secrecy of the place, together with a sense of awe before the dense impenetrable shades, will awaken in you the belief in a god."

The best way to return to the parking lot is to retrace the route you took to reach Catlin Woods. For the New England naturalist, seeing the same territory a second time is hardly a boring or repetitive prospect. After all, this environment is rich in information, too rich to absorb all in a single pass. As Henry David Thoreau demonstrated, a plant explorer can spend the better part of his or her lifetime scrutinizing one locale, extracting layer upon layer of significance. The White Foundation and other natural areas help remind us that we're locked in a loving embrace with a world of detail and embellishment.

———————————— ❧ ————————————

Leave the White Foundation by taking Route 202 East to downtown Litchfield and the junction of Route 63. Turn left onto 63 North. Along the road to

Goshen, you'll see an almost ideal upland landscape: many pastures and meadows, broad vistas, and a deciduous tree community dominated by sugar and red maples, black cherry, American beech, and white ash. In Goshen, familiar lawn trees appear, among them Norway spruce and honey locust. The main intersection in town is your next turn. At the rotary, bear onto Route 4 West. After passing a wetland area of giant reeds, drowned tree snags, and then thickets of black locust, there are residential areas planted with traditional ornamental trees and shrubs. Among these is that bushy scourge of late summer, the panicle hydrangea. Most ornamentals produce blooms that don't last long enough, at least by our own selfish human standards. The panicle hydrangea, though, lasts far too long. Its flowers, borne in pompom clusters that look like the tails of clipped poodles, start as a relatively inoffensive white, but then degrade into a sickening institutional pink that lasts in severe cases until the next spring thaw. This plant has some stiff competition from the Colorado blue spruce, but it's still the sorest thumb in New England horticulture.

At 4.0 miles beyond the Goshen rotary, all lingering thoughts of the panicle hydrangea are banished by the beautiful entrance to Mohawk State Forest. Turn left here, onto the access road. Just 1.1 miles farther is a pull-off to the right, where the low profile of the Litchfield Hills reveals itself. Shortly beyond this point, go left at the T-intersection and follow the signs for Black Spruce Bog. Park on the side of the roadway near the maintenance garage. The entrance to the short Bog Trail is just across the road, but before you start down it, take a look at the red pine grove across from the garage complex. You'll find a descriptive sign there. As the sign notes, these trees are under attack from red pine scale, an insect pest thus far confined to southern parts of New England. The symptoms of the potentially lethal infestation include foliage discoloration (in which the needles turn yellow and then red) and severe branch dieback. As is almost always the case with plant pest and disease problems, the situation is at its worst when there is a monoculture, or exclusive community, of the susceptible plant type.

Now head toward the bog. This location complements our previous wetland discussion. Having seen marshes at the White Foundation, you here have the opportunity to examine a peatland. The path down to your goal features lichens, mosses, forest mushrooms, mountain laurel, and wild sarsaparilla. Then the boardwalk onto the sphagnum mat begins. As the trail's name suggests, the dominant tree here is the spindly black spruce, one of the best indicators of true bog conditions. This sickly looking relative of the mountain-loving red spruce has in fact adapted very well to its extremely harsh life in the acid-saturated peat. For example, the tree's short stature prevents it from toppling over in the shallow, soggy substrate. If the water level rises, the black spruce puts out new roots at the appropriate level; and new trees sprout wherever the branch of a parent plant touches the surface. This sort of asexual propagation is known as layering. (By early 1993, some of the black spruces had fallen across the boardwalk. If they still have not been cleared away, be careful negotiating your way over, under, or around them.)

The Black Spruce Bog boardwalk, in Mohawk State Forest.

Even if the black spruce is the most eye-catching plant here, the least obvious plant, sphagnum moss, is still the overall determinant of conditions in the bog. Sphagnum is related to the mosses seen elsewhere in the woods and on rocks, but its evolutionary path diverged from theirs at an early stage. In common with the regular mosses and the liverworts, though, sphagnum is what botanists call a nonvascular plant—one lacking the highly developed tissues that move water, nutrients, and food substances up and down a stem. Without this advanced internal circulation system, nonvascular plants cannot afford to rise high in the air, away from their source of moisture. They are therefore subject to severe size and shape limitations. Still, if you pick up a small clump of damp sphagnum and squeeze it, you'll see that it has tremendous water-absorbing capacity. What isn't so apparent, though, is that the sphagnum is a fierce rival of the more advanced plants, often successfully out-competing them. This is partly the result of its skill in hogging the scanty nutrients available, and partly because it actively poisons the bog's water by increasing its acidity. Here, however, these nasty tricks have not prevented a few highly specialized types of shrubs and herbaceous vascular plants from getting established. Close to the boardwalk, for instance, grow three representatives of the heath family: highbush blueberry, sheep laurel, and dwarf huckleberry, all of which require acidic conditions. Somewhat more secretive is a fascinating carnivorous plant, round-leaved sundew. A red-tinted species that tends to be found where the red types of sphagnum grow, it often escapes the gaze of all but the most careful observer. It manages to extract additional nutrients by digesting the insects it traps on the sticky hairs of its paddle-shaped foliage. In the field of herbal cures, sundew has been used to treat whooping cough and even to counteract the aging process. Had it been successful, you have to suspect there would be quite a sundew-growing industry by now.

―――――――――――――― ∾ ――――――――――

The final destination of the tour lies just a few miles from Black Spruce Bog. Return to Route 4 and turn west (left) onto it. Follow the route signs carefully in this area; 2.2 miles beyond the state forest entrance, you will come to the junction of Routes 4 and 125 North in Cornwall. Turn left here onto Pine Street and drive past the curiously eccentric town library. This small-scale delight thumbs its nose at architectural good order by somehow combining cedar shingles with Doric columns. At the T-intersection, bear left and follow the road toward the hillside ahead. Be prepared for a shock.

This hillside is the site of the famous Cathedral Pines, once renowned as the tallest surviving white pine trees in New England. These giant conifers were often cited as an example of how first-growth timber must have appeared to early English settlers. But on July 10, 1989, a violent summer storm produced a tornado that ripped through Cornwall. The grand old trees were largely stripped of their branches, split in two, or otherwise weakened beyond much hope of recovery. The Connecticut chapter of The Nature Conservancy, which administers this parcel of land, has decided to let nature take its course. Accordingly, the shattered tree trunks remain, an unsettling testament to the potential fury of New England weather. Had this amazingly selective destruction of some of the region's greatest trees taken place two or three centuries earlier, no doubt it would have been the subject of more than one allegorical Sabbath-day sermon, on how in their pride the exalted shall be laid low.

Cornwall's Cathedral Pines, four years after the deadly tornado. Still discernible are the tall, straight trunks characteristic of mature white pines. There's little wonder that the species was considered ideal for sailing-ship masts.

105

TOUR 11 ⟨⟩

In Connecticut:
Bloomfield to Simsbury

Driving Distance:
3 miles

Connecting Tour:
Number 12

Starting Point:
The junction of
Routes 178 and 185
in Bloomfield

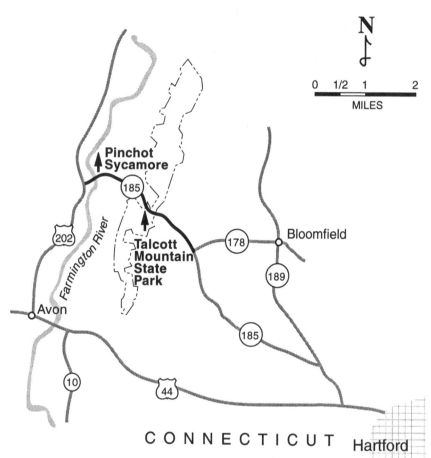

Tour 11
The Plants of Metacomet Ridge
and a Riverside Titan

HIGHLIGHTS

Traprock Considered ∿ Talcott Mountain State Park ∿ A Pumped-up Tree and Its Many Uses ∿ The Moral Equivalent of Fruit ∿ A Fine Collection of Forest Trees ∿ The Ridgetop's Spring and Summer Wildflower Shows ∿ White Baneberry: A Case of Toxic Loveliness ∿ A Deep and Shady Grove ∿ A Tower and Its Fancy Annuals ∿ The Pinchot Sycamore ∿ Connecticut's Biggest Tree ∿ A Matter of (Structural) Integrity ∿ Keeping a Stiff Upper Limb ∿ Boneless Wonders ∿ What It Takes to Be a Real Tree ∿ A Sappy Theory

T o the student of landscapes, the broad and fertile Central Lowland is New England's great anomaly. It belies the region's hard and hilly image. It is open, cornucopious, and populous. It has been the cradle of industry, agriculture, and a fine array of educational institutions; yet it also contains some of the most fascinating natural features in the entire Northeast. The Connecticut River, an uncommonly beautiful waterway, runs through the lowland from the northern border of Massachusetts to the point where it veers off to the southeast, at Middletown, Connecticut. And throughout the lowland run massive walls of rock that impart an unusual sense of vertical foreground.

In contrast to the lowland floor, these dramatic heights are formed of resistant igneous dolerite and basalt, known collectively as traprock. This term derives from the Swedish word *trappa,* meaning step, or perhaps from the related Dutch *trap,* signifying staircase. At any rate, the reference is to the rock's tendency to form platforms, or levels, on its exposed faces. By far the longest of the trap hills is Metacomet Ridge. Named for the Indian resistance leader who launched King Philip's War in the late seventeenth century, the ridge extends with only one significant break from Branford on Long Island Sound to the Holyoke area of Massachusetts. The other Connecticut traprock locations described in this book are formed of intrusive rock—that is, from large bodies of molten magma that cooled and hardened before they reached the earth's surface. But Metacomet Ridge, lying to the east, is composed of basalt, a rock that was extruded onto the surface in lava flows before it solidified. (For more on other traprock sites, see Tours 3, 4, 12, 25, and 26.)

107

*A welcome sign of win
end. The strange, mott
inflorescences of skunk
cabbage poke through (
leaf litter at the base o
Talcott Mountain.*

In this tour, we visit one of the most beautiful parts of the Metacomet heights, the section northwest of Hartford known as Talcott Mountain. The hike to the summit, along a well-defined foot trail, is good exercise, and it isn't overly taxing to most people when they choose a comfortable pace. The view from the summit's Heublein Tower makes the exertion worthwhile. From the trip's starting point, head west on Route 185 a scant 1.1 miles, past Penwood State Park and on to the entrance of Talcott Mountain State Park, on your left. Go up the paved access road to the cul-de-sac; the Heublein Tower Trail begins here. Park on the side of the road, wherever permitted. As you'll see, you will be ascending the ridge near its steep cliff face, so watch your footing.

The forest in the lower section of the hike contains a wide variety of woody plants, from Canadian hemlock, black birch, red oak, and white ash to pignut hickory, pin cherry, witch hazel, and the sugar and striped maples. One other tree worth hunting for is the American hornbeam, a birch family species that has simple toothed leaves and smooth gray bark. Its irregular, bulging trunk and branches often give the impression that the tree has been doing muscle-building exercises. The hornbeam is sometimes also called ironwood, in recognition of the exceptional strength of its timber. At one time, the wood was widely used for levers, handles, and the like. Two great naturalists living a century apart noted other applications: Pehr Kalm wrote that it was used for cart axles and as a premium-grade fuel, and Thoreau observed that Indians fashioned it into canoe paddles.

After the first, steepest segment, the trail levels off along the ridgeline. Magnificent views of the peaceful Farmington River valley open up at various spots along the cliff face—and once again, be careful of your footing. One of the more common clifftop plants is that tough customer, the juniper tree known as red cedar. Its female cones are small and fleshy; they may look like berries, but they definitely aren't. Only angiosperms, the flowering plants, produce true fruit, berries or otherwise; and red cedar is very much a gymnosperm. Not concerned with such botanical realities, however, the New England Indians treated the plant's cones as the moral equivalent of berries, and happily collected them as a food rich in sugar and starch. More recently, the species' fragrant heartwood has been used for pencils, and an oil distilled from the wood has done service in liniments, insecticides, soaps, and perfumes.

Elsewhere along the trail, you should catch sight of basswood, the native linden tree of the Northeast and the North Central states. It's joined here by American beech, American chestnut sprouts, hackberry, bigtooth poplar, red and white oaks, and American cranberry bush. Wildflowers are most common along the sides of the trail. My experience is that the summer-blooming species are the most prevalent. Some of these are the woodland sunflower (a traprock-ridge regular), Lowrie's and Schreber's asters, white snakeroot, black snakeroot, clearweed, agrimony, horse balm, and at least three legumes—round-headed bush clover and the panicled and pointed-leaved tick trefoils. But this is not to disparage the

Red cedars and other trees at the Talcott Mountain cliff edge. The valley of the Farmington River is seen in the distance.

springtime floral display. In that earlier phase, black cherry trees and viburnum shrubs burst into bloom, as do the herbaceous white baneberry and false Solomon's seal. The baneberry is a plant of special note; later, it develops striking white fruit on bright red stalks. Resist the temptation to sample the berries—they are quite poisonous. Many plants, whether wild or cultivated, have developed toxic substances for their own defense. *Never* taste or ingest any part of a plant unless you're absolutely sure it's safe. This most definitely goes for houseplants, too.

On the final approach to the summit, the trail passes an outcrop of the basalt that forms the bastion of the ridge. A bit farther along stands a darkly attractive grove of hemlocks. The wood of this familiar, shade-tolerant conifer species is usually considered too brittle for small-scale carpentry uses, but its resin has yielded Canada pitch, a substance that has a lower melting point than the pitch normally extracted from pine trees.

A moment later you're at the tower, an imposing structure built in 1914 for Gilbert Heublein and used as a summer home until 1937. It stands 900 feet above sea level, an impressive height here, but a mere one-seventh the altitude of New England's highest peak, Mount Washington. If you're in luck, you'll find the garden beds at the base of Heublein Tower well planted with fancy annuals such as Madagascar periwinkle. They provide a cheerful and exotic contrast to the hemlock stand you've just walked through.

———————————— ∾ ————————————

The hike up and down the ridge constitutes the bulk of this tour, but there's one more treat in store. As you drive out to the park entrance, turn left on 185 West and follow the route signs carefully for another 1.3 miles. This brings you to the bank of the Farmington River and a pull-off to the right, which you should take. This is the site of the awesome Pinchot Sycamore, reputedly the largest tree in Connecticut. While there are various criteria that can be applied to determine tree size, no one can quibble with the overt majesty of this titanic organism and its companion just across the way. The sycamore, one of the New World's native plane trees, is the tallest broad-leaved species in the forests of eastern North America. Curiously, its wood is not particularly rot resistant, though it is odorless and tasteless—a fact that once made it a popular material for food containers.

There is no better place than this to contemplate the great masterpiece of design that a tree really is. Consider what the Pinchot Sycamore must be, and what it must accomplish, to survive year after year. For one thing, its architectural framework must be supple yet strong enough to bear its massive branches and their foliage. Animals achieve their structural integrity by means of skeletons; but plants, even the largest, have no shells, no bones, no ligaments. They have tackled the challenges of the upright life in a wholly different way, down at the microscopic level. Plant cells are individually fortified by walls of a carbohydrate compound called cellulose. In addition, woody plants contain another extremely important organic substance, lignin. Obviously, the system works. When was the last time you saw an animal as big as this tree? The giant blue whale and its relatives don't count, since they have all that lovely salt water to buoy them up. (Incidentally, herbaceous plants lack lignin. To the extent they need vertical support, they count on turgor—the confined water pressure in their cells. One drawback of this system is evident enough in a dry spell: when a nonwoody plant's internal water level falls beyond a critical point, it begins to droop and wilt.)

Though many people have trouble realizing it, trees are as fully alive as any other being on earth. The ingenuity of their design would be all for nought if they did not undertake such complex life processes as the absorption of water and nutrients, and the transportation of food to their tissues. Interestingly, for years scientists were unsure how a plant the size of the Pinchot Sycamore could possibly get substances from its roots all the way to its uppermost leaves. No one physical process could account for more than a portion of the total rise of the sap. Finally, however, a plausible composite explanation, dubbed the Cohesion-Adhesion-Tension Theory, was put forward. Simply stated, it suggests that water reaches the top of the tree because its molecules cling together in long, continuous columns in the tubelike channels of the plant tissue. These extremely narrow water columns are pulled upward by the loss of the water already in the foliage. You won't be able to observe the workings of this subtle process, but you certainly will note its beneficial effects. As you gaze upward, you can tell by the existence of thousands of leaves responding to the sunlight that this tree has mastered the art of living on a scale we human beings can only imagine.

The massive lower trunk of the Pinchot Sycamore.

TOUR 12

**In Connecticut
and Massachusetts:**
Simsbury, Connecticut, to
Ashley Falls, Massachusetts

Driving Distance:
63 miles

Connecting Tour:
Number 11

Starting Point:
The junction of
Routes 185 and 202 in Simsbury

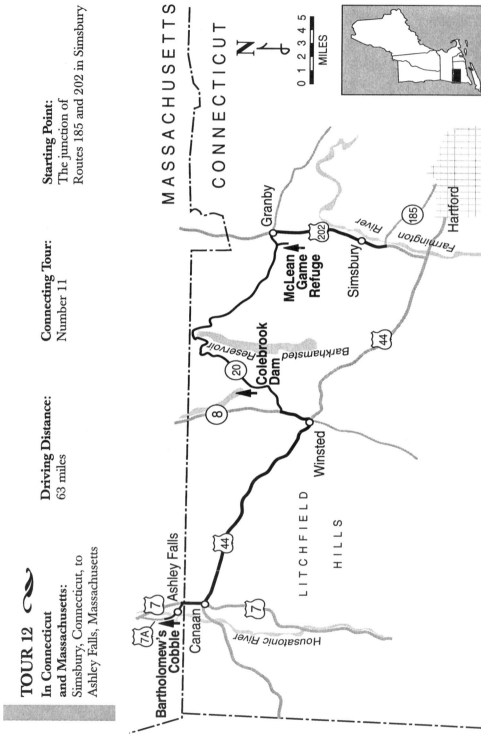

Tour 12
The Road to Bartholomew's Cobble

A Recommended Fall Foliage Tour

HIGHLIGHTS

The Sycamores of the Central Lowland ∽ McLean Game Refuge ∽
The Beautiful Barndoor Hills ∽ The Shagbark and Its Hickory Kin ∽ A
Gateway of Hemlocks and the Trailing Arbutus ∽ Two Surprising Oaks ∽ Into
the Litchfield Hills ∽ Megabotany and the Colebrook Dam ∽ A Meeting of
Hardwood Regimes ∽ The Cottonwood, the Black Locust, and the Sweet-
Splitting White Ash ∽ Bartholomew's Cobble ∽ A Rock to Resist the River ∽
A Miracle of Plant Diversity∽ A Profusion of Pteridophytes and a
Walking Fern ∽ Two Studies in Blue

This trip should be required driving for any person who thinks Connecticut is nothing more than an extended bedroom community. It's
a journey that offers a triad of gorgeous rural landscapes: first the lush
Central Lowland with its traprock hills, then a remote and forested part of the
Litchfield Hills highlands, and finally the enchanting Housatonic Valley, with a
botanical marvel just over the Massachusetts line. The itinerary takes a full day,
or pretty close to it, so you're advised to get under way at a fairly early hour.

The starting point is just across the gentle Farmington River from the mighty
Pinchot Sycamore, the concluding stop of the previous trip. Head north on Route
202. This residential area is graced by some impressive lawn trees. The prepon-
derance of sycamores is interesting; in the wild, they're rarely numerous in any
one spot, but here and in other Central Lowland towns they've often been planted
as ornamentals. Their appeal is easy to understand. They grow to be huge, stout-
trunked shade trees with bark that peels off in patches to reveal cream-colored,
light green, and orange-brown sections. The effect can be dazzling. By the same
token, the plant's wood has always been highly prized, especially for baskets,
pails, and other containers. It also has a lovely grain and a pleasant, streaked
pattern that has ensured its use in decorative paneling.

Some of the most venerable extant sycamores in this great valley date from
the Revolutionary era. Their species, often mistaken for a kind of maple, actually
belongs to the plane tree family. (Alas, there is also a totally distinct sycamore
maple. It is a European maple of minimal appeal that can be found in New

113

England parks and arboreta—see Tour 7. Don't confuse it with the genuine article.) The history of the plane tree family goes back a long way by angiosperm standards; it extends into the last part of the Mesozoic era, otherwise known as the Age of Dinosaurs. In the past few centuries, the London plane tree, a hybrid of our native sycamore and a closely related species from the Balkans, has become an exceptionally popular urban street tree. (For more on this hybrid's origins, see Tour 14.)

At 1.7 miles from the start of this tour, the road goes through Simsbury town center. Note the honey locust and pin oak plantings along the way. Farther north you enter the town of Granby, where, for the time being, the traditional nursery trade coexists with condo "development." Soon thereafter you reach the junction of Routes 202 and 20. Bear left onto 20 West and go an additional 1.3 miles to the intersection of Barndoor Hills Road. A good landmark is the stolid traprock farmhouse standing on the near left-hand corner. Turn left onto that road and head south 1.0 mile. Turn left again, down the access road to McLean Game Refuge, and proceed to its streamside parking lot. You are now in the Barndoor Hills, a part of the complex of traprock heights that runs most of the length of the Central Lowland.

This pleasant stop contains the tour's one substantial hike, along woodland footpaths to the summit of East Barndoor Hill. It should present no particular difficulty, unless you attempt the ascent at a dead run. Before heading onto the trail, pause to look at the trees fringing the parking area. Among them is an aptly named shagbark hickory, with its trunk peeling in long, curving strips. Hickories form the genus *Carya,* and they are one of the mainstays of the central hardwoods regime that makes up the vast bulk of Connecticut's forest lands. Hickory wood is extremely tough and has been used in all sorts of ways: for bows, tool handles, baskets, carriages and wagons, and even for horse-racing sulkies. And it has one of the highest thermal outputs of any native firewood.

Now cross the tranquil stream. This is a pretty spot, and local folks often come here to while away hot summer afternoons. Head over to the north end of the picnic grove, where the woodland trail begins. Generally speaking, moisture-loving Canadian hemlocks are encountered at the bases of the traprock hills, rather than in the drier upper reaches; and sure enough, here you pass through a dense group of mature specimens. As you walk under these big conifers, it is difficult to suppress the feeling that the world has already reached its end point, in a time more primitive than our own. Unfortunately, many hemlocks in Connecticut are under attack from a serious insect pest, the hemlock woolly adelgid. We can only hope that the dire predictions about the infestation will ultimately prove to be wrong.

A little farther down the trail, our second hickory of the day, the pignut, appears among the oaks and other broad-leaved species. Its handsome, vertically ridged trunk does not exfoliate the way the shagbark's does. Two evergreen pteridophytes, the Christmas fern and the marginal woodfern, dot the forest floor. As in many other central hardwoods areas, American chestnut suckers still sprout here, even though the main stems of their parent trees, attacked by the devastating

Lowly in habit, heavenly in scent: trailing arbutus flowers in the lower woodland of the McLean Game Refuge.

chestnut blight fungus, may have died decades ago. (For more on the chestnut and its tragic downfall, see Tour 24.)

This segment of the hike has a particular transcendent reason why the McLean Refuge should be visited in May. If you look carefully along the trail, you should find it—the ground-hugging trailing arbutus. Though it might not appear to be, this species is a woody plant and an ericad—that is, a member of the heath family—so its kin include azaleas, rhododendrons, and mountain laurel. The white or pink flowers it produces are not particularly showy, but their fragrance is beyond the reach of adjectives. Unfortunately, this has almost been the species' downfall. It has been picked relentlessly by people who can't understand that the plants' gentle perfume, once taken out of its forest setting, is not nearly as wonderful. In other places, I have seen thoughtless individuals bringing home sap buckets stuffed with trailing arbutus in full bloom, not to preserve and study the plants, of course, but simply to add a pretty smell to their dining room centerpieces for a day or two, which is downright shameful.

One of the chief attractions of East Barndoor Hill is that it's a terrific place to see the zonation of plant communities. Indeed, as you reach the upper half of the trail, several new plant types have replaced some of those encountered below. Mountain laurel and lowbush blueberry are common ericads here, and there are a number of new wildflowers, such as early saxifrage and yellow false foxglove. Other interesting plants include a monocot, wood rush, and a common rock-loving fern, polypody.

When the main trail comes to the top of a rise and begins to descend again, bear off to the right on the footpath that leads up to the summit. Here is the uppermost plant community. Near the cliff, shrubby common juniper and upright red cedar mix with shagbark hickory, white pine, and Canadian hemlock. (Zonation notwithstanding, these last three show that more lowland types do make it to the summit. In nature, nothing is quite as clean and distinct as the working hypotheses of ecologists.) The summit area also contains two surprising members of the oak genus. While the word *oak* usually conjures up the image of great, long-lived trees, some kinds are actually spreading, understory shrubs. Our first example of this is the scrub oak, a Cape Cod regular that also does well here, close to the drop-off, where conditions are the most challenging. (The largest known plant of this species, a modest 25 feet tall, grows not far away in the town of Windsor.) A bit farther down the slope, several colonies of the other—less frequent—shrub form, chinquapin oak, can be found interspersed among the trees.

Before you start your return trip, switch off the botanizing for a minute and enjoy the view. From this vantage point in the Central Lowland, you are facing a wholly different terrane, the Western Highlands, including the Litchfield Hills. That's where we're headed next.

--------------------- ⟲ ---------------------

As you exit the McLean Game Refuge, turn right onto Barndoor Hills Road and return to Route 20. At the junction, bear left onto 20 West. The road soon climbs out of the great valley and onto the much older landscape we saw not long ago from our traprock perch at the refuge. This route, through the forest north of Barkhamsted Reservoir, is one of the best places in the state to see autumn foliage colors: the flame orange of sugar maple, the scarlet and burgundy of red maple, the brighter yellows and reds of black cherry, the reddish brown of red oak, and the odd pale yellow and dull purple mixture of white ash. At the picturesque town of Riverton, you cross the West Branch of the Farmington River; 2.2 miles beyond it, turn right onto Route 8 North. Proceed another 3.7 miles and take the right-hand turnoff to Colebrook River Lake. Go past the U.S. Army Corps of Engineers office and across the dam, to the parking area on the far side.

This site provides a view of botany on its grandest scale, and also a stunning look at what one species of the animal kingdom can do to alter the environment. On the hillsides above the water, and also in Algonquin State Forest just across Route 8, lies a transition zone where the central hardwoods regime of southern New England blends with northern hardwoods more characteristic of New Hampshire and Vermont. The vastness of the forest community, coupled with the sheer cliffs exposing rock units more than half a billion years old, imparts a feeling of monumental timelessness. But glancing a little lower, you see a different kind of monumentality, the work of a creature obsessed with time: the Colebrook Dam and its lake, formed by the impoundment of the Farmington's West Branch. Completed in 1969, the dam serves as a hydroelectric generating station, a recreation facility, a protection against floods, and the source of 10 billion gallons of fresh water for the city of Hartford. It could be said that this is not what nature intended; but it is now the local version of nature.

The final segment of the trip takes us deeper into the Litchfield Hills of northwestern Connecticut and over the border into Massachusetts. Return to Route 8, turn left onto it, and continue south 5.4 miles to downtown Winsted, where you bear right onto Route 44 West. Here, it is a little easier to note the

The wilderness dramatically modified by a single animal species. Colebrook River Lake and a portion of its dam (right).

mixture of American beech, a northern hardwood, with its central hardwoods equivalent, red oak. By 11.0 miles west of Winsted, you've crossed the divide between the drainage systems for the Connecticut and Housatonic Rivers and are following the course of the westward-flowing Blackberry River. Cottonwoods and black locusts, both colonizers of floodplains and other bottomlands, are well established here. The black locust, a legume highly prized in colonial and Federal times for its rot-resistant wood, originally was a native of the Appalachians to the south, but now it's found in the wild as far north as the mountain watercourses of Vermont, New Hampshire, and Maine. The cottonwood, on the other hand, is a poplar, which means it's a representative of the willow family. This native is fairly prevalent in our region, but it takes a much more prominent role in the Midwest. Its common name derives from the silky white tufts attached to its seeds. One of the most vivid memories of my childhood in Illinois, where the tree is particularly abundant, is of lawns and sidewalks blanketed by drifting masses of cottonwood fluff. This must be one of the most effective modes of seed dispersal ever devised by plants.

The valley scenery in the environs of East Canaan is especially pleasing. White ash is plentiful along the roadside. The supple wood of this tree has a multitude of uses, and as every veteran woodchopper can testify, its logs practically stand up and split themselves, provided you ask them to, in the right tone of voice. As you enter Canaan proper, welcome yourself to the beautiful Housatonic Valley. At the downtown junction of Route 7 North—and make sure it is North—turn right. Go another 1.2 miles and bear left at the junction that has the "To 7A" sign. Almost at once, you enter Ashley Falls, Massachusetts, and in fact you are on Route 7A. After a mere 0.7 mile from the state line, you're at the town's center; turn left onto the road that bears you across the lush farmland of the

The three great elements of Bartholomew's Cobble: an extremely diverse plant community, the sweetening limestone beneath it, and the somnolent Housatonic. –The Trustees of Reservations

Housatonic floodplain. At this point you should notice signs directing you to Bartholomew's Cobble; 0.9 mile from the town junction, bear left onto the gravel road, and then turn into the Reservation parking lot.

Bartholomew's Cobble is one of the undisputed gems in a marvelous network of open spaces and historic sites administered by The Trustees of Reservations, a Bay State nonprofit organization supported by visitor donations and dues. Massachusetts, which also boasts an impressive array of state-run lands, is especially blessed to have this effective partner in such a good cause. The other forty-nine states should be so lucky.

The Cobble consists of two knolls of white marble and quartzite. These rock types were formed off the coast of ancient North America early in the Paleozoic era, before the appearance of land plants or animals. Later, the rocks were subjected to the intense forces of mountain building that time and again changed the eastern side of our continent so dramatically. If you've ever tried to hammer a piece of quartzite off an outcrop, you already know it is one of the toughest and unyielding substances on earth. This may help explain why the Cobble has so far survived the erosional effects of the Housatonic, which has snuggled up next to it like a patient boa constrictor.

Still, the most notable aspect of this locale is its botany. New England is blessed with a number of distinct environments, but rarely can several of them be found all in one place. That's the miracle of the Cobble. On its grounds are some eight hundred wild plant species—a mind-boggling number for such a small area. The floristic diversity reflects the diversity of the habitats themselves. Unusual lime-loving ferns roost where the marble's calcite has neutralized the acid-prone soil; there is a miniature dry upland zone and also a river's-edge wetland regime. And as if that weren't enough, this valley setting is a meeting place for southern and northern plant communities.

Make sure your first stop is at the little museum not far from the entrance. Among other interesting exhibits are samples of forty Cobble fern species, some of them extremely rare. (A collection of forty flowering plants would raise no eyebrows; but extant fern species are so few in number, relatively speaking, that this really is a treasure trove.) I heartily recommend the Ledges Interpretive Trail booklet, available here; it can be borrowed or bought, as you wish. In the museum's vicinity, you possibly might catch sight of one of the most splendid wildflowers in the mint family, wild bergamot. Its genus, *Monarda,* commemorates the name of the great Spanish botanist Nicolas Monardes. He produced the first book on the medicinal uses of New World plants.

When you're ready to explore the Cobble proper, set out on the Ledges Trail. Depending on the season, you may see wild ginger, clearweed, round-lobed hepatica, bellwort, white baneberry, and many other wildflowers as well. Of these, the hepatica is best known for its old-time use as a liver ailment cure. This was in accordance with the medieval Doctrine of Signatures: the plant's lobed leaves somewhat resemble the shape of the human liver, so, obviously, it was meant to treat that organ. (For more on the notorious Doctrine, see Tour 30.)

However interesting the flowering plants here may be, the ferns steal the show. The path named after naturalist and field guide author Boughton Cobb is the absolute high point. On an adjacent rock face, the delightfully eccentric walking fern spreads itself by forming a new plant wherever its frond tips touch the mossy surface. There and elsewhere along the main trail are a host of other ferns: maidenhair spleenwort, silvery spleenwort, Clinton's fern, purple cliff brake, wall rue, *Woodsia* species, and more. For the pteridophyte fancier, this is the Garden of Eternal Bliss.

If you have time, amble down Bailey Trail. This takes you along the river and into the lush habitat of bottomland plants. The most splendid wildflower of all, great blue lobelia, opens its coolly elegant blossoms in August. It's difficult to believe that anything this chaste-looking could have once been used in the treatment of venereal disease, but after all, the plant's taxonomic name is *Lobelia siphilitica.* Another striking angiosperm here is the woody swamp dogwood. Found growing in the water or very near it, this shrub produces clusters of pale yellow or cream-colored blossoms that later develop into the kind of fruit known as drupes. These resemble berries and ripen to a breathtaking steely blue.

It's only fitting that the last word be devoted to our friends with the fronds. One of the daintiest species of all, the maidenhair fern, shares its low, damp habitat with heartier and higher-growing companions, the cinnamon, sensitive, and interrupted ferns. If there is any one spot in the natural world of New England worthy of repeated visits, Bartholomew's Cobble must certainly be it.

TOUR 13 ❧

**In Connecticut
and Rhode Island:**
North Windham, Connecticut,
to the Great Swamp
Fight Site, Rhode Island

Driving Distance:
60 miles

Starting Point:
The North Windham
junction of Routes 6
and 203, just east of
Willimantic, Connecticut

Tour 13
From the Land of the Pequots
to a Forest Battlefield

HIGHLIGHTS

Connecticut Mahogany ∿ New England's First War ∿ Goodwin State Forest ∿
Autumn Elaeagnus ∿ An Ecotone and Abundant Wildlife ∿ The Aromatic
Spicebush ∿ Lightning Strikes and Exploding Trees ∿ Aquatic Flora and Fauna ∿
A Blight-Resistant Chestnut? ∿ From Avalonia to the Provisions State ∿
Pachaug State Forest ∿ A Symbiosis Showcase ∿ White Cedars and Rhodos ∿
Arcadia Management Area ∿ Winterberry, Maleberry, and Leatherleaf ∿
Hurricanes as a Way of Life (and Death) ∿ A Swampy Killing Ground

I n the first decades of New England settlement, European colonists faced two major military confrontations with the American Indians they were beginning to displace. The first crisis, the Pequot War, occurred in 1637 and was basically confined to southeastern Connecticut; the second, King Philip's War, lasted from 1675 to 1676, and involved most of southern and central New England. In this tour, we'll use these two conflicts as historical benchmarks, the starting and ending points of our journey through the wilds of eastern Connecticut and Rhode Island. In between them is a story about the region that is at the same time more modern and more ancient. It describes both the present-day plant life and the geological and meteorological setting it inhabits.

Head east on Route 6 from the starting point. At once red oak, that key player in the central hardwoods regime, is present along the roadside. Among the other trees are two well-known conifers, red cedar and white pine, and one familiar species in the rose family, black cherry. The latter's heavy, warp-resistant wood polishes to a deep and luminous red tone. It has long been a favorite material for making tabletops, cabinets, and paneling, and even U.S. dollar-bill engraving plates. In this state, traditional furniture making so often involved the use of black cherry that the tree's wood was nicknamed "Connecticut mahogany."

If you were making this trip some four hundred years ago, you would have found the territory claimed not by English-speaking landholders but by the people then considered the region's fiercest Indian tribe, the Pequots. Relative newcomers to this area—they had migrated from the upper reaches of the Hudson Valley in the late 1500s–they apparently didn't waste much time in transcendental meditation classes or sensitivity seminars. On the contrary, they conquered, bullied, and otherwise intimidated neighboring tribes whenever the opportunity presented itself. Indeed, it's said that the name *Pequot* was the Algonquin term for "destroyer." (Melville fans will recognize the variant *Pequod,* which was the seagoing home of Ishmael, Queequeg, Starbuck, and Captain Ahab.)

In retrospect, it's understandable that this warlike nation would react rather dramatically to the encroachment of another highly acquisitive people, the English. Within a year of the Puritans' 1636 founding of Connecticut Colony, the Pequots, led by their sachem, Sassacus, launched attacks on the new white settlements. However, they clearly underrated Old World technology, their enemy's Calvinistic determination, and the resentment of their Indian neighbors. Before long, the colonists and a force of Mohegans and other native allies counterattacked, hitting the Pequots at their fort near the seacoast, on the bank of the Mystic River. In this one engagement, practically the entire Pequot culture was obliterated. Amazingly, one small component of the Pequots survived and eventually settled on the faraway shore of Wisconsin's Lake Winnebago. Sassacus, who had once ruled a large portion of Long Island as well as the mainland from Rhode Island to eastern New York state, had to flee for his life. Later, while taking refuge among the Mohawks, he was assassinated. And so began a familiar pattern of fear, confrontation, and genocide that would manifest itself from one side of the continent to the other over the next two and a half centuries.

At 4.8 miles from the starting point, you'll see the entrance to your first objective, James L. Goodwin State Forest, on the left. Park in the small lot just up the access road, and note the large cottonwood tree in the adjacent lawn area. Mature cottonwoods have deeply furrowed gray bark, and the pattern of vertical ridges can be quite striking. This stop involves a short hike; begin by walking down the open area toward the large body of water, Pine Acres Pond. In the meadow is a group of large shrubs with eye-catching silvery leaves. This is elaeagnus, a genus of salt-tolerant plants now common in southernmost New England. My guess is that the shrubs here are the species known as autumn elaeagnus, based on the fact that their bright red fruit ripens in late August and September. Another species, cherry elaeagnus, is similar, but its fruit ripens in the first half of summer. (For the story of this plant's great escape from its intended setting, see Tour 7.)

The pond was created in 1933 when a swamp was flooded. Take the Natchaug Trail, which starts just north of the elaeagnus meadow and runs along the pond's western shore. It is marked with blue blazes. (Somehow I have successfully resisted the obvious What-in-Blue-Blazes pun. In case you're not familiar with the term, a blaze is a marking of a particular shape or color, often made with spray paint or metal tags affixed to trees at intervals along a path.) Before reaching the water's edge, you'll see a demonstration area illustrating the ecotone, or interface, of forest and field. This is a zone particularly rich in wildlife. Notice how the pioneer staghorn sumac has established itself at the edge of the taller tree growth. One important point to be made here is that periodic tree cutting, whatever its effect on the forest itself, tends to increase animal diversity. In contrast, a mature hardwood habitat is home to relatively few wildlife species.

Once you reach the main section of the woods, you'll find several notable woody plants, including such southern-range species as tulip tree and spicebush. The latter is a shrub with strikingly aromatic foliage. Crumble a leaf between your fingers, and you'll smell the connection between this North American native and its Mediterranean relative, the laurel of bay leaf fame. In the same area grow the fall-flowering witch hazel and the curious beaked hazel, which has fruit in the form of a nut covered with a long-snouted involucre, or husk. By all means seek out the white pine with the lightning-strike scar. Lightning is more common as a cause of tree mortality than many people realize. A western study revealed that one out of every three dead ponderosa pines in northern Arizona had been electrocuted by a thunderbolt. When lightning hits a tree, it travels down the outer sapwood or exterior. The sudden, huge increase in temperature causes water in the plant tissue to boil instantly, with explosive results. One characteristic sign of a strike is a long, grooved depression in the trunk where the bark has been blown away. To prevent such dire results, some large and especially fine specimen trees in botanic collections are equipped with lightning rods, in the same way buildings are.

The forest floor, meanwhile, is home to cinnamon, royal, New York, and Christmas ferns, as well as goldthread, an evergreen spring-blooming wildflower. It's also a good place to have a conversation with some familiar forest residents, the chickadees. When they feel like it, they can be amazingly tame. With a good supply of patience and sunflower seeds, I've managed to get them to perch on my hand.

At the pond's edge is plenty of evidence of another industrious animal resident, the beaver. The last time I was there, a lodge stood just offshore, and

Work in progress. This beaver-gnawed yellow birch in Goodwin State Forest is a good reminder that humankind was not the first lumber-hungry mammal to reach New England.

there were quite a few saplings that had met an untimely end by the meticulous gnawing of these large, flat-tailed rodents. Two of the bigger water-loving plants, clethra and swamp loosestrife, share the crowded bank with skunk cabbage and jewelweed, and in some places, fragrant water lily inhabits the shallow-water zone.

As you return to the parking lot, try to visualize how this woodland must have looked when it contained mature, massive-trunked American chestnut trees. This area once had many chestnuts gracing it, but they were brought low by a murderous fungus disease, the chestnut blight. Apparently, it originally came from the Orient and was first discovered in this country in the vicinity of New York's Bronx Zoo in 1904. In the following decades, it spread throughout the tree's range and struck down virtually every adult specimen of this handsomest Northeastern hardwood. Curiously, the fungus does not attack the roots. This accounts for the host of new shoots, or suckers, that keep sprouting from the rootstocks of blighted trees. Eventually, they too succumb, but not before they produce some food reserves for the plants' long-suffering underground tissues.

You may recognize some American chestnut suckers along the trail. In fact, there is a larger tree located along the roadway, a little to the north of the parking lot. To date it has managed to beat the odds, even though it has the stem cankers characteristic of the disease. According to a State Parks and Recreation official, this is due to the fact that the plant also hosts a virus that curtails the full development of the cankers. Scientists are investigating the ramifications of this potential microscopic ally. Another sign of chestnut research on the state forest grounds is the presence of disease-resistant hybrid varieties.

The second destination is another state forest with a distinctly different plant community. When you're ready to set out for it, return to the main road and turn left onto Route 6 East. After 1.4 miles, you'll come to the junction of Route 97; bear right onto 97 South. This stretch is typical of much of eastern Connecticut in modern times. Woods composed of oaks, basswood, white ash, and black cherry have reclaimed land once extensively cleared. Here's one telling statistic that points out the decline of farming: even as late as 1935, more than two-thirds of Connecticut's land was used for agricultural purposes. By 1979, only one-sixth was employed for that purpose. This is a great change from the days of the American Revolution, when the bounty of this land was gratefully acknowledged by the quartermasters of the half-starved Continental army. In fact, Connecticut was known as the Provisions State. Its main contributions included wheat and beef. After the Revolution, rye outdistanced the wheat, which had been subject to such deadly diseases as black stem rust ever since the 1600s. Still, the development of more resistant strains set the stage for one last wheat comeback in the middle of the nineteenth century.

For many years, earth scientists have realized that this region, in common with Rhode Island, eastern Massachusetts, and maritime Maine, is geologically distinct from the old oceanic and continental rock units that lie inland. One recent theory holds that this area was actually a separate minicontinent that crashed into (or, if you prefer, docked with) ancestral North America hundreds of millions of years ago. This enigmatic piece of the earth's crust, or suspect terrane, has been named Avalonia.

In another 4.8 miles, turn left onto Route 14 East. This is a good place to observe the lush, rounded look of the central hardwoods community. In summer it is the perfect hiding place for wildlife, but in winter it often provides less warmth and protection than northern woods, which have a higher percentage of evergreen conifers. As a rule, the floor of the central hardwoods forest is also drier, hence more fire prone than the soggier soils associated with the northern hardwoods.

On the eastern side of Canterbury town center, bear right onto Route 14A. Continue on 14A through Plainfield and past Interstate 395, until you reach the right-hand turn onto Route 49 South. Take that road through open dairy country, which features vistas reminiscent of colonial and Federal times. After 7.5 miles, you'll see the entrance to Pachaug State Forest on the right. Turn in there. If you're of a mind to relax for a few minutes, a pretty pond setting is on the left just beyond the entrance. Our main goal, however, is the Rhododendron Sanctuary located up the access road. Go to the Y-intersection and take the gravel-surfaced Cutoff Road into the camping area. Park there; the Sanctuary trailhead is well marked.

As you reach the head of the trail, look for two shrubby colonizers of dry, sandy soil. These are scrub oak and sweet fern. Actually not a fern but rather a flowering plant of the bayberry family, the second species has one of the most unique leaf shapes—the long, narrow blade has irregularly spaced indentations and blunt, lobed projections. One of the most intriguing aspects of the lowly sweet fern is that it has a symbiotic relationship with actinomycetes, bacteria that help its roots fix crucial nitrogen from the nutrient-starved ground. New England Indians used this plant's leaves as a tobacco additive.

Farther down the path, Canadian hemlock, white oak, shadblow, and hazel shrubs are common; then the track descends into an area where the broad leaves of skunk cabbage announce the beginning of a wetland community. The hemlocks are the dominant tree type, but you'll also see good specimens of Atlantic white cedar, a classic swamp tree of southern New England. Take a good look at this striking conifer. In times past, it was often overcut in response to the high demand for its rot-proof wood. (For more information on this species, see Tour 16.) Here, somewhat closer to the ground, grow clethra, mountain laurel, cinnamon fern, wild iris, and even pockets of sphagnum moss. All are within sight of the walkway. But then you come upon the most exciting sight of all: a stand of mature rosebay rhododendrons. Their taxonomic name is *Rhododendron maximum,* and it's easy to see why. Their size is impressive, and in the first part of July, when they break into blossom with large pink or white flower clusters borne over sprays of evergreen leaves, the sight is unforgettable.

These rhododendrons, and also the mountain laurels and blueberry shrubs found nearby, are ericads, or members of the diverse and widespread heath family. Like the sweet fern, ericads have developed a symbiotic way of extracting nitrogen from their challenging environment. In the case of these rhododendrons, the problem is soggy, acidic soil. Their solution is to enter into their own kind of association—not with bacteria, but with fungi. The result is the complex relationship known as mycorrhizae, or "fungus roots." The fungal partner vastly increases the surface area of the roots, and so improves the absorption of nutrients and water. Indeed, this bond is so important, the fungus and the rhododendron could be regarded not as two separate organisms but rather as one composite superorganism. Does it sound farfetched? Then stop to consider the lichens. They are actually associations between fungi and green algae, even though we routinely classify them as single entities.

The second half of the trip takes you into some of the loveliest parts of inland Rhode Island. Return to the Pachaug State Forest entrance and turn right onto Route 49 South. In less than a mile, you reach the junction of Route 138. Bear left onto it, heading east. In just 1.0 mile you come to a Y-intersection; go left onto Route 165 East. Make sure you note your mileage at this point. Soon, you enter Little Rhody in the vicinity of Beach Pond. Pitch pines, well adapted to the dry soils, are quite plentiful. At 6.3 miles from the last turn, go left onto

For obvious reasons, the shagbark hickory, like this one from Pachaug State Forest, is one of the easiest New England trees to identify in any season.

A peatland at Arcadia Management Area. In the foreground, low-growing leatherleaf shrubs blanket the sphagnum mat.

a gravel road that is marked with a small sign denoting Brook Trail. This is part of the Arcadia Management Area. You'll make two brief stops here: one at 0.2 mile from the entrance, and the other at 0.6 mile. Both are located just off the road to the right.

The first spot is a swampy area fringed with gray birch and red maple. The main attraction is a beautiful shrub known as winterberry or black alder. The former name is the better one for a couple of reasons. To begin with, the plant's attractive red fruit often persists through the winter, when it provides an unexpected jolt of color. And besides, the species isn't really an alder. Believe it or not, it's a holly—not the evergreen kind referred to in Christmas carols or found in Cape Cod thickets, to be sure, but a lesser-known deciduous type. Once you know what to look for, winterberry is one of the chief delights of the cold season.

Another wetland awaits you at the second spot, but it is a wetland with a difference. On its edge stand maleberry, red maple, withe rod viburnum, and some pitch pine. But the center section is given over to a large community of a low-growing ericad called leatherleaf. If you venture out to look at it, be prepared for wet feet as you encounter the substrate of soggy sphagnum moss. The prevalence of leatherleaf suggests that this peatland should be classified as a bog. In this environment, the availability of substances crucial to plant growth is at a minimum, and only the toughest customers can make a go of it.

Leave Brook Trail by turning left onto Route 165 East. Precision navigation is still the order of the day, so note your odometer reading once again. At 1.6

miles down the way, turn right onto an unmarked, two-lane blacktop. (Rhode Island will never be accused of being oversaturated with road signage.) To the south 1.5 miles lies the beautiful Browning Mill Pond, also part of Arcadia Management Area and an appealing place for a picnic under the tamarack trees. After another 1.1 miles, you'll come to a T-intersection. Turn left and continue another 3.1 miles past the parkland headquarters, to the intersection of Route 138. Go left onto 138 East.

This is one of the state's main thoroughfares. In August 1991, the trees in this locale suffered massive damage from Hurricane Bob, a lethal storm despite its unintimidating name. Some of the largest lawn trees were wrenched out of the ground or snapped like toothpicks. For days afterward, the wreckage blocked many streets and sidewalks. The campus of the University of Rhode Island, which lies somewhat to the east of this tour's track, was so littered with branches and other debris that it looked as though the gods themselves had thrown a frat party. This was just the latest reminder that great storms are an important factor in the shaping of New England's environments. It is not difficult to find people who can give firsthand accounts of the great 1938 hurricane, which knocked down an incredible three billion board feet of timber. Recalling a much earlier tempest, Plymouth Colony governor William Bradford wrote:

> This year [1635] . . . was such a mighty storm of wind and rain as none living in these parts, either English or Indians, ever saw. . . . It began in the morning a little before day, and grew not by degrees but came with violence in the beginning, to the great amazement of many. It blew down sundry houses and uncovered others. Divers vessels were lost at sea and many more in extreme danger. It caused the sea to swell to the southward of this place above 20 foot right up and down, and many made of the Indians to climb into trees for their safety. . . . It blew down many hundred thousands of trees, turning up the stronger by the roots and breaking the higher pine trees off in the middle. And the tall young oaks and walnut trees of good bigness were wound like a withe, very strange and fearful to behold. . . . The signs and marks of it will remain this hundred years in these parts where it was the sorest.

The last destination of this journey is close at hand. Turn right onto Route 2 South at its intersection with 138 and proceed 1.9 miles to the left-hand turnoff for the Great Swamp Fight site. A gravel-fill road leads through otherwise impassable terrain to the Battle Monument. If the road hasn't been graded recently, be careful—it can get rather washboardy. Along the way, note the dense growth, which imparts a sense of directionless gloom. Shrubs includes both mountain and sheep laurels and the second native holly of the trip, the evergreen inkberry. It is here that we take a concluding look at colonial history.

If the Pequot War alerted English settlers to the threat of native belligerence, King Philip's War came close to destroying their New England holdings entirely. A large confederacy of Indians, more or less under the control of the Wampanoag sachem Metacomet (called Philip by the English), launched widespread attacks. Numerous settlements from the seacoast to the Connecticut River valley were raided or burned; patrols were ambushed and many colonists were tortured or taken into captivity. In proportion to the population, it was the bloodiest conflict in American history, with one out of every sixteen male settlers killed in battle.

By winter of the war's first year, things looked extremely grave for the English. In December, the colonies of Plymouth, Massachusetts Bay, and Connecticut agreed to form a joint army. Commanded by Plymouth governor Josiah Winslow, the force was to attack the formidable Narragansetts in their Rhode Island hideout. How this was to be accomplished lacked some explanation: the soldiers were about to run out of rations, they were not sufficiently equipped with proper clothing for the unusually cold weather, and besides, no one was exactly sure how to confront or even find the wily foe. From the Narragansett perspective, however, there was plenty of reason for optimism. The bulk of the tribe, including its warriors and their families, was safely holed up with a large cache of winter supplies in a palisaded fort about 5 acres in extent, in the depth of this swamp. Normally, the high water table and tangled vegetation restricted access to the fort to one easily defended point. What sachem Canonchet and his people did not know, however, was that one of their own had betrayed their location and had offered to lead the English to the fort.

Winslow's luck was not restricted to his having a good Indian guide. The day he chose for the attack—Sunday, December 19, 1675—was as cold as the past few days had been. As a result, the surface of the swamp had frozen solid and could bear the weight of his men. They marched straight across what previously had been uncrossable. As if that weren't enough advantage, they arrived to find one part of the palisade still unfinished. After bloody fighting, they took the fort and set it aflame. Many of the Narragansetts, including women, children, and elderly noncombatants, were killed in the fracas. The massacre was a horrible strategic blow for the Indian alliance, since it crippled its supply base and removed the stigma of white helplessness.

With the exception of the monument and a roadside historical marker, there remains no evidence of the encounter. The swamp is still a swamp. But a different kind of battle, common to all forest communities, still rages: the battle between countless plants striving to get light, nutrients, and the reproductive edge that ensures the survival of their species.

TOUR 14 ❧

In Rhode Island:
Newport to Bristol

Driving Distance:
22 miles

Starting Point:
The Downtown Newport
Harbor front

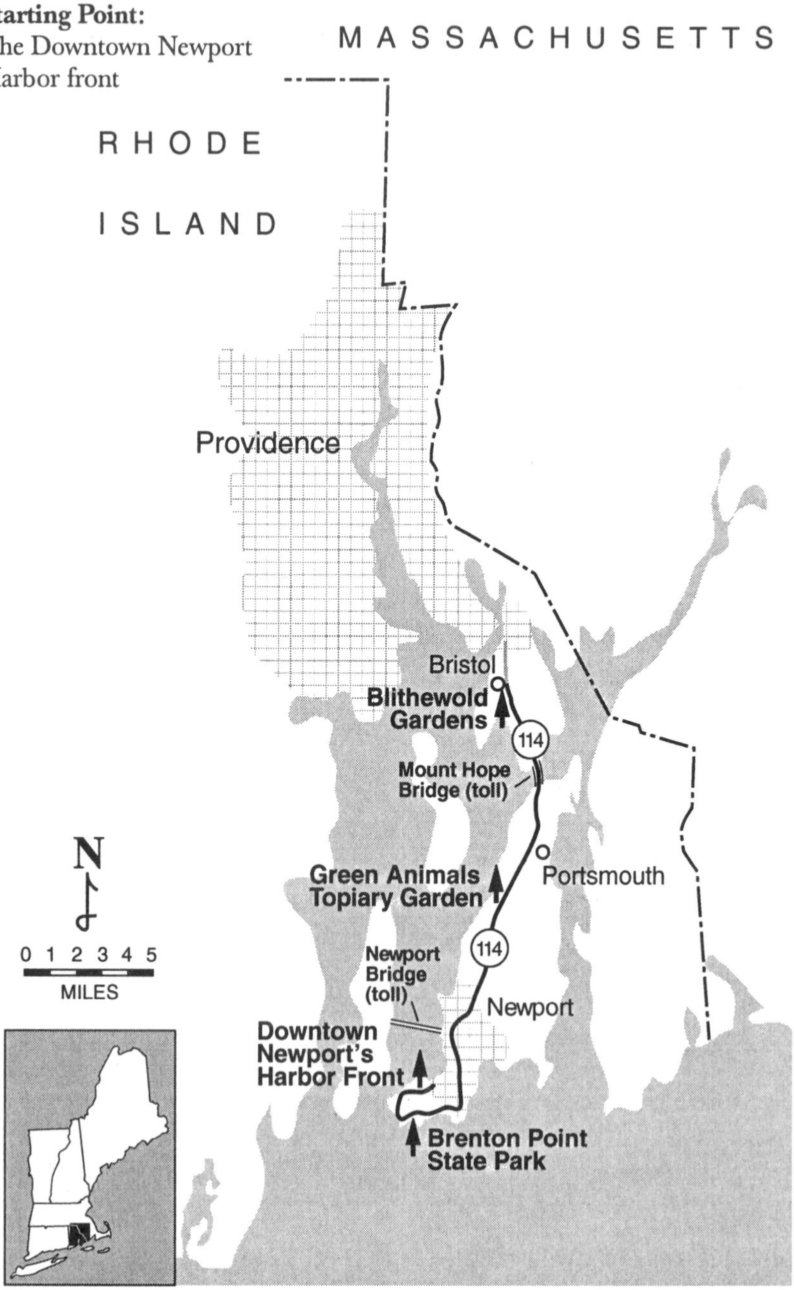

M A S S A C H U S E T T S

R H O D E

I S L A N D

Providence

Bristol

**Blithewold
Gardens**

114

**Mount Hope
Bridge (toll)**

Portsmouth

**Green Animals
Topiary Garden**

114

**Newport
Bridge
(toll)**

Newport

N

**Downtown
Newport's
Harbor Front**

0 1 2 3 4 5
MILES

**Brenton Point
State Park**

Tour 14
Ocean and Bay

*I*f you are not already acquainted with Rhode Island's Narragansett Bay and its environs, you are in for a delightful surprise. The bay, one of the world's best natural anchorages, has long been a haven for ships of every description, from the brave craft of sixteenth-century Florentine explorer Giovanni da Verrazano to modern U.S. Navy surface combatants and America's Cup yachts. One of the best things about the bay's dazzling blue surface is that it keeps popping into view. The low, rocky surface of Aquidneck Island, and all its fine plants and buildings, seem to be an ark in its midst, heading seaward. In this Ocean State tour, you'll see how beautiful garden settings benefit from having a saltwater backdrop.

When you begin the trip in downtown Newport, you'll probably find that it is the human environment rather than botany that commands your attention. Once a maritime center and now a tremendously popular tourist attraction, Newport is blessed with noteworthy examples of eighteenth-century architecture, from Richard Munday's Trinity Church to Peter Harrison's Touro Synagogue. The narrow lanes and closely spaced buildings of colonial urban centers precluded much in the way of street plantings, but in modern times Newport has boulevard trees in its more open places, especially the harbor front area. Stroll for ten or fifteen minutes around the main boulevard and surrounding commercial area, and see if you can locate the following species: the London plane tree, the ginkgo, the Japanese zelkova, and the Kwanzan cherry tree. We'll briefly consider each in turn.

If you've already had the opportunity to take either Tour 11 or Tour 12, you're familiar with the sycamore, the largest of all New England broad-leaved trees. This noble riverbottom species is a parent to one of horticulture's most famous hybrids, the London plane tree. The history of the London plane is a long and fascinating one. In 1637, John Tradescant the Younger, one of the most notable of the early British plant explorers, returned to England with sycamore specimens he'd collected in Virginia. Eventually, he planted them in the vicinity of a preexisting Oriental plane tree. The Oriental plane, a closely related species native to the Balkan region, cross-pollinated with the sycamores, and a new, extremely vigorous hybrid came into being—the London plane. Like its parents, it produced simple, lobed leaves reminiscent of a maple's; also, its bark exfoliated, or peeled, in the same attractive patchwork pattern.

Over the centuries, the hardy and undemanding London plane has been a favorite choice for park and curbside plantings in European and North American cities. But, as with the American elm, this seemingly disease- and pest-free hybrid was often planted in monocultures, practically to the exclusion of everything else. In modern times, it was ultimately hit by a fungus disease called anthracnose, and soon thousands of trees were afflicted with stem dieback and cankers. Some died, and others became badly disfigured. This workhorse of street plantings is slowly but surely giving way to more recent and less vulnerable introductions. So far, though, the London planes of Newport appear to be unaffected by disease problems. May they flourish for many years to come.

The next tree is the ginkgo, one of two "living fossils" now widely used in horticulture (the other one is described in Tour 4). Though it certainly looks like a broad-leaved angiosperm, it's actually a nonflowering gymnosperm in an ancient and distinct order of its own. For more on its lengthy pedigree, see Tour 18. Here, we'll concentrate on its role as a street tree. Its greatest virtue thus far has been its amazing resistance to pests and most diseases. The tree grows reliably in neglected soil and tolerates most bothersome city conditions well. While the London plane produces almost no autumn color at all, the ginkgo's fan-shaped foliage turns a striking golden yellow in the fall.

The two remaining selections on our list both hail from Japan. The Zelkova is a fairly recent introduction to the urban street tree setting. While it has been touted as a replacement for its relative, the disease-stricken American elm, it is not as graceful a plant, and in some locales it has suffered a great deal from the effects of the foliage-nibbling elm leaf beetle. Still, it has much to commend it in its own right. On the other hand, the fancy-flowering Kwanzan cherry is not a good choice for more stressful locations, such as unprotected street-edge pits. It is prone to both insect attack and vandalism. Fortunately, someone here knew that; you'll find the Kwanzan tucked away in the less footworn and less abused confines of the median strips.

When you've finished your tree spotting, proceed by car to Brenton Point State Park, situated at the extreme southwestern tip of Aquidneck Island. Continue past Fort Adams (currently the site of the Jazz Festival) and the Castle Hill Coast Guard Station. If you stick to the shore road, you'll soon be there.

Brenton Point is a bright and breezy place, the one stop on this journey that is not situated on Narragansett Bay. Instead, it faces the unfettered expanse of the Atlantic. Park in the long lot facing the water and proceed across the roadway to the beach.

The point sits atop rugged Cambrian period rock units some 300 million years older than the sedimentary beds underlying the rest of the island. Few ornamental plants can withstand a habitat as blustery and desiccating as this. There are some clumps of salt-resistant rugosa rose, and over by the park house stands one flag tree, its branches contorted to leeward by the force of the wind. But the real attraction here is not the plants but the protists—the separate kingdom of life that includes the marine algae. At low tide, these seaweeds can be discovered easily enough along the rocky shore and in tidal pools. Among the most common are the crisp-edged Irish moss, a member of the red algae, and such brown algae as knotted rockweed and wrack and kelp species. If you spot one of the rubbery, long-bladed kelps—they can measure 15 feet or more—take a look at its basal stem, or stipe. It's connected to what appears to be a mass of roots. This structure, the holdfast, is actually not roots at all, but a device the seaweed uses to attach itself to rocks or other surfaces. Incidentally, kelps are residents of the zone below low tide. If you find one thrown up on dry land, its holdfast must have lost its grip. In this pounding environment, that's understandable.

───────────── ∾ ─────────────

To depart Brenton Point State Park, exit at the eastern end of the parking lot and head east on Ocean Avenue. Along this shoreline are tidal marsh areas containing *Phragmites,* or giant reed. It perches on the uppermost, least saline parts of the wetland. Park managers and wildlife experts usually regard this tall-growing plant in the grass family as an overt pest, since it rapidly takes over disturbed areas. It reproduces quickly in two different ways: sexually, through flowers and seed, and asexually, through spreading underground stems called rhizomes. Once established, the species is almost impossible to eradicate with any technology available to mere civilians.

After approximately 3.5 miles, Ocean Avenue intersects Bellevue Avenue. Turn left (north) onto Bellevue. This is prime mansion territory, and the road might well be renamed the Boulevard of Conspicuous Consumption. Some of Newport's most famous estates—Marble House, Rosecliff, and Chateau-sur-Mer—are here, while everybody's favorite, The Breakers, is located just a few blocks east. Bellevue continues north into the downtown area, where it joins Route 114. Brace yourself and go north on 114 through Middletown and what at first seems to be an endless expanse of stoplights and clonescaped shopping centers. (Remember—if Verrazano could make it all the way from Tuscany, the least you can do is endure a little bit of road rash.) Before long, the smell of deep-fat french fries dissipates and you're in Portsmouth. Some farm fields persist in this area; keep an eye out for the old, low boundary walls. The stone fences of Aquidneck Island are said to be among the most finely made in the region. In the traditional New England scheme of things, many towns chose a citizen to serve in the post of Fence Viewer. This official's duties included inspecting property lines and gathering evidence in legal cases involving boundary disputes, wandering cows, and the like. As time went on, this increasingly meaningless responsibility became the butt of many jokes, if not the target of downright contempt. If anyone ever offers you the position, you might want to consider moving two or three counties away, regardless of the benefits package.

The rocks that make up these sturdy fences have their own story to tell. To geologists studying the complex prehistory of New England, the bedrock around here speaks of a time 300 million years ago when sediments were deposited in a depression we now call the Narragansett Basin. Both here and far to the west, the land was covered by the lush coal swamp forests of the Carboniferous period. These forests, however, were utterly different from the New England woodlands of today. Modern flowering trees, and even their more primitive conifer counterparts, had not yet come into existence. Their place was occupied by strange, tall-growing pteridophytes, huge relatives of the current-day horsetails and club mosses. These ancient free-sporing trees included the scale-barked *Lepidodendron* and the jointed-stemmed *Calamites*. Also present was *Cordaites,* one of the early seed-producing gymnosperms. Interpretations of all three of these plant types can be seen in the great *Age of Reptiles* mural at Yale's Peabody Museum (see Tour 4). The time of the coal swamps was poised between two of New England's major mountain-building episodes, the Acadian and Alleghenian events. The second of these marked the joining of North America with Europe and Africa into the supercontinent Pangaea.

At an elapsed distance of about 13.8 miles from Brenton Point, you reach the stoplight intersection of Cory's Lane. Turn left down the lane and go another 0.7 mile to the entrance of Green Animals. This intriguing site is not a research facility where members of the animal kingdom are genetically engineered to contain chlorophyll. Rather, it is a showplace for the grand old art of topiary, or plant sculpture. Topiary enjoys a long tradition, especially in European formal gardens. In this magnificent bayside setting, eighty capricious animal forms have been sheared and pruned from a variety of hardy shrubs. Another section displays more formal, geometrically shaped plants, where by rights we should catch a fleeting glimpse of the Red Queen in pursuit of Alice.

There are other, less anthropomorphic, sights on the Green Animals grounds. In front of the main house stands a huge copper beech, a cultivar (cultivated variety) of the European beech. Its name refers to the reddish or purple hue of the leaves. It would be difficult to imagine a finer lawn tree for a situation such as this. Nearby stand scaly-barked and confusingly named sycamore maples: they are true maples, but not true sycamores. They belong to a Eurasian species that has been planted in some Northeastern parks and arboreta.

One more horticultural high point awaits you a short distance to the north. Return to Route 114 and turn left onto it, heading north once again. Along this stretch, you'll probably notice a number of conifers with reddish orange upper trunks and a distinctive blue-green cast to their foliage. These are Scots pines, widely used as ornamentals both in this region and across much of the northern United States. At 2.9 miles up from the Cory's Lane stoplight, bear left onto the graceful Mount Hope Bridge. Have your toll change ready. This single-span suspension bridge is by no means the largest structure built over the bay, but it certainly is the prettiest. On the far side, you enter the mainland and the town of Bristol. At the three-way intersection, bear left onto Ferry Road (this is still Route 114), then turn at the left-hand entrance to the Blithewold Gardens and Arboretum, 0.1 mile farther up. This too is a former private estate now preserved and open to the public. The parking area is close by.

A large portion of Blithewold's 33 acres is taken up by the Great Lawn, which discloses another lovely view of Narragansett Bay. If you stop by the visitor center, pick up a copy of the self-guided tree tour brochure. Hurricane Bob hit this area in force late in the summer of 1991, and a number of plants on the grounds were damaged, but according to staff members, none of the trees listed in the brochure were seriously damaged.

While you're still in the vicinity of the visitor center, take a look at the rose garden. There, rising above the rest, is the remarkable chestnut rose, a vintage variety that has been allowed to grow to tree size over the course of many years. In front of the mansion, you'll see other eye-catching woody plants. The most arresting conifer there is the Sargent weeping hemlock, a cultivar of the Canadian hemlock species so plentiful in the forests of New England. Nearby stand beautiful mature examples of Japanese cedar, two European beech cultivars, and the rare, almost extinct Franklin tree described in Tour 6.

At Green Animals. Here plants serve other ends, as topiary figures.

But the very best has been saved until the end. Go into the Enclosed Garden to the north of the mansion. Here you behold a breathtaking sight, a conifer that has the potential to be one of our planet's largest organisms. Be respectful. Before you stands the largest sequoia, or *Sequoiadendron giganteum,* east of the Rockies. Even though it is a preadolescent by sequoia standards, it's pushing the 100-foot mark. Early English settlers in New England reported seeing white pines two and a half times the height of this tree, but it's difficult to imagine how even those lofty plants could match the effect of the massive symmetry of this specimen. Its species' native habitat is the western slope of California's Sierra Nevada range. While the tree does not grow quite as tall as the equally famous coastal redwood, it is stockier, heavier, and more imposing.

The most celebrated older sibling of the Blithewold tree is the General Sherman Tree, in Sequoia National Park. At last report it was 102 feet in circumference at its base and weighed about 6,200 tons. One other statistic about it would warm only the heart of a lumber company lobbyist: the tree's timber could make 20 billion toothpicks. (Let's hope the day never dawns when someone seriously suggests using it for such a purpose, even if it fits that grand excuse for wrecking forests, Saving Jobs.) The General Sherman Tree is estimated to be between 2,300 and 2,700 years old. It's thrilling to think that this sequoia might still be here, towering over the bay, in A.D. 4600.

A beautiful baby monster. The eye-catching sequoia at Blithewold Gardens. It is definitely a juvenile form, even though it is already almost 100 feet tall.

TOUR 15 ◠◡

In Massachusetts:
Sagamore to Plymouth

Driving Distance:
74 miles

Connecting Tour:
Number 16

Starting Point:
The junction of
Routes 3 and 6, just
north of the Sagamore
Bridge to Cape Cod

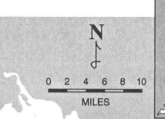

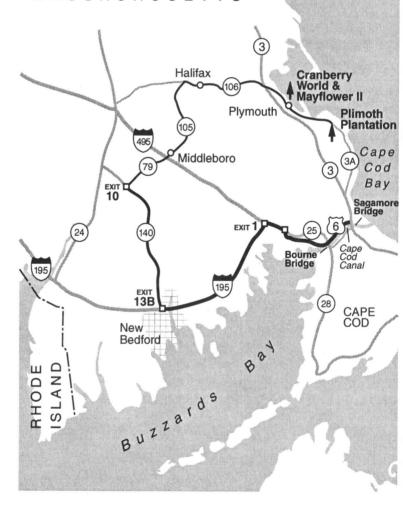

Tour 15
Cranberry Land and a Trip
to the Seventeenth Century

*I*f you glance at a Massachusetts road map, you'll notice that the driving distance cited above seems much too long for a trip from the northeast fringe of Cape Cod to nearby Plymouth. This isn't a misprint. We'll be taking a roundabout, clockwise swing through the Bay State's southeastern mainland. This is a different sort of place, with its own look, its own agriculture, its own geology, and its own weighty contribution to the early chapters of American history. As we'll see in this tour's final two stops, Plymouth and its surroundings are a mecca for tourists. But by and large, the region has a fairly lonely feel. Its uniquely numinous atmosphere, apparent to the earliest English settlers, still lingers in the air.

Our starting point, the Sagamore Bridge junction, is an excellent example of that great winnower of souls, the traffic rotary. This seems to be Massachusetts' special contribution to automotive mayhem. Other states have them, but not in the same profusion. After one, two, or twenty orbits around this one, extract yourself from the maelstrom as safely as possible and exit onto Route 6 West. You are now paralleling the Cape Cod Canal, which was completed in 1914. There are several scenic pull-offs to the left where you can take a less hurried look at this feat of civil engineering. The local tree community is composed of stands of pitch pine, black cherry, and black locust, as well as oaks and tall-growing sumacs. (The pitch pine is described in Tour 16.)

By the 2.7-mile mark, you will see Bourne Bridge. Not far from this span, on the Cape side, is the reconstructed Aptucxet Trading Post, established by the

Pilgrims not long after the founding of Plymouth Colony. As you pass under the bridge, make sure you stay on Route 6 heading west toward New Bedford. At 5.0 miles from the trip's beginning, you cross over a tidal estuary and pass through a popular American landscape, where mere natural wetland and beach periphery have been magically transformed into souvenir stores and peewee golf courses. After another 3.3 miles, follow the "To 195 West" sign and go across the overpass. This puts you on Route 25 North. At Exit 1 bear right onto I-195 West.

The interstates of New England often provide an excellent overall view of the megaflora—large-scale plant communities and their ecological settings. The big highways are often built in relatively unpopulated areas where introduced ornamental trees and shrubs are less prevalent than in more settled areas. In the first few miles of I-195, you'll find that white pine has replaced the sand-loving pitch pine. Another gymnosperm, the red cedar, takes up its post in unmowed sunny strips and fields. Its narrow, almost columnar habit makes it an easy species to spot, even from a great distance.

As noted in other tours, the red cedar is actually a close relative of both the common juniper, a broad-shouldered shrub, and the prostrate creeping juniper. Why do these three members of a single genus, so similar in other ways, vary so grossly in shape? The answer has much to do with the plants' hormones. The severely upright red cedar is apically dominant—the growth of side branches is suppressed in favor of vertical development. One theory suggests that hormones in the red cedar actively inhibit the side growth; another explanation is a simple lack of other hormones that would encourage the side branches to elongate. In the shrubby common juniper, more of a balance exists between horizontal and vertical growth, while the "hormonal programming" of the ruglike creeping juniper ensures that vertical growth literally never gets off the ground.

In roughly fifteen or twenty minutes, you reach the environs of New Bedford, one of the great seacoast cities of the region. In its heyday, the economy was based not so much on the exploitation of plants brought from nearby inland harvests but rather on the exploitation of the world's largest animals, hunted in the farthest reaches of the sea. Once the whaling center of the world, New Bedford has suffered an inevitable decline from maritime and manufacturing prominence into the long and rusty twilight of postindustrial times. Its legendary aura persists, though. Nowadays, it is one of New England's "gritty cities," but it still inspires the imagination. Any place trafficking in the nineteenth century is bound to be alluring.

Next, swing inland by leaving I-195 at Exit 13B in downtown New Bedford. Head north on Route 140. If any freeway deserves the adjective "bucolic," this is it. It's a lovely introduction to interior southeastern Massachusetts. Note the size of the surrounding trees; they're taller than many others in modern New England, thanks in part to the apparent lack of logging and real estate encroachment. Red maple is especially numerous. This adaptable species, so well represented in the region, has a remarkably wide range. It is native from Newfoundland to Minnesota, and from Florida to the bayous of Louisiana and farther westward.

While it can be found in many woodland situations, including dry, windy hilltops, red maple is especially at home in waterlogged soil, and is often associated with swamps and the fringes of bogs. There it begins a dazzling fall color display, featuring just about every gradation of red and even some yellow, weeks earlier than elsewhere. Still more welcome to the New Englander, perhaps, is its blossoming, when the tree produces dangling flowers that impart a haze of reddish tones to the glum browns and grays of early spring.

At 9.3 miles north of the turn onto Route 140, an extended granite road-cut dates from the era geologists call Proterozoic Z. The name has an oddly poetic ring to it. While this was the last phase of Precambrian time, it was still a very long time ago—roughly one billion years. The relatively simple forms of life in the sea were still showing a pronounced lack of desire to evolve into anything more complex. But the gigantic game of continental bumper pool, the result of the forces of plate tectonics, was already well under way. The ancestral core of North America, to the west and north of here, collided with a section of what would later be northern Europe. The granite seen at this location was part of a different, mysterious piece of continental crust geologists call Avalonia.

Leave the highway 5.3 miles farther north, at the Exit 10 turnoff to Route 79. Proceed north on 79. Before long, you pass through Lakeville, where a common lawn tree, the purple-leaved Norway maple, adds its dreamlike effect to the surroundings. Among the various cultivars that fall into this type, two have been most generally used—the Schwedler and Crimson King varieties. The striking foliage color of the Crimson King persists throughout the summer, while the more traditional pick, the Schwedler, usually fades to a somber dark green by August. On the far side of town, stands of wild pitch pine reappear—a sign that the soil here is sandy and nutrient poor. Continue on Route 79 through the junction with Route 18; just before the pond, note the plantation of big white pines.

Turn left when you reach the T-intersection where Route 79 becomes Route 105. Continue through the I-495 underpass and stay on Route 105 through Middleboro. Be careful to follow the route signs. This is the southern edge of the Narragansett Basin, a zone filled with sedimentary beds deposited at the time of the great Carboniferous coal swamps. It extends southwestward to Newport, Rhode Island. (For more on the plants that grew in this area in those lush coal swamp days, see Tour 14.) Before you reach the town, a native vine, fox grape, can often be seen scuttling over the stems of other woody plants. The broadly lobed, or "entire," leaves of the fox grape have brown or rusty undersides; this is especially evident as summer wears on. The species is the most important source of several cultivated varieties, including the famous Concord grape.

Beyond Middletown, a good deal of land is still under cultivation. Most is given over to a crop plant known in general parlance as corn. Its more accurate common name, however, is maize. Corn is actually a catchall term; in Europe, for example, it has been used to denote other cereal grains altogether. Without a doubt, maize is the one most important agricultural gift bequeathed to the English settlers by the New England Indians. It was apparently developed from a wild Mexican grass, *teosinte*. While it rapidly exhausts soil fertility when care

A commercial cranberry bog to the west of Kingston town center.

is not taken, it is a dependable plant adaptable to a wide range of climates. It's also resistant to most diseases, including those that beleaguered New England wheat almost from the beginning. After its original development far to the south, maize spread from one American Indian community to the next, until it was introduced into this region, some five hundred years before Christ, by the migrating people of the Adena culture. Its arrival triggered New England's first great agricultural revolution, long before the arrival of the whites.

A few moments more and you've reached the intersection of Routes 105 and 106. Bear right onto 106 East, and continue through the pleasant town of Halifax. You are now entering a zone where another form of farming, more characteristic of this specific area, prevails. Keep a sharp lookout around 6.4 miles after the turn onto Route 106; on both sides of the road you should see large, sunken fields that may be flooded, depending on the time of year. This is a small portion of the 12,000 acres of commercial cranberry bogs to be found in the Bay State. Of course, these are not bogs in the ecologist's sense of the word, but carefully regulated environments where one humble plant species, native to New England peatlands, has been put to work on a massive scale.

In pre-Columbian times, the local Indians were well aware of the cranberry's worth, and it was an important ingredient in their pemmican. Still, they apparently restricted their harvest to the fruit of wild plants. According to a later Wampanoag legend (perhaps influenced by Christian symbolism) the first cranberry plant was a gift from the Great Spirit, and was brought from heaven in the beak of a dove. Cultivation began in earnest in the Federal era, when Captain Henry Hall of East Dennis, Cape Cod, determined that the large cranberry, *Vaccinium macrocarpon*, was the more promising species for cash crop use. Before long, the basics of the new trade were established. The best soil for it was found to be coarse sand, replenished periodically and underlain by a rich layer of peat or muck. Hall's idea caught on as soon as the shrewd Yankee mind realized that this corner of New England best supplied the cranberry's needs for acidic soil, reliable rain, and a relatively long growing season. By the 1850s, the region was experiencing "cranberry fever," the local equivalent of the California gold rush. Plantings covered thousands of acres, including many high and unproductive sites that in retrospect were bound to fail. In those days, retired sea captains were especially drawn to this watery form of agriculture. Some of their nautical terms still persist in the trade. On leaving a bog, for instance, you are "going ashore." The cranberry itself saw plenty of use at sea, at least on American clippers and whalers, where, like the Britons' limes, it was taken to combat scurvy, long before its vitamin C content was identified.

Over the years, the mixtures of ethnic groups that have made up cranberry harvesting crews have presented some interesting juxtapositions. In the decades around the beginning of the twentieth century, for instance, Finns often worked with Cape Verdeans; more recently, Puerto Ricans and Scots have been employed side by side. Perhaps the most exotic group to hit the bogs, though, was the 150 German Wehrmacht POWs detailed to help bring in the crop during World War II. The traditional method of gathering the berries by hand, with toothed scoops, must have been grueling. Since the 1940s, mechanical picking devices of ever increasing sophistication have been used. Now there are two modes of harvesting. The majority of cranberries, destined for juices and sauces, are harvested by what is known as the wet method. In October, when the berries are ripe, the bog is flooded; then an amphibious machine called a water reel loosens the fruit from the stems. When the berries float to the surface, workers herd them into an enclosure, then pump them into a waiting truck. The dry harvesting method is reserved for berries destined to be sold fresh or frozen: a mechanical picker something like a lawn sweeper is pushed along the unflooded bog surface. In either case, the picked berries are hauled to a special facility where they are screened by other workers, who patiently remove damaged and other unwanted cranberries.

Cranberry growers once used dangerous insecticides, including DDT, sodium cyanide, and lead arsenate, but they now resort to less toxic substances and to IPM (Integrated Pest Management). This more enlightened and efficient system of pest control subjects insect foes to a broad-based, well-timed attack that takes the place of the traditional wasteful application of a single poison based more on an arbitrary schedule than on real need. No one familiar with the world of plants would doubt that some sort of careful protection is necessary; and the cranberry has been a favorite target for insects ever since the fire worm hit town in the early days.

In addition to IPM, the growers use a combination of old-fashioned and high-tech tricks to protect and improve their investments. With the help of a system of ditches, dikes, and flumes, the bogs are flooded at intervals to kill pests and provide wintertime protection. Also, low-temperature sensors alert growers to turn on sprinklers, which quickly sheathe exposed plants in a layer of ice. This ploy, ironically, keeps them from getting frostbite during a bad cold snap. And in warmer weather, a thoroughly modern cranberry farmer may be seen regrading a bog's sandy surface with a bulldozer fitted with a laser-guided blade. It's a bit more than either Captain Hall or the Great Spirit's dove could have possibly imagined.

———————————— ❧ ————————————

Our track next takes us to the shore of Cape Cod Bay, and what was once the heart of Pilgrim territory. Resume your eastern heading on Route 106; when you arrive at its junction with Route 3A in Kingston, continue straight. This puts you onto 3A South. This is now a heavily settled area, full of ornamental plants almost totally unknown to the founding fathers and mothers of 1620. Another 3.8 miles brings you to downtown Plymouth. Turn left at this intersection; at the Water Street rotary, go north along the shoreline 0.3 mile to the Cranberry World visitor center, on your right.

This facility is operated by Ocean Spray, the growers' cooperative that has become a household name across the country. Currently included on the Fortune 500 list, Ocean Spray markets about 85 percent of the entire United States cranberry crop, which includes other production centers in New Jersey, Wisconsin, Washington, and Oregon. Cranberry World features interesting displays that complement the description given above. The best part of the whole show is the miniature cranberry bog outside the exit. (If you patronize the snack bar counter at the end of the indoor segment, think twice before you order orange juice.)

When you've had your fill of cranberry products, either literally or figuratively, return to the Water Street rotary and continue south to the parking area for the *Mayflower II* exhibit. Since you will be going to Plimoth Plantation next, take advantage of the lower-priced combined admission ticket available here.

This may be your first time in the seventeenth century, so take a moment to get yourself in the right frame of mind. The moment you board the full-sized replica of the original *Mayflower,* you will be greeted by its crew members, who are as eager as starving used-car salespeople to make eye contact. They'll acquaint you, in appropriate British dialects of the era, with the early history of the Pilgrims: the sixty-six-day voyage across the Atlantic, the theological disputes with the Anglican church, and practically anything else that pertains to the period. But don't bother to engage your new friends on such topics as global warming or the movies. If you do, they're likely to tell you that verily, they know not whereof you speak.

The ship itself is a priceless example of the wood-based technology of preindustrial Europe. Captain John Smith, an early New England explorer better known for his role at Jamestown, aptly described the sailing vessel of this epoch: "Of all fabricks a ship is the most excellent, requiring more art in building, rigging, sayling, trimming, defending, and moaring, with such a number of severall termes and names in continual motion, not understood of any landsman, as none would think of, but some few that know them." If you can, find the master (the captain) and ask him to explain the basics of how forest trees from Scandinavia, the Baltic countries, and Russia were turned into this seaworthy ship and others like her. Also note the masts, formed from single straight conifer trunks imported from Norway.

———————————— ❧ ————————————

The second part of your sojourn in the 1600s, Plimoth Plantation, is also the tour's concluding stop. Drive south from the *Mayflower II* parking area, along the waterfront and past mythic Plymouth Rock. At the first intersection, turn right and go up the hill one block to Route 3A. Turn left onto 3A South and proceed through Plymouth and its outskirts, where many of the houses still sport traditional cedar shingles. Then 2.6 miles south of the turn onto 3A, you'll see the Plimoth Plantation entrance on the right. Proceed to the parking lot 0.3 mile down the access road.

With its labyrinth of entrances, exits, signs, finger-wagging ticket takers, and orientation movies, the Plantation is one of those latter-day historical attractions that trembles on the verge of being a theme park. Once past the nonprofit grandeur, however, you meet more of those fascinating people who are so well versed in their period roles. For that, the management deserves the highest praise.

The Mayflower II *at her accustomed berth in Plymouth.*

A full description of the points of interest here would take up an entire book; but there are several plant-related things that serve as a focus. If possible, first visit the small native settlement called Hobbamock's Homesite. American Indians, including Wampanoag descendants of the Pilgrims' hosts, display here a fascinating array of house-building and domestic skills. Ask the interpreters to explain how wetland monocots—cattails, rushes, and bulrushes—were transformed into woven mats and shelter coverings, and how bark from cottonwoods and other trees was used for waterproof roofing. Also, you may be able to observe a large maize-grinding mortar being made. This process involves patiently hollowing out the heartwood of an elm log section with glowing coals. (As an aside, it's also interesting to note that the Wampanoag people here are the only staff members on the grounds who speak modern American dialect. After all, how many visitors today would follow a weaving demonstration spoken in seventeenth-century Algonquin?)

Having seen some of the canny Indian uses of the native vegetation, you're now ready to see how European settlers strove to transplant their own culture onto the shore of a new continent. The first sight of the 1627 Pilgrim Village is a revelation: with its utter dependence on local plant materials, it looks like an earth-daddy's vision of organic heaven. Several times, the people who lived in the original community had a serious struggle surviving, but their sense of purpose and religious zeal carried them through. The early Plymouth governor, William Bradford, wrote of the colony: "Thus out of small beginnings greater things have been produced by His hand that made all things of nothing, and gives beginning to all things that are; and, as one small candle may light a thousand, so the light here kindled hath shone unto many, yea in some sort to our whole nation; let the glorious name of Jehovah have all the praise."

No one would doubt the great courage of these early settlers, yet when you enter their community, you see that it does not refer to local conditions so much as it duplicates the late medieval town planning and building practices of the Pilgrims' original homes. One example is the overall disposition of the settlement. It's a good demonstration of what geographers call the nucleated community. The houses and associated buildings are clustered together and surrounded by outlying communal fields. In the years to come, this customary European practice would succumb to population pressures and give way to the more familiar New England pattern of private fields and separate farm houses.

Another example of the Pilgrims' conservatism can be found in one of the architectural styles here. Many of the immigrants came from one particular part of England, East Anglia; accordingly, a common East Anglian feature, the thatched roof, is conspicuous. The demographics of the situation expressed themselves in other ways, too. Wooden buildings were more frequent than stone ones in early colonial villages, even in already cleared areas like Plymouth, because the economy and ecology of East Anglia produced few masons but many carpenters, sawyers, and joiners. In addition, an absence of lime for making mortar in this and many other parts of the region precluded brick or stone houses in the first colonial phase.

146

The upper portion of a Plimoth Plantation house. Note the clapboard sides and chimney, the thatched roof of Phragmites *stems, and the ridgetop made from cattail plants.*

Plymouth Colony's first inhabitants had to rely on the plants nearby to replicate the timber-intensive, Old England look. Notice how the thatched roofs (made, incidentally, of *Phragmites* reed) are surmounted by ridgetops made of the more supple cattail. Also, see if you can find the strips of maple wood used to keep the thatch in place. The settlers twisted and shaped the maple into large staples while it was still green; once it dried out, it would retain the desired shape permanently.

The earliest English arrivals weren't too picky about which wood was chosen for which house-building task; their first priority was to get a roof over their heads as quickly as possible. On the homes with exterior wooden planking, for instance, the vertically set weatherboards were placed flush against one another, and not installed in the more artful board-and-batten style, with intervening protective strips. The other type of board siding, clapboard (pronounced "*clab-*erd"), is mounted horizontally. The clapboard style later became prevalent in New England and elsewhere.

Dependence on European traditions can also be seen in the early Pilgrim attempts at agriculture. In retrospect, you wonders why the ranks of the first Plymouth settlers contained so few people with farming experience or why, for that matter, the leaders failed to bring with them livestock and other essentials. As every American schoolchild used to learn, the first attempts at growing wheat and peas the Old World way were a dismal failure. This, of course, set the stage for Squanto.

It would be difficult for any novelist, even Charles Dickens, to invent a life story more implausible than Squanto's. The sole Patuxet survivor of a European-introduced plague, this supremely adaptable native was abducted by early white traders and taken to Spain. Eventually, he made his way to England, and then back to the New World, where he arrived just in time to be the self-appointed, English-speaking welcoming-committee-of-one for the Pilgrims. A shameless opportunist, he was a sharp-tongued gossip and a political foe of the great Wampanoag sachem, Massasoit. This caused a good deal of tension in the alliance between the powerful Wampanoags and their uneasy white neighbors; but nonetheless, Squanto was a godsend. As William Bradford acknowledged, Squanto single-handedly taught them the native planting practices that ensured their survival. Nowadays, his crash course might be offered as Indigenous Agriculture 101; his students, none too adept even in the farming techniques of their home-land, needed the basics. The syllabus included how to clear underbrush and vermin from fields by controlled burning; how to hoe the compacted soil; how to manure the soil with small fish; and, finally, how to plant and harvest a mixed crop of maize, squashes, pumpkins, gourds, watermelons, and beans in small hills rather than in furrows.

There has been considerable debate among modern scholars concerning Squanto's use of fish to fertilize the soil—which tells you more about modern scholars than it does about Indian agriculture. At any rate, some experts believe that Squanto actually learned the technique from Europeans in the course of his travels; others tenaciously defend the older view that this was a widespread Indian practice before the advent of the whites. Perhaps you can clear the whole matter up by asking one of the village elders, or better yet, by asking the Wampanoag interpreters at Hobbamock's Homesite.

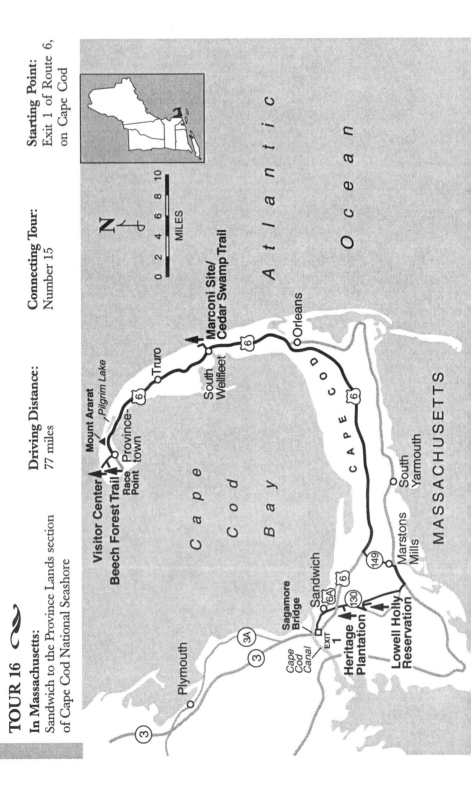

TOUR 16 ∿

In Massachusetts:
Sandwich to the Province Lands section
of Cape Cod National Seashore

Driving Distance:
77 miles

Connecting Tour:
Number 15

Starting Point:
Exit 1 of Route 6,
on Cape Cod

Tour 16
The Plant Lover's Cape

HIGHLIGHTS

Heritage Plantation ∿ The Horticultural Wizardry of Charles
Dexter ∿ Rhododendrons, Day Lilies, and Handsome Hostas ∿ Lowell
Holly Reservation ∿ Ma and Pa Kettles ∿ Boughs of Holly ∿ Pitch Pine, the
Flame-Loving Conifer ∿ Cape Cod National Seashore ∿ The White Cedar
Trail ∿ The Rare Broom Crowberry ∿ Black Oak and Ship Bottoms ∿ The
Realm of the Atlantic White Cedar ∿ Swamp Azaleas and a Yankee Spanish
Moss ∿ Salt Meadows and Sand Dunes ∿ An Ark Reaches the New Ararat ∿
Beech Forest Trail ∿ Beach Grass, Reindeer Moss, and Hidden Oases

*N*o place in New England presents a greater set of contrasts than Cape
Cod. Look in one direction and you'll see motels, campgrounds, and
sprawling real estate tracts awash in the sea of tourism that is the
economic lifeblood of the area. Look in the other, and you'll find surf rolling
onto deserted beaches, an endless expanse of woodlands, and the strange windy
peace of the dunes. On a planet as dynamic as ours, perhaps there is no better
symbol of permanence than the ocean; yet this land fronting it is one of the
youngest and most transient features on the continent.

The most persistent image in the New England landscape is one of rugged
surfaces that yield only grudgingly with the passing of the ages. But while the
rocky mainland's legacy stretches back hundreds of millions of years, on Cape
Cod there's not a single square yard of genuine bedrock, young or old. A hike
through the low, resin-scented forests convinces you that this must be the most
ancient place on earth. The sunlight itself seems older. In truth, the Cape is just
a temporary barrier of loose debris left behind by the most recent wave of
continental glaciers. So brief is its history that it postdates the origin of our own
upstart species. How long the Cape will last, no one can say, but even in recent
times, its fragile boundaries have shifted, and its surface has been reshaped in
significant ways.

This tour spans the entire peninsula. Just as the local folks refer to the area
as the Lower Cape and the Upper Cape, so this trip can be easily divided into
two sections, in case you wish to linger at the early stops, then finish on another

day. The Lower Cape is first. The starting point is located at the southern end of a landmark familiar to thousands of tourists, the Sagamore Bridge. This span, built in the 1930s, is one of two road links with the mainland. It soars over another, slightly older human artifact, the Cape Cod Canal.

Take Exit 1 off Route 6. Once you reach the stoplight at the base of the ramp, turn right onto Route 6A East. This takes you under the bridge and gives you a chance to admire it from another perspective. After going 1.4 miles past the stoplight, turn right onto Route 130 and proceed 1.0 mile more, to the right-hand entrance of your first destination, the Heritage Plantation. The parking lot is down the lane and to the left.

Heritage Plantation is one of the Cape's most important attractions. Once the private estate of prominent industrialist and plant breeder Charles Dexter (1862–1943), the land was eventually bought by Josiah K. Lilly III and turned into a showplace of American artifacts. The plantation's horticultural showpiece bears witness to Dexter's avocation—which might better be described as a consuming passion—for hybridizing rhododendrons. This is by no means the worst consuming passion the human mind has come up with. If you visit in late May or early June, during the rhododendrons' peak flowering, this will be especially apparent. The Dexter hybrid rhododendrons, produced by their namesake in

Lowell Holly Reservation, framed by Mashpee and Wakeby Ponds.
–The Trustees of Reservations, airview by Kelsey

152

almost spendthrift fashion, are now legendary in the world of horticulture. Just a century before Dexter's time, few people would have thought of using rhododendrons ornamentally, one reason being that the plants are often finicky about soil and climatic conditions.

As with many other innovators, Dexter preferred new experiments to the patient development and documentation of his earlier work. As a result, the quality and viability of his cultivars varied considerably. On top of that, the plants were widely distributed before they were evaluated. Fortunately, Dexter's genius was recognized and rewarded when, a few years after his death, a committee of experts tracked down his best varieties and saved them from oblivion. The splendid rhododendron display area, located between the art and military museum buildings, is a sort of museum itself. It gives visitors the invaluable opportunity to evaluate the hybrids' competing traits. It is a testament to the committee's efforts, and to Dexter's.

Elsewhere on the meticulously tended plantation grounds, other displays feature important ornamental plant types. The day lilies, themselves the subjects of endless hybridization, are particularly fine; and perennial fanciers will be delighted by the substantial collection of hostas. Herbaceous members of the august lily family, hostas are one of the best things that ever happened to gardening. In their cultivated varieties they too illustrate the human craving for variation. Some have variegated, white-edged leaves, while others sport foliage with a pronounced blue or golden cast. Various leaf shapes, from narrow and lancelike to very broad, have been developed as well.

Depart Heritage Plantation by turning right onto Route 130 South. Continue 3.4 miles through the Route 6 underpass to the junction of Cotuit Road. Go left on Cotuit. Here you see the typical outwash plain of the Cape interior—low, sandy ground once dominated by pitch pine barrens but now threatened with real estate development. (Wouldn't it be a more accurate use of language to switch the two words, and talk about pitch pine development and real estate barrens?) After another 3.4 miles, turn right onto South Sandwich Road. Our next stop is the Lowell Holly Reservation, located just beyond the Ryder Conservation Lands. At 0.7 mile from your last turn, bear right on a road marked with residence signs; you'll spot the reservation entrance almost at once. Carefully drive down the narrow reservation access road to the parking area at its end. You are now on a small peninsula between Mashpee and Wakeby Ponds. These bodies of water, technically known as "kettles," were formed at the end of glacial times, when large masses of ice, separated from the retreating main sheet, became embedded in the sediments. These big chunks melted slowly, and when they were finally gone, they left large depressions that became the lakes you see before you.

Unfortunately, Lowell Holly Reservation was badly hit by Hurricane Bob in August of 1991. Among the tree victims were several large white pines, common enough elsewhere but rare in this locale. These plants were snapped in two about 15 feet aboveground. Our primary objective, the American holly trees, can still be found growing in a diverse community that also contains rosebay rhododendrons, white oaks, American beeches, and poison ivy vines. Although they never attain the size of their larger neighbors, the hollies are of tremendous interest. For one thing, they remind us that broad-leaved plants can be evergreen—an attribute that is by no means restricted to pines and other conifers. For another, these American hollies represent one of their species' most northern populations. Generally, they are more typical of lowland habitats farther down the Atlantic Coastal Plain. If you take a close look at their foliage, you might feel the sudden urge to start singing Christmas carols. For what it's worth, however, the holly most associated with the winter holidays is the English species, which, in truth, is a more attractive plant, with shinier leaves and better fruit set. The bark of the American holly was once used as a malaria cure; the plant's compact, smooth wood has been employed in engraving, inlay work, and handles. Another of its holly genus relatives is maté, a South American native plant widely used in a beverage similar to coffee or tea.

The reservation is a wonderful spot to have a picnic or to simply relax for a while. When you're ready to push on, return up South Sandwich Road to the Cotuit Road junction. Turn right and take Cotuit until it merges with Route 130. After exactly 1.0 mile on 130, turn left onto Route 28 South at the T-intersection. After another 2.1 miles, make a second left, this time onto Route 149 North. On this leg, pitch pine shares the limelight with oaks, red cedar, and black cherry. Roadside wildflowers are rather sparse here, but in the first half of summer, you might spot the numerous yellow blossoms of wild indigo, a native legume that has yielded a traditional remedy for two formidable diseases, typhus and scarlet fever. Before long, you're back at the Cape's main highway, Route 6. Turn left onto 6 East.

In this area, the highway runs along high ground called the Sandwich Moraine, yet another glacial feature. Barring traffic tie-ups, the next 25 miles is ideal for honing your high-speed tree-identification skills. You may not be able to differentiate all the hardwoods, but you should at least recognize the Cape's most characteristic softwood, the pitch pine. As a rule, it is smaller and scragglier than its heftier relatives, the white and red pines. One of the pitch pine's best telltale features is its cones. Produced in great number, they tend to persist on their branches a long time. In fact, they often don't open to disperse their seeds until they're triggered by the heat of a passing forest fire. Botanists call such delayed-action cones "serotinous." In a dry environment like this, fire-activated cones are an important reproductive advantage. Most conifers are especially vulnerable to burning, but the pitch pine profits by it. And in addition to reseeding a flame-cleared environment more quickly than most other trees, the parent trees actually survive too, thanks to their thick, fireproof bark and deep roots. They even possess the ability to resprout from their bases, which is a rare trait in softwoods. Given all this admirable fire-protection equipment, it's ironic that human beings have long used pitch pine stems as easy-to-light kindling, and as a substitute for wax tapers, which accounts for the tree's alternative common name, candlewood.

———————— ❧ ————————

The Lower Cape segment of the tour ends in the vicinity of Orleans, where Route 6 begins its north-south trend. To begin the second half of the trip, continue up Route 6 as it parallels the Cape Cod National Seashore. Your first stop at this splendid and multifaceted public holding is the Atlantic White Cedar Trail in the South Wellfleet area. To reach it, turn right onto Marconi Site Road; then bear left past the headquarters complex toward the Marconi Station Site (do not bear right toward the beach). Signs at the parking lot will direct you onto the White Cedar Trail.

The trail, approximately a mile long, is a botanical wonderland chock-full of unusual seaside and dunal vegetation. At the beginning, you pass through a pygmy forest, a sort of East Coast equivalent of New Mexico's chaparral country. Here, an oak predominates, though its true identity might escape your notice at first, since it only grows to dwarf-tree size at most. This is scrub or bear oak, a diminutive but hardy colonizer of poor soils. Also present along the way are stunted versions of our old friend, pitch pine, and black huckleberry, a shrubby

The Atlantic White Cedar Trail leads visitors to an unexpected world lorded over by rare and highly prized conifers.

ericad much loved by Thoreau for its fruit. And the plants that hug the ground are every bit as intriguing. The gloriously fragrant-flowered trailing arbutus puts in an appearance, as do checkerberry, bearberry, and sweet fern. The real highlight, though, is the rare, mat-forming broom crowberry. Its tiny, needlelike leaves somewhat resemble a conifer's, but in reality the plant is an angiosperm in good standing. The Cape is one of the few places in the world where you're likely to find it growing wild in such abundance.

As you progress farther inland, a subtle change takes place. The pitch pines are no longer dwarf forms; this signifies a somewhat less rigorous environment. Oaks are still common, but they now largely belong to two tree-sized species. White oak, with its rounded-lobed leaves, is readily identified, but black oak can be easily mistaken for its inland look-alike, red oak. Black oak is best distinguished by its stubbier acorns and by its habitat. While red oak is most at home in more fertile forest soils, black oak prefers sandy, well-drained conditions. In the days of sail, its wood was used for the submerged sections of ships' hulls, since it was more resistant to marine boring organisms than other oak types. In the seventeenth century, John Josselyn noted that black oak wood was also a favorite for wainscoting. And as if all that wasn't enough, its inner bark yielded a fine yellow dye. If you visit this site in the dormant season, you'll no doubt notice a common oak trait: the leaves tend to persist on their branches in winter long after they've turned brown. The explorer Pehr Kalm, who was also an ordained Lutheran pastor, took this as proof of God's wish to protect the birds.

Not long after you enter the realm of the bigger oaks, you come upon yet another habitat change, and this time it's a dramatic one. A boardwalk loop takes you through a low, wet area, part swamp, part peatland. Though it might not look like it, this is a miniature version of Wakeby and Mashpee Ponds, seen at our earlier stop at the Lowell Holly Reservation. It too is a kettle, which owes its existence to a detached block of glacial ice. In the intervening centuries, organic matter, including layers of sphagnum moss, has accumulated and formed a waterlogged zone in the midst of the sandy terrain.

The undisputed king of this little peatland is a conifer with characteristic cat-scratch bark, the Atlantic white cedar. The wood of this tree has been prized for centuries for its incredible capability to resist rotting. Josselyn wrote, "This tree the English saw into boards to floor their Rooms, for which purpose it is excellent, long lasting, and wears very smooth and white; likewise they make shingles to cover their houses with instead of tyle, it will never warp." Later, its timber was pressed into service as telephone poles, railroad ties, and even pipe organs for New England's churches. In this wetland setting, the tree performs its best role—living and growing and furnishing visitors with a sight both rare and uncommonly beautiful.

In the shade of the great Atlantic cedars reside a variety of understory plants: highbush blueberry, the delightful white-flowered swamp azalea, and two dissimilar hollies, the deciduous winterberry and the evergreen inkberry. (If you can possibly visit in early July, do so. That is when the swamp azaleas, the very best of their kind, burst into bloom.) Weird *Usnea* lichens, draped over some of the tree branches, resemble Spanish moss and give the place the feel of a bayou. These lichens are not plants but complex associations of fungi and green algae. In contrast, the sphagnum moss on the damp surface is a true plant. What sets it apart from the trees and shrubs above is its lack of a well-developed internal plumbing system. Unlike the Atlantic white cedar, for instance, it cannot pump water, nutrients, and food products to its different tissues; each cell on its surface must more or less fend for itself.

Other wonders await us along the National Seashore. After returning to Route 6 and heading north, 1.8 miles up the road a good example of a salt meadow appears on the left. Giant reed occupies the fringes, wherever it doesn't have to come into direct contact with the salt water. Four miles later, large sand dunes illustrate how vegetation can "fix," or anchor in place, the otherwise shifting ridges of windblown sand. On the final approach to Provincetown via Route 6, the dunal scenery becomes even more arresting. It was here, in the Pilgrim Heights area, that a reconnaissance team from the newly arrived *Mayflower* came ashore in November 1620, some time before the storm-tossed settlers chose their permanent Plymouth site across Cape Cod Bay. At the northwestern end of Pilgrim Lake stands the impressive Bowl of Mount Ararat. At times, the sand in this neck of the dunes makes a genuine effort to take over the roadway. A bit farther down, you come to an intersection; turn right toward Race Point and the Province Lands section of the National Seashore. At 0.5 mile from the previous turn, bear left into the parking lot for the Beech Forest Trail.

At Race Point: beach grass colonizes the seaward dunes (foreground), *and pitch pine stands to leeward* (background).

This brief stop is a refreshing foretaste of northernmost Cape Cod. In this protected location there is yet another pond, but it was not glacially formed. Instead, it is one of the small bodies of water in this vicinity that have collected in low points in dunes or other high ground. Known as a water table pond, it is fed by the pressure of groundwater underneath, in basically the same way an artesian well is recharged. On its shore stands the remnant of a once extensive forest of American beech. It is a testament of earlier days, when a better soil profile had developed atop the sandy substrate. Closer to the water, tupelo trees have established themselves a few feet from the floating pads and white blossoms of the fragrant water lily. New England Indians made wide use of water lily roots to cure sores; white settlers relied on them to treat dysentery and diarrhea.

Just 1.0 mile farther up the access road lies your final destination. Park in the visitor center lot. The visitor center offers plenty of information, free and otherwise, as well as a good view of the dunelands that extend outward to the ocean's edge. See if you can identify three familiar seaside shrubs in the parking lot area and near the building entrance: beach plum, bristly locust, and Scots broom. Of these, only the much-beloved beach plum is a true Cape Cod native. In Britain, the Scots broom has been used herbally as a diuretic and a cathartic, but it supposedly also contains potentially poisonous compounds. Both it and the bristly locust are legumes; they can thrive in this sandy ground and even enrich it with the help of their associated nitrogen-fixing root bacteria. (For more on this symbiotic arrangement, see Tour 31.)

If you're still in an exploratory frame of mind, take a stroll into the dune area, down the paved bicycle path that runs close by. Keep a sharp eye out for speeding cyclists and crawling ticks, both of which can cause long-term grief to the unwary. (See the Mounting Your Expedition section for advice on avoiding disease-bearing ticks.) In addition, equip yourself with a hat, sun screen, and a canteen full of some pleasant, nonbefuddling beverage (water works well). As you leave the unadventurous crowd far behind, you encounter an otherworldly dunescape where straight lines have no meaning. The plant hero of this environment is beach grass, the first species to colonize the raw surface of the sand. Also surviving on the bare spots are reindeer moss and other lichens, as well as a tough little shrub known as woolly hudsonia or beach heather. You may even catch sight of small marshes nestled between the dunes. These secluded oases support such unexpected inhabitants as bulrushes, royal ferns, and dwarfed bigtooth poplar.

TOUR 17 ❦

In Massachusetts:
Norwell to Hingham

Driving Distance:
13 miles

Starting Point:
On Route 123,
in Norwell town center

Arnold
Arboretum

Atlantic

Ocean

World's End
Reservation

Whitney &
Thayer Woods

(123) ○ Norwell

Norris
Reservation

(3)

Plymouth
○

M A S S A C H U S E T T S

(3A)

(3)

N

0 2 4 6 8 10
MILES

Tour 17
A Trip to World's End

*B*oston residents don't have far to travel if they want to experience the wonders of the plant kingdom in a number of beautiful settings. Besides having their own incomparable hometown arboretum (see the next tour), they can visit an excellent assortment of parks and other natural areas that ring the urban core. This trip describes three such open spaces, situated close to one another on the South Shore. All three are properties administered by The Trustees of Reservations, the nonprofit organization that maintains a constellation of nature preserve spaces and historic sites throughout the Commonwealth of Massachusetts.

Once you reach the town of Norwell, you're close to our first point of interest, the Albert F. Norris Reservation. To reach it, turn off Route 123 in the center of town and head down West Street. Go just 0.3 mile, past the cemetery to the entrance of the small reservation parking lot. What a delightful place this is to start a tour. The path from the parking lot takes you through a short stretch of woods to the edge of a sleepy pond. This quiet water served a saw and grist mill constructed in the closing years of the seventeenth century. Beyond it lies a more extensive forest, with a dominant central hardwoods community of shagbark hickory and red oak mixing with a few northern hardwoods, such as American beech and yellow birch. See if you can also locate white oak, red cedar, and hop hornbeam. White oak was considered the best all-around timber tree in colonial times. It was utilized for all sorts of things, from barrel staves and hoops to fencing, furniture, and premium-quality firewood. European shipwrights

The tranquil mill pond at Albert F. Norris Reservation.

were of the opinion that all American oak was inferior to their own, but this didn't prevent Yankee builders from using white oak in many stout and long-lived vessels that made it from New England to China and back on more than several occasions. Still earlier, the Indians of this region had a much different use for the tree, as a sort of agricultural timer. In spring, they waited until the newly emerged leaves appeared, marking the beginning of the growing season, when they could safely plant their crops. Some New Englanders today still follow this rule of thumb, planting corn when the oak leaves are the size of a mouse's ear.

In many places, you will notice that the forest floor is not as overgrown with brush as you might expect. The pleasant openness in the understory is the result of controlled burning, an effective and time-honored method of suppressing unwanted thickets. The practice predates the arrival of European settlers; the Indians frequently set fire to large areas, especially near the seacoast. They had several good reasons for doing so. For one thing, burning was a quick way to clear agricultural land. In addition, the fire helped limit the spread of pests and diseases, and it facilitated travel and hunting in heavily vegetated areas. When the English first came to these shores, they often established themselves in the spots cleared by previous Indian burning, even if they weren't always aware why the broad, parklike expanses were there.

Only in recent times have we begun to understand how significantly the indigenous people altered the natural landscape. Once portrayed as "noble savages" who lived in total harmony with the preexisting natural order, the Indians in fact induced significant ecological change. While nothing can match the wholesale environmental impact of modern industrial civilization, the first native New Englanders demonstrated how even "primitive" societies can transform the landscape when they have a mind to.

Farther along, the woodland path comes within view of the salt marsh of the North River. At the edge of the woods, two classic wetland trees, tupelo and swamp white oak, are fairly common. The latter, a close relative of the regular white oak, was even more highly prized by boat builders. Beyond the trees, however, a whole new community of plants takes over. One of the few woody types in this zone is marsh elder, a shrub in the composite family. This gigantic group of dicots also includes the dandelion, ragweeds, daisies, and sunflowers. Two other composites grow here, white-blossomed perennial salt marsh aster and seaside goldenrod. They are joined by a goosefoot family representative, the fleshy-leaved orache. But the most important herbaceous plants by far are the grass family members of genus *Spartina*. Also known as the cord grasses, they dominate vast tracts of New England's tidal zone and are crucial to the maintenance of healthy coastal wetland habitats.

On your way back to the parking lot, take the other leg of the trail back to the pond, so that you can see the large white pines and excellent specimens of American holly. At this location, the holly is near the northern tip of its natural range.

--------------------- ❧ ---------------------

On leaving the Norris Reservation, return to Norwell town center and turn right onto Route 123 East. One popular ornamental conifer you'll probably notice along the way is the Norway spruce, a European evergreen easily identified by its large size, its Christmas-tree shape, and its long, dangling branches. After 3.2 miles, you come to the junction of 123 and Route 3A; bear left onto 3A North. As you pass through Scituate, you will see a crazy-quilt mixture of wild and cultivated plants, with the somber green of white pines clashing with red-leaved maples and the pale, irreconcilable color of Colorado blue spruces.

At 5.4 miles from the last turn, look for the junction of Sohier Street, to your right. Directly across from it, on the left-hand side of the road, is the gravel driveway entrance to the Whitney and Thayer Woods Reservation. Turn into the driveway and proceed to the parking lot at the far end. This too is a woodland preserve; it merits a visit primarily because of its splendid assortment of forest wildflowers. Take the path that leads through the picnic area sheltered by the stately pines (notice the fire clearing here, too) and follow the triangular white blazes of Bancroft Trail. On the right, as you enter the heart of the woods, is an excellent example of what foresters call a windthrow—a tree that has been toppled over, roots and all. Note the large hole once filled by the massive root system. Depressions that form this way are known as "cradles." In the oldest forests, where human interference has been minimal, there are many cradles that alternate with pillows of slightly higher ground, to form a typically hummocky surface. In contrast, forest floors that are quite flat probably were cleared and farmed by human beings at one time or another.

The wildflower populations change as you descend from the path's high point toward the brook. In the drier upland, among sassafras and witch hazel shrubs, there are a number of herbaceous spring bloomers: starflower, Canada Mayflower, Indian cucumber root, jack-in-the-pulpit, wood anemone, checkerberry, and partridgeberry, as well as those deep-shade showboats of late summer, the asters. Closer to the water, though, the scene changes. Jewelweed, a native member of the *Impatiens* genus, forms sizable colonies among the ferns and sweet-smelling clethra shrubs. Another woody plant worth identifying is spicebush, a laurel family representative with aromatic leaves that give off a refreshing fragrance when crushed.

From this secluded woodland setting, you now proceed to another sort of landscape, featuring sea views, pastures, and rows of old specimen trees standing in full sunlight. To reach this final stop of the trip, turn left onto Route 3A North and stay on it for 2.5 miles, to the intersection of Summer Street. Bear right onto Summer and cross Rockland Street. Before you know it, you'll be on Martin's Lane; continue straight until you come to the entrance of World's End Reservation.

Simply stated, this is one of the loveliest spots on the New England seacoast. It doesn't offer the rocky headlands of Cape Ann or the pine-covered dunes of Cape Cod; instead, it has a sort of breezy, tree-studded dreaminess that can turn the worst day imaginable into something much better. To fully enjoy the peaceful beauty of the place, it helps to know how its landscape came into being. First of all, the terrain of World's End bears the indelible stamp of the Ice Age. The reservation's hills, in common with many other high points in the greater Boston area, are drumlins—huge, humpbacked piles of glacial sediments plucked and sculpted into their present form by a great ice sheet. Originally, the outermost hill here was an island, but a connecting neck of fill, known as the Bar, was added in colonial times. For many years, the land was private farmland. In the late 1800s, America's most famous landscape architect, Frederick Law Olmsted, was contracted to design what today we would call a housing subdivision. Olmsted, the designer of much of the Boston park system, as well as Manhattan's world-renowned Central Park, provided the plan; fortunately, the project never went much further than the planting of the trees. In the twentieth century, World's End was considered a worthy site for a number of additional development schemes, including one for a nuclear power plant and another for nothing less than the headquarters of the fledgling United Nations. Still, the land must have been protected by a guardian angel. It survived intact, and in 1967, The Trustees of Reservations stepped in and succeeded in raising the funds necessary to purchase and permanently preserve it.

The full walking circuit of the reservation is roughly 3.5 miles, but even if you choose to do an abbreviated version of the complete hike, you'll see some terrific scenery and impressive trees. Begin by taking the trail from the parking lot across the western neck of the marsh, not far from the main entrance. In earlier days, this human-induced wetland was called the Damde Meddowes because it was blocked, or dammed, at its eastern end. If you're lucky, you'll see some mute swans gliding over the water's surface.

Once past the marsh, the trail gently climbs the first drumlin, and the showcase of trees begins. Genuine New England plants such as red oak, tulip tree, and black walnut stand at various points along the way, but in general, imported European species predominate. Both in Olmsted's day and our own, landscape architects have often emphasized exotic vegetation at the expense of the native. America's first great horticultural writer, Andrew Jackson Downing, had something to say about this.

> How many grand and stately trees there are in our woodlands, that are never heeded by the arboriculturist in planting his lawns and pleasure-grounds; how many rich and beautiful shrubs, that might embellish our walks and add variety to our shrubberies, that are left to wave on the mountain crag or overhang the steep side of some forest valley; how many rare and curious flowers that bloom unseen amid the depths of silent woods, or along the margin of wild water-courses. . . . Nothing strikes foreign horticulturists and amateurs so much as this apathy and indifference of Americans to the beautiful sylvan and floral products of their own country.

As is obvious here, Olmsted was much too canny to ignore the native plants he thought were good, but Downing's complaint rings true with many other designers before and since. Whatever one's views on the subject, the fact remains that this is a splendid place to get acquainted with some of the finest Old World trees.

One of these Old World species, the deciduous conifer known as the European larch, could easily be mistaken for our own tamarack. The two are very closely related; the key to differentiating between the two is that the European tree has cones that are usually much larger than those of its Yankee next-of-kin. In its native locale, the European larch has been used to treat eczema and other skin conditions. Other transatlantic selections along the way include sycamore maple, horse chestnut, and Scots pine. After you pass a large conglomerate boulder (probably an example of the Roxbury Puddingstone familiar to Bostonians) you reach the Bar. A geologist from out of town might reasonably hypothesize that this narrow neck of sediments is a tombolo, or naturally occurring connection between the mainland and a former island. That hypothesis would not stand in the light of local records, though. The Bar was definitely constructed by human hands.

Walk close to the water's edge and take a good look at some of the herbaceous plants managing quite nicely in this salty, challenging environment—orache, cord grasses, the incomparable sea lavender, and beach pea. The beach pea has the intriguing taxonomic name of *Lathyrus japonicus*—intriguing because it translates to "Japanese lathyrus." This is a rather perplexing tag for a species native to the New England seacoast. Actually, the beach pea is a fairly rare example of a circumglobal flowering plant. Not only is it indigenous here and a thousand miles inland, on the western shore of Lake Michigan, it also claims the Far East as its home. In an era of dramatically varying climates and widely separate land masses, the beach pea's huge geographic range is impressive indeed.

On reaching the outermost drumlin, you have the opportunity to examine two well-known imports in the beech family, the English and turkey oaks. The first of these has small leaves reminiscent of the white oak's; one difference is that the English version has leaves that are sessile—that is, they're directly attached to their branches without connecting petioles, or stems, at their bases. This noble species is one of the most widespread forest trees in Europe. For untold centuries, its rot-resistant wood has been cut for fuel, shipbuilding, and many other industrial and domestic applications. Its galls have been a source of ink, and its bark, a provider of dyes. The turkey oak, meanwhile, has seen a good deal of service in the eastern United States as a park shade tree. (The two great Olmsted parks in New York City, Central and Prospect, also feature impressive turkey oak specimens. The master certainly liked the species.) It has one of the most outrageous acorns known to botany—the nut is enclosed by a shaggy cup that resembles a miniature Halloween fright wig.

The intricate architecture of the broad-leaved tree, revealed at World's End. –The Trustees of Reservations

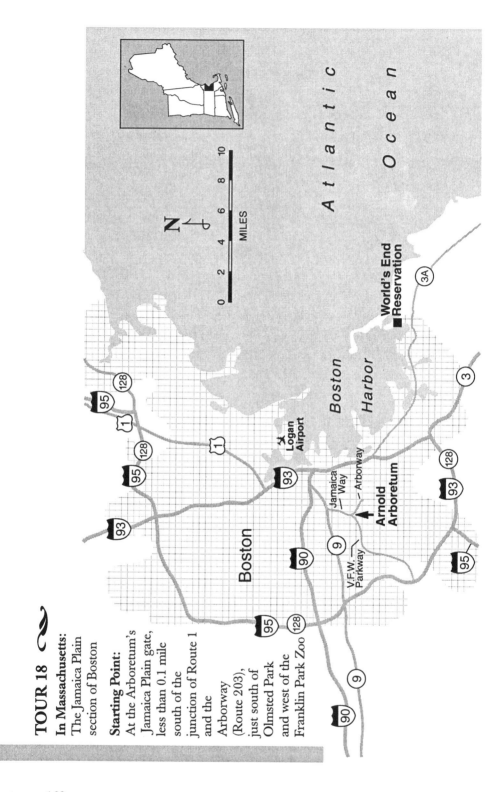

TOUR 18

In Massachusetts:
The Jamaica Plain
section of Boston

Starting Point:
At the Arboretum's
Jamaica Plain gate,
less than 0.1 mile
south of the
junction of Route 1
and the
Arborway
(Route 203),
just south of
Olmsted Park
and west of the
Franklin Park Zoo

Tour 18
The Arnold Arboretum

HIGHLIGHTS:

The Visitor Center ∾ Magnolias and a Lofty Relative ∾ Trees with Roots Deep in the Past ∾ The Dawn Redwood and the Magnificent Katsura ∾ A Broad-Leaved Gymnosperm? ∾ The Ginkgo's Tale ∾ New England's Mulberry Mania ∾ The Osage Orange and Its Brainy Fruit ∾ The River Birch and Its Far-Flung Kin ∾ A Tree That Traffics in Handkerchiefs ∾ Fancy Flowerers ∾ The Conifers and a Transpacific Hybrid ∾ The Reason for Resin

*A*ccording to *Merriam Webster's Collegiate Dictionary,* Tenth Edition, an arboretum is "a place where trees, shrubs, and herbaceous plants are cultivated for scientific and educational purposes." Several fine botanical institutions in New England fit that definition admirably, but one, the Arnold Arboretum of Harvard University, can lay claim to something more. This great Boston landmark is imbued with a distinct, almost grandfatherly personality. It would be impossible to write the history of American horticulture without citing its role and the efforts of its founders. For all of its venerableness, it's an extremely easy place to love. From the first few minutes of your first visit, you'll probably feel, as most people do, that the arboretum is home turf. The forest within a city has a peculiar attraction.

When you arrive at the arboretum's Jamaica Plain gate, enter by way of the Hunnewell Visitor Center. Inside, you'll find maps, historical highlights, and information on blooming dates. If by any chance you're moved to become a member of the arboretum (and I respectfully suggest you should be) you will soon be receiving *Arnoldia,* a delightful quarterly magazine that always manages to hit upon some fascinating topic or other, from the plant-collecting habits of Henry David Thoreau to the life and times of the dawn redwood tree.

The grounds cover a fairly compact 265 acres, including two main hills that provide sweeping panoramas of the Boston skyline and its environs. The success of the arboretum's design is due partly to the intrinsic lie of the land, and partly to what its designers, Charles Sprague Sargent and Frederick Law Olmsted,

An essay in floral primitiveness. One of many magnolia blossoms to be seen in spring at the Arnold Arboretum.

managed to do with it. Every season here has its own special charm. A visit in midwinter can be as engrossing as one at a peak flowering time. The plants are often grouped by taxonomic family or genus, so the botanically motivated person can easily compare far-flung plant relatives. For example, you can view North American birches next to their Eurasian equivalents. The staggering breadth of this collection is the legacy of various stouthearted plant explorers, including the legendary E. H. "Chinese" Wilson, who probably deserves to be remembered as botany's closest approach to Indiana Jones.

The first trees you'll see around and behind the visitor center are members of a family thought to represent one of the oldest in the lineage of angiosperms, or flowering plants. Here, its best-known members, the magnolias, share their home with their less showy but much grander relation, the tulip tree. This straight-trunked plant is one of the loftiest inhabitants of the woodlands of the eastern United States. In New England, regrettably, its wild range is restricted to the southern part of the region. (Should you wish to see other tulip tree examples, refer to Tours 1, 13, and 17.)

The magnolia family and its relatives are of particular significance because they occupy a pivotal position in the development of the flowering plants. They appear in the fossil record early in the Cretaceous, the period that was both the first boom time of angiosperms and the swan song of the dinosaurs. Were you to dissect a magnolia blossom (please don't do it here), you would see that its internal structure is arranged in a helical or spiral pattern, in much the same way that a pine cone is. This anatomical detail and others have led evolutionary theorists to postulate that the magnolia group is a primitive type indeed, with

170

traits that hearken back to the gymnosperms. This is not to say that pines or other conifers were the direct ancestors of flowering plants. The issue of the origin of the angiosperms remains an "abominable mystery," as Darwin put it. Many ancient gymnosperm candidates have been proposed as the ancestor of flowering plants; perhaps the most plausible of these, in light of modern research, is a long-extinct type known as *Caytonia*.

As it happens, the area around the visitor center is rife with other striking hangers-on from the Cretaceous. If I didn't mention the deciduous conifer known as *Metasequoia* or dawn redwood, this book would be banned in Boston; after all, the Arnold Arboretum has been this fascinating tree's preeminent exponent since its arrival in this country in 1948. (For a description of its amazing story and a look at prehistoric dawn redwood fossils, see Tour 4.)

Another ancient angiosperm, the katsura tree, is located just across the roadway. To me, this species is the loveliest of the bunch. It has been used as a specimen tree for decades, but is still unknown to most people. That's a distinct shame, because it has tremendous ornamental potential. Its heart-shaped leaves are flushed with purple when they emerge in the spring; as they mature, they

The lower trunk of a living fossil: a dawn redwood at the Arnold Arboretum.

171

become a pleasing blue-green; and in fall, they turn a clear yellow or a refined pinkish orange.

But the most famous symbol of ancient times is unquestionably that other "living fossil" species, the ginkgo, or maidenhair tree. Head over to the spot behind the Dana Greenhouses where you can see this magnificent loner of the plant world. Its dramatic fan-shaped foliage might suggest that the ginkgo is a broad-leaved angiosperm; in fact, it's a gymnosperm with certain characteristics that link it, however incompletely, to cycads and conifers. Its heritage goes back at least a hundred million years further than that of the magnolia or katsura families. Some paleobotanists believe it is the oldest genus of plants still living. Take a good look at the leaves. They exhibit a wide variation in shape, and an unusual network of equal, branching veins. In contrast, many angiosperms have a more hierarchical network of major and subsidiary veins. From an ornamental standpoint, the ginkgo's foliage is a tremendous asset, especially in autumn, when it turns a rich yellow.

The ancient geographic distribution of this amazing veteran is truly impressive—it extended across several continents, including North America. A 60-million-year-old fossil ginkgo leaf in my collection, for instance, comes from North Dakota. In later times, though, the tree's native range shrank down to one region, the hilly country around China's Yangtze River. Clearly, ginkgo individuals can be exceedingly long-lived; one Chinese tree is reportedly three thousand years old. A wonderful tale relates how the careful attentions of Buddhist monks saved the last remnants of this unique order of plants from utter extinction. Like most wonderful tales, it lacks much in the way of real evidence. What is indisputable, however, is the ginkgo's latter-day return to the other continents. Thanks to human intervention, it is now seen far and wide as a street, park, and lawn tree.

Many aspects of this remarkable messenger from the past can excite one's curiosity. The ginkgo is not particularly resistant to air pollution, but it is amazingly free of pests and diseases, excepting the Texas root rot of the South, which is lethal to the extent that it's also known as the "take-all" disease. The tree's tolerance of bad soil is legendary; it survives so handily because its roots are mycorrhizal (see Tour 13 for more on this symbiotic relationship). Initial research indicates that the ginkgo may prove useful medicinally, as a treatment for asthma, toxic shock, Alzheimer's disease, and blood problems. The ginkgo species is dioecious, with separate male and female trees. Like many other gymnosperms, it has a drawn-out reproductive cycle—five months pass between pollination and fertilization, and it takes approximately nine months after that for the seeds to mature. However much the seeds on female trees may resemble plums or other fruit—the Chinese term *ginkgo* means silver apricot—the fact is that they are not fruit, but simply seeds enclosed by fleshy coats (the seeds are highly esteemed in Oriental cookery). These seeds are responsible for the tree's one objectionable trait: when they fall to the ground and begin to rot, they smell like rancid butter or worse. They also trigger dermatitis in some people.

The first European to describe the ginkgo was the botanical explorer Engelbert Kaempfer, who found it growing in Japan in 1690. The tree was introduced to

North America by the latter part of the following century. Thoreau saw a ginkgo at the Plymouth home of his friend Marston Watson; at roughly the same time, the enterprising Yankee jack-of-all-trades Asa Sheldon was asked to move and warrant for one year a "gingo tree, an exotic, and I expect the only tree of the kind in this country." He declined the offer, since it was made in June, hence too close to the heat of full summer. However, his ever practical eye estimated that the plant would yield about 2 feet of good cordwood. Later, some other contractor moved the plant from its original Boston site at Pemberton Hill to a location near the State House. Apparently, it survived—at least in the short term. If you're in a mood for a little research, you can see if it's still there.

Another type of tree with a long record of cultivation in the Far East is the mulberry. It's the next focal point of our stroll through the arboretum. Walk over to the north side of Bussey Hill, where you'll find a collection featuring plants of this genus. Though few people realize it today, New England was a center of American silk production from approximately 1785 to 1840. The silkworm feeds on a strict diet of mulberry leaves, taken most usually from the *multicaulis* variety of the white mulberry species. This fact led to the planting of thousands of trees in such far-flung locations as Northford and South Manchester, Connecticut; Northampton, Massachusetts; Walpole, New Hampshire; and even selected sites in Maine. In many cases, the state governments were at the forefront of advocating the new form of agriculture, even though no one really had enough experience in the matter to know whether the industry could survive over time. No matter—it soon became a genuine craze, and many Yankee farmers, normally so resistant to the unknown, surrendered to mulberry mania. For several decades it looked quite promising. But disaster struck in 1839 and 1840. Severe storms killed or ruined many trees; then a severe winter was followed by a serious blight. That triple whammy pretty well ended the region's production of raw silk.

When you examine the foliage of the white mulberry here, you'll note that it tends to be dimorphic—it manifests itself in two different forms. The toothed leaves can have entire margins or they can be distinctly lobed. Mulberry fruit is a composite affair, with each "berry" composed of a number of tiny drupes. Birds find it irresistible, even though it works on them as a powerful laxative. If a tree is planted near a sidewalk or parking lot, the results can be horribly messy. Though you might not notice the connection, the genus *Ficus* also belongs to the mulberry family. Its members include the common fig of agriculture and the weeping or Benjamin fig of indoor horticulture. The Boston climate may be too brisk for the *Ficus* crowd, but you will find another mulberry relative nearby. This is the unique North American thicket tree, Osage orange. Native to a small area in the South Central states, it has often been used by Midwestern farmers as a windbreak planting. It has without a doubt one of the most bizarre-looking fruits imaginable. When ripe, it matches a hardball in size and specific gravity. On top of that, it has an unsettling chartreuse color, and convolutions that bear a striking resemblance to the naked human brain. I have seen youngsters in

Brooklyn's Prospect Park use it to pelt passing police cars; personally, I'd rather be hit by a cruise missile.

While it's impossible to give anything resembling a complete description of the arboretum's other wonders, I hope the mention of a few more highlights will pique your curiosity. One additional point of interest is the suite of birches located on the Bussey Hill slope facing the Dana Greenhouses. Look for the river birch, an especially attractive eastern American species. In New England it is not at all common as an ornamental, but in the Midwest it is rapidly approaching cliché status. Notice its beautiful peeling or shredding bark. The color scheme varies on different trees from pink or orange-brown to silvery tones.

In spring, many plants bathe the arboretum landscape in scent and color. Perhaps the most exotic is the dove tree, a member of the tupelo family, located on the southern side of Bussey Hill. Its insignificant flower clusters are draped in large, white bracts, invariably reminding people of drooping handkerchiefs. And among the summer-flowering species are two trees more typical of somewhat warmer climes, the silk tree, close to the Bussey Hill summit, and the golden rain tree, situated a little more than halfway down Meadow Road from the visitor center. The silk tree, a legume, is a native of southern Asia; it produces finely dissected compound foliage and pink, puffball flowers. It is now a common lawn tree in the Sunbelt region. The golden rain tree, another Asian selection, is a representative of the soapberry family. It, too, has striking compound leaves, though they are less feathery. The flowers, borne on long, branched spikes called panicles, are a bright yellow. By fall they give way to green or brown lanternlike seed capsules. I've seen the golden rain tree perform admirably as a curbside planting in such dissimilar places as New York City, Springfield, Illinois, and the waterfront slums of Palermo, Sicily. The sight of it cheerfully blooming in the squalor of Palermo was both heartrendingly ironic and somehow redeeming.

Finally, don't miss the outstanding collection of conifers that is spread out in various locations in the arboretum's middle section. Yews, junipers, hemlocks, and even a sequoia are clustered in the vicinity of Bussey Brook. If you choose to ascend the Woodland Hill Path paralleling South Street, you'll come upon a botanical curiosity—examples of a successful hybridization between the white pine, so near and dear to every Yankee soul, and the stately if languid Himalayan pine. The trunks of these large gymnosperms are sometimes streaked with an important substance, resin. This compound is secreted by special canals in the woody tissue of conifers and serves to deter fungal diseases and some insect pests. It can be broken down into two basic constituents: a solvent, turpentine, and a waxlike material called rosin. If you're a fan of that other great local institution, the Boston Symphony Orchestra, you might have noticed the members of its string section rubbing rosin on their bows.

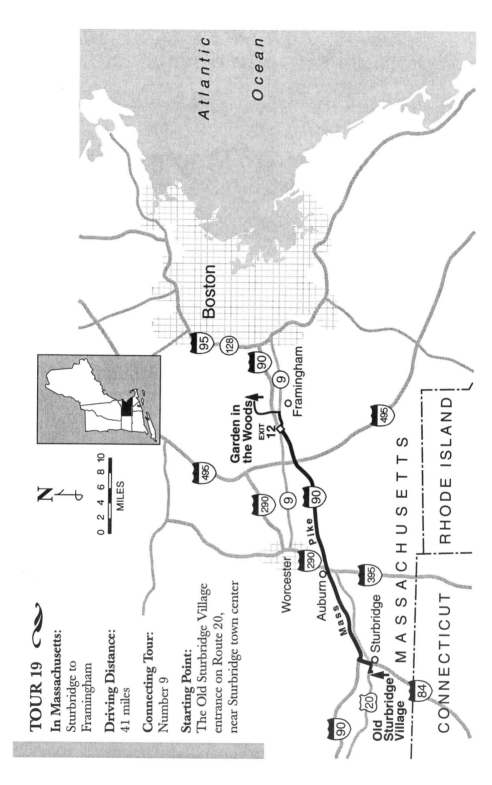

TOUR 19

In Massachusetts:
Sturbridge to
Framingham

Driving Distance:
41 miles

Connecting Tour:
Number 9

Starting Point:
The Old Sturbridge Village
entrance on Route 20,
near Sturbridge town center

MILES
0 2 4 6 8 10

N

Garden in
the Woods

Boston

Atlantic
Ocean

Framingham

Worcester

Auburn

Sturbridge

Old
Sturbridge
Village

Mass Pike

MASSACHUSETTS

RHODE ISLAND

CONNECTICUT

Tour 19
Gardens Past and Present

*I*f you've had a chance to visit Plimoth Plantation and its seventeenth-century inhabitants (see Tour 15), you've had one taste of time travel already. Old Sturbridge Village, our first stop on this trip, is another famous Bay State tourist attraction that traffics in the past—though the historical period it portrays is two centuries closer to our own. While the Plymouth site demonstrates colonial life in its earliest phases, Old Sturbridge Village (often "OSV" on signs) re-creates New England farm and town life in the period that could arguably be called the region's golden age, the early 1800s.

The Route 20 entrance to OSV is situated on the southern side of the road. It's difficult to miss. When you get there, proceed to the facility's main public parking area. Various signs will direct you from the parking lot to the admissions window at the visitor center. You may wish to see the chief horticultural attraction, the herb garden, first. If so, go up the path running northward behind the visitor center and museum shop. You'll soon see the large, attractively designed planting on your right.

Naturally, this is the perfect place to consider what actually makes a garden a herb garden. Strictly speaking, the term *herb* refers to any plant without woody tissue—grasses, cattails, sedges, and all garden annuals and perennials are herbs, at least in the overall botanical sense. But generations of horticulturists have used the word in a more restricted sense, to signify plants grown specifically for culinary, domestic, or medicinal uses. New England, with its strong cultural ties to the Old World, has a particularly rich tradition of herb growing. As you can

A planting of grace and utility. The handsome herb garden at Old Sturbridge Village.

discover at the Plimoth Plantation reconstruction, even the earliest English settlements in America had kitchen gardens. These were customarily the domain of those family members who were responsible for both food preparation and the practice of home remedies—the womenfolk. While the Indians' use of plant products and materials is often emphasized, it isn't always understood that the Anglo-Saxon settlers and the other European immigrants also had a rich herbal lore, passed on from mother to daughter, all the way from medieval times or earlier. In an era of few doctors and rudimentary medical techniques, the role of the female home practitioner assumed an importance in the life of the family that we can barely comprehend today.

One of the best things about this herb garden is the comprehensive labeling. This is no small matter. Many splendid plant collections suffer from a lack of good signage, with the result that their public education value is undermined. If you survey each bed in turn, you'll probably notice one of the most important aspects of the cultivated herbs. To a large extent they are members of just a few plant groups, in particular the mint family (Labiatae), the carrot family (Umbelliferae), and the composite family (Compositae). This is not a coincidence. Many of the species in these families produce volatile oils in their stem and leaf tissues. These oils are distinctively fragrant, sometimes pungently so. Botanists speculate that this sort of chemical specialization, thoroughly exploited by the herbalist and delightful to the passerby, came into being as a sort of biochemical warfare system to defend plants against insects and other pests. (For two extreme examples of plants that fight back, see the discussions of poison sumac in Tour 22 and of stinging nettle in Tour 37.)

Once your senses of sight and smell have had their fill at the herb garden, take a stroll down the pleasant byways of the rest of the complex. Don't be shocked if members of the interpretive staff are a little less worried about their total image than the Plimoth Plantation folks are; here you may see men and women shamelessly wearing their wire-rimmed aviator glasses along with their bonnets and top hats. Also, my own research conclusively proves that at least some of the farmers and tradespeople who inhabit this Emersonian daydream have strong opinions about the Red Sox, especially when the team drops a doubleheader. But there is no doubting the sense of authenticity. The workmanship here is impressive.

Make sure you stop by the sawmill building. It is an utterly fascinating example of pre-industrial America's dependence on wood products. Practically everything except the saw blade itself originally came from the forest. Farther along, you come upon the Freeman Farm, with a trim and tidy apple orchard and wooden fences in both the straight split rail and zigzag "worm fence" styles.

These boundary barriers, which represent a general type of enclosure that antedates the more famous and enduring stone fence, illustrate one important cause of the wholesale deforestation of New England that took place in the seventeenth through the early nineteenth centuries. Even when their posts were made of the most enduring wood, such as black locust, cedar, American chestnut, or black walnut, wooden fences usually deteriorated quite quickly. Over the course of generations, countless thousands of trees had to be felled to keep the enclosures in fairly good repair. Interestingly, foreign visitors, including the property-minded British, were quick to point out the strange and wasteful Yankee obsession for fencing. In Europe, property lines were often left open or planted with hedgerows. Fabricated barriers, including wooden fences, were usually relegated to military uses. As a result, when Old World travelers saw each and every parcel of New England property carefully closed off, they thought the local landowners must be fortifying their holdings in a sustained fit of collective paranoia. In fairness, though, it must be said that New England farmers had plenty of incentive to join in the craze. Residents who did not build and maintain good fences often

A freckle-faced plant called lungwort—one of the residents of the Old Sturbridge Village herb garden.

became the objects of scathing ridicule. Even more important, they might jeopardize a portion of their property rights if they left their fields open to strangers and straying animals.

On the way back to the parking lot, pause by the creek to admire the flowering dogwoods. These indigenous understory trees bloom in early to middle May. If you visit then, look closely at the lovely blossoms. In fact, they are inflorescences, or clusters of tiny individual flowers. The four showy white appendages are not petals but bracts—highly modified leaves that help the otherwise drab blooms attract pollinating insects. The common name of this particular species is misleading, because all dogwoods are flowering plants, even if some are subtler than others. The general term *dogwood* owes its existence to the old English practice of washing mangy dogs in an infusion made from the plant's bark. According to Pehr Kalm, the bark taken specifically from flowering dogwood was used as a colonial substitute for quinine in the treatment of malaria. In addition, it yielded a fine black ink.

A little farther along, on the grounds of the Quaker Meetinghouse, stands an august white oak, a truly impressive specimen of one of the region's most highly prized timber species. If this visit has spurred your interest in tree types and their uses, don't leave without browsing through the bookstore in the Museum Shop. It offers the single best selection of titles on woodworking and related subjects you're likely to find anywhere. It also has a number of books on herbal topics.

———————————— ⌬ ————————————

Depart Old Sturbridge Village by returning to Route 20. Turn right onto 20 East and proceed to the nearby entrance to Interstate 90, the Massachusetts Turnpike. As you approach it, get into the left lane. Once you're through the toll plaza, make sure you switch to the right lane and follow the signs for I-90 East, to Worcester and Boston. This stretch of the "Mass Pike" is bordered by a key indicator of the central hardwoods regime, the red oak. You should be able to discern red maple, black cherry, and white pine as well. Poplar, better known out west as quaking aspen, shares the sunnier spots with large colonies of staghorn sumac. Of this plant, seventeenth-century botanist John Josselyn wrote, "Our English Cattle devour it most abominably, leaving neither Leaf nor Branch, yet it sprouts again next Spring." Josselyn was very observant; the species spreads most effectively by suckers rising from the roots. If its mature stems are cut back, it seems as though twice as many take their place, Hydra-like. Named for its branching pattern, which does indeed resemble a stag's antlers, the plant has been an important source of tannin. It also provides the motorist with a wonderful show of flame oranges and reds in autumn.

In the vicinity of Exit 10 to Auburn, there is an impressive road-cut of the metamorphic rock known as schist. It's tough stuff; the same sort of stone holds up mighty Mount Washington, the highest peak in the Northeast. This exit is the highway's main link with Worcester, New England's second largest city and the home of a number of educational institutions, including Clark University, Worcester Polytechnic, and Holy Cross. As you pass the Auburn Mall, note how its ornamental tree plantings contrast in shape and texture with the aggressive wild vegetation nearby. Scattered red cedars come into view farther east. These conifers, in the cypress family, are among the first trees to invade untended open land.

Now the directions get a bit involved, but the next destination is well worth a little extra attentiveness to navigational details. At 20.9 miles from Exit 10, bear off to the right at Exit 12. Note your mileage here. After the toll booth, turn right onto Route 9 East. Stay in the right-hand lane; 2.6 miles from the Exit 12 turnoff, go up the ramp, and get over to the left as soon as you can. Turn left at the stoplight. You're now on Edgell Road heading north. Stay on Edgell for 2.1 miles; this takes you through a highway underpass. Then turn right onto Water Street. At this point you should start seeing signs for your goal, the Garden in the Woods. After an additional 0.3 mile, turn left onto Hemenway Road and proceed 1.1 miles through the residential neighborhood until you see the Garden in the Woods entrance on the left. Signs will lead you to the parking lot. When you get there, take a deep breath and pat yourself on the back. Believe it or not, there was no easier approach to this splendid but rather secluded facility. In a moment, when you start your stroll, you'll be in a better position to appreciate the merits of the garden's seclusion.

The Garden in the Woods is the showplace of the New England Wild Flower Society, a nonprofit conservation organization dedicated to the preservation of indigenous plant species. This 45-acre preserve is sectioned into a number of large plantings, each of which demonstrates a particular type of habitat, from wild woodland and manicured gardens to peatland and desert ensembles. Each of these locales lie along the main walking path. There are also secondary paths, one of which leads you to an area where pink lady's slipper orchids bloom in the latter part of spring. You'll obtain more precise directions when you check in at the admissions counter in the gift shop. Don't forget to pick up a visitor's map.

Once you're under way along the main trail, you'll pass the nursery complex and then come upon the woodland garden area. This section is not a free-growing natural forest habitat but, instead, a carefully tended display area. Among its intriguing plant specimens is a youthful example of the umbrella magnolia, an unusual and eye-catching tree that sports huge leaves up to 2 feet long. This is not a species you're likely to see in the wilds of New England; its range is largely restricted to the Southeast and the Central Atlantic states. (For more on the story of the magnolia family, see Tour 18.)

Look in the same locale for the herbaceous groundcover plant known as Oconee bells, or, in taxonomic lingo, as *Shortia galacifolia*. This is a particularly precious species that grows in the wild only in a few remote sites in the southern Appalachians. Harvard's great nineteenth-century botanist, Asa Gray, first became aware of the plant in 1839, when he was in Paris studying the herbarium specimens of André Michaux, the renowned French plant explorer who named many North American species. Michaux had collected the *Shortia* in North Carolina, but because it was not in flower, he could not determine its family affiliations. When Gray returned to the United States, he tried to find this humble yet mysterious plant himself. His 1841 expedition to the Southeastern mountains

failed to locate it, and an additional thirty-six years had to pass before the elusive wildflower was rediscovered by a North Carolina native. The by now elderly Asa Gray, accompanied by family and friends, made one last pilgrimage to see it. This time he did. The best time to see these Oconee bells is middle to late April, when they put forth their delicate white flowers.

As you continue along the trail, you will come upon a tranquil lily pond and a rock garden where, as you'd expect, herbaceous plants hold sway. A bit farther, past the glacially formed esker ridge, lies the second main complex, composed wholly of North American species. Bog plants are especially well represented, with a large collection of carnivorous pitcher plants in one section. Of these, the only species you're likely to see in our region's peatlands is the standard pitcher plant, *Sarracenia purpurea*. In the same part of the garden try to find the Hartford fern, a rare New England native that actually climbs like a vine. This is a most unfernlike thing to do, at least in this part of the world. A close relative of this species, the Japanese climbing fern, has escaped into the wild in Southern states and has become quite rampant in places. I've seen it practically take over streambanks in central Louisiana. These two vining pteridophytes belong to an ancient family that first entered the fossil record during the Carboniferous coal swamp times, more than 300 million years ago. Their own genus, *Lygodium,* appears to have had a worldwide distribution in the balmy days of the Eocene epoch, some 50 million years ago. Now most of the family, with the notable exception of the doughty Hartford fern, is restricted to tropical or subtropical climes.

One other plant type of note, even though it is probably not part of the garden's intended design, is a stemless, leafless, and rootless species that forms dark green mats in front of some of the beds. This is a liverwort of the genus *Marchantia.* It is a nonvascular plant with a particularly primitive appearance. At winter's end, it develops tiny umbrella-shaped structures that constitute the liverwort's spore-producing generation. The flat portion below it is the gamete-producing generation. This clearly distinct alternation of two different reproductive types for each species is also found in the true mosses and the pteridophytes.

In my experience, one of the biggest attractions at the Garden in the Woods is the collection most antithetical to New England's native flora, the western garden. It can be most entertaining to watch crusty Yankee visitors cast a typically skeptical and suspicious eye on these otherworldly forms from far-away deserts. To the unprepared, the display of various cacti seems incongruous in this treesy Massachusetts setting; but surprise quickly yields to delight. These strangely shaped succulent plants are well labeled, and it's fun to discover how the different specimens are related to one another. Incidentally, there is one cactus, a species of prickly pear, that is actually found growing wild in New England. Its taxonomic name is *Opuntia humifusa.* Do you see it here? For an account of its preferred habitat, see the description of Connecticut's West Rock in Tour 3.

When you've taken in all these unusual sights, return to the parking lot by way of the Hop Brook Trail. Along the way, a natural stand of native stalwarts—skunk cabbage, witch hazel, Canadian hemlock, American beech, and many others—returns you to the familiar world of the New England woodland.

Native New England it isn't. Yuccas of fearsome aspect bristle in the western garden, at the Garden in the Woods.

183

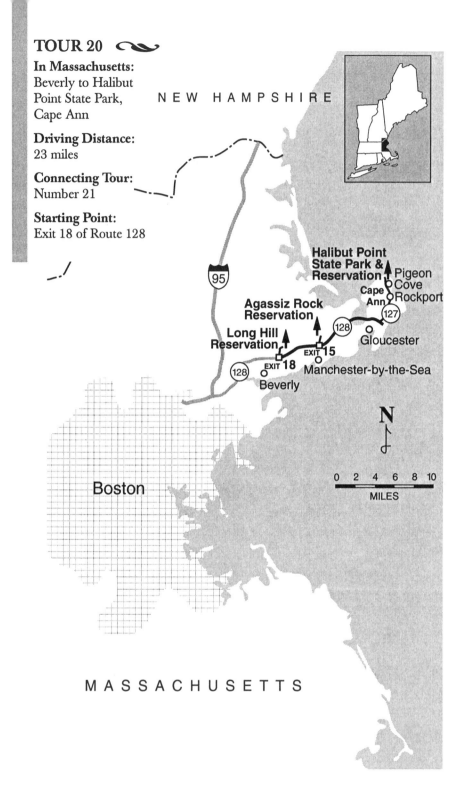

TOUR 20

In Massachusetts:
Beverly to Halibut
Point State Park,
Cape Ann

Driving Distance:
23 miles

Connecting Tour:
Number 21

Starting Point:
Exit 18 of Route 128

NEW HAMPSHIRE

Halibut Point
State Park &
Reservation

Pigeon
Cove
Rockport

Cape
Ann

Agassiz Rock
Reservation

Long Hill
Reservation

Gloucester

EXIT 15

EXIT 18

Manchester-by-the-Sea

Beverly

Boston

N

0 2 4 6 8 10
MILES

MASSACHUSETTS

Tour 20
Exploring the North Shore–Part I

HIGHLIGHTS

Long Hill Reservation ∾ Mabel Sedgwick Begins ∾ Marjorie Sedgwick
Continues ∾ A Search for Four Viburnums ∾ Tree Peonies, the Court of
China, and a Mysterious Mode of Development ∾ Weeping Trees and Dead
Nettles ∾ Agassiz Rock Reservation ∾ An Erratic Hike ∾ Bearberries, Rock
Tripe, and Reindeer Moss Plants That Won't Stand Up ∾ Where the Land Is
Twice as Old as the Ocean ∾ A Purple Invader That's a Sight for Sore Eyes ∾
Halibut Point ∾ Vines and a Seacoast Quarry ∾ Public Enemy Number
One ∾ A Seaside Heath ∾ Entering Another Kingdom

*T*his is the first of a pair of tours devoted to the lovely section of the
Massachusetts seashore that lies north of Boston. The trip features the
best of the area's plant life: a gardener's dream of a landscaped estate,
a secluded woodland site that combines Ice Age wonders with rock-loving lichens
and wetland plants, and a sunlit heath on the windy Cape Ann shore.

We begin in the leafy suburb of Beverly, located just north of Salem. Leave
Route 128 via Exit 18. At the top of the turnoff ramp, turn left onto Route 22
North and proceed 0.9 mile to the fork in the road. At this point, make sure
you stay on 22 by bearing left. Just 0.4 mile beyond, look for the brick gate
posts and the sign for Long Hill Reservation, and turn down the long access
drive to the estate's main parking area.

Long Hill is a site of special interest to horticulturists and landscape architects. Now serving as a public display area and the administrative headquarters
of The Trustees of Reservations, a private, nonprofit land conservation organization, the estate was originally the home of *Atlantic Monthly* editor Ellery
Sedgwick and his wife, Mabel, a talented plantswoman who laid out the gardens.
After Mabel's death, Ellery married another skilled horticulturist, Marjorie
Russell, who added many unusual trees and shrubs. The Sedgwick home was
based on an early nineteenth-century house in Charleston, South Carolina. Its
interior features woodwork detail brought from the original site, as well as
transatlantic wallpaper from nowhere less literary than Shakespeare's own
Stratford-upon-Avon.

Mabel Sedgwick's design of the grounds incorporated a blend of native and exotic plants, with an informal woodland area contrasting with the more-or-less formal garden groupings. As you head down the path from the parking lot to the house, you first see the "natural" section, where some of New England's best trees and shrubs, including sugar maple, pignut hickory, and winterberry holly coexist with woody plants from Europe and Asia. Try to find four North American members of the *Viburnum* genus: nannyberry, black haw, arrowwood, and withe rod. You probably will find it difficult to conclusively distinguish each of these species, but you may at least succeed in identifying them at the genus level. One of the characters, or identification traits, they hold in common is the opposite arrangement of their buds and leaves—the foliage is set in pairs.

However significant Mabel Sedgwick's design was, her successor deserves proper credit for her additions. Marjorie's hand is especially evident in the almost spendthrift assortment of exotic plant species she added to the scene. She had good ties with the Arnold Arboretum, and must have been well aware of promising new introductions from the Orient and elsewhere. There is such a breathtakingly diverse selection of plants types in this relatively small locale that you'll probably overlook some of your personal favorites unless you purchase the inexpensive Sedgwick Gardens guidebook available inside at the Trustees office.

Perhaps the richest section of the grounds is the terrace and garage areas to the south and east of the house, where, among many other things, there is an excellent display of tree peonies and hybrid rhododendrons. Take a good look

An ornament *azalea bursts* *forth, part of* *the sumptuou.* *plantings at* *Long Hill* *Reservation.*
—The Trustees of Reservations, Dorothy Kerper Monnelly photo

at the tree peonies. Unlike their herbaceous counterparts, the more common garden peonies, these plants are quite persistently woody, though their maximum height of 6 or 7 feet hardly qualifies them for real tree status. Cultivated in the Far East for well over a millennium, tree peonies reputedly enjoyed the special favor of the Chinese imperial court. To the microbiologist and evolutionary theorist, the peony genus is an intriguing one because it exhibits a quite unusual type of reproductive development. When it begins to form, the peony's embryo undergoes a period when its nucleus divides repeatedly. This trait, called poly-embryony, is characteristic of many gymnosperms, but is unknown in other flowering plants.

Before leaving Long Hill, make sure you track down another of Marjorie's focal points, situated just off the main terrace—the Sargent weeping hemlock, which is set among ornamental cherry trees that also sport weeping habits. However unrelated and alien it might appear, this striking evergreen conifer is in fact a cultivated variety of the Canadian hemlock so prevalent in the moist woods of New England. Also, see if you can locate nearby plantings of two attractive herbaceous plants. The first is dead nettle—not really a nettle at all, but a perennial in the mint family with yellow, hooded flowers handsomely offset by variegated green and silver leaves. This distinctive foliage provides a longer-lasting means of identification than the blooms themselves. The other species has the poetic name "honesty." This dramatic biennial has four-petaled, rose-purple blooms that indicate its membership in the mustard family. If you're too late for their flowering period, you will be able to look for the flat, round fruit, which is often likened to silver dollars.

Depart Long Hill by returning to Exit 18 of Route 128. Then head north on 128. Here a familiar collection of oaks, birches, red maple, and black cherry fringes the highway. At Exit 15 bear right at the turnoff. At the head of the ramp, turn left onto School Street North; you are now in Manchester. Go 0.7 mile and carefully pull off the road on the right-hand side, at the Agassiz Rock Reservation, another Trustees property.

This site takes its name from one of the giants of nineteenth-century science, Louis Agassiz (pronounced "ag-uh-see"). Born in Switzerland in 1807, Agassiz eventually became the most eloquent proponent of continental glaciation, the concept that large portions of the Northern Hemisphere were recently covered by thick sheets of ice. In 1846, Agassiz emigrated to America and began his eminent teaching career at Harvard. During that period, he visited this site and found additional evidence for his Ice Age theory in the form of two large boulders, or erratics. Thousands of years ago, these detached masses of rock had been hefted from more northerly locations by the movement of glaciers, eventually to be dumped in their present resting place.

Both of these erratics can be reached today by well-defined footpaths. Take the trail into the wooded area inhabited by red oak, black birch, American beech, hobblebush, and maple-leaved viburnum. After a while, the trail splits; the right-hand fork takes you to the hill summit (home of the smaller of the two boulders), and the left-hand fork leads to a small wetland area and the larger erratic. Both locations are unique ecological settings. The dry, stony surface of the summit location hosts a large colony of prostrate bearberry shrubs. In May, these produce numerous white or pinkish flowers that resemble tiny bells or inverted urns. Bearberry is an ericad, or member of the heath family; its relatives include blueberries and rhododendrons, though taxonomists believe it is even more closely linked to the manzanita shrubs of the Arizona uplands. Its dried leaves have been used herbally in an infusion said to ease urinary problems.

On other rock surfaces, including the boulder's, there are sizable communities of slow-growing lichens—composite creatures, part fungus, part algae or blue-green bacteria. Here, lichens are the easiest to spot; they can be distinguished by the central point of attachment on their lower surfaces, and often by their tan or olive-brown color. Some of the most common umbilicate lichens, both here and elsewhere in the region, are known as rock tripes—a rather unflattering reminder of what they resemble. Lichens are frequently the hardiest colonizers of all. Slowly but surely, they break down the rock into more rudimentary components needed for the creation of soil. To get some idea of the broad range of shapes that lichens assume, look for another type, reindeer moss, which lives not on the boulder faces but on the ground nearby. It forms a mat of crinkly, interlocking, and minutely branched stems.

The lower and larger erratic occupies a much different environment. As you approach the big rock, keep a lookout for goldthread, a low-lying evergreen wildflower, and also for soggy green masses of sphagnum moss. Both plant types

The lower and larger glacial erratic at Agassiz Rock Reservation. Note the young tree snuggled against it. –The Trustees of Reservations, Anna Abbott photo

never rise above the level of your shoelaces, yet they are fundamentally unalike. Goldthread, a member of the buttercup family, is one of the countless vascular species that now dominate the plant kingdom. To be a vascular, a plant must have well-developed internal tissues that conduct water and nutrients from the soil upward and photosynthesis-produced food downward from the leaves or stems. As fundamental an arrangement as this seems, there are actually some plants that thrive without this circulation system. These are the nonvascular types, usually considered more primitive for this and other reasons. Sphagnum and other mosses are excellent examples of nonvascular plants. Since they can't transport water any distance internally, they must either float in it or hug the ground where moisture is especially plentiful. However primitive sphagnum may appear, it has developed a number of advanced survival techniques, including the remarkable ability to absorb and retain up to twenty-five times its own weight in water.

———————— ❧ ————————

The final leg of this trip takes you to the rocky headlands of Cape Ann. Once again, return to Route 128 North and head toward Gloucester. At 4.9 miles from your previous stop, you'll come upon a tidal inlet, complete with salt-meadow cord grass, which signals the proximity of the great, 200-million-year-old ocean. As you pass onto Cape Ann proper, you will see many examples of the granite bedrock, which is twice as ancient as the Atlantic itself. Stay on 128 through the first and second traffic rotaries; at the junction of Route 127 North, take that road toward Rockport. In this vicinity, patches of freshwater wetland still exist, where a beautiful but aggressive invader, purple loosestrife, is very much in evidence, especially during its flowering time in late summer. In its native range in Europe, it has served as a treatment for sore eyes and even for blindness.

After 3.0 miles, bear left toward Pigeon Cove; you are still on 127 North. This area is largely devoted to summer homes. The black locust trees that grow in this location are doing just fine without human help. While they are not really native to this region at all, their species, prized for its rot-resistant wood, has long been a common New England sight. At 2.5 miles from where you made the previous turn, you'll see the sign for Halibut Point State Park. Go right, into the parking area.

It seems perfectly reasonable to name such a watery place after a fish, but actually *halibut* is a creative corruption of the nautical phrase *haul about*—which is what mariners had to do when they neared the point. The preserve we'll visit here is partly state park property and partly an adjoining parcel administered by The Trustees of Reservations.

Proceed on foot down the trail toward the shore and Babson Farm Quarry. The first part of the path gives you the opportunity to identify several important tree species of the Northeast, including American elm, paper birch, red and white oaks, and black cherry. Additionally, notice how certain plants such as bull brier and Oriental bittersweet have adopted a vining habit that allows them to spend less energy maintaining structural support and more on finding the sunny spots in and above the forest canopy.

In a few moments, you'll reach the long-inactive, flooded quarry. Be careful here, especially if you have children along. A stroll around the quarry hole gives you a good chance to examine the hard Cape Ann granite, which so stubbornly resists the onslaught of North Atlantic gales. Woodland trees give way to other woody plants: red cedar, the beautiful spring-flowering shadblow, chokecherry, and that supreme seacoast delight, bayberry. In addition, it isn't difficult to find specimens of European barberry. This unfortunate was once the plant pariah of New England, much hated for its inadvertent role as alternate host of the wheat-destroying black stem rust fungus, also known as "the blast." In the early years, no one associated the economy-threatening disease with the barberry. As John Josselyn wrote in the late 1600s, "Our Wheat . . . is subject to be blasted, some say with a vapour breaking out of the earth, others, with a wind North-east or North-west, at such time as it flowereth, others again say it is with lightning."

By the middle of the eighteenth century, though, the shrub's role was correctly identified. The crown colony of Massachusetts, one of the major wheat-producing centers of the time, ordered all barberry shrubs to be destroyed. No doubt thousands of plants were eradicated, but as you can see here, it's one thing to introduce a new type of organism to a virgin land, and something quite different to extirpate it.

Take one of the footpaths that lead from the quarry down to the rocky shore. The view here is splendid. In clear weather, you can see the coastline curving up at least as far as New Hampshire. If you head over onto Trustees of Reservation property, you'll find yourself in another fascinating plant community, often referred to as a seaside heath. Here only dwarfed versions of woody plants (sassafras, Virginia rose, and black chokeberry) and naturally short shrubs (lowbush blueberry) can survive in the blustery, desiccating coastal environment. A few wildflowers, especially the late-blooming seaside goldenrod, also manage to do well. Being herbaceous, they avoid the harshest of the seasons by dying back to the surface in the fall, or by completing their entire life cycle in the space of several months. As you near the water's edge, you come upon fractures in the granite, called joints. The surf has taken good advantage of these zones of weakness in the otherwise tough rock. At this most fundamental of the earth's boundaries, the plant kingdom gives way to another grand scheme of chlorophyll-bearing creatures, the kingdom Protista. It is represented here by green algae and also by the brown algae known as knotted rockweed. This rubbery olive-tinted organism has mastered the art of survival in the surging, ever-changing zone between high and low tides.

The seaside heath at Halibut Point.

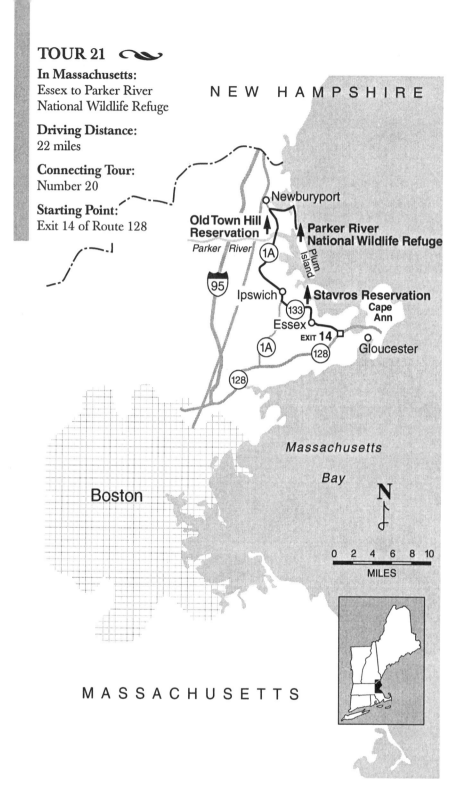

TOUR 21 ～

In Massachusetts:
Essex to Parker River
National Wildlife Refuge

Driving Distance:
22 miles

Connecting Tour:
Number 20

Starting Point:
Exit 14 of Route 128

NEW HAMPSHIRE

Newburyport

Old Town Hill Reservation

Parker River National Wildlife Refuge

Parker River (1A)

Plum Island

95

Ipswich

Stavros Reservation

133

Essex

EXIT 14

Cape Ann

(1A)

128

Gloucester

128

Massachusetts

Bay

N

0 2 4 6 8 10
MILES

Boston

MASSACHUSETTS

Tour 21
Exploring the North Shore–Part II

*T*he Bay State's North Shore presents a superb study in ecological contrasts.
This second half of our exploration of the upper Massachusetts seacoast
demonstrates these contrasts, and complements the preceding tour
beautifully. While it always parallels the shoreline closely, it offers an excellent
variety of terrains and plant communities. As its concluding stop shows best of
all, the saltwater fringe of our continent contains both the lush and the harshly
unforgiving, with windblasted, almost barren zones counterpoised with calm
waters and rampant growth.

Begin the journey by leaving Route 128 at Exit 14. Head north on Route
133. In the first 2.8 miles, you pass through a built-up area surrounding downtown
Essex. Here and there, tidal inlets are visible. Spirelike red cedars have managed
to establish themselves in patches on drier ground, and meadow-forming cord
grasses predominate in the lower expanse, within reach of the tides. Cord grasses
are the primary constituent of New England's great seaside agricultural product,
salt hay. Traditionally, salt marshes were often reserved for the summertime
grazing of cattle. But when harvested instead, the salt hay was used for fodder.
It still has important uses today; as every thrifty Yankee gardener knows, it makes
a terrific mulch, considerably less expensive than fancy bark chips or plastic
sheeting. Throughout the trip, keep an eye out for a particular fixture of the
old salt meadows, the hay-drying supports known as staddles.

North of Essex town center, the scenery bears testimony to one of the worst
indignities ever to hit roadside ornamental trees: the local power company and

At the Stavros Reservation. Red cedars atop one drumlin frame another (center background).

its contracted tree-surgeon minions. These chainsaw-wielding artistes consider it a point of honor to turn healthy, well-shaped plants into topiary-from-hell. All sorts of monstrosities result. Some of the most common forms resemble upright guillotine victims or giant doughnuts with telephone wires running through them. Utilities officials are quick to point out that all this brutal sculpting is necessary to protect the power lines from falling branches. You can't help wondering whether the most merciful thing would be to refrain from planting any trees within striking distance of the lines or the folks in the cherry pickers. Part of the problem could be allayed by the judicious selection of low-growing tree species.

As you near the point 2.0 miles north of downtown Essex, watch for a golf course on the right-hand side of the road. After briefly noting the presence of the obligatory weeping willows, the *sine qua non* of water hazards, turn right onto Island Road, at the course's far end. Immediately after this turn, you'll see the entrance sign for the James N. and Mary F. Stavros Reservation. This Trustees reserve, named for its donors, is our first stop.

The reservation is one of the many drumlins clustered in and around Boston. These curious high points on the otherwise flat landscape are elongated, often humpbacked features that were formed when the motion of an Ice Age glacier molded deposits of fine-grained sediments into swarms of hills that locally all point in the same direction. This drumlin is a fairly large one, more than 100 feet high, but the climb up the trail to the summit is a quick one, and not at all taxing for most people. Near the base are May-blooming flowering dogwoods and a colony of staghorn sumac. This sumac species is a familiar sight on

194

roadsides and along railroad embankments, where it often forms large communities. Its ability to grow so successfully in massed formation is due in large part to mastery of the art of asexual reproduction. As an angiosperm, it produces true flowers and fruit sexually; but it also clones itself rampantly by producing suckers, or new sprouts from the rootstock, that are genetically identical with the parent plant. If you visit in spring before these large shrubs have put out their leaves, you may additionally find the lily-family wildflower known as starry false Solomon's seal in full bloom at their feet.

Farther up the path, arrowwood and chokecherry are prevalent. Just before you reach the top, you'll find a lofty Norway spruce. This conifer is a common sight in the region, but it's definitely not native. The specimen here was probably planted as an ornamental many years ago. At the summit, more red cedars stand among the vegetation that commands the foreground. Take a good look at their foliage and curious fleshy cones. If these remind you of the characters that distinguish junipers, there's a good reason, which is revealed by the tree's taxonomic name, *Juniperus virginiana*. In fact, it's not a true cedar at all. Before you descend, take in the sweeping view of Ipswich Bay. Imagine the time some twenty thousand years ago when this landscape was buried in an ice sheet that marched all the way out to the sea.

━━━━━━━━━━ ∾ ━━━━━━━━━━

Depart the Stavros Reservation by returning to Route 133 and continuing northward on it. Both red and white pine are readily visible along this stretch, as are black cherry and red maple. On the very fringes of the salt marshes look for the waving plumes of another large plant, giant reed. This is New England's largest native grass. It is the cosmopolitan plant species par excellence, and includes all but one of the earth's continents in its natural range. (If you can't figure out the one reedless continent, think of Roald Amundsen and flocks of snowbound penguins.) Giant reed is an aggressive colonizer of disturbed land and has profited greatly from wetland "reclamation" schemes, where direct access to the sea has been cut off. When this happens, ecologically crucial cord grasses are stripped of their saline habitat, and this fast-spreading giant moves in.

Within 3.0 miles of the Stavros Reservation, Route 133 merges with Route 1A. For the next 8.5 miles, until you reach the turnoff for the next stop, follow the signs for 1A North. After passing through Ipswich and Rowley, 1A crosses over a broad tidal inlet of the Parker River, where small islands of oaks and red cedars stand just a little bit above the wetland surface. To your right lies a long natural barrier of sand, Plum Island, which protects the mainland from the worst ravages of the sea; to the left is the pine-crowned Old Town Hill, your next destination. Once you reach the northern side of the river, turn left onto Newman Road and proceed 0.1 mile to the entrance of Old Town Hill Reservation, on the right.

This is the second drumlin of the trip, and though its dimensions and origin might be similar to the one you saw before, both its plant community and its view are special and well worth another short climb. No sooner have you reached

Staddles (the posts driven into salt marsh, foreground*) were used by local farmers to dry salt hay. The drumlin of Old Town Reservation rises in the background.* –The Trustees of Reservations

the entry gate than the botany, or rather the horticulture, gets interesting. At once, you see clues that the hill was originally cleared farm or residential land. For example, ornamental Norway spruces stand sentry duty by the entrance. In the same location, you may come upon an enduring old garden escape, common tansy. This late-blooming species of the composite family is easily identified by finely dissected, fernlike foliage and clusters of buttony yellow flower discs. It was an extremely popular herb of the colonial kitchen garden. Its name is said to have come from the Greek *athanaton,* or "immortal," which perhaps refers to its long-lasting inflorescences. Tansy cakes were traditional Easter fare, and the plant was also a remedy for nerve disorders and intestinal worms.

On the way up the path, see if you can locate the abandoned apple trees that now contend with the rampant thicket growth. Also note the numerous examples of what seem to be sugar maples. If you snap one of their leaves off its twig and check the oozing sap, you will see it is milky rather than clear—these specimens are really Norway maples, probably the naturalized offspring of an old lawn tree. Along the same stretch of the trail stands an example of the rose-family plant type called mountain ash. New England's flora contains two indigenous mountain ash species, but this isn't exactly their habitat of choice. They much prefer the mountain heights miles to the north or west of here. A

196

little sleuthing with a field guide confirms that the species is actually European mountain ash, yet another introduced ornamental from an earlier time. One clue to its true identity is the buds, which are distinctly fuzzy.

Still, not all the trees here are cultivated varieties. It's quite easy to find pin cherry, a second member of the rose family that is a certified native. This pioneer species is among the first to appear on burned or cleared land. It can't take much shade, and after a while, larger and more enduring trees outcompete it. Take a look at its striking reddish bark; it almost seems to be lacquered. The sides of the path are also home to summer grape, one of the most common indigenous vines. At the summit, you'll come upon a meadow with a pair of oaks standing at one end. Determining which oak species is which can sometimes be frustrating, because the different types may hybridize quite freely. Happily, the U-shaped sinuses on the leaves of these trees suggest they're probably standard pin oaks. Here on the hilltop once stood a mighty American elm of great significance to the region's maritime commerce—it served as a marker for ship masters sailing in this vicinity. The famous landmark is gone, but the view is as splendid as ever. Take a moment to study Plum Island and its sound. It is this fragile but vital coastal environment that you'll visit next.

To reach Plum Island, resume the northward trek on Route 1A into Newburyport, a quaint though busy town of cedar-shingled houses. At 2.6 miles from Old Town Hill Reservation, turn right at the stoplight; you should see a sign for Parker River National Wildlife Refuge. At the T-intersection 0.6 mile farther, turn right. This coastal road takes you south past salt marsh and stands of speckled alder and beach plum, and over the Plum Island bridge. At the far end of the bridge, bear right again and drive up to the official entrance of the Parker River National Wildlife Refuge. You will have an opportunity to meet a federal government professional who will ask you to invest in America, at least a little, by paying the required entrance fee. When you hand her or him the cash, get your money's worth and ask for the free list of refuge plant species.

Before proceeding with your exploration of this beautiful preserve, note the following driver's advisory. In all probability, you will encounter the alarming behavior of one particular member of the animal kingdom that is seen virtually everywhere within the refuge. The creature in question is best identified by its bizarre stalking habits and its erratic and distracted way of getting from one place to another. Of course, this is *Homo sapiens* subspecies *birdwatcherensis,* and it should go without saying that driving a vehicle anywhere near it requires a mixture of attentiveness and gentle pity. After all, it's missing the really important life forms, the plants. Whatever you do, do not imitate a piping plover, or you'll be followed by squadrons of telephoto-studded Land Rovers wherever you go.

With this in mind, muster your last ounce of fortitude and proceed resolutely up the road. On your left are sand dunes and their highly specialized vegetation; on the right lie the upper portions of the tidal wetland, called salt pans. The dunal plant species include the splendid spring-flowering beach plum, source of

Beach heather and seaside goldenrod thrive in the dune sand at Parker River National Wildlife Refuge. Note the tracks of the animals that share this harsh, high-energy environment with them.

the planet's best breakfast jelly, and the ruglike woolly hudsonia shrub, often called beach heather. The plant life of the salt pans is something quite different: here salt-meadow cord grass predominates, with the ruddy stems of the odd-looking, succulent slender glasswort also apparent in many spots. The name glasswort refers to the plant's high soda content, which ensured its use in glassmaking (and also in soapmaking). Cattle, always attracted to salt, are especially fond of its saline taste.

The salt pan environment is worth some scrutiny. It's situated just above the main extent of the salt marsh, and consequently, seawater floods into it only during the spring tides. (And just in case you're not the nautical type, spring tides are not to be confused with tides that occur only in springtime. A spring tide is one of higher-than-normal range that occurs near the time of the full moon.) Since evaporation is more of a factor here, salt precipitates out more substantially. The predominant underlying sediment of the salt pan is clayey silt; it accounts for the characteristic mud cracks with upturned corners.

Continue down the road to Lot Number Four, at Hellcat Swamp Nature Trail. A stroll down this well-maintained boardwalk path is a must. Begin by taking the Dune Loop Trail, which leads across the road and over the low, sinuous hills of white sand. For all its beauty, this is an incredibly demanding environment, poor in soil nutrients, rich in toxic salt, and exposed to the whims of the world's great energy machine, the ocean. Interestingly, one of the common conifers in the backdune area is the introduced Austrian pine. More to seaward, there are large clumps of bayberry shrubs, and woolly hudsonia miraculously manages to form large mats where nothing else will grow. In somewhat more sheltered locations, other species take hold: poplar, sassafras, Morrow honeysuckle, Virginia rose, poison ivy, and black cherry. In one swampy area, clethra, arrowwood, winterberry holly, and common buckthorn are all well established.

Taken as a whole, this forward dunal area is a high-energy environment, a fact made obvious by the force of the wind and surf. As you look about this undulating terrain, speculate. What is the sand, really, and where did it come from? It is often composed of a number of minerals, but chief among them is one of the hardest: silicon dioxide, or quartz. The sand that forms these dunes and beaches comes from two main sources: sediments delivered to the coast by rivers, and older reworked sand deposits. Geologists call shorelines such as this "progradational"; that is, the beach is building outward, rather than being eaten away.

Before you return to your car, take in the connecting Marsh Loop on the inshore side of the island. The boardwalk passes through a seemingly endless expanse of giant reed and common cattail, where the ravishingly beautiful but extremely troublesome alien purple loosestrife is also widespread. In direct contrast to the dunelands just to the east, you'll come into a protected low-energy environment, one of the richest, most productive ecosystems anywhere on earth. Tidal wetlands such as this often produce a greater mass of vegetation than the most carefully tended and artificially fertilized agricultural land. While each salt-marsh community has its own peculiarities reflecting local conditions, it's possible to recognize a general pattern in the way they form. Their uppermost edge is usually marked by giant reed. Below it, heading toward the heart of the marsh, is salt-meadow cord grass (*Spartina patens*). Its associates include salt-marsh aster, seaside plantain, glasswort, and sea lavender. In the lowest, most saline section, only the taller saltwater cord grass (*Spartina alterniflora*) really thrives.

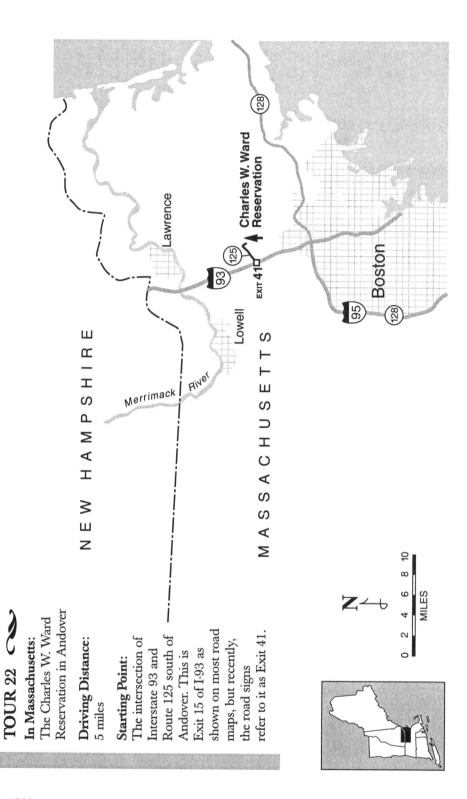

TOUR 22 ⁓

In Massachusetts:
The Charles W. Ward
Reservation in Andover

Driving Distance:
5 miles

Starting Point:
The intersection of
Interstate 93 and
Route 125 south of
Andover. This is
Exit 15 of I-93 as
shown on most road
maps, but recently,
the road signs
refer to it as Exit 41.

NEW HAMPSHIRE

MASSACHUSETTS

Merrimack River

Lowell

Lawrence

**Charles W. Ward
Reservation**

Boston

EXIT 41

N

0 2 4 6 8 10
MILES

Tour 22
The Terra-Not-So-Firma Tour

HIGHLIGHTS

Mayflowers, Starflowers, and Late Summer Asters ∾ Bees Nabbed
in Major Orchid Scam ∾ Farewell to Land: The Boardwalk Begins ∾
A Sphagnum Review ∾ Bog or Fen? ∾ Swamp Horsetails and Royal Ferns ∾
Poison Sumac, Its Observers and Victims ∾ The Little Executioner ∾ The
Famous Crane-Berry in Its Native State ∾ The Arum Family: The Good,
the Bad, and the Smelly ∾ Another Orchid Delight ∾ The Web-Footed
Black Spruce and the Water's Edge

O ne August day in 1856, Henry David Thoreau paid a visit to a peatland
     near his home in Concord, Massachusetts. What he saw impressed him:
     "I seemed to have reached a new world, so wild a place that the very
huckleberries grew hairy and were inedible. . . . What's the use of visiting far-
off mountains and bogs, if a half-hour's walk will carry me into such wildness
and novelty?"

This tour involves the same sort of wildness and novelty. The peatland at
the Charles W. Ward Reservation, like Thoreau's hometown example, is tucked
in amid the familiar. It is surrounded by suburbia and is only a few minutes
away from a commuter-choked interstate. But it too is a new world, an offbeat
and isolated community, preserved for the enjoyment and understanding of the
nature lover. It is a priceless place.

To reach this sole stop of the tour, proceed north on Route 125 from its
junction with I-93. Go 4.8 miles to the intersection of Prospect Road. Turn right
on Prospect; the reservation entrance and parking lot are in a residential area
0.4 mile down the road and to the right. Much of the Ward preserve is dry
woodland, but the focus of our visit is a small but splendid peatland and pond
that are accessible by boardwalk. One word of warning is in order at the outset:
If you are particularly susceptible to poison ivy, be aware that you will be passing
near its even more toxic relative, poison sumac, once you're on the boardwalk.
Make sure you read Pehr Kalm's description, which follows, and maintain a
respectful distance from the plant.

After you've parked your car, proceed across the small meadow to its far end (in the direction away from the Prospect Road–Route 125 junction), and pick up the path that leads through a lowland woods. This is an ideal place to see some classic New England wildflowers: starflower, Canada mayflower, and pink lady's slipper, in the spring, and wild sarsaparilla, enchanter's nightshade, and whorled aster in middle to late summer. As you might expect, it is the lady's slippers that usually provoke the most interest. As with many other orchids, this plant depends on a complicated, even deceptive pollination scheme. When a bee visits the flower, it enters the pouchlike area but can't exit the same way. Instead, it must brush against the bloom's stigma and a pollen-dispensing organ situated near the true exit. The trickery works like a charm. The bee departs with a dose of pollen, which it dutifully deposits on the next lady's slipper's stigma.

Indians knew this species well and used it in a treatment to ease the strains of childbirth. More recently, it has suffered at the hands of people who dig it up and try to transplant it into their gardens without realizing that they're destroying the soil habitat that's crucial to the plant. If you do see orchids on this visit, please don't try your luck at the relocation game.

To your right is a thicket that suggests the presence of wet soil, and sure enough, before long you reach the boardwalk entrance to a dramatically different habitat, where the earth's surface quakes and the plants fight back against animal intruders. Before you proceed down the boardwalk, take a moment to consider

The peatland boardwalk at the Charles W. Ward Reservation, seen from the nearby ridge. The "dry land" between the boardwalk and the open water is actually a floating mat of sphagnum moss, which supports other vegetation. –The Trustees of Reservations

this peatland environment. It is one major type of wetland, and as its name suggests, it's characterized by the presence of peat-forming organic soil. In many parts of New England, the primary contributor to this substrate is a bryophyte, or nonvascular plant, known as sphagnum moss. There are a number of different sphagnum species that grow in this region, but all share a basic form, which you can see when you carefully pull a small individual plant from a damp sphagnum mat. To begin with, each plant has a central stem with tiny branches attached to it. At the top, the youngest branches form a compact cluster known as the capitulum (Latin for "little head"). Farther down the stem, the branches are fully grown and arranged in discrete groups, or fascicles. These are the main water-absorption devices. If you look more closely, you'll see the branches are made up of minute, triangular or wedge-shaped leaves. Somewhat harder to spot, though, are the other leaves, which cling almost invisibly to the stem itself. It should be mentioned, for the sake of botanical purists, that since sphagnum and other mosses aren't vascular plants, they lack true leaves and stems. Still, the structures they've developed are externally so similar that practically everyone, including bryophyte experts, uses those terms anyway.

Sphagnum has played a significant role in human affairs for many years. Nowadays, it is used as a horticultural peat moss; most of this product is mined, so to speak, in Canada. In addition, it has been employed as a medical dressing since the Middle Ages; its peak use in that capacity was during World War I. It also served the American Indians as an absorbent material for diapers.

Ecologists recognize two overall classes of peatlands, bogs and fens, though the dividing line between them can be indistinct at times. In general, bogs are the more acidic and nutrient-starved environments, with less surface water cir- culation and less floral diversity. Fens are somewhat richer in nutrients and somewhat less acidic, because they have some surface water flow. Both bogs and fens have their characteristic indicator plant species. For instance, black spruce thrives best in bogs, while tamarack prefers fen conditions. When there is a mixture of these or other characteristic types, as is the case at this site, chances are that the regime is a transitional one, lying in the gray area between the two definitions. From now on, however, we'll refer to this peatland simply, and not too inaccurately, as a bog.

The following description of the plants on view from the bog boardwalk incorporates surveys made in more than one season, so don't be disappointed if you can't find every species listed here during your first visit. Incidentally, this is one of the best places cited in this book to visit a second and third time. Never-ending change is common to all New England vegetation, of course, but there is so much going on here, in such a small area, that this is a perfect place to see transformations in their most dramatic form.

At the very start of the walkway, you are at the edge of the wetland, where bog conditions are not particularly manifest. The plant community here contains species more indicative of a straight swamp habitat—common cattail, royal fern, and swamp horsetail. The last two are pteridophytes—nonflowering, vascular plants that reproduce by free spores. In Carboniferous times, some 300 million years ago, relatives of the horsetail achieved tree size, but most modern members of its group are considerably smaller. Take a good look at the strange stems of this most unusual plant type—they may have whorls of narrow, jointed branches, or no branches at all. The leaves are extremely reduced and are no more than small, toothlike projections that hug the stem in ringed clusters. The royal fern, on the other hand, has extremely prominent, compound leaves, often called fronds. This handsome and aptly named plant, one of the most notable members of the ancient osmunda family, has an impressively broad geographic distribution. It is a genuine native of the six warmer continents, and its fossil remains have even been discovered in Antarctica—proof that it flourished there, too, when climatic conditions permitted. No other plant species, not even the *Phragmites* giant reed, is so widespread.

Even in this first section of the boardwalk, however, you'll see one tried-and-true bog species, the distinctive sedge relative known as cotton grass. This plant produces flower heads that resemble tufts of tan cotton borne aloft on a thin stalk. As a whole, the sedge family is overlooked by all but the most diehard naturalists. What a pity; though it isn't known for its showy growth forms or colorful blossoms, it is New England's second largest plant family.

A little beyond stands an assortment of shrubs. WATCH OUT. While there are plenty of inoffensive speckled alders here, there are also several poison sumac plants, which can be distinguished by their long, compound leaves, each of which has seven to thirteen large leaflets. Thoreau, ever the iconoclast, declared this species "one of the chief ornaments of the swamps." Even if you agree, maintain a respectful distance. This species should not be touched. Pehr Kalm, the great Swedish botanist and explorer who visited North America in the eighteenth century, bravely tested his own reaction to this plant's chemical defense system and also investigated its effect on others. His excellent description bears repeating:

> The tree is not known for its good qualities, but greatly so for the effect of its poison, which though it is noxious to some people, yet does not in the least affect others. One person can handle the tree as he pleases, cut it, peel off its bark, rub it or the wood upon his hands, smell it, spread the juice upon his skin, and make more experiments, with no inconvenience to himself. Another person on the contrary dares not meddle with the tree while its wood is fresh, nor can he venture to touch a hand which has handled it, nor even expose himself to the smoke of a fire from this wood without soon feeling its bad effects: for the face, the hands, and frequently the whole body swells excessively and is affected with a very acute pain. Sometimes bladders or blisters arise in great numbers and make the sick person look as if he were infected by leprosy. . . . Nay, the nature of some persons will not even allow them to approach the place where the tree grows, or to expose themselves to the wind that carries the effluvia or exhalations of this tree with it, without letting them feel the inconvenience. . . . Their eyes are sometimes shut up for two or more days by the swelling.

Play it safe and resist the temptation to examine these specimens at anything closer than pistol range. (I admit I say the foregoing in the spirit of total hypocrisy. I was once foolhardy enough to experiment with this toxic plant. Much of my misdirected courage was based on my immunity to poison ivy. On a warm summer's day, I rubbed some fresh poison sumac sap on my lightly perspiring upper arm and waited for the scientific martyrdom to begin. When nothing happened, I relaxed into a state of smug self-satisfaction that lasted two days— at which point my skin suddenly erupted into an itchy, widespread rash that persisted for about two weeks. It would not have been even vaguely amusing, even in the long run, if the symptoms had been truly serious.)

In the same vicinity, you will find another kind of formidable plant if you look closely at the level of the sphagnum mat. This species is not dangerous to large creatures, to be sure, but it certainly is to insects. It is the diminutive, red-tinted round-leaved sundew. This amazing organism, like the more famous Venus fly trap, is a carnivorous plant. Its method of seizing prey is quite different than the fly trap's, however. It is covered with small, gluey, hairlike strands, and when the victim gets stuck, the hairs wrap about it, and the digestion process begins. Apparently, this allows the sundew to add essential nitrogen to its diet.

Another hard-to-spot peatland regular that lurks in this location is the large cranberry. This is the same species that is used to such advantage by commercial cranberry growers in southeastern Massachusetts and elsewhere. (For a discussion of a very different kind of cranberry bog, see Tour 15.) Though you'd hardly know it from a single glance, this trailing plant is a close relative of the hefty highbush blueberry, which grows near the far end of the walkway. Its common name is supposedly a modification of "crane-berry"—a reference to its flower, which somewhat resembles the head of that long-billed bird.

Poison sumac shrubs near the Ward Reservation boardwalk. Note the compound leaves that resemble those of an ash tree.

As you continue along, you will find some splendid examples of water-loving plants in the arum family. Relatives of such houseplant favorites as philodendrons and dieffenbachias, these northland natives are the calla lily, arrow arum, and the stout-leaved skunk cabbage. Our friend Pehr Kalm had experience with the last-named species. He wrote, "Among the stinking plants this is the most foetid. . . . Its nauseous scent was so strong that I could hardly examine the flower; and when I smelled it a little too long my head ached." While practically every human being finds the skunk cabbage's carrion smell at least a little bothersome, its odd, hooded flower receptacle is a heavenly delight to the first flies of early spring. Research has shown that the newly emerging flower structure actually generates heat to help melt the snow lingering around it, and to protect itself from frost damage on chilly nights. Later, after flowering, the plant's coarse foliage matures and often persists through at least part of the summer. Much more aesthetically pleasing are the greenish arrow arum, with its elegant, narrowly triangular leaves, and the white-blooming calla lily. Like the skunk cabbage, they produce a spiked cluster of tiny flowers, collectively known as a spadix, that is surrounded by a highly modified leaf, the spathe.

Given its role in New England wetlands, as well as in florists' shops and tropical rain forests, the arum family occupies a place of honor in the monocots, one of the two grand groups of flowering plants. Monocots include such other plants as the palms, the grasses, the true lilies, and the rushes and sedges. Perhaps the most flamboyant and rapidly evolving monocots of all belong to the orchid family; and here you may come across another of their kind. This is the splendid white-fringed orchis, a bog species that puts forth its chaste floral display in middle or late summer. Again, this beauty will not survive transplanting.

The boardwalk comes to an end not far from the point where the sphagnum mat gives way to open water. The ever adaptable red maple grows nearby, as does a rather wan-looking evergreen tree, the black spruce. This frequently scraggly but always remarkable conifer yields an essence used in the brewing of spruce beer. It sends out a shallow but extensive network of roots that often interconnect in a meshed grid. This provides enough stability for an upright woody plant of considerable height to survive on the undulating sphagnum mat. Should you decide to step off the solid boardwalk planks onto this deceptively smooth surface, be prepared for waterlogged shoes at the very least. Your curiosity has brought you a long way from terra firma.

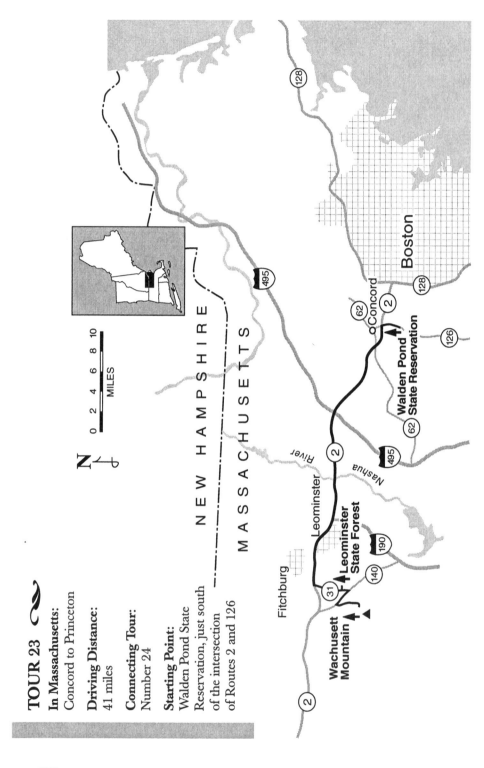

TOUR 23 ∾

In Massachusetts:
Concord to Princeton

Driving Distance:
41 miles

Connecting Tour:
Number 24

Starting Point:
Walden Pond State
Reservation, just south
of the intersection
of Routes 2 and 126

N

0 2 4 6 8 10
MILES

NEW HAMPSHIRE

MASSACHUSETTS

Boston

Concord

Walden Pond
State Reservation

Nashua River

Leominster

Leominster
State Forest

Fitchburg

Wachusett
Mountain

Tour 23
In the Footsteps of Thoreau

*N*ew England's history is awash with intriguing characters who studied and commented on the region's abundant natural treasures. In the early colonial times of the seventeenth century, such entrepreneurs and gentleman adventurers as Captain John Smith, Martin Pring, and John Josselyn first described (and sometimes shamelessly invented) the wonders of the thinly settled continent for a British public eager to learn of its new empire. By the eighteenth century, the Age of Enlightenment had dawned, and a new breed of explorer, typified by Swedish botanist and social commentator Pehr Kalm, busied itself comparing the merits of New World plants and cultures with those of the Old. It was the nineteenth century, however, when homegrown American naturalists finally came to the fore in their own land. Among these were the great Vermont polymath and environmental theorist George Perkins Marsh, and Connecticut's fascinatingly complex poet, botanist, field geologist, and recluse, James Gates Percival.

Out of this diverse cast of characters only one person has achieved long-lasting cult status. This transcultural hero, who seemed in fact a rather judgmental and frosty soul to those who knew him, was Henry David Thoreau. It was his good fortune to be born in 1817 in Concord, Massachusetts, and to spend most of his life in that town when it was the historically rich but not especially populous home of one of the most famous brain trusts the United States ever produced. Thoreau's journal entries abound with references—some of them rather unsparing—to such luminaries as Ralph Waldo Emerson, Margaret Fuller, the Alcotts,

Nathaniel Hawthorne, and Ellery Channing. In turn, these prime movers of the Transcendentalist movement regarded Thoreau as the group's most gifted naturalist. The Olympian Emerson, always his best mentor, made special note of his uncanny powers of observation. It would have been a wonderful thing to see Thoreau leading the rest of the bunch, plus their entourage of visiting city slickers, out on one of their customary huckleberry-picking expeditions.

Our own expedition begins at the very center of Thoreau country, at mythic Walden Pond. Here, Thoreau lived in his small cabin, with some interruptions, from 1845 to 1847. Now a state reservation, the pond is situated a little south of Concord town center. The reservation parking area is located on the east side of Route 126. Be forewarned that a maximum of one thousand visitors are allowed here at any one time. On a balmy summer's day, therefore, you may be waved away from the already brimming parking lot. The best bet is to arrive early, before the crowds descend.

The pond is located across the road, but before you head over to it, make sure you take a good look at the replica of Thoreau's cabin, which stands near the parking lot entrance. This was not the site of the original; you'll see that in just a few minutes. Thoreau was by all accounts a crackerjack carpenter, who, like most enterprising Yankees of his era, made the most of materials derived from local forests. The extremely modest dimensions of the place were all to his liking: "If one designs to construct a dwelling house, it behooves him to exercise a little Yankee shrewdness, lest after all he find himself in a workhouse, a labyrinth without a clew, a museum, an almshouse, a prison, or a splendid mausoleum instead." He may often have railed against the wanton destruction of woodlands, but he certainly knew how to use the end products of the timber trade, as this little gem of a home proves.

After carefully walking across Route 126, take the trail down to the bathhouse. You'll come to the extremely popular public beach that nowadays is the main attraction. There is more than a little irony in the fact that Walden, which has stood as a symbol of solitude for generations of readers, has been turned into a subsidized swimming hole. Where Thoreau's elusive loon once glided over the sparkling surface, now drift flotillas of sunbathers, catching some photons on their radio-equipped air mattresses. But pity the poor park officials, who must continually wrestle with the competing issues of recreation and preservation, and give them credit for at least restricting both the number of visitors and the length of the beach.

Head up the waterfront trail that runs roughly westward past the beach and along the near shore. In geological terms, Walden is a "kettle," a feature that owes its origin to a detached piece of a melting Ice Age glacier. (For more on how kettles form, see Tour 16.) The slope in this locale has been subjected over the years to intense erosion and soil compaction. The Massachusetts Department of Environmental Management has started to restore it, but it's obviously no simple task. Despite the problems, some of the plant species noted in *Walden* and Thoreau's journal entries can still be found here. Look for the sweet-scented clethra, shadblow, sweet fern, red maple, paper birch, and sassafras, the curious

A reconstruction of Thoreau's cabin, in the parking area of Walden Pond Reservation. For all its simplicity, it incorporates the main elements of a nineteenth-century New England dwelling: a pitched roof, a carefully shingled exterior, a full-sashed window, and a woodshed.

shrub that has leaves sometimes entire and sometimes one- or two-lobed. This representative of the laurel family was from the earliest colonial days considered something out of the ordinary. As a matter of fact, for a long time it was thought to be a cure for practically everything, including the nastiest diseases. Martin Pring described it as "a plant of sovereigne vertue for the French Poxe, and as some of late have learnedly written good against the Plague and many other Maladies." Two and a half centuries later, Thoreau noted, "I am always exhilarated, as were the early voyagers, by the sight of sassafras. . . . The green leaves bruised have the fragrance of lemons and a thousand spices." No doubt both Pring and our Concord naturalist would been more than a little surprised to learn that modern science has discovered the plant also contains safrole, a toxic compound that is reputedly carcinogenic.

Another denizen of the waterfront slope is a conifer that especially piqued Thoreau's curiosity, the pitch pine. This species, a prime constituent of Cape Cod forests and New Hampshire's inland pine barrens, is most happy growing in poor, sandy soils. It can be distinguished from the loftier white pine by its persistent cones, by its scalier bark, and by its needles, which are bound together in bundles of three, instead of the white pine's two. Some notable herbaceous plants are in this vicinity, too: hardhack, fern-leaved false foxglove, wild indigo, lance-leaved goldenrod, boneset, and the ubiquitous bracken, the dandelion of the fern world.

As you round the bend into Thoreau's Cove, the trail crosses a low spot that separates the pond proper from a small, sedge-filled marsh called Wyman

Meadow. This is a breathtaking place in autumn, when the foliage of the highbush blueberries and tupelo trees that fringe the open center flush a flame red. The tupelo is not especially beloved of the woodchopper because it is so tough to split; but no matter. It is one of the glories of the eastern wetlands.

Continuing up just a bit, you come upon the actual site of Thoreau's cabin. The monument that now stands here marks its foundation and not its owner's grave, as many visitors first surmise. Thoreau was buried in the family plot in Concord's Sleepy Hollow Cemetery. However this parcel of land looked when it was owned by Emerson and occupied by his ornery bean-planting friend, it is now woodland, with a smattering of pignut hickory, white pine, red maple, and scrub oak. There is also at least one tree-sized oak, which seems to be quite appropriately transcendental in that it possesses some traits suggestive of a scarlet oak and other traits more in line with those of a black oak. Alas, it may be symbolic, but it isn't magical. Oak species hybridize freely to produce intermediate types whose special function, apparently, is to confound taxonomists. The latter can use a little unpredictability in their lives.

Before you return to the parking lot, either through the woods or back along the shoreline path, take one last look at the body of water that has so helped to define the landscape of the American soul. From this vantage point especially, the long-suffering pond retains the rights to its legend:

> Of all the characters I have known, perhaps Walden wears best, and preserves its purity. . . . Though the woodchoppers have laid it bare, first this shore and then that, and the Irish have built their sites by it, and the railroad has infringed on its border, and the icemen have skimmed it once, it is in itself unchanged, the same water which my youthful eyes fell on; all the change is in me. It has not acquired one permanent wrinkle after all its ripples. It is perennially young, and I may stand and see a swallow dip apparently to pick an insect from its surface as of yore.

When you depart the Walden Pond parking lot, turn right onto Route 126 North. If you'd like to see downtown Concord, and its points of historical interest juxtaposed with specialty shops for the yuppie gourmet, continue straight at the Route 2 junction. If you'd rather forge ahead, turn left onto 2 West and continue on it, past the Route 62 intersection. Route 2 is one of the state's main east-west arteries. Despite all the associated development and congestion, the land here is actually much more forested than in Thoreau's day. Then, only one-sixth of Concord was forest; the rest was hayfields, pastures, or cropland. From this neighborhood a person could actually discern distant Wachusett Mountain, the final stop of this tour. Pictures of the region in the Federal era reveal how thoroughly open the ground was. The agriculturally based population was spread much more evenly across the landscape. It was not until industrialism, urban concentration, and the abandonment of many farms that the trees returned with a vengeance.

At a point 15.0 miles past the Route 62 junction, the road crosses the small northward-flowing Nashua River. In this vicinity, there are some good views of

Wachusett Mountain to the west. Wachusett is a clear example of what geologists call a monadnock, a term lifted from the famous high point of southern New Hampshire. It denotes any detached mountain or large hill that rises above much lower surroundings. Not surprisingly, the ever observant Thoreau was struck by this lonely prominence:

> Especial I remember thee,
> Wachusett, who like me
> Standest alone without society.

This may not be the best formulation of the appropriate geological concept, but it does show why our hero's literary reputation is not particularly dependent on his poetry.

If you take this trip in late summer or fall, you may behold another pleasing sight much closer at hand. If the low vegetation along the shoulder has a distinctly reddish tint to it, you're probably looking at purple love grass. In contrast to many of the herbaceous plants that thrive in tough roadside conditions, this is an indigenous species that grows from Maine to Arizona. It has a reproductive strategy reminiscent of tumbleweed: when it's through flowering, the upper portion breaks off and rolls away, propelled by the passing breezes. In this way, the purple love grass distributes its seeds, and its next generation, far and wide.

About 4.1 miles beyond the Nashua River, the city of Leominster (pronounced, by the way, something like "*lem*-inster") is visible in its pretty valley

A lovely stand of white pines in the Leominster State Forest picnic area. The open forest floor is reminiscent of the fire-cleared groves the English settlers found when they first arrived in New England.

setting. Both red and white pines are common in this stretch. After passing Leominster, keep a sharp lookout for the turnoff to Route 31. Exit there, bearing left at the end of the ramp onto 31 South. In the early 1990s, this area was one of several New England locales heavily hit by gypsy moth infestations. As almost every northeasterner knows, it is the voracious leaf-eating larva of this pest that does the damage. The gypsy moth was introduced into this country in 1869 by a Medford, Massachusetts, entomologist named Leopold Trouvelot. Mr. Trouvelot reasoned, rather plausibly, that the insect might make a good domestic silk producer, but in fact, it did not. It proved successful as an escape artist, however, and soon it was loose in the wild–defoliating whole forests. In the decades since, it has wreaked havoc over thousands of square miles in the East, in the Midwest, and, more recently, in the Far West, where it even attacks the quintessential lumber conifer, the Douglas fir. In the Northeast, the choicest foods of the 3-inch caterpillar are broad-leaved species, including oaks, alders, basswoods, and apple trees. If you've ever picnicked in or walked through a woodland where the little monsters are feeding and dangling from their threads, you probably have a keen sense of the disgusting.

In the course of its eight-week life, it is said that each larva eats more than a square yard of leaf area. Human beings have fought back in recent decades with a variety of weapons, from the justly banned DDT to a number of introduced parasitic insects and the bacterial agent *Bacillus thuringiensis*. Known in the trade simply as Bt, it can be dispensed by ground-based or airborne spraying rigs. These invariably provoke suspicious glances from a public that has grown all too accustomed to hearing about chemical side effects twenty years too late. But one of the good things about using this sort of biological agent is that it is target specific–or, in real English, it only affects gypsy moths and other related caterpillars. Some naturalists fear that Bt has the potential to harm native moth and butterfly populations in certain habitats; but it definitely does not damage car paint jobs, contribute to global warming, or mutate the household parakeet. Its effectiveness is based not on poisonous chemicals but on a microscopic organism that, with tremendous poetic justice, does unto the larvae what they would do unto the trees.

At 1.4 miles from the turn onto Route 31, you enter the confines of Leominster State Forest, your next destination. Continue on Route 31 an additional 0.8 mile, past the state forest headquarters building, to the left-hand entrance to the Crow Hill Pond picnic area. Turn in here. This is an excellent rest stop, the perfect place for lunch under a handsome grove of white pines. At first, the setting might seem quite similar to Walden, but in fact, the vegetation here is much more an upland community, composed of black birch, red oak, witch hazel, striped maple, wild sarsaparilla, and the ground-hugging checkerberry. All of these can be seen on the Rocky Pond Trail, which leads uphill from the parking lot. The trail is marked with rectangular blue blazes. Whether you choose to take this hike or not, don't miss the impressive mountain laurel stand along the low trail beyond the bathhouses. Colonies of this great ericad shrub are particularly splendid in this part of the Bay State. They bloom here and in many places along the roadside in June. (For more on the mountain laurel, see Tour 24.)

———————— ∾ ————————

When you're ready to leave Leominster State Forest, return to Route 31 and turn left (south) past the aptly dubbed Paradise Pond to the junction of Route 140. Here, bear right onto 140 North, through more prime mountain laurel country. Before long you'll spot Wachusett Mountain looming nearby, to your left. Just 1.8 miles from your turn onto 140 you will come to an intersection at a reservoir; turn left onto the blacktop that leads toward the mountain. Before reaching the ski lodge, you'll come to a Y-intersection. Bear left and you'll be on your way uphill to Wachusett Mountain State Reservation.

By all means stop at the reservation's visitor center, where you can get a map and a guide to the scenic views. You'll also see a delightful collection of tourism kitsch featuring representations of local Native Americans dressed, for some murky reason, in the traditional outfits of Plains Indians. Perhaps a hopelessly lost contingent of the Dakota peoples passed this way at just the right moment. However scrambled the image in these vintage souvenirs, the fact remains that the mountain was indeed an important rallying point for real native tribes who participated in King Philip's War from 1675 to 1676.

If you've had enough hiking for one day, you can drive all the way to the summit on the paved road, provided you visit between mid-April and the end of October. Fronting the road on the way up is a large stand of poplar, a species called "popple" by old-timers and quaking aspen by people with more western affinities. This small representative of the willow family is a pioneer of disturbed ground, and is well loved for its shimmering leaves. They tremble in the slightest breeze. You'll also see larger trees: red oak, Canadian hemlock, American beech, red maple, and basswood, the indigenous linden of the Northeast. Wildflowers are none too visible to the passing motorist, but in summer you may catch a glimpse of the pale yellow blooms of common evening primrose. Colonists applied the crushed leaves of this plant to wounds, to help them heal better; also, herbalists used the plant as a sedative. The species is a good example of a North American native that has escaped and naturalized widely in Europe. Two other examples of this counteroffensive, now found all over the Mediterranean area, are the Indian fig cactus and the century plant agave.

Just short of the summit is a turnoff with a magnificent view of the peneplain (meaning "almost a plain"), which stretches away from this great monadnock of resistant rock. Except for the slow and steady processes of erosion, and the much more dramatic advance and retreat of glacial ice sheets in the past few thousand years, this region has been remarkably well behaved for a long stretch. There was a time when this corner of the planet was a hotbed of geologic activity, with great mountain ranges being formed and various land masses crashing together and tearing apart again. Now, however, that kind of crust-wrenching action is more fashionable on the western side of the continent, since it has become the leading edge of the North American plate.

At the mountain's very top are a newly constructed parking lot and a large array of communications antennae. This is the highest point in the Bay State east of the Berkshires. The plant community that clings to this windy perch is surprisingly diverse. American mountain ash is here, and one good alpine plant, the lowly three-toothed cinquefoil, puts in an appearance. The red oaks are definitely dwarfed, but in more sheltered spots, several shrub species attain something like their normal size. These include hobblebush viburnum, witch hazel, and black chokeberry—a member of the rose family also found in fens and other wetlands. New England Indians used the chokeberry as a source of food and medicine.

One more botanic detail should tickle your sense of curiosity. The paper birches here are not the variety with rounded leaf bases most often associated with New England uplands, but the more common lowland version. How would you explain this fact? Here's one clue: Don't look just at the summit's vegetation, look at the peneplain all around you. The fact that this grab-bag assortment of plants does not exactly duplicate the purer montane community of the ranges to the north and west confirms that both Thoreau and the geologists are right: Wachusett mountain stands very much apart.

A clump of three-toothed cinquefoil, nestled between moss and bare rock, at the summit of Wachusett Mountain.

TOUR 24

In Massachusetts:
Westminster to Petersham

Driving Distance:
23 miles

Connecting Tour:
Number 23

Starting Point:
The intersection of
Routes 2 and 140 South,
in Westminster. This is
slightly to the east of the
Route 2–Route 140
North intersection.

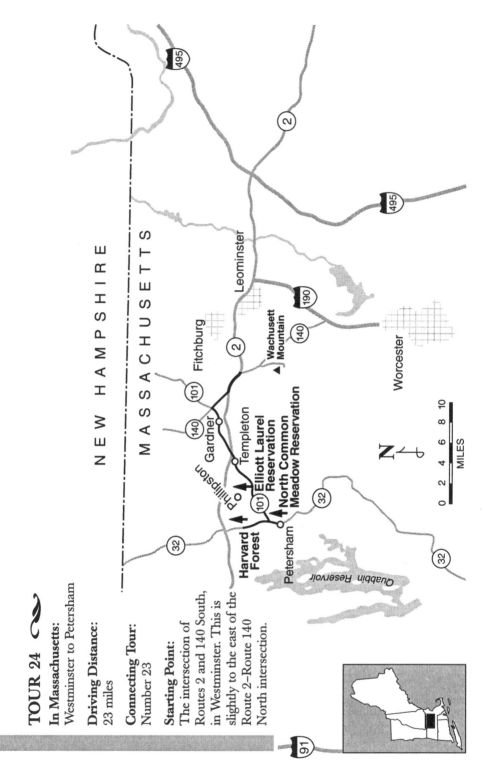

Tour 24
The Fields and Forests of Central Massachusetts

HIGHLIGHTS

A Plant for the Ecotones ∾ Dead Trees That Give Life ∾ The White Delight
of Elias Carter ∾ Treesy Overburden and a Plant Pest ∾ Elliott Laurel
Reservation ∾ New England's Most Glorious Shrub ∾ Pehr Kalm Reflects on
Kalmia ∾ The Tree of Diamonds ∾ Where Thorns Came From ∾ The American
Chestnut: In Memoriam ∾ North Common Meadow ∾ Open Land, an Artifact
of Heavy Toil ∾ The Harvard Forest ∾ A Winter Sunset Masterpiece ∾
Pitch Pine Plumbing ∾ Pulldowns, Windfalls, and Conservative Roots

*I*f your avocations include either ecology or regional history, you'll find
this trip through one of the most rural sections of Massachusetts of
special interest. It will help you better understand the patterns of human
use and abandonment that have so changed the face of New England. By the
same token, if you're an architecture buff, you'll see some prime examples of
early nineteenth-century buildings. Yet again, if you harvest your own firewood
or manage a woodlot, this section, especially its concluding stop, will give you
a deeper appreciation of the art and science of forestry.

We begin near the end point of the previous tour, just 2.4 miles north of
the Wachusett Mountain access road. Head west on Route 2, here a major divided
highway. At first, the roadway is at the level of the lower treetops. Staghorn sumac
ekes out an existence in the rock outcrops. Believe it or not, this familiar plant
of the cashew family is a close relative of poison ivy. It is a common sight on
roadsides, and also forms large monotypic stands at the ecotone, or ecological
boundary, of field and forest.

Take the right-hand turnoff for Route 140 North, which is 1.8 miles from
the tour's starting point. After going about 2.1 miles on 140, you come to an
area where there are many dead trees, or snags, as foresters call them. Can you
guess how these large trees died, apparently all within a short span of time? One
plausible hypothesis involves the arrival of a lethal insect pest or disease. However,
there's a clue that suggests another cause: common cattails now grow in abundance under the dead hulks. Cattails are restricted to the wet, silty soils found

219

in ponds, marshes, or drainage ditches. So perhaps the water table rose to the surface after the trees reached maturity. This process of wetland formation is known as paludification. If that's really what happened here, what was its cause? Here's another clue: Sometimes the construction or widening of roads, or the clogging of culverts, alters drainage patterns and water runoff. Does that make sense here? In any case, these snags, however unsightly they might seem to some people, are wonderful homes and food sources for a variety of animals. Even after death, the trees continue to be leading players in their habitat, and they certainly aren't as ugly as some high-rise human dwellings we could mention.

Approximately 0.7 mile beyond the first of the snags is a stoplight intersection. Turn left onto Route 101 South. Soon you enter downtown Gardner, where you must keep a sharp lookout for the signs that will keep you on 101. Not long thereafter, you enter the adjacent town of Templeton. Before you reach its center, there is a good example of wetland competing with real estate expansion on the left; in summer, pickerel weed sends up spikes of dark blue flowers from the still and shallow waters. Once you enter the heart of town, you'll come upon an architectural delight, the handsome Federated Church, built by Elias Carter in 1811. Seven years after it first opened its doors to the community, an almost exact clone of this building was constructed a little to the north, in Fitzwilliam, New Hampshire—proof, should you need it, that the Yankee mind usually gets its money's worth from a good idea. Happily, the Fitzwilliam church still stands, too; you can see it along the track of Tour 38.

Once out of Templeton, Route 101 enters the town of Phillipston, though there doesn't seem to be anything townlike about it. This truly is the Massachusetts outback, with abundant vegetation everywhere, giving the traveler just about the severest hemmed-in feeling, even by New England standards. Geologists, who'd much prefer a good outcrop to anything vaguely protoplasmic, glumly describe this sort of terrain as treesy; and anything that separates them from the bedrock, be it plants, soil, the singing birds, or first-rate Federal architecture, is just so much overburden. These earth scientists have had the rotten luck to be born in the wrong century; one hundred and fifty years ago, this territory was so open and forest free that it would have been the envy of Agent Orange distributors everywhere. In those days, mass deforestation was not the result of war or any other government policy; it was brought about by the thousands of individual farmers who lived and worked long before the concept of conservation had dawned in the public consciousness. Nor was the subsequent return of the woods due to any particularly enlightened policy. Rather, it represents what nature itself will do in this moisture-rich region when the farms are deserted and the fields left untended. Lest some shoot-from-the-hip anti-environmentalist think this spontaneous regeneration justifies clear-cutting the world over, I should note that this region is a special case. In many places—the Amazon Basin for one—the forest is not so resilient.

On this stretch of road, it seems as though the native plant species of the oak-dominated central hardwoods regime are in total control, until masses of bamboolike Japanese knotweed come into view on the pavement's edge. Even though it rapidly shoots up to shrub size, Japanese knotweed is not really a woody plant at all; it dies back each fall. What makes this buckwheat-family pest so difficult to eradicate is its ability to spread rapidly by horizontal underground stems, called rhizomes. In some parts of the country, notably in the New York City park system, the plant has taken over large areas and shouldered out more desirable plantings. It could be renamed, without too much exaggeration, the kudzu of the North.

Pull off the road to the right at the entrance of Elliott Laurel Reservation, located 1.7 miles beyond the sign marking the Phillipston town line. You should see a Trustees of Reservations marker and donations can there. Other than that, there's just an unadorned gateway in the stone fence. This is not one of the most frequently visited public lands in the Commonwealth, but then, the sense of remoteness is part of its appeal. The person willing to take thirty minutes or an hour on the loop trail marked by the yellow dot blazes will find a fine display of woodland flora.

Begin the trail at the stone fence opening, then bear left at the first fork. North-central Massachusetts has a reputation for being premium mountain laurel territory, and now you see why. This evergreen shrub is represented here by excellent specimens. They usually come into full bloom around the middle of June, and the show is magnificent. Individually, the small flowers are pure white, pink, or even marked with purplish tints. They are borne in dense clusters that stand out from the brooding background like American Legion Fourth-of-July floats. The result is flamboyant and hopeful. This is New England's plant epiphany. Just a week before it happens, you can't believe it's possible, but when it does, nature restores all faith.

Even the taxonomic name of the mountain laurel, *Kalmia latifolia,* is heart-warming. This was bestowed by the great Linnaeus himself, to honor his friend and pupil—a man who looms large in this book, Pehr Kalm. In the account of his travels in North America, Kalm proudly referred to the mountain laurel, describing its wood as very hard and best fit for pulley parts and weavers' shuttles. He also noted that chimney sweeps used its branches for their brooms. A later commentator, the herbalist Mrs. Grieve, pointed out a darker use for the species. The Indians, she said, once used its toxic leaves to brew a suicide potion when life became unbearable.

Elsewhere on the trail nearby, two other evergreens can be found. Both are conifers—the tall-growing Canadian hemlock and the broad-spreading common juniper. The juniper is one indication that this land was originally a pasture or other field. Browsing cattle leave its unpalatable stems and foliage alone, and they actually help the plant by nibbling away at its competitors. Common juniper is a good circumpolar species, because it is native to large parts of Asia and Europe as well as our own continent. Its fleshy cones, often mistaken for berries, contain a volatile oil that gives gin its characteristic smell and flavor. They've also been used for kidney ailments and veterinary applications.

Somewhat harder to notice along the trail's first section are the ground-running dewberry plants of the genus *Rubus.* These blackberry and raspberry relatives are identified by their prostrate habit, bristly stems, and glossy, three-parted leaves. The genus as a whole is a notorious hybridizer. Often, after struggling to reconcile a particular plant's conflicting characters, I've consigned it to my own specific epithet of *Rubus dubius*. For what it's worth, the plants here apparently are swamp dewberry, even though there's no wetland within spitting distance.

The trail next leads through the metaphysical gloom of a Canadian hemlock grove. Subsequently, it climbs to higher ground; in some places the path is rendered almost invisible by the refreshing green fronds of the hay-scented fern. Soon the trail descends again, through a stand of black birch, highbush blueberry, hazel shrubs, and American cranberry bush; then it passes into the adjoining meadow. A venerable, many-trunked white ash tree rises up in the meadow like a giant candelabra. I've always been tempted to call it Yggdrasil, after the ash tree in Norse mythology that knits heaven and earth together. Note the lichens that grow on its bark, and take a good look at the bark itself, since it's characteristic of the species. The ridges and furrows arrange themselves into a pattern of narrow diamond-shaped features.

This striking, lichen-dappled white ash on the edge of the Elliott Laurel Reservation woodland owes its unique shape to the loss of its upper central stem, or leader.

As you return to your car by way of the meadow, note the hawthorn tree planted near the stone fence and the road. Its thorns, actually modified stems, constitute a very effective method of defense, so be careful. Different plants have derived similar defensive gear from different starting points. Cactus spines, for instance, are modified leaves. This genus of the rose family has a number of species, most of which are not native to this region. It has been widely planted as an ornamental. Please heed its correct spelling; the only Hawthorne New Englanders should admit to is Nathaniel.

On departing the Elliott Laurel Reservation, resume your southwesterly track on Route 101. The next turn is 3.7 miles down the road, at the junction of Route 32. Before you get there, you pass through terrain more heavily wooded and distinguished by a high density of American chestnut stump sprouts. A good time to see these is in the autumn, when their long and coarsely toothed leaves turn a distinctive golden brown.

The sad story of the once glorious American chestnut is one of unintentional plant genocide. According to such trustworthy observers as Pehr Kalm and Thoreau, this grand member of the beech family was a common forest species in the Northeast. Besides being an excellent shade tree, it was highly prized for its rot-resistant wood, which was extensively employed in fencing, house siding and shingles, and barn construction. Additionally, its fruit was an important source of food for both wild creatures and roaming farm hogs. Thoreau noted in *Walden:*

> When chestnuts were ripe I laid up half a bushel in winter. It was very exciting at that season to roam the then boundless chestnut woods of Lincoln. . . . Occasionally I climbed and shook the trees. They grew also behind my house, and one large tree which almost overshadowed it, was, when in flower, a bouquet which scented the whole neighborhood, but the squirrels and the jays got most of its fruit. . . . I relinquished these trees to them and visited the more distant woods composed wholly of chestnut.

How shocked he'd be to learn what befell the species only a few decades later. In 1904, the American chestnuts close to New York's Bronx Zoo suddenly died and revealed the presence of a new plant disease, chestnut blight. This fungus probably originated in the Far East. Chestnut trees native to the Orient, which have had many generations to co-adapt with the fungus, are not severely threatened by it; but the American species, which had evolved in total isolation from the malady, proved defenseless. In the next half-century, the plague spread over the American chestnut's entire range. Curiously, the fungus kills the cambium layer and sapwood of the trunk, but does not affect the roots. As a result, old chestnut stumps that have been killed back to ground level continue to produce leaf-bearing suckers. Invariably, these grow for a few years, only to be struck down when they reach a certain point in their development. There have been various attempts to produce a blight-tolerant variety of American chestnut, but most of these have involved crossing the native type with Oriental species. It's ironic that a tree once famous for its imperviousness to rot-producing fungi should be so helpless in the face of one of their relatives. (For a look at one apparently resistant American chestnut specimen and a new glimmer of hope in treating the problem, see Tour 13.)

When you arrive at the Route 32 T-intersection, turn left (south) toward Petersham town center. Your next destination, North Common Meadow, lies less than 1.0 mile down the road, on your left. Pull off to the side. Next to the meadow stands the tiny Aaron Brooks law office building, which dates from 1830. Across the road is the Orthodox Congregational Church, a good Gothic Revival edifice with a rather busy facade.

The meadow is a Trustees of Reservations property. As your trip today has shown, this kind of pleasant vista is quite difficult to come by in this area nowadays. In times past, such an expanse was more actively used as pasturage, but its modern role as public open space seems, to our modern sensibilities at least, no less important. As you peer past the sugar maples and the delightful little orchard, imagine the effort involved in originally uncovering this rock-strewn, densely vegetated parcel of land. It was a time before brush hogs and backhoes. The people who did this had at their disposal a technology little better than that of the late Middle Ages. The process of clearing the land so completely, here and across the breadth of New England, took an unusual bent of mind—one that suspected the worthlessness of everything but enterprise. If there is one thing that makes us vastly different from those earlier Americans, it is our own notion of leisure as the prime source of happiness.

———————————— ∾ ————————————

To reach the final destination on this tour, retrace your path on Route 32 northward. When you've gone 2.0 miles past the 101 junction, start looking for the driveway entrance to the Harvard Forest, on your right. This facility, dedicated to the study of forest biology, is an absolute gem, so plan to spend some time here. Begin indoors by touring the Fisher Museum, located in the main brick building. If there's no one else around, you may have to switch on the lights. When you leave, respect the sense of thriftiness that hangs in the air, and turn them off again.

The highlight of the museum is the diorama display on the main floor. These models, painstakingly assembled, have as their central theme the history of New England forests from colonial times onward, using the Petersham area as the focus. A careful sequential study of these dioramas will give you an excellent introduction to the basics of forestry. As you pass from one scene to another, make sure you see the exquisite rendering of a New England winter sunset, which illustrates the principles of wildlife management. If you have ever experienced the gray and sleety hopelessness of a Yankee February, and then finally come upon one day that was clear, cold, and supernaturally still, you'll recognize the scene; you'll also understand what a masterful work of art it really is.

The upper floor of the museum houses a number of other worthwhile exhibits. There is a fascinating section of pitch pine piping that was once part

Orchard trees at North Common Meadow.

of Boston's first municipal water system, around 1790. The characteristics of this native conifer are described in Tours 16, 23, and 48. As you leave the building, pick up the descriptive material and self-guiding tours for the Sanderson's Farm and Black Gum Trails. The two together can be completed in an hour or two. Sanderson's Farm is the shorter, and is an excellent example of the interaction between agriculture and wildlife. In contrast, the Black Gum Trail is a pleasant woodland hike that emphasizes the ecological relationships of forests, both those harvested for timber and those left more or less to themselves. Black gum, by the way, is another name for tupelo; that tree species is found in one small peatland section where there are specimens more than three hundred years old. Their relatively small size comes as a surprise. How could any tree, already extant for a century when Paul Revere jumped on his horse, be so unimposing? Consider its surroundings. What does the presence of sphagnum moss suggest? Plants often react to stressful environments by growing more slowly.

Elsewhere on the trail, observe the effects—both real and simulated—of hurricane damage and renewal. One grove of red pines has trees that were actually tipped to one side by the awful lashing of the great 1938 storm. The growth they put out afterward sought the new vertical, with the result that the trunks look somewhat bent, like used soda straws. On the far end of the trail are artificially induced pulldowns, where mature oaks were toppled to check how the surrounding vegetation reacts to the altered environment. If you look closely, you'll get a fairly good feeling for the extent of the toppled trees' root systems. It is a popular misconception that tree roots plunge straight down like giant carrots. In most species, the roots fan out widely within a few inches of the soil surface. Plant morphologists point out that these underground structures are remarkably conservative in their internal design. In contrast to stems above ground, root tissue is often downright uncomplicated. One reason for this might be that the complex vascular system of the trunk and branches is largely the result of their special functions. After all, even though roots perform some remarkable functions, they don't have to produce buds, leaves, flowers, or fruit.

The sight of the pulldowns in this section of the Harvard Forest inspires one last observation. In colonial times, white pines that were toppled by storms or other natural causes were called windfalls. While the British government reserved intentionally felled pines for its own uses—primarily for Royal Navy warship masts—windfalls were considered fair game for the local sawmill operators. Woodsmen who nabbed such trees received a sizable, unexpected boost to their income. Eventually, the business world at large appropriated the lumber trade term, and now the most any corporation or investor can hope for is windfall profits.

TOUR 25 ✺

In Massachusetts:
Deerfield to
Northfield

Driving Distance:
32 miles

Starting Point:
Historic Deerfield,
off Route 5 and
about 2 miles
south of
downtown
Deerfield

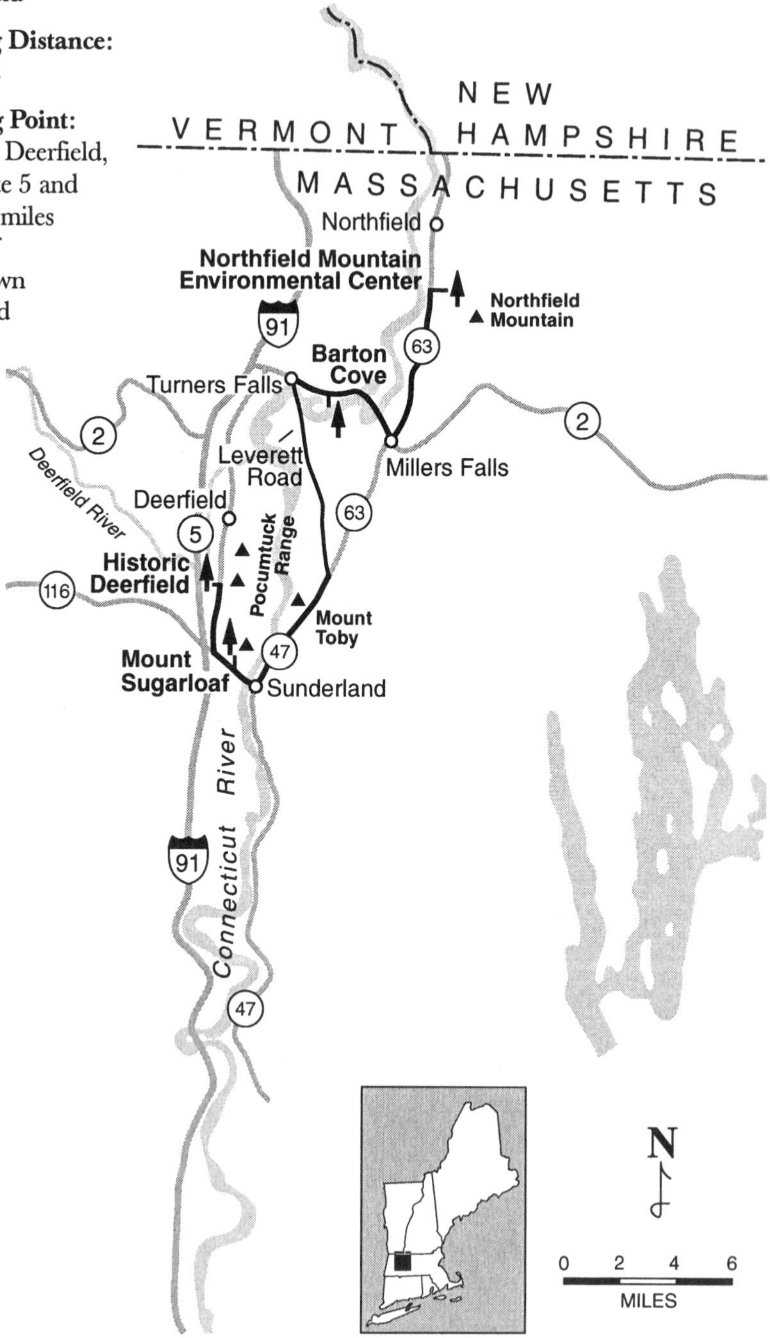

NEW

VERMONT HAMPSHIRE

MASSACHUSETTS

Northfield ○

**Northfield Mountain
Environmental Center**

91

**Barton
Cove**

63

**Northfield
▲ Mountain**

Turners Falls ○

2

**Leverett
Road**

Millers Falls

2

Deerfield ○

5

**Historic
Deerfield**

116

**Pocumtuck
Range**

63

**Mount
Toby**

**Mount
Sugarloaf**

47

Sunderland

Deerfield River

Connecticut River

91

47

N

0   2   4   6

MILES

Tour 25
Pioneer Valley Plant Life

*I*n Massachusetts, practically every road that leads from the Eastern and Western Highlands down to the lush Connecticut River valley floor has a sign proclaiming the beginning of Pioneer Valley. Indeed, this great Central Lowland corridor must still be the home of the ghosts of early pioneers who first brought its fertile banks within the sphere of British dominion. From the moment they first beheld it, those agricultural folk must have known that the valley was the best cropland New England could offer. To hold it, however, was no easy task. For decades, they bore the brunt of a terrible guerrilla war waged by the French and their Indian allies. This region, now settled and placid, was first and foremost a battleground in the conflict between two European empires.

Perhaps the best place to understand the turbulent history of Pioneer Valley is where this tour begins, at Historic Deerfield. When you turn onto Main Street from Route 5, proceed down to the parking lot located behind the visitor center, on the east side of the thoroughfare. It's a lovely spot for a botanical stroll.

In 1665, English colonists purchased 8,000 acres of the surrounding land from the local Indians, and four years later, the settlement at Deerfield began. It's important to note that at this phase the colonists were careful to strike what they considered a legally correct, square deal with the native peoples. Indian ideas about what constituted ownership differed from those of the whites; still, fairness was the general intention. Nevertheless, Indian discontent with the scale of English encroachment throughout southern New England soon erupted into a major conflict. Known as King Philip's War, this bloody confrontation raged in the years 1675 and 1676. During this period, Deerfield was on the very edge of the English-speaking world, and had to be abandoned. Resettlement took place in earnest early the following decade, but times were still hard, due to sporadic but deadly attacks by marauders from the north. Not until 1763, when the British had conquered French Canada and the Treaty of Paris had officially ended the hostilities, could the inhabitants of Deerfield enjoy the full fruits of peace.

The buildings now on display at Historic Deerfield span the period from 1717 to 1824, and feature Connecticut Valley versions of the colonial, Georgian, and Federal styles. If you wish to tour them, stop by the visitor center and purchase a ticket. A simple walk up and down the street is free, and you'll meet some interesting if not always historically appropriate trees along the way. While there are some fine specimens of indigenous species—sycamore, red oak, pin oak, sugar maple, green ash, and black walnut—other types occur that would have baffled the best eighteenth-century woodsman. For instance, Norway maple is here, including one of its purple-leaved varieties; and so is the European littleleaf linden. Oriental imports include the ginkgo, the katsura tree, and everybody's favorite latter-day street planting, the Callery pear, which is the one with the leathery, heart-shaped leaves. For more information on the wonderful katsura and the ginkgo, refer to Tour 18. The Callery pear is discussed in Tour 39.

Take a good look at two other North American trees planted here. The first, the feathery-leaved honey locust, has a natural range considerably west of New England, but it has been widely employed as an ornamental for years. For example, an immense honey locust specimen far up the Connecticut River valley is featured in the Saint-Gaudens site discussion, in Tour 32. Here, the species is represented by its popular thornless variety. The other plant is the most

A horticultural anachronism at Historic Deerfield: a ginkgo tree graces the front of the salt-cellar Allen House. Though a very ancient species indeed, the ginkgo would have been unknown to Deerfield's founders.

attractive tree in the whole neighborhood. It's a magnificent multiple-stemmed example of river birch, located in the yard directly north of the parking lot. This beautiful tree of many colors is said to be native to the Bay State, but in my experience it is not a common sight in the wild, as it certainly is in the magnificent Kankakee River stands in northern Indiana. Its horticultural use is limited in northern portions of New England because it isn't as cold-hardy as its paper birch kin. Still, in other parts of the country, particularly the Chicago area, the river birch has crossed the line into flagrant overuse. It seems to be absolutely irresistible to happy-go-lucky homeowners who equate landscaping with reflecting globes, windsock geese, and recirculating color-wheel waterfalls.

As you leave Historic Deerfield, you'll get a better sense of its geographic position. Depart the parking lot by turning south onto Main Street. To your right runs the Deerfield River, a brisk-flowing waterway that has its source in Vermont. Though it is generally a southward-trending stream, it here approaches its junction with the Connecticut by swinging quite contrarily northward. Note the flat, slightly elevated surface the town is built on. Obviously, the original contractor did a good job of constructing a well-graded terrace that would protect the inhabitants from all but the severest floods. That contractor, incidentally, was the downcutting Deerfield River itself. This terrace is one of many up and down the Connecticut Valley that mark former, higher floodplains, glacial lake bottoms, or shorelines.

Turn left at the south end of town and head east toward Route 5. Near this turn, on the left, look for a graceful American elm that displays its species' much admired and characteristic vase-shaped habit. With Dutch elm disease still rampant, it's impossible to say how long this fine tree will survive. In the summer of 1991, it appeared to be doing well, with no sign of the rapid branch dieback, or "flagging," that is the heartbreaking symptom of the deadly infection. This lone survivor is a reminder that Historic Deerfield's roadway was once lined with American elms, as were those of many other towns across the United States. In a journal entry that both praised this species and took a swipe at Commonwealth politics, Thoreau wrote:

> I have seen many a collection of stately elms which better deserved to be represented at the General Court than the manikins beneath,—than the barroom and victualling cellar and groceries they overshadowed. When I see their magnificent domes, miles away in the horizon, over intervening valleys and forests, they suggest a village, a community, there. . . . They battle with the tempests of a century. See what scars they bear, what limbs they lost before we were born! Yet they never adjourn; they steadily vote for their principles, and send their roots further and wider *from the same centre*. . . . Their one principle is growth. They combine a true radicalism with a true conservatism.

———————————— ❧ ————————————

When you reach Route 5, turn right (south). On your left, the low profile of the Pocumtuck Range separates the road from the Connecticut River. The Pocumtucks are not exactly in the same league with the higher traprock ridges farther down the valley, but they are still an impressive part of the landscape. One common roadside plant along here, especially apparent in middle and late summer, is the fancy alien called Queen Anne's lace. These white, flat-topped flower clusters are familiar to practically everyone, but few realize it is actually a wild carrot, as its taxonomic name, *Daucus carota,* suggests.

By the time you reach a point 3.4 miles south of the turn onto Route 5, North Sugarloaf Mountain, made of soft sedimentary rock, is close at hand on the left. Below it is a familiar tree, enjoying the rich bottomland soil. This is our old adaptable friend, red maple. It seems as happy in this bastion of Yankeedom as it is in the bayous of southern Louisiana, 1,500 miles to the southwest.

In another 1.4 miles, turn left onto Route 116 South. As we get closer to the Connecticut River, the poplar species known as cottonwood makes its entrance onto the scene. It is a good bottomland tree and is quite common throughout the Central Lowland. On the right is an area of ancient dunes. These surprising features came into being some twelve and a half thousand years ago, when the waters of glacial Lake Hitchcock drained out of the valley and left sand and other sediments free to blow about. In modern times they are largely fixed by vegetation. Without the restraining effect of plant roots, these dunes would tend to move in response to the prevailing winds.

At 1.1 miles from your turn onto Route 116, bear left again and then quickly right, to enter Mount Sugarloaf State Reservation. Drive up the road that takes you to the southern summit. Along the way, you can see quite a variety of woody plants: Canadian hemlock, American beech, mountain laurel, black birch, paper birch, sugar maple, striped maple, and staghorn sumac. In late summer, the predominance of somber greens is alleviated somewhat by the white blooms of asters.

When you reach the parking area at the top, pause a moment and take a look at the plants on the periphery. On the west side are some American chestnut sprouts rising from the surviving root systems of older trees stricken with chestnut blight. Eventually, these sprouts too will be attacked by that lethal fungus. (For more on the chestnut and its demise, see Tours 13 and 24.) An often unheeded tree in the birch family, the hop hornbeam, turns up here and there. Its simple, toothed leaves are similar to those of the black birch, but its bark is different, with a "cat-scratch" pattern of narrow, vertical strips or flakes. One of the few species really tolerant of the dense summertime shade of an established broadleaved forest, hop hornbeam produces dangling flower clusters, called catkins, and papery fruit reminiscent of commercial hops—hence, its common name. In earlier times, hop hornbeam was the preferred wood for bows and harrow and hayrake teeth.

As you proceed up the path, you come upon some examples of red pine, a conifer native to New England but most often found cultivated in plantations. Its rot-resistant trunks are used as dock pilings and telephone poles. Old-timers sometimes call it Norway pine. Theories abound about how this certified North American species got that particular nickname; one is that the early English settlers confused it with the Old World's Norway spruce. This is a distinct possibility, because they had trouble telling conifers, and practically everything

232

else, apart. One example of the colonists' lack of identification skills involved the intrepid seventeenth-century botanist John Josselyn. Once, while visiting Massachusetts, Josselyn espied an interesting object hanging in a tree. After due consideration he determined it was a pineapple:

> I made bold to step unto it, with an intent to have gathered it, no sooner had I toucht it, but hundreds of Wasps were about me; at last I cleared my self from them, being stung only by one upon the upper lip, glad I was that I scaped so well; But by that time I was come into the house my lip was swell'd so extreamly, that they hardly knew me but by my Garments.

One other explanation about the red pine, considerably more charitable to our forebears, is that the tree was first seen growing in abundance near the town of Norway, Maine. Take your pick. Also, watch out for killer pineapples.

When you reach the lookout tower, your thoughts will probably not be on plants, but on the excellent vista. In truth, the view from the top is not sublime. The elevation doesn't compare with that of the loftier vantage points of the valley or uplands. But the panorama before you *is* supremely beautiful: it's a vision of abundance, fertility, and peace. Even the busier towns in the distance seem enchanted. From here the Nile of New England snakes onward to the ocean. On a soundless summer's day, the fields have a drowsy agricultural grandeur. In this rocky corner of the country, such yielding openness and unrestraint are nothing less than a miracle.

Several landmarks will help orient you. To the west are the streets and buildings of South Deerfield; to the southeast, the high-rises of the University of Massachusetts campus in Amherst. Almost due east looms the forested flank

Rock laid down in the Age of Dinosaurs now outcrops at the Mount Sugarloaf summit.

of Mount Toby, which, like Sugarloaf, is not capped with harder igneous rock. On the floodplain on both sides of the river stand tobacco-drying barns that bear evidence of what was once one of the valley's most important and famous crops. You'll be taking a closer look at that industry in a few minutes.

If you'd like, walk down the access road a short distance, but be mindful of passing cars. The crumbling sandstone road-cuts are the remains of a long-vanished streambed that existed here during the Age of Dinosaurs. One of the most eye-catching things about this and other Central Lowland sedimentary rocks is their red hue. The color is due to iron in its oxidized form. For many years, geochemists have debated the genesis of these red beds. It's still a controversial subject, but it appears as though red beds can originate in both deserts and humid tropical climates. In this case, it's likely that the Massachusetts of 200 million years ago had a generally humid and tropical climate that was punctuated by a yearly dry season. Keep in mind that the region was then only about 10 degrees north of the equator, according to continental-drift experts.

Nestled among the outcrops is an assortment of wildflowers: harebell, colts-foot, fleabane, yellow hawkweed, wild geranium, and common St. Johnswort. Coltsfoot is an alien of the composite family with a flowering strategy unique for a colonizer of disturbed places. Its yellow blossoms resemble dandelions and rise out of the ground alone, in early spring; the leaves emerge later. Ancient writers recommended that its foliage be smoked like tobacco to cure coughs; Linnaeus reported that the Swedes of his day did just that. The coltsfoot and the other flowering plants present here today would not have been around when the old stream was running by, some 200 million years ago. As far as we can tell, the angiosperms had not yet appeared. Instead, the gymnosperms, or naked-seed plants, still reigned. (For a more extensive rundown on the prehistoric plant life of the Central Lowland, see Tour 8.)

———————————— ❧ ————————————

Depart Mount Sugarloaf by taking the exit that leads downward from the summit parking lot. As you leave the base of the hill, turn left onto Route 116 South and cross the Connecticut River. On its banks is a tree capable of surviving deep and sustained flooding, the silver maple. For many decades, this species, so indicative of major waterways, has also been used as a street and lawn planting in New England and the Midwest. It can develop into an impressive shade tree that cheerfully puts up with poor soils, but the first strong wind or ice storm reveals its Achilles' heel. It is weak wooded, and consequently, it drops its branches the way a rosebush drops its petals. This trait has earned the tree the alternative common name of soft maple.

Turn right at the next intersection onto Route 47 South, and go down it roughly 0.5 mile until you find yourself in the middle of tobacco country. Of this ever controversial form of agriculture our hapless friend Josselyn wrote, "It is observed that no one kind of forrain Commodity yieldeth greater advantage to the public than Tobacco, it is generally made the complement of our enter-tainment, and hath made more slaves than Mahomet." No doubt. In the earliest years of the Massachusetts Bay Colony, some settlers were eager to plant tobacco for profit, but the Puritan leadership initially decided against it, presumably because it might provide some enjoyment for its users. Not long thereafter, though, the equally straight-laced elders of Connecticut Colony permitted its production, and by the end of the 1600s, settlers downriver were busy exporting it. Perhaps a whiff of the old Calvinism can still be detected in the hyperbolic Victorian prose of Vermont's nineteenth-century environmentalist, George Perkins Marsh: "I wish I could believe, with some, that America is not alone responsible for the introduction of the filthy weed, tobacco, the use of which is the most vulgar and pernicious habit engrafted by the semi-barbarism of modern civilization upon the less multifarious sensualism of ancient life." The English herbalist Mrs. Grieve commented more succinctly on tobacco's onetime prestige, which even extended into the realm of medicine: "Kings prohibited it, Popes pronounced against it in Bulls, and in the East Sultans condemned Tobacco smokers to cruel deaths. Three hundred years later, in 1885, the leaves were official in the British Pharmacopoeia."

Actually, the two main types of tobacco that have been grown in the Connecticut Valley, the shade and outdoor varieties, are not used for the deadliest form of consumption, cigarettes. Shade tobacco—the kind grown outside, but under a protective white cloth netting that simulates a tropical environment—is used for cigar wrappers, and the unprotected outdoor tobacco is grown as a cigar binder and filler. Tobacco farming became a major regional industry in the early 1800s and experienced a tremendous boom during the Civil War (maybe in small part because Yankee soldier boys were emulating the cigar-chomping habits of their grizzled supremo, Ulysses S. Grant). In more recent times, the industry here has trembled on the brink of extinction. Shade tobacco production, the more pre-

A tobacco-drying barn (right foreground) *and shade tobacco growing under cloth netting* (background), *in Sunderland.*

dominant of the two, requires a great amount of increasingly expensive labor. That, coupled with the general decline in cigar smoking, makes it more and more difficult for tobacco farmers to make ends meet. There are still quite a few tobacco-drying barns to be seen up and down the valley, but very few still serve active farms such as this one.

Carefully reverse direction on Route 47 and head back toward the junction with Route 116. Note the stately sugar maples in front of the houses in Sunderland town center. Besides being a tobacco-growing area, Sunderland was at one time a center of flax production. Flax was introduced to the region by Scots-Irish settlers in the early eighteenth century, and it remained a major factor in New England agriculture throughout the 1800s. While its soil-exhausting nature meant it could be grown consistently only in the richest soils of river bottomlands, it was nevertheless quite the wonder plant in its day. From it came stalk fibers that were the source of linen and thread; in addition, its seed yielded linseed oil as well as oil cake, widely used as feed for hogs and cattle. Without the boom in flax production, American towns would have worn a distinctly different look, too, since house paint didn't become a widespread commodity until large quantities of linseed oil could be provided. Broomcorn, a less well-known crop, was another local specialty. This grass-family plant was harvested for its seed (still used in bird feeders today) and for its broom-making bristles.

This time as you pass the intersection of Routes 47 and 116, continue straight on 47 North. A stupendous sycamore stands just above the crossing, with a commemorative plaque stating that the tree was already growing here when the United States Constitution was signed. In fact, it is reputed to be the oldest extant tree east of the Mississippi—though such claims are difficult to prove. According to the American Forestry Association, the largest, though not necessarily oldest, plant of this species is located in faraway Jeromesville, Ohio. (See Tour 12 for more on the sycamore, its extensive pedigree, and its use in the Central Lowland.)

A little farther along, wild vegetation including cottonwood, black locust, and American elm is evident. Much of the farmland along this stretch is now planted in maize, a large grass plant developed by American Indian cultures and better known to us simply as corn. About 6.0 miles north of the 47-116 intersection, you'll come to the junction of Route 47 and Leverett Road. Continue straight on the latter; this is the way to Montague Center and Turners Falls. After another 6.2 miles, go straight down the hill at the Y-intersection in Turners Falls. At the downtown stoplight, turn right and cross the river at one of its most dramatic points, where it makes two sharp turns in the space of about 3 miles. A hydro-electric facility is downstream; to your right is the falls, the low dam built to raise the water level, and Barton Cove, your next stop.

In 1676, the falls was the scene of a decisive battle in King Philip's War. On May 19, colonial troops under Captain William Turner surprised a large gathering of Indians who were fishing here to build up their badly depleted stock of food. It was a stunning setback for the natives, though Turner himself was eventually killed, and his forces retreated in disorder as Indians from surrounding encampments entered the fray. In more recent times, the appearance of this great natural feature has been significantly altered by our energy-hungry culture. The main force of the water has been diverted into a canal that leads to the power station's turbines. This leaves the falls high and dry for most of the year, though the spillway is often open during the freshet in spring. Local fossil collectors may prefer to see the old riverbed bare, but many other people come specifically to see the spectacle of the torrents rushing by—water free, however briefly, from the meddling human hand.

When you've crossed the river, turn right at the first intersection onto Route 2 East. You are now in the town of Gill. Continue 0.9 mile past the public boat ramp, to the turnoff on the right to Barton Cove and its nature area. Take this turn. Just inside the entrance, you'll see the still waters of the cove. The plants that fringe it constitute a botanical treasure chest. If you're visiting in summer, take a few extra moments to explore this section. Purple loosestrife, an alien, blooms copiously here, as does the indigenous pickerel weed. The latter can be distinguished by large arrowhead leaves rising directly out of the muck, and by handsome blue flower spikes. A bit inshore from these are the woody dwarf or shining sumac and the herbaceous blue vervain. These take second seat to yet another species, the rank and red-stemmed pokeweed, which attains the size of a big shrub, though it's really herbaceous. The botanical explorer Pehr Kalm noted that European settlers used its roots as a source of red dye, despite the fact that the leaves and stems can be extremely poisonous. It is also said that during the Civil War, when the Union blockade of the Confederacy was having its effect, desperate Southerners used the juice from its berries for ink. I have talked to people who claim that pokeweed gives them a skin rash similar to that produced by poison ivy.

Next, drive up the road a bit farther, until you reach the parking lot for the first campground area. Park here. The peninsula that juts into the river is about a mile long. If you'd like to take an easy and pleasant hike, walk up the paved road, which is bordered by Canadian hemlock, wild grape vines, purple-flowering raspberry, and basswood. Take a moment to examine the basswood. It is New England's native linden tree. In early summer, its small, yellow blossoms perfume the air. Note the unique leaves: they're unusually large by linden standards, but they have the typical lopsided, asymmetrical shape. The red or bright brown buds of the basswood are quite distinctive, too; they have been compared to miniature mittens or teardrops.

Stop off at the first walkway overlook on your right for a pleasant view of the cove, and good examples of bloodroot, black birch, wild geranium, the small rock-hugging fern known as polypody, and the larger marginal woodfern. The rock outcrop here is the brown Turners Falls Shale, which dates from the Jurassic period. Return to the roadway and continue up it; at the next gate, take the right fork. A little farther along, a nice picnic grove on the left has an excellent vista upstream; on the right is the beginning of the nature trail. Look for diamond-shaped metal tags marking the route. This is a favorite place for bird watchers, and for good reason. In recent years, the peninsula has become home to fish-hunting bald eagles.

As you continue down the trail, you'll soon come to the old dinosaur footprint quarry. These days, you probably won't find any of the three-toed impressions, but another rare and wonderful sight is in abundance—maidenhair spleenwort, a delicate fern that clings to the rock face. Nearby, colonies of the low-growing Canadian yew, as well as shade-loving striped maples and hop hornbeams have become established. Past the quarry, the trail leads through more woodland. Look for pteridophytes in the form of regular maidenhair fern and the evergreen Christmas fern. They are joined by such herbaceous angiosperms as false Solomon's seal, white baneberry, tick trefoil, and spotted wintergreen. Wintergreen, in particular, has been used herbally—to treat urinary problems. Above these wild-flowers grow a variety of woody species: sassafras, white oak, witch hazel, mountain laurel, and maple-leaved viburnum.

──────────── ∾ ────────────

Depart the Barton Cove area by returning to Route 2. Turn right onto 2 East. As you head toward French King Bridge, see if you can identify the pines in this area. At least two species, white and pitch pine, are visible from the road. As you cross the bridge, glance up the narrow gorge. The Connecticut threads its way through this constricted channel just before it makes its first big turn. If the water is low enough, you may catch sight of the large boulder known as French King Rock, which sits in middle of the stream. Before the Turners Falls dam was built, this landmark was a much more prominent feature. The gorge itself follows one stretch of another, more geologically significant feature, the Eastern Border Fault. This fault marks the boundary between the Central Lowland and the older highlands to the east. The downthrown valley and its associated faults are one place the supercontinent Pangaea began to split up, midway through the Mesozoic era. As things turned out, however, the main breach ultimately formed some distance to the east. If you want evidence of that breach, take a good look at the Atlantic Ocean.

Very soon after you've crossed the bridge, you'll see the turnoff for Route 63. Take it, then turn left onto 63 North and proceed toward the town of Northfield. About 2.5 miles up this road and on the right is the entrance to the Northfield Mountain Recreation and Environmental Center, a facility operated by Northeast Utilities, the local power company. Turn up the drive and pay the center a visit. Inside are nature displays, printed material and information on river cruises, plant identification hikes, and much more. One book on sale here, the paperbound *Northfield Mountain Interpreter,* gives children and adults alike a helpful overview of Pioneer Valley ecology. It even contains nifty continental-drift diagrams that show you where Northfield has been all these millions of years relative to Africa, Europe, and everywhere else. When I last visited the center, it was open Wednesdays through Sundays, but you might want to check ahead to confirm its current schedule.

In case you've been wondering, the power company people are not running an environmental education institution on this spot for the sheer joy of it. Just uphill, at the top of Northfield Mountain, is one of the most unusual electricity-generating facilities you're likely to encounter. An artificial reservoir 300 acres in area has been excavated and filled with water pumped uphill from the nearby river. In times of peak electrical use—for instance, when everyone is running air conditioners at full blast—the water is allowed to flow downhill through turbines, which helps generate more juice. Later, when there is energy to spare, the reservoir is recharged with more river water.

Your final stop is just a minute or two away. As you go back down the driveway, notice the trees planted in front of the building, along the bus road. These are balsam poplars, close relatives of the cottonwoods and common poplars. They are more frequently encountered far upstream, in the northern-most reaches of New Hampshire and Vermont. When you're back at the entrance to Route 63, turn right, and then take the next left. Go through the railroad overpass and into the Riverview Picnic Area, another amenity administered by Northeast Utilities. In the parking lot is a circle that contains a good shagbark hickory specimen. True to its name, the bark looks as though it has been pried apart by some vandal wielding a claw hammer. Actually, this kind of peeling, or exfoliation, is perfectly normal, and it makes simple the task of identifying the species.

A short stroll takes you to the bank of the big river that has been the focal point of this journey. The greatest inland waterway of New England presents an especially gentle aspect here. Nearby, sycamores, red oaks, and sugar maples cast welcome shade on picnickers, and wildflowers such as fringed loosestrife, black-eyed Susan, and bird's-foot trefoil nod in the light breeze. It seems appropriate on this tour for Thoreau to have the last word:

> Rivers must have been the guides which conducted the footsteps of the first travellers. They are the constant lure, when they flow by our doors, to distant enterprise and adventure, and, by a natural impulse, the dwellers on their banks will at length accompany their currents to the lowlands of the globe, or explore at their invitation the interior of continents. They are the natural highways of all nations, not only levelling the ground, and removing obstacles from the path of the traveller, quenching his thirst, and bearing him on their bosoms, but conducting him through the most interesting scenery, the most populous portions of the globe, and where the animal and vegetable kingdoms attain their greatest perfection.

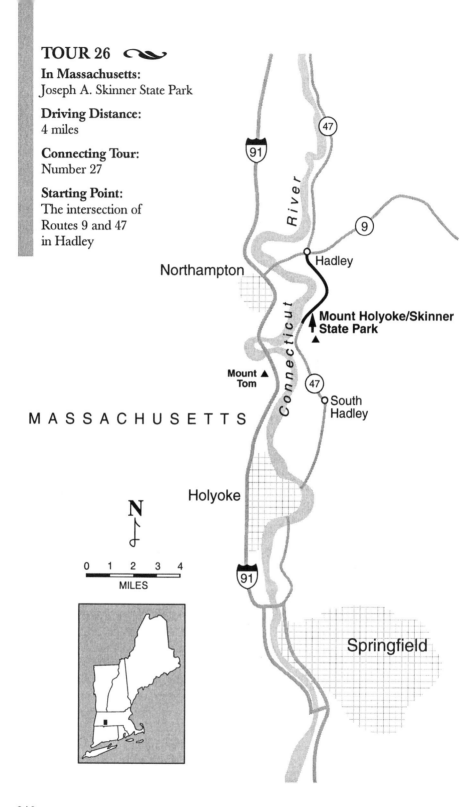

TOUR 26

In Massachusetts:
Joseph A. Skinner State Park

Driving Distance:
4 miles

Connecting Tour:
Number 27

Starting Point:
The intersection of
Routes 9 and 47
in Hadley

Northampton

Hadley

Mount Holyoke/Skinner
State Park

Mount
Tom

South
Hadley

MASSACHUSETTS

Holyoke

N

0 1 2 3 4
MILES

Springfield

Tour 26
Mount Holyoke

O f all the traprock ridges in New England's Central Lowland, perhaps Mount Holyoke occupies the most dramatic position. As one of the higher points in an impressive east-west structure capped with ancient lava flows, the ridge is situated less than a mile from the bank of the Connecticut River, in a highly populated area famous for the number and quality of its colleges and universities. This setting is the result of the river having done something remarkable. Instead of taking an easier route, the Connecticut has bored a water gap through what otherwise would have been a continuous barrier of resistant basalt. On the far side stands Mount Tom, now isolated from the Holyoke Range by the river's narrow passage. As it is, the four lanes of Interstate 91 can barely squeak through the opening.

Two prevalent theories help explain how the river made it through such a formidable obstacle. The first states that the stream actually predates the ridge. As the landscape in this area slowly eroded, the Connecticut gradually cut down through the emerging Mount Holyoke rock. The other explanation, considerably more complicated, holds that a stream on the south side of the range captured one on the north, when their headwaters met. This created an ever deepening notch in the ridge through which, eventually, the main river was pulled. Whichever hypothesis is correct—and it could be that the truth lies in a synthesis of both—the water gap and its surrounding heights make for terrific scenery.

Mount Holyoke is special in a botanical sense as well. Along with the other traprock ridges in Massachusetts and Connecticut, it is a haven for its own distinct

The approach to a traprock ridge. Mount Holyoke is home to a diverse forest community.

plant communities. As you'll see, it's a wonderful place to find some relatively rare woody plants, as well as many different species of wildflowers and pteridophytes. To appreciate the abundant flora fully, it's best to plan on walking up a fairly steep but steady grade for an hour or so. In other words, you'll be hiking rather than driving up the paved summit road. If you're interested in observing plant life close up, in great diversity and detail, this tour is ideal.

To reach Mount Holyoke's Skinner State Park, head south on Route 47 from Hadley for 4.1 miles. The park entrance is on the left. Turn down this access road. In summer, it's possible to take your car all the way to the top after paying an admission fee. But if you do, you'll miss the majority of fascinating plants. Besides, pedestrians pass the checkpoint free of charge. So, if you're feeling frisky and want some exercise, pull off to the side just shy of the parkie with the cash box. (Please note that it is not my intention to deprive the Commonwealth of Massachusetts of much-needed income, nor am I writing a guidebook on how to beat the system. If you feel moved to contribute anyway, by all means do. With all due respect to the whales, it would be difficult to imagine a better cause.)

The fun begins the moment you pass through the gate. Keep a sharp lookout for a substantial shrub with opposite, three-part leaves. This species, the American bladdernut, is a member of a small family of angiosperms better represented on other northern continents. It has close relatives, for instance, in Japan, China, and the Caucasus Mountains. It produces small, drooping, white flowers in spring;

242

by August, these develop into unique pouchlike pods. As you'll discover, American bladdernut is well established most of the way to the summit.

The lower segment of the road, the portion below Halfway House, has an overstory that includes sugar and red maples, black birch, and Canadian hemlock. Each of these trees has served human needs. The sugar maple, besides providing the famous syrup, has hard wood that is widely used for furniture, bowling alley surfacing, and dance hall flooring. The wood of the red maple is not so highly esteemed, but the Indians used a part of the tree to provide relief for sore eyes. Black birch wood is prized for cabinets and pianos. Hemlock timber has been employed in bridge construction, building beams, and papermaking. Native peoples used its roots for cordage; and, as John Josselyn recorded, colonial fishermen tanned sails and nets with dye extracted from its bark.

Also present nearby is the pignut hickory. This tough-timbered species can be distinguished from the other hickories by its combination of foliage and bark characters. Its leaves usually have only five leaflets, and the trunk is vertically striped or ridged, not shaggy or exfoliating. Lower down, in the understory, you'll notice prickly clumps of blackberries; but see if you can spot the less rampant mountain holly. This shrub has deciduous leaves that are simple and mostly untoothed, though they do sport a tiny tip, or mucro. The mountain holly fruit is a red, berrylike drupe poised on a long stalk.

There is no lack of wildflowers along the lower part of the road, either. Their show begins by early April. Among the first wave are the lurid red trillium and the bloodroot. The best treat of the spring, though, is that red-and-yellow marvel, columbine. It's a representative of the buttercup family, a group often cited as being among the most primitive flowering plants. But to the untrained eye, the columbine blossom looks anything but primitive as it reveals long, graceful spurs and unusual design. One of its other common names, meetinghouses, refers to the fact that the spurs somewhat resemble the elegant spires of Yankee churches.

From midsummer until the first hard frost, another set of herbaceous flowering plants have their turn. A member of the orchid family known as downy rattle-snake plantain has attractive, seersucker leaves; it puts up spikes of small, white blossoms. It is joined by a host of other species. You should be able to locate the humble but ubiquitous partridgeberry, which manages to insinuate itself into practically every woodland from northern Maine to southwestern Connecticut. Also in evidence are yellow false foxglove and the coarse and sprawling pokeweed. The latter is a red-stemmed native with sap often regarded as poisonous. Notwithstanding that, the Indians used it medicinally, and the herbalist Mrs. Grieve noted that "the young shoots make a good substitute for asparagus." Personally, I'd leave them alone. (For more on pokeweed, see Tour 25.)

Among the late-flowering asters, goldenrods, and other representatives of the composite family are small colonies of the tall-growing sweet joe-pye-weed. Its flower heads are a curious, pale pinkish purple that has an almost metallic sheen. It's rather odd that a plant with such a unique and pleasing color hasn't been introduced into the garden, the way its smaller relative, the ageratum, has. Sweet joe-pye-weed has been used herbally, though, as a diuretic and tonic.

Sweet joe-pye-weed, a tall, late-summer wildflower, on Mount Holyoke's lower slope.

Remain on the paved roadway as you pass Halfway House. As the woods resume, you'll see some of your friends from below, and some new plant types as well. Woody plants more tolerant of drier upland soil begin to show themselves. Among their number are mountain laurel, lowbush blueberry, paper birch, and red and black oaks. The wildflowers here include one of the heftiest you'll run across: spikenard, an indigenous aralia with huge, compound leaves and flowers arrayed in a long-stemmed structure called a panicle. This species has served as a stimulant and a home remedy for heart problems. Another plant with compound leaves is white baneberry; its white fruit with a single black dot is eye-catching but definitely poisonous. Also in this part of the walk are harebell, tall meadow rue, and the small, pink-flowered geranium called herb Robert.

Ferns are also well represented in the roadside habitat. Farther down, the quick-spreading hay-scented fern was most noticeable, but now the little polypody and mountain woodfern have become more prevalent. Polypody loves nothing so much as a barren rock surface to perch on, and if left to its own devices, it will form large colonies on top of glacial erratics and other rocks. These look for all the world like giant boulder toupees.

One other pteridophyte here does not equip itself with the fancy, dissected foliage of ferns. It's the shining club moss, a small, vascular plant known as a lycopsid. The extremely lengthy pedigree of the lycopsids stretches back to the Devonian period, one of the earliest chapters in the story of plant evolution. At a somewhat later point in earth history, when the Carboniferous coal swamps covered much of North America and Europe, one branch of the lycopsid group attained tree size. Interestingly, its members didn't have the tough internal structure that modern trees, such as oaks or palms, have today. Larger specimens of these ancient giants had lower trunks estimated to have been more than half spongy outer tissue. On top of that, the upper portions of the stem were largely pith—which all in all must have made these lycopsids the Gumbies of the plant world. A determined woodchopper might have felled one with a Swiss Army knife, or perhaps just with a well-placed nudge of the elbow. Still, there's no doubt that such trees not only survived to maturity but also dominated their habitat for millions of years.

In the final approach to the summit, where the road snakes about in a series of switchbacks, more new plants come into view. The herbaceous species include the distinctively formal-looking bottlebrush grass, woodland sunflower, and the diminutive clearweed. Clearweed usually stands only about 4 to 6 inches tall and grows in large mats. Look for its glossy, simple, and oppositely arranged leaves. This is the native member of the *Pilea* genus, which also contains such houseplant favorites as the aluminum and artillery plants. While clearweed opts for the flatter ground of the roadside, another wildflower, early saxifrage, prefers to snuggle in the basalt crevices. This white-flowered gem is well named, since it blooms in early April. (Not long after I saw this species here, I traveled to northernmost Maine. At one point, not far from the Canadian border, the road rises onto a basalt upland. When I checked the road-cuts, I saw the early saxifrage there as well, blooming roughly three weeks later than its Mount Holyoke relations.)

One other wildflower present in this stretch deserves mention—the white snakeroot. A close relative of the sweet joe-pye-weed seen farther down the road, it's considerably shorter and has white flower heads. This inoffensive-looking species, when browsed by dairy cows, causes the lethal cattle disease known as the trembles. The toxins in white snakeroot can even make it into commercially sold milk, which in turn produces the so-called milk sickness that has been responsible for the death of some children in other parts of the country.

The best indicator plant of the upper heights of this traprock ridge is one of the largest, the chestnut oak. The leaves of this striking tree do not have deep sinuses but rather are fringed with large, rounded teeth. The sweet acorns of this true oak are a favorite food for a number of animals, and its deeply furrowed bark gives it a great deal of ornamental appeal. You would look for it in vain in the nutrient-rich soils of the bottomlands. It demands thin, stony ground.

You'll find your efforts in climbing the hill well rewarded when you reach the Summit House. On the cliffs below, you may see the orange blossoms of the introduced tiger lily swaying to and fro in the updrafts. In the Old World, this plant was sometimes used to soothe the discomforts of pregnancy. But even its showy flowers are upstaged by the breathtaking view beyond. To the northwest lies Northampton, the starting point of the next tour; to the southwest, the Connecticut River scrolls about the foot of Mount Tom; to the northeast is Emily

Dickinson's Amherst; and to the south, a little less than 20 miles away, stand the great river cities of Holyoke and Springfield. This is the perfect spot to study the megaflora, or the grand, overall patterns of the vegetation. Notice how there are more trees in the towns than on the agricultural lands.

Imagine this scene three hundred years ago, when the valley was North America's first major wheat belt. Picture how thousands of acres of this one cultivated grass type, developed in ancient times and brought here from the Old World, must have looked from this height on a breezy day. Then travel back further, some thirteen thousand years. By then, the latest wave of Ice Age glaciers had retreated, leaving a great body of still water and settling sediments that, at its maximum extent, stretched from central Connecticut to Vermont's Northeast Kingdom. It was Lake Hitchcock—sullen, frigid, with drifting chunks of rotten ice; the landscape beyond was slowly turning from treeless tundra into forests of white spruce. This was the setting, possibly, when the first human beings reached this region. One wonders if our forerunners made it up Mount Holyoke, perhaps to hunt the biggest beast of their woodlands, the mastodon. Or perhaps they made the climb, like us, just to enjoy the view.

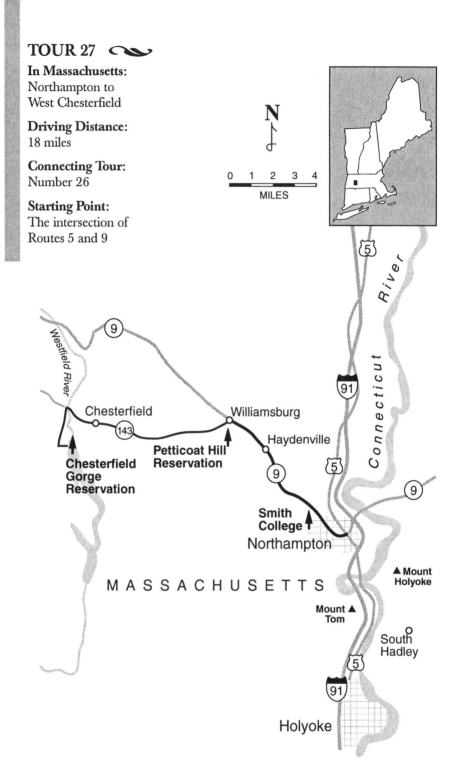

TOUR 27 ∾

In Massachusetts:
Northampton to
West Chesterfield

Driving Distance:
18 miles

Connecting Tour:
Number 26

Starting Point:
The intersection of
Routes 5 and 9

N

| 0 | 1 | 2 | 3 | 4 |
MILES

Westfield River

⑨

Chesterfield

⑭⑶

**Chesterfield
Gorge
Reservation**

**Petticoat Hill
Reservation**

Williamsburg

Haydenville

⑨

**Smith
College**

Northampton

Connecticut River

⑤

⑨①

ⓢⓤ⑤

⑤

M A S S A C H U S E T T S

▲ Mount
Holyoke

Mount ▲
Tom

○
South
Hadley

⑤

①⑨①

Holyoke

Tour 27
From the Valley to the Hills

A Recommended Fall Foliage Tour
HIGHLIGHTS

Fervor, Northampton Style ∿ The Botanical Campus of Smith College ∿
Houseplant Heaven ∿ When Ferns Decide to Be Trees ∿ The Silk Oak's Leaves ∿
Colossal Cycads and 2-Foot Cones ∿ The Heyday of Herbaceous Plants ∿
Heath and Heather ∿ A Mill Town's Trees ∿ Petticoat Hill Reservation ∿ The
Beauteous Bloodroot ∿ A Shiplike Forest ∿ First Growth, Second Growth ∿
Nasty Bracken and an Essay on Primitiveness ∿ The K-T Fern Spike ∿
Chesterfield Gorge Reservation ∿ A Chasm Tree and an Acrobatic Fern

*T*o travel westward from the fertile plain of the Connecticut River to
the Berkshire highlands is to take a retrograde journey through time,
from a landscape that has its origins in the age when dinosaurs prowled
the earth to one that began its turbulent and complex history even before their
ancestors and ours emerged from the sea. But the trip from the lowland to the
hills is also an excellent study in the New England sense of scale. In some ways,
the proportions and distances of this region are more European than American.
The traffic of busy towns suddenly gives way to isolation; and woods, fields,
and housing lots succeed each other, time and again, in the space of a few miles.
There is a sense of intimacy and filigreed detail, even in the wildest places, that
in some ways is more reminiscent of Old England than of the slowly changing
vastness of the Midwest or West.

We begin in Northampton, one of the most famous towns along the Connecticut River corridor. Always prominent in the valley's history, Northampton
was one of the centers of the religious revival movement that swept New England
in the 1730s. This movement was largely the result of Jonathan Edwards, a
spellbinding preacher who delivered sulfurous sermons with such heartening titles
as "Sinners in the Hand of an Angry God." Under his influence, the whole town
was so swept up in things spiritual that, according to Edwards himself, it was
brimming over with both joy and distress.

You can be hopeful that the only distress you'll encounter here will be caused
by the irritatingly long stoplight that marks our starting point. To reach your

A tree fern rises above other fern species, in Smith College's Lyman Plant Houses.

first destination, the Smith College campus, head west on Route 9 and follow its signs through town for 0.6 mile. Then turn left onto College Lane and pull off in front of the Lyman Plant Houses, located across from Paradise Pond. (College Lane may perchance be blocked off. If so, park where permitted, along Route 9, and walk down.)

Smith College wisely chose Frederick Law Olmsted to supervise the design of the campus. Olmsted's credentials were truly impressive: he helped create Manhattan's Central Park, Chicago's glorious waterfront, and the loveliest parts of the Boston park system. To the delight of plant lovers ever since, the entire Smith campus was laid out as a botanic garden, with labeled specimens arranged by their taxonomic affiliations. Needless to say, this concept predates our own age of enlightenment. Now we realize higher institutions of learning exist primarily to build stadiums, parking structures, and big-name sports teams.

Before you take in the delights of this unique outdoor setting, visit the Lyman Plant Houses, a teaching facility open to the public. This small and immaculate greenhouse complex is a must for every houseplant fancier, and also for anyone who wants to see some of the plant types that inhabited this area 200 million years ago.

While most of the specimens are self-explanatory and immediately engaging without textual preening, a few extraordinary plants deserve mention here. Magnificent staghorn fern specimens lurk overhead like dozing vultures. These grotesquely appealing plants are epiphytes—that is, they normally grow upon the branches of other plants, but without being parasitic. Quite a few tropical species are epiphytic. Can you figure out why? In warm, humid climates, the battle for available light is intense because of the extremely high density of the vegetation. Many herbaceous plants, such as the trunkless ferns and the orchids, would be relegated to the full gloom of the forest floor were it not for their ability to ride piggyback, so to speak, on the trees that can reach up into the sunlight.

Another pteridophyte type not often encountered by New Englanders is the tree fern. A number of species have adopted this upright habit, though most are tropical or subtropical in their distribution. The examples in this collection have developed distinct trunks, to be sure, but their huge fronds still develop in the classic fern way, by unrolling from a coiled position. In the arcane terminology of plant morphologists, this method of uncoiling is known as circinate vernation. More generally, people call the undeployed fern leaves fiddleheads, croziers, or monkeytails.

Another specimen that could almost be mistaken for a tree fern, given its exceptionally beautiful and lacy foliage, is the silk oak. Originally from Australia, it is now commonly planted as a yard tree in southern Arizona and other warmer parts of the United States. In fact, it's a genuine flowering plant, a member of the intriguing protea family, which is well established in arid parts of the Southern Hemisphere.

The most fascinating plants of all, though, are the cycads, which look something like squat palms on steroids. The chief of these is a burly specimen that goes by the jangling taxonomic name *Encephalartos altensteinii*. It's also known by the common names of bread tree, Kaffir bread, and prickly cycad. However much it may outwardly resemble a palm, it actually represents one of the main groups of the gymnosperms. Whereas a palm reproduces by flowering, this plant develops huge cones, either male or female, depending on the plant. The female cones, the larger of the two, can reach a length of roughly 2 feet. Cycads are of more than passing interest because they were an important part of this region's flora in the Triassic and Jurassic periods. Many of their kind apparently were under-story trees or shrubs, which made them the mountain laurels of the early Mesozoic forest. It was at this point that rifting first opened the Atlantic Ocean and created the Central Lowland.

Once you've wended your way through the potted citrus trees and camellias and returned to the entrance, head outside and walk toward the small artificial pond. Look there for the native, blue-blossomed pickerel weed among the other aquatics. This section is the Herbaceous Garden, where annual and perennial species are grouped by family. As you look at these relatively low-growing angiosperms, keep in mind that the nonwoody flowering plants evolved from woody precursors, and not the other way around. In fact, we are currently in the heyday of highly advanced herbaceous plants; much of the evolutionary action nowadays is centered in the grasses, orchids, and members of the composite family. So, don't think a tree is necessarily more advanced simply because it's learned how to grow big. The process of evolution sometimes involves the

The feathery foliage of a dawn redwood tree on the Smith College campus.

development of complex structures and complicated mechanisms; at other times, it calls for their reduction or elimination.

Nearby are some of the best campus trees. A booklet on sale at the greenhouse office provides a map and list of the plantings campuswide. Also, as noted above, individual plants are tagged, so you can identify the more exotic types without resorting to guidebook keys. The ginkgo specimens are particularly fine. Their venerable ancestors, like those of the cycads, are not represented in the fossil record of the valley's Mesozoic sedimentary rocks, but they do form part of the Mesozoic fossil flora elsewhere. (For more on their species, see Tour 18.)

The other famous living fossil, the dawn redwood or *Metasequoia,* is represented by two individuals this side of nearby Neilson Library. They should be called Baby and the Beast, though Baby will hit adolescence and even adulthood soon enough, if it displays its species' typical frenetic growth rate. A relative of these, the pond cypress, stands on the west side of Wright Hall, next to the library. It is a variety of the bald cypress, the evocative conifer of Southeastern swamps that is mentioned more fully in Tour 2. Like the dawn redwood, this tree is deciduous. If you see it leafless in the dormant season, don't be regretful–it will put out a new set of needles come spring.

252

The trees are certainly the main attraction, but don't leave without seeing the ornamental shrubs of Azalea Bank, situated along the path leading to Neilson Library. Among the ericads planted here are good examples of both heath (genus *Erica*) and heather (genus *Calluna,* consisting of many cultivated varieties of one species, Scots heather, or *Calluna vulgaris*). Both instantly conjure up images of the windswept moors of northern Britain; but in fact, some of the less cold-hardy heaths used commercially hail from southern Europe and Africa. As appealing as these low-growing plants are, especially when they're in full bloom, they oftimes drive their owners to conniption fits. They're very finicky and won't tolerate poorly drained, nonacidic soil.

Now you leave the Connecticut River and its lowland behind. Return to Route 9 and turn left (west) onto it. The road follows the Mill River upstream. The landscape in the first part of the 7.5-mile stretch to Williamsburg is mostly residential. As you pass through Haydenville, note how the town's old mill complex has been turned into a shopping center—a common turn of events in postindustrial New England. The nineteenth century was the boom time for mills in the region, though their prototype, the Slater textile facility in Pawtucket, Rhode Island, was completed just a little earlier, in 1793. As a rule, mills erected in Massachusetts and New Hampshire were bigger than those built to the south, because their sources of power, the waterfalls, were larger. In the lower New England states, entire families, fresh from their failed farms, might be employed; farther north, the work force was usually restricted to young women, who lodged in dormitory quarters on the company grounds.

Haydenville boasts quite a selection of street trees. After your Smith College visit, you should be able to recognize the ginkgo. It is considered one of the most problem-free selections for stressful locations. A bit surprisingly, there are also some examples of European mountain ash. This species normally is much more vulnerable to problems resulting from stressful environments, but it does sport highly prized flowers and showy fruit clusters. Another veteran of Northeastern curbsides, the pin oak, has been placed here, too. In more demanding locations, parts of New York City, for example, many pin oaks have suffered from chlorosis, a condition in which the foliage turns a sickly yellowish green. It's most likely to occur when soil acidity is neutralized by nearby cement or other factors. As a result, the trees can't extract the iron they need to keep their leaves dark green. One other problem with this species is not so easily corrected. As pin oaks mature, their lower, drooping branches tend to die back. The dead wood can be pruned off, but it often gives the plants a rather leggy look.

When you reach Williamsburg town center, a little less than 2.0 miles north of Haydenville, bear left at the main intersection and go up Petticoat Hill Road 0.2 mile to a small parking turnoff on the left-hand side, which is the entrance to Petticoat Hill Reservation, a Trustees property. This is a showplace of the central hardwoods regime found throughout most of southern New England, and it's also a good stop for wildflower enthusiasts. In early spring, a large colony of bloodroot can be seen just inside the entrance. Easily identified by its unique leaf shape, this member of the poppy family, a native, produces striking white-petaled flowers that rise out of its still partly unfurled foliage. Its common name refers to the orange or reddish sap found in both the roots and stems. This sap contains poisonous alkaloid compounds; when ingested, they trigger intense thirst, vomiting, vertigo, and impaired vision. As usual, though, herbalists rush in where executioners fear to tread, and they have used the plant—very carefully, one has to hope—to cure asthma and the croup.

Go up the summit trail marked with white blazes. The slope is fairly steep, and the hike, though quite short, does require some exertion. Along with those characteristic central hardwoods, the red oak and the shagbark hickory, there are other familiar broad-leaved tree species: paper birch, black cherry, and sugar maple, as well as white pine and Canadian hemlock. On a breezy day, the hemlocks creak wonderfully, like the masts of a square-rigger under full sail. Less imposing, no doubt, but still appealing in their own way, are the shrubs that form the understory. Among these are witch hazel, striped maple, mountain laurel, and American fly honeysuckle. The fly honeysuckle is by no means the fanciest flowerer of its genus, but its greenish or faint yellow blooms are a heartening sight when they appear in April.

April is also the month the wildflower show begins in earnest. Early species, some of them spring ephemerals, include red trillium, trout lily, wild ginger, foam flower, false Solomon's seal, at least one yellow violet species, and, best of all, the chaste sharp-lobed hepatica. In summer, other blooms take over—wild sarsaparilla, partridgeberry, fragrant bedstraw, white baneberry, and wood aster. The partridgeberry, one of the most familiar sights on dry forest floors across New England, was once used by Indian women to ease the pains of childbirth.

One other understory plant is obvious along the way. It's a fern. Can you identify it? Its common name refers to the fact that it is still green around the 25th of December. As you approach the summit, the trail markings are less comprehensible (at least that was the situation the last time I was there). Whether you go on to the top or turn back while there are blazes to guide you, take a moment to ponder the fact that this hillside was once cleared land and even a main part of town. This reflects the common colonial practice of placing settlements high above the flood-prone valleys. Fields situated near the summits of hills did not enjoy the optimum soils typical of the bottomlands, but they received more sunlight and had longer growing seasons. Where they're not bare ledge, the hillsides are cloaked in till, glacial debris ranging from clay and silt to cobbles and boulders. Despite its irritatingly high rock content, till makes a fairly good medium for crop raising.

The trees that have taken the place of the farmland comprise a community that foresters term second-growth. However stately they may seem to us, to the experienced woodlot manager they are distinctly inferior in shape and wood quality to the ancient first-growth trees that the early colonists saw. First-growth timber was said to have the almost miraculous quality of increasing in strength and durability as it aged. Second-growth wood is more likely to split, warp, and rot. The issue of this particular woodlot's profitability is deferred indefinitely, thanks to The Trustees of Reservations.

⸻ ☙ ⸻

The last stop lies in the heart of the Berkshire highlands. Return to the main intersection in town. Turn left onto Route 9 West, go past the Williamsburg post office, then go left again onto Route 143. As the road gains altitude, red oak and Canadian hemlock mix with paper birch. In summer, a succession of common milkweed, black-eyed Susan, common mullein, spotted joe-pye-weed, rabbit's-foot clover, and a variety of goldenrods are visible as well. In the vicinity of 3.3 miles from the previous turn, wetland appears on both sides of the roadway. Gray birch and poplar are nearby and, as you'd expect, so is the swamp-loving red maple. Also keep an eye out for large communities of bracken not far from the shoulder's edge.

Of the roughly eleven thousand extant fern species, bracken must certainly be on the short list for overall nastiness. An aggressive colonizer of disturbed ground, this one species is native to North America, Central America, Europe, Asia, and Africa. For a primitive plant, it definitely knows how to look after itself. According to one recent account, its stem fibers inflict deep cuts, and animals with the bad luck to browse it receive a stiff dose of toxic chemicals that destroy vitamin B reserves, kill bone marrow and blood cells, and even cause blindness. The various clergymen of past centuries who tried to find signs of divine helpfulness in the natural world would have been hard pressed with this mean-spirited plant; they'd be better off attributing it to the creative powers of Old Nick instead.

The modern-day success of bracken in particular, and ferns generally, illustrates an important point. Long-term survival is by no means a simple issue of being more evolutionarily advanced—whatever that means—than other competing forms of life. In each environment on the earth's surface, it's a matter of which complicated set of factors works, and which doesn't. Nature refuses to limit its options to simplistic notions of evolutionary progress. Ferns represent a relatively cumbersome mode of sexual propagation that predates the efficiencies of seed-borne reproduction. Not sensing their own inferiority, however, they're doing just fine; here on this roadside, for example, they often manage to shoulder their so-called betters out of the way completely.

Nor is the ferns' nimble opportunism an entirely recent state of affairs. One of the hottest topics for paleontologists has always been the lurid subject of the mass extinctions that have occurred time and again in geologic history. And no extinction gets more press than the one that signaled the demise of the dinosaurs at the end of the Cretaceous period. This episode, known in the trade lingo as the K-T Boundary Event, had as a little-known sideshow an interesting development in the world of plants. Some time before and after the catastrophe, the fossil record was dominated, as you might expect, by the pollen of angiosperms, the flowering plants. At the boundary itself, however, the pollen was largely supplanted by fern spores. This fact, dubbed the K-T Fern Spike, suggests that the ferns were best able to exploit the opportunity provided by a deadly meteorite impact, climatic change, or whatever else it was that wiped out so many other organisms.

The story of the Fern Spike brings up one other interesting point. Millions of years later, in the more gradual transition induced by the cooling climate worldwide, another primitive plant type, the conifers, claimed the huge cold tract of the earth's surface known as the taiga. So, in certain cases at least, ferns and conifers have proven capable of dealing with change better than the angiosperms. It's possible to say that angiosperms still dominate the most favorable climates, but tell that to a white spruce, which finds the winters of Labrador to be perfection. From its perspective, it occupies the best set of conditions; and one can imagine it pitying the poor plants here that must endure the hellish heat of summertime in the Berkshires. In other words, our concepts of primitiveness, success, and dominance may define our own biases more than they define the way nature itself proceeds.

--------------------------- ∾ ---------------------------

The center of the beautiful upland community of Chesterfield lies 2.6 miles beyond the wetland area. The road descends on the far side of town, affording a grand view of rounded hills cloaked in second-growth timber. When you have driven 2.2 miles past the heart of town, just over the Westfield River bridge, turn left onto Ireland Street and go an additional 0.9 mile. Take another left, onto River Road. This is the access to the final point of interest, Chesterfield Gorge Reservation. If you visit in summer, you may spot naturalized tiger lilies blooming on the side of the road. (See the end of Tour 26 for their role in herbal medicine.)

This book features two Chesterfield Gorges; the other one, in New Hampshire, is treated in Tour 38. This breathtaking spot is a Trustees site that offers a close-up view of what the Westfield River has managed to do to the Devonian period metamorphic rock beneath it. The gorge is narrow and deep. The best time for dramatic splashing effects, naturally, is the spring freshet. The predominant tree species here is Canadian hemlock, though American beech, red oak, paper birch, and white ash are about, too. The hemlock is this region's number-one chasm tree. As you can also see at Vermont's Quechee Gorge (Tour 32), it loves to hang over the rising mist of a plunging stream. Since ferns have received our attention at every other stage of this journey, we'll finish the trip by noting one more—the widespread but ever delightful polypody. This little acrobatic pteridophyte has provided a home cure for coughs and other bronchial complaints. True to its nature, it insists on adorning the bare cracks and surfaces of the stone along Chesterfield Gorge.

Chesterfield Gorge at the spring freshet.

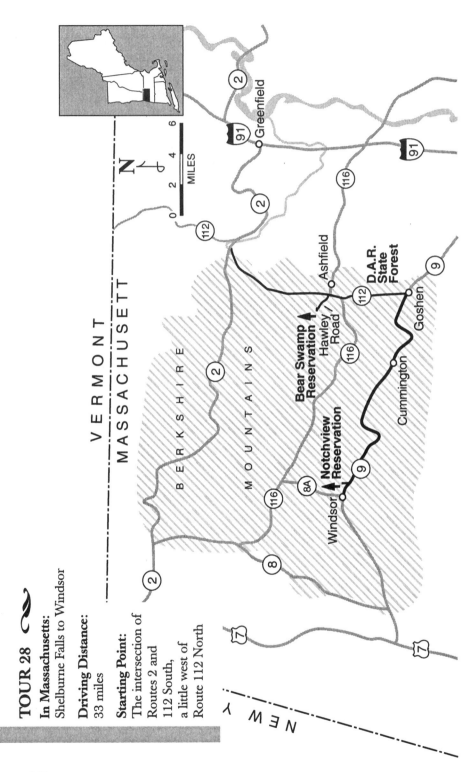

TOUR 28

In Massachusetts:
Shelburne Falls to Windsor

Driving Distance:
33 miles

Starting Point:
The intersection of
Routes 2 and
112 South,
a little west of
Route 112 North

VERMONT

MASSACHUSETT

NEW Y

Greenfield

Ashfield

D.A.R.
State
Forest

Goshen

Hawley
Road

Bear Swamp
Reservation

Cummington

Notchview
Reservation

Windsor

BERKSHIRE

MOUNTAINS

N

MILES

258

Tour 28
Exploring the Northern Berkshires

A Recommended Fall Foliage Tour

HIGHLIGHTS

A Land of Farms and Pastures ✶ An Amatory Willow ✶ Bear Swamp
Reservation ✶ Franco-American Rock Tripe Stew and Its Swedish Critic ✶
The Ebb and Flow of the Northern Hardwoods ✶ Hummingbirds and
Bulrushes ✶ The Flower the Field Guides Forgot ✶ Netherlandish Trousers ✶
A Delicate Cure for Consumption ✶ An Appreciation of Hay, Its Plants and
Its Mowers ✶ The Godzilla of Shade Tree Horticulture ✶ Black Locusts Go
Italian ✶ Rising into the Hoosacs ✶ Notchview Reservation ✶ Soothing
Valerian ✶ The Acquisitive Cocklebur ✶ A Glimpse of the Montane Taiga

*I*n some ways, this tour is good preparation for a journey through the
more northern parts of New England. After all, the Berkshires are the
Green Mountains by just another name. They are geologically identical
with Vermont's famous range; and in this part of Massachusetts, they offer the
same sense of upland remoteness. The trip begins at the junction of a heavily
traveled thoroughfare, but it's devoted to one of the Commonwealth's most
beautiful and least populated areas. Better yet, it culminates on high ground, at
the crest of the Berkshires, where the woodland community is the Bay State's
answer to Canada's forests.

If you approach the starting point along Route 2, the busy Mohawk Trail,
you may not initially realize how rural the surrounding countryside is. When
you reach the intersection of Routes 2 and 112, turn onto 112 South, toward
Buckland and Ashfield. Now the landscape bares its lonely soul. There are still
some working farms, and handsome ones at that, but the overall impression is
one of forested isolation. White pine, black cherry, poplar, sugar maple, American
elm, black locust, and red oak can all be found along the way. In the vicinity
of Clesson Brook, there are two classic bottomland trees, sycamore and black
willow. The latter has seen herbal duty as both an aphrodisiac and a venereal
disease cure—which makes it, I suppose, an all-in-one product. Before long, a
second conifer, the Canadian hemlock, appears. Its preferred habitat is cool, moist
ground, though it can be found on dry slopes as well. This terrain, with small,
secluded valleys and low, rounded hills, definitely has a Vermontish look to it.

The road rises as you pass the boundary into Ashfield. In the nineteenth century, this town was a center for the production of essences, or herbal extracts used in perfumes and various domestic and medicinal applications. (For more on this most aromatic of agricultural pursuits, see Tour 46.) A total of 7.9 miles from the tour's starting point, you'll come to a four-way intersection, with Route 112 continuing straight, Route 116 going off to the left, and Hawley Road heading to the right. Take Hawley Road. Proceed about 1.2 miles and then bear right onto the lower road at the Y-intersection. Continue 0.7 mile farther until you see the Trustees sign marking the entrance to Bear Swamp Reservation, on the left side of the road. Pull off and park here.

Bear Swamp Reservation has that wonderful in-the-absolute-middle-of-nowhere feel to it. Whether your interest is wildflowers, tree identification, lichens, or wetland ecology, this is a tempting place to pay a lingering visit. My own favorite seasons at this site are early spring, say, mid-April or a little sooner, and the goldenrod days of late summer. If you come here in mud season, by all means wear ooze shoes, and don't plan to attend a wedding or the opera directly afterward. Your pant legs will probably accrete a wide assortment of nature's finest camouflage materials along the way.

As you pass through the reservation entrance, note the cedar posts that have been driven in there, and speculate on why that wood, and not maple or pine, was used. If you don't have a clue, check the discussion of fencing materials in Tour 19. Just beyond, the understory contains striped maple and one of the best of the native viburnums, hobblebush. Above them rise sugar maple, American beech, yellow birch, Canadian hemlock, and, a little farther in, basswood. Taken as a whole, this community is a good northern hardwoods ensemble. Elsewhere on the property, the oak-dominated central hardwoods species, so prevalent at lower elevations, mix with these. Keeping track of the complex ebb and flow of plant regimes can be a fascinating pastime. In northern Massachusetts and southern New Hampshire, for example, central hardwoods may be more common on the warmer, southern sides of hills, while the northern hardwoods tend to congregate on the cooler slopes opposite.

Also near the entrance are two familiar pteridophytes, the hay-scented and evergreen Christmas ferns. Nearby and to the right stand boulders that provide the kind of lifestyle irresistible to rock tripe lichens. According to eighteenth-century explorer Pehr Kalm, rock tripe was a food source for the French Canadians. They boiled it in salt water mixed with suet until the lichen swelled up. As a native of Sweden, Kalm was perhaps not in the best position to criticize strange cuisines; nevertheless, he came down hard on this Franco-American concoction: "It is a food that neither nourishes nor satisfies hunger, nor has it a pleasant taste. It serves only to sustain life." At least, he noted afterward, it wasn't so bad when served with fish.

After a short trek through the forest, you'll come upon a beautiful upland pond on the left. It emanates a sense of unsurpassed tranquillity. If you walk out to the point, you will find a distinctive plant community: red spruce, sweet gale shrubs, various asters and goldenrods, sensitive fern, the orange-flowered

The spring-blooming corn lily, a common sight in the forests of the Massachusetts Berkshires. —The Trustees of Reservations, Jack Swedberg photo, Massachusetts Division of Fisheries and Wildlife

jewelweed so attractive to hummingbirds, and, at water's edge, a colony of low-growing bulrushes. These representatives of the sedge family might be the common species known as three-square bulrush, but certain minute details, including the fact that their stems are fairly round in cross section rather than triangular, suggest they might belong to another closely allied species, weak-stalked bulrush, instead.

As you leave the pond, follow the white-dot blazes that lead farther into the woods. There is no ultimate objective, and you'll be returning by the same route, so go as far as you like. In places, the trail snakes along steep slopes, so watch your step. The flower show at winter's end is spectacular. Through the mist, drizzle, and melting snow appear a who's who of spring ephemerals: red trillium, foam flower, trout lily, Carolina spring beauty, and assorted blue and yellow violets. The violets have a curious reproductive strategy: they produce both insect-pollinated flowers that open fully and others that never open and pollinate themselves.

Among these early herbaceous species are two others, the plantain-leaved sedge and Dutchman's breeches, which illustrate different endpoints of flower design. If you're not already familiar with the first species, you may find it difficult to locate in a basic field guide to wildflowers. Sedges are often considered to be in a world of their own. Because they're numerous, often tricky to tell apart, and of little ornamental value, they're frequently ignored. Their obscurity is also the result of their being wind pollinated. Since they've dispensed with insect help, they've no need of the colorful petals and fancy patterns that attract bees, flies, or amateur naturalists. In brazen contrast, there is the unequivocal flamboyance of Dutchman's breeches. This plant, a member of the fumitory family, develops intricate, dangling blossoms that are invariably likened to a pair of baggy white trousers. If you come across any Northeastern wildflower guide that doesn't mention this one, find a trash receptacle and throw the book in it. Sedges are doomed to be excluded for the heinous sin of subtlety, but Dutchman's breeches makes great copy. Incidentally, this charming little showboat is a close relative of a splendid old New England garden perennial, bleeding heart.

Later in the spring, another characteristic woodland plant, Indian cucumber root, blooms in these woods; later, in summer, wild sarsaparilla, tall meadow rue, and white snakeroot follow suit, only to yield the floor to the ubiquitous shade-loving asters. Throughout the warm months, nonflowering plants hold their own, as well. Look for marginal woodfern, polypody, and that miracle of delicacy, maidenhair fern. In Kalm's time, the mid-1700s, this dainty species was used to cure consumption, which today is called tuberculosis.

──────────────── ❧ ────────────────

Bear Swamp Reservation is indeed an excellent upland site; but now you head toward still higher ground. Return to the junction of Hawley Road and Routes 112 and 116, and turn right onto 112 South—which, for the puzzlement of tourists and outsiders in general, has also been designated 116 North on this stretch. There are wonderful views in this area, thanks to the surviving hayfields and pastures. While the eye naturally passes over these open spaces to the woods beyond, the foreground itself is worthy of close scrutiny. As with land planted in potatoes, corn, or wheat, these painstakingly cleared sections are largely given over to introduced species. Among the traditional pasture plant favorites are timothy, red top, orchard grass, the red and white clovers, and sometimes lucerne, another name for alfalfa. In early colonial times, each farming family tended to use communal town pasture; later, private land carefully fenced off became the rule. While pastures are devoted to grazing livestock, hayfields grow high and then are harvested for fodder. Even today, mowing the hay is a wearying and tricky business, and an unexpected rain can ruin hours of hard labor. It's not a good occupation for the half-attentive. Thoreau described it this way:

> This haying is no work for marines, nor for deserters; nor for United States troops, so called, nor for West Point cadets. It would wilt them, and they would desert. Have they not deserted? and run off to West Point? Every field is a battle-field to the mower,—a pitched battle, too,—and whole winrows of dead have covered it in the course of the season. Early and late the farmer has gone forth with his formidable scythe, weapon of time, Time's weapon, and fought the ground inch by inch. It is the summer's enterprise. And if we were a more poetic people, horns would be blown to celebrate its completion.

In this same stretch of Route 112 South, the broad-leaved trees, including both northern and central hardwoods, prevail in the forested sections. But about 5 miles from the last turn, you come upon a massed stand of hefty conifers flanking the entrance of D.A.R. State Park, on the left. These are introduced Norway spruces and native red pines. Though this simple fact has escaped general comprehension, the Norway spruce is the Godzilla of shade tree horticulture. Owners of very small houses almost invariably plant one or two cute little specimens, then pay for it the rest of their shadebound lives. This common domestic tragedy could be called *The Christmas Tree That Ate the Front Lawn*.

Just 0.7 mile past the state park, in Goshen, bear right onto Route 9 West. Now Canadian hemlock is abundant; red oak, bigtooth poplar, and some white pine also put in an appearance. Before long, the road has entered the valley of the Westfield River. At 5.6 miles from the turn onto Route 9, good examples of bottomland tree species are in evidence: black willow, cottonwood, and black locust. As noted in other tours, the black locust was originally a native of more southern parts of the eastern United States, but it was introduced into New England at an early date in the colonial period. It was also one of the first North American trees to be used in the Old World, perhaps because its wood is so strong and durable. Today, it is well established, even in central Italy; I have seen it form dense thickets along the road from Naples to Rome. The black locust is quite easy to identify, even from a moving vehicle. Its pinnately compound leaves have relatively broad, oval leaflets, and in early June it produces dangling clusters of white flowers. By late summer, these have turned into large, dark brown pods that undeniably mark the plant as a legume.

This hardwood-dominated region is also a fine place to study the obvious but still significant seasonal changes that take place in New England's deciduous forests. In winter, the landscape has a more luminous and delicate texture, thanks to the visible intricacies of the branching treetops. After leaf break, however, the landscape balloons into an agglomeration of rounded, large-scale forms. These create the impression of massive uniformity. That feeling is dispelled quickly enough when you learn to distinguish individual tree species and the different plant regimes.

Next, the road climbs toward the crest of the Hoosac range. A total distance of 15.5 miles from the turn onto Route 9, you'll see the entrance to the final stop, Notchview Reservation, on the right. Pull off and park in the lot near the Budd visitor center. By Massachusetts standards, this is high ground. This Trustees property comprises 3,000 acres—roughly four times the size of New York City's huge Central Park. In addition, it has miles of paths for hikers and cross-country skiers. We'll begin by taking a look at the short Hume Brook Interpretive Trail located directly across Route 9. As always, it's best to get a map, if possible, before you set out. If the visitor center is closed, maps and trail booklets should be available at the information sign just uphill from the building.

Carefully cross Route 9. The Hume Brook Trail emphasizes forest management, an important topic in these parts and indeed across all New England. It is also another good wildflower station, even though many of the species here are aliens, which accompanied new settlers across the Atlantic. Among these are spotted knapweed, a purple-flowered composite (see Tour 7 for more on its genus), and valerian, a highly prized herb whose reputed powers are indicated by its alternative common name, all-heal. The twentieth-century herbalist Mrs. M. Grieve wrote that this handsome, pink-blossomed plant was administered to British air raid victims and other people suffering psychological traumas. If it's truly as good as advertised, it would be perfect for botany writers on a tight deadline.

Another wildflower here, agrimony, also has enjoyed a herbal reputation—in this case, as a low-budget Mickey Finn especially popular in medieval times. To quote a Middle English description of the plant, as given in Mrs. Grieve's book:

> *If it be leyd under mann's heed,*
> *He shal sleepyn as he were deed;*
> *He shal never drede ne wakyn*
> *Till fro under his heed it be takyn.*

As a more practical matter, this species produces some of the most exquisitely designed, and therefore most totally bothersome, cockleburs. Close examination reveals that they are armed with many curved bristles, which allow them to stick to practically anything. The efficiency of this seed-dispersal system was especially irritating to farmers, who often called agrimony by the name "harvest lice."

Farther up the trail, you enter a pleasing meadow whose inhabitants include many of the fodder plants discussed earlier, in the haying section. Among these is the grass known as timothy. A member of the genus *Phleum* (pronounced "*flee-um*"), its common name refers to one Timothy Hanson, an eighteenth-century farmer who first promoted its commercial use. Its green or pale brown flower panicle looks like a narrow cylinder borne aloft on its culm, or stem. If you carefully dissect one of the individual flowers from the cluster, you'll see it resembles, in miniature, the head and horns of a charging bull. This identification tool has been the pretext for one of the vilest botanical puns of modern times, to wit: "What do you do when you see a charging bull? You *Phleum*."

In common with white and red clovers, timothy tolerates the oxygen-poor soils found in pastures trampled by browsing animals. People acquainted with the basic life processes of plants are aware that vegetation requires atmospheric carbon dioxide to conduct photosynthesis, and that animals correspondingly need free oxygen. It's less well known that plants require oxygen, too. Its presence in the soil is especially crucial. When plants (overly coddled houseplants in particular) get too much water, the excess moisture drives the oxygen out, and the root tissue begins to rot. In the case of compacted soil, there's simply not enough room for sufficient oxygen in the first place. Timothy's patient capacity to withstand compaction definitely gives it a competitive edge.

This meadow setting is the proper locale to ponder more generally the role of the grass family in human affairs: of the eleven plant species that provide the vast bulk of our food, ten are grasses. One of the most impressive aspects of these monocots is their extensive root development. One rye plant that was carefully studied proved to have an astounding fourteen billion root hairs. Put end to end, these would have extended from Notchview Reservation to the California seacoast and back again—more or less. In other words, the water- and nutrient-absorbing potential of the grasses is remarkable, at the very least.

When you've returned across the road to the vicinity of the visitor center, take one of the trails up into the forest above. After you've passed the fringe commu-

264

nity of sugar maple, yellow birch, and American beech—northern hardwoods all—you'll realize that something has changed. There are still plenty of broad-leaved trees about, yet the red spruce has become prevalent. In effect, you've entered a transitional zone where the montane version of the taiga regime is making its presence known. By reaching this elevation of approximately 2,000 feet, you have done the equivalent of traveling to a sea-level habitat about 500 miles to the north. In that corresponding locale in Canada, another spruce species, the white, would take the place of our red. Other than that, the conditions are much the same. As you stroll through this evergreen wonderland, congratulate yourself on finding a patch of the north woods conveniently located in Massachusetts.

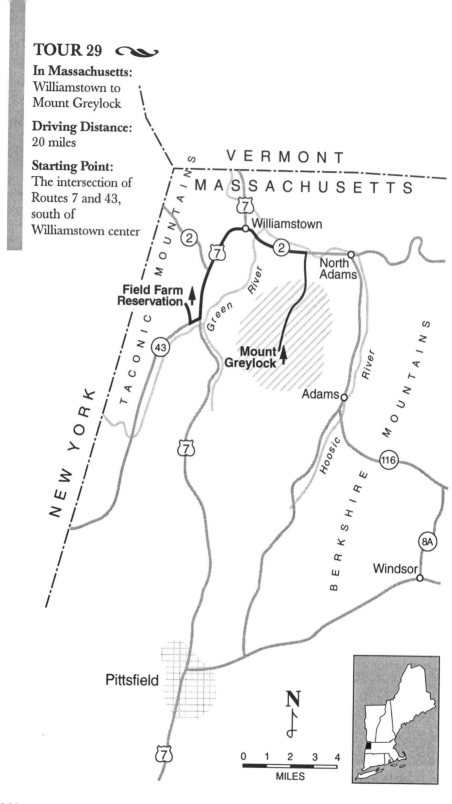

TOUR 29 ～

In Massachusetts:
Williamstown to
Mount Greylock

Driving Distance:
20 miles

Starting Point:
The intersection of
Routes 7 and 43,
south of
Williamstown center

VERMONT

MASSACHUSETTS

7

Williamstown

2

2

North
Adams

Green River

**Field Farm
Reservation**

43

**Mount
Greylock**

Adams

TACONIC MOUNTAINS

Hoosic River

BERKSHIRE MOUNTAINS

NEW YORK

7

116

8A

Windsor

Pittsfield

7

N

0 1 2 3 4
MILES

Tour 29
Great Greylock

A Recommended Fall Foliage Tour

HIGHLIGHTS

Field Farm Reservation ∾ A Footloose Mountain Range ∾ Healing Comfrey
and Its Scorpion Blooms ∾ Gray Dogwood and the Snakes' Own Herb ∾
Considering Penstemons ∾ Fort Massachusetts and What the French and
Indians Thought of It ∾ Mount Greylock Reservation ∾ New England's
Largest Pteridophyte and the American Beech ∾ An Egg-Laying Artillery
Battery ∾ The Ague and the Mountain Ash ∾ Entering the Taiga ∾ Flags
and Banners ∾ Structures Good and Bad ∾ The View of a Blessed Land

Mount Greylock, the highest point in the Commonwealth of Massachusetts, is located in a portion of the state undeniably remote from its great cultural and historical centers. Nonetheless, this magnificent mountain is a center in itself. Framed by beautiful river valleys and thus far preserved from the large-scale encroachments of money-making schemes, it stands intact, a high point in the landscape of the human imagination. Described or eulogized by Thoreau, Hawthorne, William Cullen Bryant, and Oliver Wendell Holmes, Mount Greylock probably received its most fulsome tribute from another man—part sailor and wanderer, part bureaucrat, part master novelist—by the name of Herman Melville. Melville lived in nearby Pittsfield while he finished writing *Moby Dick;* later, he dedicated his rather perplexing book *Pierre* to the mountain, as though Greylock were a monarch whose patronage was essential to the success of the humble artist.

Two substantial rivers run past Greylock. To the west is the northward-flowing Green; to the north and east is the Hoosic (its valley, preserving an alternative spelling, is the Hoosac). The Green is a tributary of the Hoosic and meets it east of Williamstown center. The tour begins on the Green River side, and the first stop is the Trustees property called Field Farm Reservation. To reach it, proceed west from the starting point on Route 43, then take a quick right onto Sloan Road. Go 1.1 miles to the clearly marked reservation entrance on the right.

Field Farm and its surroundings are situated in a particularly interesting place. The Green River flows into the Hoosic north of Greylock; the Hoosic continues on through the southwestern corner of Vermont and eventually meets the Hudson. This lush lowland is actually a southern continuation of the broader Valley of Vermont. It is underlain by dolomitic marble that formed off the coast of ancient North America some 500 million years ago. Geologists refer to these rock units as "autochthonous"; in other words, they've remained just about where they were created. That is not the case, however, with Greylock or the main body of the Taconics, which rises directly to the west. According to one widely accepted geological interpretation, the rock units that comprise these heights are "allochthonous"—that is, they were pushed from their original position miles to the east, and came to rest, quite contrarily, on top of younger rocks In the course of time, they were isolated by erosion. This kind of topsy-turvy feature, typical of the Taconics, is known as a "klippe."

However complex its geologic past, this setting is now supremely peaceful. As you start along the reservation's pond trail, keep a lookout for buckthorn shrubs and arrowwood, and also for a large, coarse, hairy plant that is probably an escape from an old kitchen garden—comfrey. One of its most important traditional uses was as a vulnerary, or wound-healer. Since its roots contain a great deal of mucilage, it has also been used as a cough medicine and even as a treatment for intestinal problems. Its flowers are also of special interest, because they reserve the right to be practically any color they want to be, from off-white to yellow, pink, or even purplish. Like many other members of the borage family, comfrey's blooms are arranged on a partly coiled stem that botanists describe as scorpioid, since it resembles a scorpion's arched tail.

The pond is home to arrowhead, a highly successful aquatic plant indigenous to both the Old and New Worlds. Cultures as disparate as the Chinese and the American Indians have used its rhizomes as a source of food. Arrowhead's common name refers to the shape of the leaves, which can vary in form from extremely narrow to quite broad. The species is a representative of the monocots, one of the two main groups of the angiosperms. Monocots generally have flower parts arranged in threes or multiples thereof; if you catch arrowhead in summer, when it's blooming, you'll note that each flower has three white petals. The other group of angiosperms, the dicots, normally have flower parts in fours or fives.

And speaking of dicots, there are plenty of interesting ones where the path reenters the thicket growth on the pond's far side. By mid-June, gray dogwood panicles of small, white blossoms stand out among the other shrubs, and two nonwoody *Lysimachia* species show how different members of a single genus can be in overall form. The first, an upright plant called whorled loosestrife, has four tapered leaves at each point of attachment, while the second, moneywort, is a prostrate scrambler, with rounder leaves arranged in pairs. The key to their close affiliation is their similar, five-petaled yellow flowers. Moneywort's common name refers to its coin-shaped foliage. The traditional English *wort* suffix, originally *wyrt,* means "herbaceous plant." This species' herbal use as a wound treatment parallels comfrey's. It was said that even snakes recognized moneywort's healing powers, and would seek it out when they were injured.

Another *wyrt* worth searching for is the beardtongue penstemon. This relative of the snapdragons has striking tubular flowers that are zygomorphic, or irregularly shaped, so that they're symmetrical only along one plane. Penstemons aren't exactly rare here in the East; but in the Southwest, they're at their greatest glory. One of the most beloved wildflowers at Organ Pipe Cactus National Monument, on the Arizona-Mexico border, is Parry's penstemon. It has the temerity to bloom just about the time the frost heaves return to New England.

The path leads to a farm field that gives you a splendid view of this trip's second and main objective, Mount Greylock. Return to your car and drive back through the lovely farmscape dotted with holstein cows to the intersection of Routes 7 and 43. Turn left onto 7 North. As mighty Greylock looms to the east, you pass through an area where the white pine, appreciative as always of unobstructed sunlight, has established itself in previously cleared spots. Continue straight on Route 7 past the junction of Route 2 West; 2.3 miles beyond that junction, bear right onto 2 East. Immediately before this turn, you pass by the Williamstown green, site of the historic 1753 House.

To the settlers of the seventeenth-century Plymouth and Massachusetts Bay colonies, the Williamstown area must have seemed a horrible wilderness halfway to China. But by 1745, the British had built a frontier outpost here, which they named Fort Massachusetts. Three years later, the Indians allied to the king of France registered their opinion of this idea by burning the place down. Colonial true grit prevailed, and it was rebuilt a year later.

Route 2 takes you past the airy and elegant Williams College campus and then over the Green River. Continue into North Adams and note the odometer reading when you pass the airport; 2.2 miles beyond that point, turn right onto Notch Road—you should see a Mount Greylock Reservation sign. Half a mile up the road, bear left at the fork; after another 0.6 mile take a sharp left. Once again, there should be a sign to guide you. Along this roadway are nice pastures and an interesting mixture of trees: Canadian hemlock, red pine, cottonwood, and a plantation of Norway spruce. After 1.1 miles, yet another sign will direct you to the right and through the official entrance of the state-owned reservation.

As you ascend the paved 9-mile road, pay attention to the plant regimes. You'll definitely see a change take place. At first, the northern hardwoods prevail, quite common in the Berkshire uplands, with an overstory of yellow birch, paper birch, and American beech. The humbler striped maple skulks beneath them. Due to extensive logging, the American beech is rarely seen in its full glory, but when it attains maturity, it is one of the most impressive trees in the world, massive and ghostly. Its three-sided nuts are enclosed in a prickly husk, or involucre; they're an important food for wildlife. The tree's wood is prized as fuel and also serves as flooring and veneer. For a description of the bark disease that is now threatening it, see Tour 1.

You should be able to recognize some of the roadside wildflowers as you pass by. Yellow jewelweed, a species frequently seen in this kind of setting, is here. It's one of two native New England members of the *Impatiens* genus familiar

The upper portion of Mount Greylock's well-maintained Notch Road, with its taiga conifers.

to gardeners. The handsome tall meadow rue pops up as well; look for its dangling white or green flowers in late spring. Elsewhere among the abundant vegetation is the magnificent ostrich fern. Its graceful, plumelike fronds are capable of growing almost 6 feet long—when they're not collected as they emerge by fanciers of fiddlehead cuisine—making this species New England's largest pteridophyte. (For more on fern cookery, see Tour 38.) In contrast to the vase-shaped habit of the ostrich fern foliage, the leaves of the smaller hay-scented fern rise singly from their far-spreading rhizomes. Whatever their scale, fern fronds are usually impressive structures. In fact, ferns are the only pteridophytes that have substantial leaves.

This is prime porcupine country, and you may have the good fortune, as I once did, of having your progress slowed by an ambling and unflappable character who refuses to yield to faster traffic. Early New England naturalist John Josselyn (the one who thought a wasp or hornet nest was a pineapple) reported to his seventeenth-century English readership that porcupines lay eggs and fire their quills in salvoes at unwary human passersby. Should you see one of these lumbering creatures, be assured that you won't take a broadside unless you insist on making direct contact.

At approximately 2.5 miles above the reservation entrance, mountain maple becomes evident, especially in early June, when its long, yellowish flower clusters stand out against a dark green foliage background. Mountain maple is a small tree with leaves that are sometimes confused with those of the striped maple, even though they are much more coarsely toothed. Soon thereafter, American mountain ash, a tree in the rose family with pinnately compound leaves, shows up. In pre-Revolutionary times, its bark was used as a quinine substitute to treat the ague, or what we would now call malaria.

One herbaceous plant that is well adapted to roadside drainage ditches and other wet spots is American white hellebore, one of the very first species to sprout at winter's end. Its broad, pleated leaves may reach a foot in length. Long after these appear amid the mud and melting snow, the plant shoots up a large spike of flowers that are so thoroughly green they must be U.S. Army issue. Josselyn noted the white hellebore in his delightful guidebook, *New-Englands Rarities Discovered,* and remarked that it "is the first Plant that springs up in this Country, and the first that withers; it grows in deep black Mould and Wet, in such abundance, that you may in a small compass gather whole Cart-loads of it." He also mentioned that the Indians he knew used it for healing wounds and soothing toothaches. In more modern times, the plant has been described as highly poisonous, though it has been used to treat arteriosclerosis.

On the final approach to the summit, the look of the vegetation changes dramatically as you enter the montane equivalent of the taiga regime. Two kinds of conifers, red spruce and balsam fir, are more capable of dealing with desiccating wind, cold, and impoverished soil than most of the broad-leaved trees, and they've clearly taken over at this elevation. Of the two evergreens, balsam fir is somewhat the tougher, and it will hang on even after red spruce quits the field. These two gymnosperms present good examples of what easterners call "flag" trees and westerners know as "banner" trees. These are deformed woody plants, often dwarfed, with contorted branches pointing only in one direction in response to the prevailing wind.

There is a spacious lot up at the top where you can park and take one of the more elevated strolls of your life (the summit is at 3,491 feet, more or less). The striking building just across the way is Bascom Lodge. It was built in 1937 for visitors who choose to stay overnight; nowadays, it is jointly administered by the Appalachian Mountain Club and the state. Constructed appropriately enough of rough stone and red spruce, it fits into the surroundings admirably. The gift shop offers maps and books that describe the wonders of this mountain more fully.

The other structure at the summit defies rather than accepts its location. This is the War Memorial Tower, originally designed as a Boston lighthouse but engrafted here instead, like a truck-stop salt shaker glued to the back of an elephant. Its beacon, often cited for its powerful beam, must be turned off during the seasonal bird migrations. This granitic sore thumb notwithstanding, the scenery facing outward in all directions is spectacular. Peering down to the east, you'll spot the lime-producing area of Adams and the Hoosac Valley; and to the north and west, you will espy the Taconics, including Vermont's Mount Equinox, the goal of the next tour. This is also a good spot to take a close look at the dwarf black cherry and pin cherry thickets, the mountain ash and mountain maple, and the wind-blasted balsam firs. These conifers, like their magnificent relatives, the white firs of the Rockies, can be differentiated from spruces by their round-tipped needles and by their cones, which stick up, rather than hang down, on the branches.

While you're here, make sure you pause from your botanical pursuits long enough to silently thank the dedicated men and women who over many decades have saved this unspeakably important place from the ravages of human greed cloaked in the form of tramways, ski lodges, and the like.

The view eastward from the Greylock summit; the heavily settled Hoosac Valley lies in the background.

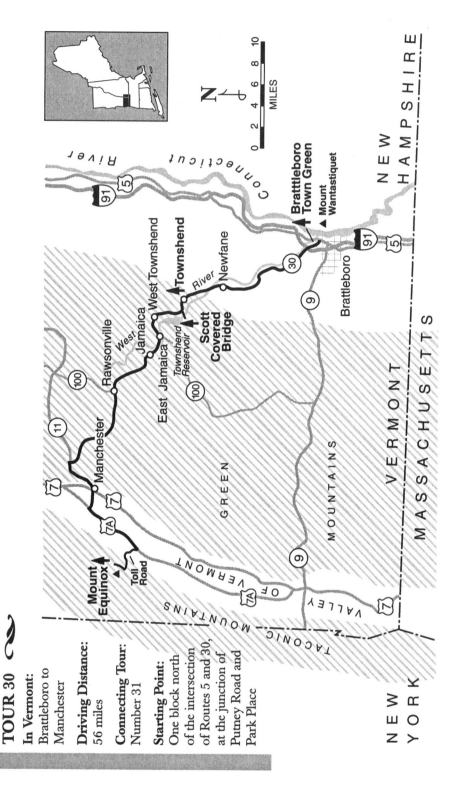

TOUR 30 ⁓

In Vermont:
Brattleboro to
Manchester

Driving Distance:
56 miles

Connecting Tour:
Number 31

Starting Point:
One block north
of the intersection
of Routes 5 and 30,
at the junction of
Putney Road and
Park Place

Tour 30
A Trip to the Valley of Vermont and the Realm of the Mountain Maple

A Recommended Fall Foliage Tour

HIGHLIGHTS

An Everlasting Mountain and Fort Dummer ∾ Brattleboro's Town Green ∾
A Meeting of Maples ∾ The Basswood's Heavenly Scent ∾ Classic Yankee
Towns ∾ A Visit to Scott Bridge ∾ A Beginner's Guide to Covered Bridges ∾
The Versatile Indian Hemp ∾ The Land of a Billion Years ∾ Over the Crest of
the Greens ∾ *Facilis Descensus Averno* ∾ The Sweet Soil of the Valley of Vermont ∾
Wild Parsnips ∾ Mount Equinox ∾ A Herb of Awesome Aspect ∾ The Doctrine
of Signatures∾ A Montane Birch ∾ The Path to Lookout Point ∾ Conifers,
Lichens, and Mountain Woodferns ∾ The Gods of the Hills

*T*here is a general understanding about New England: The farther north
you go, the wilder the scenery is, and the less civilization you have
to contend with. It makes sense until you reach Vermont. For one thing,
the population of the Green Mountain State is concentrated in the northwestern
and north-central sections; for another, the southern reaches contain some of the
most untrammeled upland terrain, preserved as national forest.

This journey takes you through a beautiful lower portion of the state—over
the spine of the Greens, into the broad western lowland known as the Valley
of Vermont, and up to the highest point in the Taconics, the range that forms
Vermont's boundary with upstate New York. You'll begin at the historic Connecticut River town of Brattleboro.

In contrast to the beautiful broad-valley towns of Massachusetts just 20 or
30 miles downstream, Brattleboro is situated in a region of hard-rock geology.
Highly resistant metamorphic bedrock ensures that the great river will not manage
to carve a spacious lowland setting for itself. In fact, New Hampshire's Wantastiquet
Mountain, just across the water, is a powerful reminder that the landscape of
northern New England only grudgingly gives way to even the most powerful
forces of erosion. Thoreau, who visited Brattleboro in 1856, noted in his journal:
"Above all, this everlasting mountain is forever lowering over the village, shortening the day and wearing a misty cap each morning. You look up to its top
at a steep angle from the village streets."

Brattleboro's history reaches back to the early eighteenth century, a comparatively late founding date by Connecticut or Massachusetts standards, but an early one in the chronicles of the colonization of the Upper Valley. In 1724, the outpost called Fort Dummer, the first permanent British settlement in Vermont, was erected on a site south of the modern town center. That spot now rests under the waters of the river.

The first stop on this tour is located at its starting point, where Putney Road meets Park Place. Turn west onto Park Place and pull over to the curb where permitted. Take a turn on foot around the small but charming town green. This is a good place to study ornamental maple trees, especially if you're visiting in summer, when the trees are bearing their clusters of two-seeded, winged fruit known as samaras. These are one of the more helpful identification features of the large genus *Acer*.

One species native to the bottomlands of our region's larger streams is the silver, or soft, maple. Its ability to grow quickly into a hefty shade tree is matched, unfortunately, by its weak and easily snapped branches. One of the silver maple specimens here, at the park's western end, has highly dissected leaves and may be the cultivar, or cultivated variety, that goes by the name of Laciniatum. Silver maple samaras have large wings spread at a fairly broad angle; they form two sides of a more or less equilateral triangle. The other native here is one of the chief glories of Vermont, as well as one of the best indicator species of the northern hardwoods regime—the famous sugar maple. Its broad-lobed, taper-pointed foliage is the standard pattern for maple leaf symbology everywhere, most notably on the national flag of our great northern neighbor. The samaras of this substantial tree are surprisingly small; their wings are positioned to form a small angle, and in some cases they almost touch each other.

Two other maples grow on the green. The first has leaves that closely resemble the sugar maple's. The distinction is that the leaf stems of this species exude a milky sap when broken. This is the Norway maple, originally a European import, but now used so widely in this country that it has naturalized in some areas. One other way of distinguishing it from the sugar maple is to check its samaras. These are often likened to aviator-wings insignia, because the two halves are pretty nearly arranged in a straight line. The other introduced species, the amur maple, is less well known, though it has been used as a dependable, often shrub-sized ornamental. The leaves of this species are lobed, too, but the central lobe is generally much longer than those on the side. Its samaras resemble miniaturized sugar maple fruit. The original home of the amur maple is the colder parts of the Far East.

Before you leave the green, pay your respects to the other trees that faithfully provide shade for the townsfolk year after year. These include a purple-leaved crab apple, thornless honey locust, horse chestnut, the inevitable Colorado blue spruce, and basswood, the predominant indigenous linden of the American Northeast and Midwest. Its yellowish white flowers, always a favorite of bees, produce one of the most heavenly scents on earth when they open in early summer.

———————————— ❧ ————————————

Depart Brattleboro by continuing on to the end of Park Place and turning right onto Route 30 West, otherwise known as Linden Street. After passing the Brattleboro Retreat, the road descends the bluff slope and skirts the West River where it broadens into a basin of slack water before entering the Connecticut. This wetland, the Retreat Meadows, is now partially under cultivation; elsewhere, the monocot common cattail forms large communities. On the low, midchannel islands, a much more substantial flowering plant, the arching American elm, continues to hold out against the continuing onslaught of Dutch elm disease. In the wild, these stately trees often indicate habitats exposed to shallow flooding of short duration, whereas the silver maple is a good sign of long-term inundations.

The West River is one of the most scenic streams in Vermont, and is one of the larger tributaries of the upper Connecticut. Route 30 takes you up its course for the next 20 miles or so. In this first stretch, it is a wide and boulder-strewn watercourse. The banks support a wealth of woody plants, from hefty, mottled-bark sycamores and black willows to red oaks and black locusts. On the hillsides to the left, basswood, our now familiar friend, grows quite happily without human intervention among bigtooth poplar, staghorn sumac, and other determined competitors. At 6.4 miles from the starting point, you'll see the first covered bridge of the day on the right. This is the West Dummerston bridge, the longest surviving structure of its kind in the state. A few miles up the road, you'll be stopping to look at another bridge of almost exactly the same dimensions.

Townshend's church on the green, a chaste commingling of Greek Revival and Gothic Revival styles.

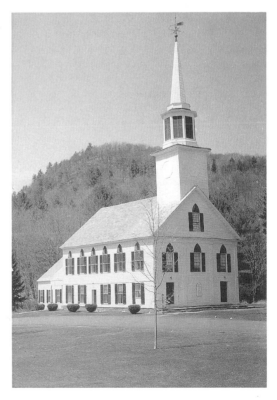

Townshend's Scott Covered Bridge, as it appeared in 1991.

In a few minutes' time, you've reached the placid heart of Newfane. Continue past its bemapled green and onward into the heart of Green Mountain country. Notice the pleasant little floodplain the West River has built in this typically narrow valley. Floodplains are usually premium agricultural territory, and here much of the bottomland is still actively farmed. Glacial features known as kame terraces are a common occurrence in this neighborhood. They resemble shelves extending partway up the valley sides, but it often takes a trained eye to detect them.

The center of Townshend lies almost exactly 5.0 miles from downtown Newfane. Lorded over by a prim church, Townshend is so quintessentially Vermont that it provokes one's suspicions—is it the two-dimensional set from an old Bing Crosby movie? Was it built by Madison Avenue for a thirty-second bread commercial? Fortunately, its white and green-shuttered austerity is absolutely authentic. If you can stop here for a moment and amble past the tour buses to the aptly unadorned green, by all means do so.

Afterward, continue west on Route 30 for a distance of 1.5 miles and then carefully pull off to the left, near the historical marker that describes Scott Bridge. This covered span, a mere 4 feet shorter than the West Dummerston structure already seen, is a wonderful spot for a brief stop and an excellent place to consider one of the classic symbols of old-time Yankee architecture and culture. Scott Bridge is not the epitome of successful bridge design nor does it embody one particular style of construction better than its companions elsewhere in northern New England. It significance, for this book at least, lies in its rather battered charm and its pleasant location.

The first reasonable assumption an uninitiated visitor can make about covered bridges is that they were roofed over to protect travelers from falling snow and slippery stream crossings in winter and mud season. In fact, for many years bridge owners or town governments actually hired boys or men to "snow" the bridges in winter—that is, to shovel the white stuff *into* the bridges, where it would be spread over the flooring. The rationale for this would be clear enough to you, were you a nineteenth-century Vermonter who depended not on one type of vehicular transportation, but two: wheeled buggies in the warm seasons, and runnered sleighs in winter. The bridges were closed in, therefore, for other purposes. The roof and siding apparently added to a span's structural integrity; at the same time, they deflected water from the main load-bearing elements, allowing them to season properly and resist deterioration longer. (It's easier to replace detachable roofing, panels, or snow-rotted floorboards than an attached truss.) In addition, a well-tended superstructure gave a bridge a more stolid and impressive look, which contributed to its resale value. This was an important consideration in the last century, when many of these spans were privately owned toll crossings, bought and sold frequently by entrepreneurs.

Another widely held assumption is that covered bridges are quaint artifacts from New England's earliest colonial days. In fact, they came into full fashion in the United States only in the nineteenth century. The first one in Vermont dates from the 1820s. (The first covered bridges, incidentally, were constructed in Germany, Austria, and Switzerland.) The Scott Bridge is a relative newcomer; it was built by Harrison Chamberlain in 1870, a full five years after the end of the Civil War. Surprisingly, the vast majority of extant covered bridges are located in states outside the New England region. Pennsylvania tops the list with (depending on your source) two hundred or even three hundred surviving examples. The totals for Ohio, Indiana, and far-distant Oregon are close to Vermont's— each hovers close to one hundred.

If, given that sort of competition, the Scott Bridge can lay claim to any special distinction, it is that it was the first covered span in Vermont to be deeded to the state and preserved by it. Its design is an interesting hodgepodge of truss styles, combining one Town Lattice section—the part that has diagonal, crisscrossing beams—with two kingpost segments. On top of that, beautiful but ultimately ineffective bow-arch bracing was added. When it failed to help the sagging structure, a concrete pier support was put in underneath. Notice how the arch system is laminated: it's composed not of one great beam but of successive layers of wood designed to help distribute the stress better than a solid piece could.

Covered bridges were most often made from familiar New England tree types: white oak, white pine, balsam fir, or Canadian hemlock. In modern steel construction, the strongest points are where the beams are joined together; in wood, the joints are the weak spots. Still, wooden spans are superior in one respect. Being somewhat elastic, individual wooden beams actually get stronger when they're compressed.

As you walk across this fascinating product of timber technology, take a moment to investigate the living plants on both riverbanks. Red oak and sycamore are main components of the overstory, and beneath them are poison ivy—the walker beware!—and communities of sensitive and ostrich ferns. One wildflower worth hunting for is Indian hemp, a species in the dogbane family with oppositely arranged, smooth-margined leaves. This is one plant that actually has an accurate common name; native peoples used the tough-stemmed herbaceous plant for sacks, cordage, and fishing nets. English settlers found it handy, too; some thought the thread it provided was better than cotton.

———————————— ❧ ————————————

Depart Scott Bridge by resuming your westward trek on Route 30. Almost at once, you pass the Townshend Reservoir, where the progress of the West River is partly checked by an earthwork dam. Here the scenery is particularly exhilarating. Four miles beyond the dam, Route 30 meets Route 100; stay on 30. Another 1.2 miles farther along, an outcrop of metamorphic phyllite shimmers and sparkles on a sunny day.

After passing through Jamaica, take heed of the upland forest community, with its mixture of American beech, paper birch, white pine, and Canadian hemlock. Here at last we bid farewell to the West River, which swings due north toward its source in Weston. You've reached the Precambrian core of the Green Mountains. The rocks that underlie this splendid landscape came into being one billion years ago, roughly speaking, which makes them more than twice as old as the earliest record of plants on earth. Subsequently, these rock units were subjected to the tremendous forces of mountain building, as the margin of North America came into contact with other land masses.

Remain on Route 30 when Route 100 splits off from it. Here, foliose lichens are a common sight on the damp tree trunks. Red spruces, marking the lower limit of the montane taiga forest, are easier to spot at this elevation. Another 5.9 miles past Bondville Center is a rare open vista. From this distance, the ski slopes of the far hillsides could almost be mistaken for rock slides. In another 3.0 miles, you reach a T-intersection. Bear left. Nearby, the legendary Appalachian Trail crosses the roadway. In a moment or two, your mountain journey is rewarded by the first view of the glorious Valley of Vermont. Now, as the Roman poet Virgil once noted in a similar context, the descent is easy. Proceed into Manchester, the local mecca of tourism, past the fern bars with cute names and the cloned factory outlets, which, as usual, have no factories anywhere near them. At the downtown intersection, bid Route 30 and its coy whims a fond farewell and turn left onto Route 7A South. If all the happy jaywalking shoppers permit, take in the view that intermittently presents itself between the various examples of consumer-friendly architecture. Congratulate yourself on having two separate mountain ranges at your beck and call: the Green Mountains of recent memory on your left, and the quirky mass of the Taconics on the right. Their mightiest representative, Mount Equinox, is close at hand. (For a brief discussion on how the Taconics came to be, see the first part of Tour 29.) Hildene, the estate of Robert Todd Lincoln (Abraham Lincoln's son), is located just off the road at 1.9 miles south of the last turn. Check your odometer reading when you reach it.

In contrast to the highlands that border it, the Valley of Vermont is underlain by such yielding rock types as dolomitic marble. The marble, similar in composition to limestone, sweetens the soil by neutralizing acid content. Acidity is a common problem in much of the rest of New England, thanks to the region's high rainfall. You can see at least one good marble outcrop along 7A. One of the most frequent roadside wildflowers you'll see is wild parsnip, which raises spreading clusters of yellow flowers throughout most of the summer. It is an indicator of calcareous soil. In earlier times, wild parsnip served as a popular treatment for all sorts of diseases, including asthma, tuberculosis, and cancer.

Our final destination, Mount Equinox, now looms to the right, and at 3.3 miles south of Hildene the entrance to the toll road leading to it comes into view. Turn in here, pay the piper, and begin your ascent of this private, fully paved road to the highest point in the Taconics. (The Equinox summit is about 400 feet higher than that of lofty Mount Greylock in northwestern Massachusetts.) At first, the vegetation is as lush as anything the valley floor can offer, with poplar, American elm, and purple-flowering raspberry mixing with wild grape vines and an assortment of wildflowers—ox-eye daisy, bladder campion, and many more.

After summer's start, keep an eye out for a formidable borage-family alien named viper's bugloss. Resembling something out of Darth Vader's garden, this bristly beast of a herbaceous plant can be a pest, but it's actually quite beautiful when you examine the flowers closely. Their long, red stamens are offset by dark blue corollas—striking yet complementary colors. This is another demonstration of nature's predilection for getting it right the first time around. When, after all, was the last time you saw a clashing forest or a badly composed wetland community? It takes the human element to come up with ceramic white-tailed deer standing in a bed of orange marigolds.

As a herbal remedy, viper's bugloss has been used to treat nervous disorders and headaches; supposedly, the lowest leaves are the most effective in making the soothing infusion. In addition, it was believed by some to neutralize snake venom—because its flower structure resembles a snake's head in miniature. This is one illustration of a system of healing widely accepted by doctors and herbalists for centuries. This concept, known as the Doctrine of Signatures, states that the aspect of a particular plant indicates its curative properties. For instance, for disorders of the mind, a healer might use walnut meats, because they look like small brains, or the herb skullcap, since its blooms remind you of the human skull. By the same token, Vermont's state flower, red clover (originally an Old World species), was used to purify the blood.

This logic might seem strained to us nowadays, but if you'd voiced your reservations to the great academics of medieval times, the learned professors would have pitied your weak grasp of science. It's important to realize that botanists still take the same basic approach: they look for clues, often fairly simple ones, in the structure of plants. Today, however, those clues lead them not to corresponding parts of the human anatomy but to the greater anatomy of evolutionary theory—viper's bugloss points not to a snakebitten hand but to a particular twig on the profusely branched tree of the history of life. One day, these striking blossoms will probably help justify yet another predominant view of existence. Stay tuned.

Stop at the roadside pull-off at the 2.1-mile mark from the entrance. Northern hardwoods in the form of American beech and sugar maple are evident here, as are red maple and some escaped tatarian honeysuckle. The latter is a shrub much loved by cedar waxwings and other fruit-eating birds. As you continue upward, you soon enter the zone dominated by a small and often overlooked relative of the first trees we examined on the Brattleboro green. This is the mountain maple, a species that normally confines itself to the upland heights. Its leaves superficially resemble those of the red and striped maples, but the plant can be differentiated from the other two by a moment's examination of the foliage, flowers, or fruit.

If you want to enjoy one of the best views to be had from the roadway, stop in the area known as Skinner's Hollow. Just below you is what geologists call a cirque, a feature resembling an amphitheater, carved into the upper reaches of the mountain by an alpine glacier during the Ice Age. Behind you is the summit of Little Equinox, with an array of electricity-generating wind turbines. This is also a good place to see another veteran of the heights, American mountain ash, which has compound, pinnate foliage with toothed leaflets. While you're at it, take a close look at the paper birches. Their leaves have rounded, almost heart-shaped bases, a sure sign that this is the montane variety of the species. In the lowlands, paper birches produce leaves with tapered bases.

In summer, the final approach to the main summit features still more wild-flowers, most of them members of the gigantic and highly complex composite family. Among these are devil's paintbrush, oyster plant, and the splendid rag-wort, a yellow-flowered member of the widely distributed genus *Senecio*.

Park in the lot in front of the inn and enjoy the sweeping view. The summit of Equinox is a little more than 3,800 feet above sea level. In clear weather, you can see Mount Greylock to the south, past the Green Mountains to Mount Ascutney in the east, and up the Taconics to the impressive mass of the Adirondacks in the north. For an excellent view of the Valley of Vermont and its river, the Batten Kill, take the 0.4-mile trail to Lookout Point. Conifers, including the toughest mountain species of all, balsam fir, predominate. At their feet grow Canada Mayflower, trout lily, golden Alexanders, and large communities of mountain woodfern. Also, foliose lichens, ever appreciative of highland mists, have staked their claim on many tree trunks and branches. Although they are among the most cold-hardy organisms on the planet—their Antarctic kin grow within 4 degrees of the South Pole—they are for the most part markedly sensitive to air pollution, especially sulfur dioxide. That they thrive here reinforces the fine feeling that this high point is the best place in a beautiful but heavily traveled region. To repeat Ethan Allen's famous line, "The gods of the hills are not the gods of the valleys."

Mount Equinox. Skinner's Hollow, a glacier-formed cirque, is visible on the mountain's flank to left of center.

TOUR 31 ⤷

In Vermont and New Hampshire:
Brattleboro, Vermont, to Springfield, Vermont, via New Hampshire

Driving Distance:
31 miles

Connecting Tours:
Numbers 30 and 32

Starting Point:
Exit 3 of Interstate 91, the junction of Route 9 East

Tour 31
The Upper Valley–Part I

*I*n this tour, and the one that follows, our focal point is one of New England's most significant natural features–not a seacoast or a range of mountains, but its largest inland waterway, the Connecticut River. You'll see the utterly distinctive terrain it has created north of the expansive Massachusetts Central Lowland. This is an area rich in history and abundant in striking scenery. It is also a magical place for the botanist and geologist. It is known, simply enough, as the Upper Valley.

In both this tour and the next, our route will take us back and forth across the river, which serves as the boundary between the Green Mountain and Granite States. There's something especially fitting in this, because in many ways the Upper Valley comprises one cohesive community, almost a state in itself. Family, educational, and economic ties across the water are strong, and have been for two and a half centuries. As a matter of fact, in the late 1700s, thirty-eight towns along New Hampshire's western fringe tried to secede from their own government and join their neighbors in the fledgling republic of Vermont, which was an independent nation, however reluctantly, from 1777 until 1791. It took the Continental Congress, along with veiled threats of political dismemberment and military intervention, to persuade the Vermont leaders to disavow the defection.

The starting point of the journey is Brattleboro's northernmost interstate exit. This is the link with Route 9 East to New Hampshire. If you want to orient yourself in terms of the previous tour, it's 2.1 miles north on Route 5 from the town green on Park Place. When you're ready, head north on I-91. This superhighway is one of the most beautiful in the nation, and it provides superb views of the surroundings. Geologists love the huge road-cuts that go with interstate construction, and here in the Upper Valley, it's easy to understand why. The glistening masses of phyllite and other metamorphic rocks add a sense of timelessness and monumentality entirely appropriate to a landscape as enduring as this.

The roadside tree population in this area is largely broad-leaved, though it won't take much effort to spot some red and white pines. Look for one classic bottomland species, the cottonwood, a type of poplar that has a natural range extending westward almost halfway across the continent. Should you ever decide to travel by car from New England to Chicago, for instance, the cottonwood species will gladly accompany you the whole way; and, in fact, it will become more prevalent the farther you go. True to its willow-family heritage, the tree tolerates deep and lengthy flooding. And yet, surprisingly, it is also the first to occupy the sandy beaches of the Great Lakes; it resists burial by shifting sand dunes by growing new roots higher up its stem, as necessary. The Algonquin nations used the tree's roots for cordage and the bark for wigwam exteriors. This organic version of aluminum siding was best peeled off cottonwood trunks in spring, when the sap was flowing. In modern times, this species is preferred for pulpwood, which makes high-grade paper for fashion magazines and other sleek publications.

Also along the way is one of the interstate's most pleasing herbaceous species, though few motorists ever fully appreciate it. It shows up in late summer and fall when it imparts a red or maroon blush to the roadside. This is purple love grass, a true grass indeed, and one that must be thankful to the road-building habits of twentieth-century humankind. In creating dry, sandy shoulders for the pavement, we have extended the plant's favorite habitat by leaps and bounds, which is not a cause for complaint; no true connoisseur of natural color would underrate its role in beautifying the local environment. (For more on purple love grass and its clever way of spreading, see Tour 23.)

At a point about 2.9 miles from the start of this tour, an especially good view of the New Hampshire side of the valley, where the hills trend north-south, can be seen on the right. Closer at hand, on the left, are cornfields still actively farmed. This is a region with plenty of American elms, red oaks, and black locusts. By the second half of summer, the yellow tones of common mullein and various goldenrods predominate in the more open spots. In an additional 3.4 miles, you've reached Exit 4, to Putney. Leave the interstate here and head north on Route 5 by turning left at the head of the ramp, and by following the signs through the town center.

As you pass through Putney, see how many lawn trees you can identify. Consider yourself in need of remedial instruction if you can't distinguish the Colorado blue spruces from the plants that actually match and complement each other. One arresting broad-leaved lawn tree here is the catalpa. Horticulturists describe this plant as "coarsely textured," and one glance will tell you why. The leaves aren't just broad, they're gross; their off-green color and heart-shaped outline makes you suspect the tree has recently escaped from an Henri Rousseau

A catalpa tree in Putney. This photo, taken in early spring, shows last year's seed capsules still hanging from the branches.

painting. The catalpa reaches its state of maximum outrageousness by late summer, when its tubular, white flowers have long since given way to seed capsules resembling giant mutant string beans. These frequently have the supreme bad taste to keep dangling from their branches long after the foliage gives up and flops exhausted to the ground. I have heard some folks in the Midwest refer to this fruit as "Indian cigars," though no one has had the temerity to suggest that Native Americans, usually so canny in their use of nature's bounty, would waste much time puffing away on these flameproof eyesores. To maintain a little perspective, though, catalpa wood does make some of the most durable lumber, and the springtime blossoms that precede the pods exude a sweet perfume.

Mutant string beans notwithstanding, Putney is a lively and attractive river town. One of its biggest draws is a wicker outlet, irresistible to locals and leaf-peepers alike. Since wicker is a plant product, it's worth a few seconds' pondering. Its name derives from the Middle English term *wiker,* which apparently meant willow. This makes good sense; after all, the traditional material for baskets and other wicker items is a kind of European willow known as osier. This species is a shrub often cultivated in the Old World; the stems are harvested on a yearly basis.

On your way out of town, take a look at the big black locusts on the left, just before the cemetery. When last I saw them, the trunks were covered with wild grape vines. Wild grape species are aggressive pests in this region, and while the vines here probably won't be allowed to do much harm to these trees, they can kill large plants in the wild by blocking off access to sunlight.

Beyond this point another 1.7 miles, the road passes a spot on the left where gray clay is exposed on the bank. The clay, also visible in other locations up the road, is a deposit of sediments that slowly settled to the bottom of Lake Hitchcock. This body of glacial meltwater filled the Connecticut River valley all the way from Rocky Hill, Connecticut, to the Northeast Kingdom of Vermont roughly thirteen thousand years ago. One of the indications that the clay is a lake deposit, and not one made by the river itself, lies in the tiny size of its individual particles. Usually, very fine sediments such as this don't drop to the bottom until the water is still, or at least sluggish. Sediments deposited by a flowing stream are generally larger and heavier, which is why much of the river's contribution to the valley floor consists of sands and gravels. There is one important exception to this rule: When the river floods and overtops its levees, the water velocity decreases, giving the surrounding floodplain a new layer of nutrient-rich, fine-particled mud. This fact jumps out of the geology textbooks straight into human history, for it is in the self-renewing fertility of river systems, the gift of floods, that civilization has its deepest roots.

Geology isn't the only attraction at this point. On the right is a good example of blueberry farming, with row after row of fruit-bearing shrubs of the highbush type. This sight will trigger the salivary glands of anyone who has ever experienced the supreme hedonism of a bowl of fresh blueberries topped with an overkill sweetening of sugar. Truth to tell, this form of agriculture is not a particularly big operation in the Upper Valley. (To see the industry on a much larger scale, where lowbush blueberries are used, trek over to Down East Maine and take Tour 54.)

Only 0.6 mile past the blueberry fields, the clay exposure is a massive road-cut into gray Paleozoic era schist. Interestingly, this metamorphic rock began its career as a clay, too, or something similar, but it is of marine origin, and is about forty thousand times as old as Lake Hitchcock. Long before the Ice Age hit, this rock was subjected to mountain-building forces that altered it profoundly.

In midsummer, one wildflower, bird's-foot trefoil, does a particularly good job of colonizing the local roadside. Look for clusters of bright yellow blossoms sticking up above the more or less prostrate foliage. If you get a chance to examine the flowers of bird's-foot trefoil at one of your stops, you'll notice that their form is reminiscent of a pea blossom, and for good reason: both peas and trefoils are members of the huge legume order. These plants have developed the ability to thrive in nitrogen-poor soils (sandy shoulders, for instance) because they have a special type of bacterium that lives in nodules on their roots. This bacterium, *Rhizobium,* fixes nitrogen from a much more abundant source, the atmosphere itself. For this reason, legumes such as bird's-foot trefoil are among the first to establish a foothold in impoverished locations.

From now on, keep a lookout for a roadside shrub known as winterberry or, more misleadingly, as black alder. It isn't an alder at all, but a native holly. This one plant all by itself makes the tour a worthwhile endeavor in the dour months before the onset of spring, because it has persistent, bright red fruit that adds a much needed splash of color to the cold-season landscape. While its "berries" (actually drupes) do indeed remind one of the evergreen hollies used in holiday decorations, the leaves of this species are quite different. For one thing, they don't have a shiny luster, and for another, they're not evergreen. As a result, the drupes stand alone and unobstructed on the naked branches in winter. It's a wonder that such fruit-nibbling feathered friends as migrating cedar waxwings don't make short shrift of this high-visibility target. As far as the winterberry's herbal uses go, the Indians derived an astringent from it; in the pre-Revolutionary era, white settlers made an extract from its bark to use in treating malaria.

As you go up the rise on the far side of the outcrop, you come upon the best thing cultivated land has ever been used for—apple orchards. There is something about the sight of a field of well-tended trees that evokes the husbandman in each of us. Perhaps it is the combination of meticulous order and unbridled abundance. In the full sunlight of summer, a broad-spreading apple tree seems to be the most truly living organism the world has ever produced. In his encyclopedia of A.D. 1470, Bartholomeus Angelicus noted that the apple tree "makyth shadowe wythe thicke bowes and branches," and that it had "floures of swetnesse and lykynge: with goode fruyte and noble."

This good fruit and noble has been an important component of New England agriculture since the earliest colonial days. Vermont in particular is prime apple country. While the apple was always used for vinegar, molasses, and pie fillings, its primary traditional role was in the production of home-brewed happiness. When the first English settlers found that barley grew poorly in the New World, while apple trees prospered, they switched their loyalties from beer to cider and cider brandy. Consumption of these relatively stiff forms of cheer was widespread in farm country. Given the rigors of hard work and the soul-testing climate, no one worried about the regional dependence on alcohol until the temperance movement swept the country in the 1800s. Ironically coupled with the introduction of mass-marketed hard liquor, it resulted in a serious decline in orchards, which was checked only several decades later, when the apples themselves became a promising cash crop.

To the systematic botanist, the genealogy of the apple tree presents a bit of a mystery. It may be descended from one species native to the Near East; or its primal parent may have been a cross between that and a European species. Today, there are literally thousands of cultivars. Some of the most famous varieties employed in this region are (or were) Fameuse or Snow Apple, which was first brought to Vermont by the French in the early 1700s; the superb cider selection Roxbury Russet, developed in Massachusetts in 1647; and such other poetic entries as Hubbardston Nonesuch and Seek-No-Further. As far as Vermont's apple production goes, it remains substantial, though there are now fewer than one-tenth the number of trees there were a century ago, when well over a million and a half were under cultivation in this small state alone.

Just past the biggest orchards, you'll spot a sugarbush, which is not a sweet-stemmed shrub but a stand of maples tapped for sugar production. If you happen

An Upper Valley sugarbush during the sap run.

by here in late winter, you may see buckets hanging from the trees. This is the traditional way of collecting sap. Notice how the trees are well spaced to permit easy access for tappers, horse-drawn sleds, or more modern vehicles—the result of either good planning or good luck. I once had the grim duty of hauling brimming sap buckets up a steep, overgrown, muddy ravine. Occasionally, the odd co-worker would go rolling past. But that was in northern Indiana, in what deserved to be called Satan's Own Sugarbush. The original owners of that parcel were obviously not transplanted New Englanders.

Elsewhere in this region, you may notice the high-tech sap-collection method, which involves plastic milk jugs or tubing that drains a number of trees into a central reservoir. For a more extended discussion of sugaring and its economic significance, and for a visit to a Vermont maple sugar factory, see Tour 45.

———————— ∾ ————————

Continue north on Route 5. After another 2.0 miles, you'll come upon a magnificent view of the valley. In summer, vegetation swarms over every available surface. Among this multitude of plants is a wildflower common in these parts, bouncing Bet. Its dramatic, five-petaled blossoms are white or light pink. Also present is a handsome alien legume, white sweet clover, also known as melilot. It is one of the tallest herbaceous plants you're likely to see; it often reaches a height of 6 feet or more. In Europe, it has seen herbal application as a digestive, and even as an ingredient in "a bath for melancholy."

The next pleasing sight is Westminster. Note the lie of the land as you approach the center of this handsome settlement. You rise to a plateau, which gives the town its expansive setting. This is one of the finest river terraces in the entire Connecticut River drainage system. The early white settlers were wise to choose this relatively flood-free surface for their dwellings and to devote the more exposed lower terraces and floodplain to pasturage and crops. Here and there along Route 5, people who aren't convinced of the river's slumbering power have built homes right at the water's edge. This is much the same as having a picnic on top of Old Faithful: 99 percent of the time, the experience is pretty enjoyable, but when Mother Nature finally cuts loose, you learn that the catastrophic 1 percent was all that really mattered.

Interestingly, the Indians who lived in New England's river valleys for centuries before the advent of the Europeans actually had the same daredevil mentality. They also settled close to the bank and reversed the whites' pattern by planting their crops on the higher ground. But in their case, it made sense: their wigwams were either disposable or easily moved during a flood—something that can't be said about the modern waterfront homes. The Indians' different mode of settlement actually proved to be a boon for the early white colonists. Since the more elevated terraces were usually already cleared of heavy vegetation, thanks to the native practice of controlled burning, the whites did not have to devote much time or energy to that task before they raised their permanent structures.

This is a splendid place for a drive on a lazy summer afternoon, when the corn is high, the sand bars have appeared midchannel in the river, and the cicadas have worked up their unearthly drone. If you have any Midwestern blood in you, even unto the third and fourth generations, you'll find the farm country of Westminster a sort of spiritual ground zero. Here, at least, New England has a sky that is really an expanse.

———————————— ❧ ————————————

After descending from the Westminster terrace, you pass through a pleasant landscape of dairy cows, truck farms, and highly recommended produce stands. The road winds upstream and eventually leads across a newly constructed bridge over Saxtons River. At the stoplight, continue on Route 5 by bearing right. Welcome to Bellows Falls, the part of the town of Rockingham that occupies the site of the largest natural falls on the Connecticut River. The brooding mass of rock on the New Hampshire side is Fall Mountain; it's composed of the same Devonian period Littleton schist that forms mighty Mount Washington, far to the northeast. On September 10, 1856, a solitary tourist by the name of Henry David Thoreau climbed this hill with a heavy valise on his back. It was a detour on his way to visit Bronson Alcott and family, then residing in Walpole. When he reached the summit, he noted the great red oaks growing there, as well as the rocky narrows below. On the way down, however, he fell several times because he was wearing smooth-soled shoes. Fortunately, the only serious injuries were those sustained by the valise.

At the T-intersection in the heart of town, turn right (toward the river) and then park where permitted in the vicinity of the post office. Several points of interest are within easy walking distance. The hydroelectric station is one of them; like the facility at Turners Falls, Massachusetts (see Tour 25), it converts the energy of dammed and diverted river water into electricity. Close by is the narrow Bellows Falls Canal, the first artificial waterway of its kind begun in the United States. When you've had a chance to inspect these, walk down to Vilas Bridge and turn right onto the dirt road that skirts the gorge on its near side. Almost immediately, you'll see the locally famous Indian petroglyphs—carvings—on the rock surface below. These were probably made by the Pennacook people, a branch of the Algonquins, in precolonial times. The bright yellow paint that has been liberally used to highlight the petroglyphs is a later artifact, however, dating from what experts call the Age of Rampant Tourism.

To resume the journey, reclaim your vehicle, cross Vilas Bridge, and say hello to the Granite State. The span you've just used is probably a less impressive affair than the original one that connected Bellows Falls and Walpole. It was an open wooden structure, constructed in 1785 by Colonel Enoch Hale. In fact, it was the very first bridge across the Connecticut at any location.

At the stoplight on the New Hampshire side, turn left onto Route 12 North. As you head through North Walpole, look to your left and notice the bed of the falls, which is dry except at the spring freshet and other times of especially high river levels. Even there, vegetation constantly tries to gain a foothold, though it is mostly kept in check by periodic torrents. On the slopes on Fall Mountain, to your right, the situation is naturally quite different. Besides birches and poplars, you'll see numerous oaks—descendants, no doubt, of the ones Thoreau noted.

Above the town, the river turns into a broad expanse like a lake, the result of the Bellows Falls dam. For the first time on this tour, the conditions are right for plants characteristic of sluggish-water environments. At various locations, pickerel weed, purple loosestrife, and fragrant water lily are visible close to the bank. The water lily's white flower is an intriguing essay on floral primitivism. When carefully dissected, its components prove to be arranged in a rising spiral pattern rather than in a flat ring, as is the case with flowers thought to be more advanced. This spiral arrangement, also found in magnolias and buttercups, strongly suggests the structure of pine cones and other more ancient forms of seed production.

However peaceful this setting may be now, it was once the scene of one of New England's biggest commotions. Each spring, from the mid-1700s to 1915, annual log drives came downstream past this point. The logs were those of north country conifers; being softwoods, they floated readily. Lumber companies took advantage of this fact to turn the Connecticut and its tributaries into the region's greatest conveyor belt. Two of the largest business concerns, International Paper Company and Connecticut Valley Paper Company, effected an arrangement: the I.P.C. drive was held up at the mouth of the Ammonoosuc River, upstream at

Woodsville, while the C.V.P.C. timber came down the main waterway first. The combined complement of logs took from three to six weeks to pass Bellows Falls.

───────────── ❧ ─────────────

By the point 2.9 miles north of Vilas Bridge, you've reached the Great Meadow, part corn fields, part wetland. Notice the farmhouses, some of which have passageways linking the living quarters with barns and other buildings. This is a common New England design that precludes the need for wading through barnyard snow on frigid winter days—even if it does give the house proper all the arresting aromas of the barn. Usually, the structures are oriented in a line, but here the river terrace topography has stipulated that the joining be done sideways.

The wetland scenery is especially good in one marshy zone just up from the closest farmhouses. This spot, a favorite with local fishermen, is also a breeding ground for ducks, some of whom insist on nesting on the wrong side of the road. It isn't all that uncommon to see a mama mallard leading her squad of ducklings across the pavement, totally oblivious to the cars swerving around them. This would be a good place for the kind of Duck Crossing signs posted at the Missisquoi Wildlife Refuge in northwestern Vermont.

The next community, Charlestown, features a classic main street setting with hefty shade trees and spacious lawns. It was from this location in 1777 that New Hampshire's Revolutionary War hero, John Stark, began his army's march against the invasion force of General "Gentleman Johnny" Burgoyne. Eventually, this resulted in the Battle of Bennington. With the help of Seth Warner's Green Mountain Boys, Stark's troops managed to keep that Vermont town from being plundered by the Hessian mercenaries attached to the British. The victory was the prelude to Burgoyne's complete disaster at Saratoga. New Hampshire was the only one of the thirteen original states that did not suffer the direct effects of war within its own borders. But as this episode shows, its citizens were more than willing to join in the fray elsewhere.

Continue on Route 12 through town. About 1.0 mile north, bear right onto the turnoff for Route 11 West. Take Route 11 about 0.5 mile to the left-hand entrance to the Fort at No. 4. Head down the drive and park in the lot. This stop is an excellent demonstration of early settlement architecture and timber uses. It will also acquaint you with one of the earliest chapters of colonial history in the Upper Valley.

In 1740, the area's first white settlement, prosaically named No. 4, was established a mile south of here, in what is now Charlestown center. Its founders were three families from Lunenburg, Massachusetts. Population pressure was already beginning to mount in the English-speaking colonies to the south. Relocation to places as remote and dangerous as this was often the only way that men from large families could hope to have their own land. Still, the risks were of a magnitude we can hardly comprehend today. Susanna Johnson, one of the first to arrive here, wrote of No. 4: "During the first twenty years of its existence as a settled place, until the peace between Great-Britain and France, it suffered all the consternation and ravages of war; not that warfare which civilized nations wage with each other, but the cruel carnage of savages and Frenchmen."

Imagine yourself staking a claim in this densely wooded land when it was without roads or reliable communications. Your enemy is an extremely capable guerrilla fighter with a system of tactics and honor totally alien to your own background. The ethics of his culture, sometimes more laudable than yours, nevertheless condone the use of ambuscade and torture. He can move to your north, east, west, and south with complete freedom; furthermore, he understands this land and its resources much better than you do. You may possess the property and independence you desired, but you've purchased these things at the cost of a seemingly endless siege in the loneliest place imaginable. Once again, Mrs. Johnson:

> A detail of the miseries of a "frontier man," must excite the pity of every child of humanity. The gloominess of the rude forest, the distance from friends and competent defense, and the daily inroads and nocturnal yells of hostile Indians, awaken those keen apprehensions and anxieties which conception can only picture. If the peaceful employment of husbandry is pursued, the loaded musket must stand by his side; if he visits a neighbor, or resorts on Sundays to the sacred house of prayer, the weapons of war must bear him company; at home, the distresses of a wife, and the tears of lisping children often unman the soul that real danger assailed in vain.

However melodramatic her description might sound to us, Mrs. Johnson knew the situation firsthand. Captured by the Indians when she was in the last stage of pregnancy, she gave birth on the second day of a forced march to French Canada. She and other family members were subjected to various privations and adventures, until they were released or ransomed one by one. By the way, her child born en route, Elizabeth Captive Johnson, survived to adulthood.

The reconstruction of the community here is based on a sketch by a British Army officer who visited at the height of the French and Indian troubles. As you enter the grounds through the gift shop, you'll notice that the buildings are constructed of squared timbers. These were made from two native conifer types, pine and hemlock. Keep in mind that this complex wasn't an expertly engineered military base, but merely an improvised fortified settlement. The palisade logs on the perimeter were deliberately spaced about 5 inches apart so that the defenders could see their foes approaching. This meant the barrier's main purpose was to prevent the entry of burning wagons used as incendiary devices. It would slow down attackers on foot, but it wouldn't absolutely prevent their getting in. Due to space limitations, crops had to be planted outside the palisade, and it was in the exposed fields, as Mrs. Johnson implied, that the danger was most extreme.

Be sure you chat with the interpreters here, who will gladly acquaint you with many other fascinating aspects of the fort. By all means climb the stairs to the watchtower, and enjoy the gorgeous view of the river. Also, don't leave without seeing the Indian dugout canoe, formed from a single tree trunk, in one of the exhibit rooms. As you look about from the center ground of the settlement, you can understand the vital role timber products played in colonial life. For

The Cheshire toll bridge, seen from Hoyt Landing.

good reason, it wasn't long before the residents here erected a sawmill. When it was completed, they held a celebratory dance on the first run of newly cut boards.

———————— ∾ ————————

As you exit the No. 4 parking lot, turn left onto Route 11 West for one last bit of Upper Valley history, the venerable Cheshire toll bridge. This is a lonely survivor of an age when many river crossings in northern New England were privately owned. Return to the Vermont side after paying the toll taker the requisite two bits. This steel span, built in 1930, is the fourth in this location; the first two were wooden. A tough survivor of the awesome 1938 hurricane, the present bridge takes its name from the fact that Charlestown used to be in Cheshire County, before a readjustment of the boundaries put it in Sullivan County.

The Vermont riverbank belongs to the town of Springfield, a famous center of the machine tool industry. Complete the tour by turning left immediately at the far end of the bridge, into the Hoyt Landing parking area. This is where the Connecticut absorbs the Black River, a tributary that rises from lakes and brooks high in the Green Mountains. In this setting, you can better acquaint yourself with the bottomland trees of the valley: cottonwood, black willow, American elm, silver maple, and its close relative, box elder. In the wetland facing the mouth of the Black River, common cattail and pickerel weed thrive in the slack water. If you look back toward the bridge at just the right moment, you may see a kingfisher darting over the rippling surface for prey. With the possible exception of a passing motorboat, the river and valley these days are an image of placid contentment. Not long before her death, our heroine Mrs. Johnson noted the remarkable change that had come over the region in her later years. It is largely the same setting today: "The gloomy wilderness that fifty years ago secreted the Indian and the beast of prey, has vanished away; and the thrifty farm smiles in its stead. . . . The tomahawk and scalping knife have given place to the plough-share and sickle, where the terrors of death once chilled us with fear."

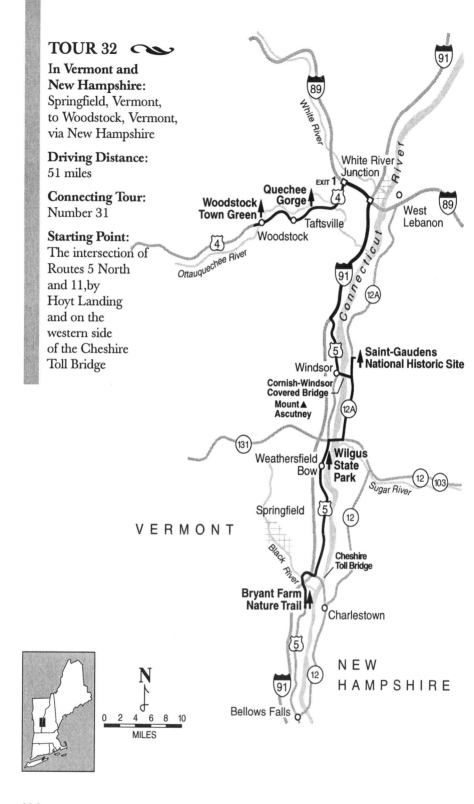

TOUR 32 ∽

**In Vermont and
New Hampshire:**
Springfield, Vermont,
to Woodstock, Vermont,
via New Hampshire

Driving Distance:
51 miles

Connecting Tour:
Number 31

Starting Point:
The intersection of
Routes 5 North
and 11, by
Hoyt Landing
and on the
western side
of the Cheshire
Toll Bridge

Woodstock
Town Green

Quechee
Gorge

White River
Junction

EXIT 1

West
Lebanon

Taftsville

Woodstock

Ottauquechee River

White River

Connecticut River

Saint-Gaudens
National Historic Site

Windsor

Cornish-Windsor
Covered Bridge

Mount ▲
Ascutney

Weathersfield
Bow

Wilgus
State
Park

Sugar River

Springfield

VERMONT

Black River

Cheshire
Toll Bridge

Bryant Farm
Nature Trail

Charlestown

N E W

H A M P S H I R E

N

0 2 4 6 8 10
MILES

Bellows Falls

Tour 32
The Upper Valley–Part II

I f you've taken the preceding tour, you are already well aware that the Connecticut River's Upper Valley is a landscape combining many aspects of botany, horticulture, agriculture, geology, architecture, and history in a context of overtly beautiful scenery. This tour features the same appealing terrain and subject matter, but its individual points of interest are quite distinct. This tour can be enjoyed either as a continuation of the previous one, or as a wholly separate itinerary.

The first stop of the journey is Springfield's Bryant Farm Nature Trail, situated just across the mouth of the Black River from Hoyt Landing. From the starting point, proceed west on Route 11 for 0.8 mile, then turn left onto Route 5 South. The parking area for the Bryant Farm Trail is located on the east side of Route 5, just 0.4 mile down from the last turn.

The trail offers a short hike through the lush world of the river's bottomland vegetation. Take the gravel path at the northern end of the parking pullout and proceed downslope. Not far from the beginning, you should be able to spot hackberry, a tree with simple, toothed leaves that have an asymmetrical outline. It's growing in the thicket to your left. Hackberry belongs to the elm family; while it never seems to be a dominant species anywhere, it is well adapted to a wide range of conditions, from damp woods to dry, rocky uplands. The hackberry genus, *Celtis,* also includes the sugarberry tree of the Deep South and a thorny desert shrub of the Southwest.

Another tree of interest along the way is the box elder. It's the umpteenth example of the inaccuracy of common plant names. It is not an elder–it's a maple,

albeit the most unusual one native to North America. Defying the usual maple pattern of simple, lobed foliage, the box elder has a compound pattern with three, five, or even more leaflets. The native range of the species is gigantic—almost all of the United States and part of Canada—so it makes even the far-flung red maple seem a local yokel in comparison. This is a characteristic site for box elder; it is almost always found slouching about streambanks. And speaking of streambanks, linger a moment or two at the edge of the torpid Black River. This it is not much more than a stone's throw from the place where it loses its own identity in the larger being of the Connecticut.

Staghorn sumac forms a magnificent monotypic stand nearby, demonstrating how the species spreads rapidly by suckering. In addition, you'll see American elm, poplar, Virginia creeper, cottonwood, and two trees adapted to the deep flooding of big rivers, black willow and silver maple. See if you can track down an old spreading plum tree, perhaps the last survivor of a farmhouse garden or orchard.

Wildflower fanciers are in for a treat. At various times in the summer, you can find motherwort, jewelweed, spotted joe-pye-weed, goldenrod species, blue-eyed grass, Virginia knotweed, bird's-foot trefoil, and another alien legume, red clover, which just happens to be Vermont's state flower. If you locate motherwort, examine its sharp-lobed leaves and feel its stem. Is it square in cross section?

This is one of the most common characters, or identification traits, of the mint family. The motherwort plant has been a favorite herbal cure for nervousness, and a few centuries ago it was even thought effective against evil spirits.

Ferns are practically everywhere you step. As always, bracken is in the forefront of the action, and in its typical mean-spirited way, it tries to obliterate every patch of open ground, especially the official trail, which it obviously resents a great deal. Ages will come and ages will go, but bracken will always be with us; probably, we should just be thankful it hasn't taken control of the state government. Two other common species, sensitive fern and interrupted fern, grow a little less rampantly in the moist spots. (For a discussion of the odd frond structure of the latter species, see the end of Tour 38.)

When you've completed the trail circuit and are ready to press on, retrace your route on 5 North and 11 East back to the vicinity of Hoyt Landing. While you're on Route 11, be careful not to inadvertently end up on Interstate 91. Just before the Cheshire toll bridge, turn left; this is the continuation of 5 North. You are now in some of the loveliest terrain in the Upper Valley. On the left, woods composed of basswood, white ash, American elm, red oak, and Canadian hemlock have taken over the eastern face of Skitchewaug Mountain. Wherever the slopes are less extreme, though, the land has been preserved for agricultural pursuits. For example, on the broad floodplain 2.0 miles north of the bridge, a Christmas tree farm comes into view, with row after row of young conifers.

To the innocent among us, this might seem an easy way to make a fast buck—the horticultural equivalent of chinchilla ranching. As with any commercial

The demise of a graceful giant. An American elm (center) *in Claremont, New Hampshire, shows the first sign of Dutch elm disease infection, which will swiftly kill it. The introduction of Old World plant pests and diseases is one unintended legacy of the European settlement of North America.*

operation involving huge numbers of plants, however, nature poses plenty of problems. The vigor and attractiveness of baby Christmas trees can be threatened by grasses and other competing plants. In addition, if the ground between the rows of trees isn't mowed regularly, mice and other rodents move in and nibble the tender branches and bark. One other potential pest is arguably New England's most beautiful bird, the evening grosbeak. This chunky black, gold, and white finch has plenty of visual appeal, but it's also an avian Hell's Angel. It devotes almost every waking moment to its two favorite pursuits, arguing violently with its relatives and trashing its surroundings. Since it travels in large flocks, the mayhem can reach monumental proportions; and this can spell substantial trouble for the Christmas tree farmer. In the eye of an evening grosbeak, there's no better place to spend a nice, long, contentious winter than in a grove of dense evergreens, which is its perfect source of food and shelter.

North of this spot the road continues through rolling farmland. If you pass by at the right time of day, the sunlight glitters on the river's surface wonderfully. Notice the shade trees sheltering some of the farmhouses. As their planters no doubt knew, the trees are not just ornamental additions to the yard. In a sense, they are also heat regulation devices that moderate the temperature extremes the house must endure. In a few moments more, you enter the small settlement of Weathersfield Bow. It's difficult to tell nowadays that this was once a center for one of the country's great agricultural movements.

In 1810, the American consul in Lisbon, a Vermonter by the name of William Jarvis, managed to ship some four thousand prize Merino sheep out of Spain for sale in the United States. In the process, he sent a substantial flock to his own farm here in Weathersfield Bow. This feat of entrepreneurship was more significant than it might appear. Apparently, the Merino breed was first developed by the Moors of the Atlas Mountain region; later, Merinos were introduced into Spain and widely bred there for their excellent wool, which compared favorably to the finest English broadcloths. Ownership of the Merinos was expressly reserved for the Spanish nobility, and their export was forbidden under penalty of death. In the chaos of the Napoleonic Wars, however, both the Spanish crown and the French invaders began selling them to foreigners, including Jarvis and other interested Americans.

In terms of domestic animals, the goat is unquestionably the supreme ecological terrorist. But the sheep, with its cleft lip and pronounced lack of concern for environmental issues, is no sluggard either and has caused the destruction of many preexisting habitats. In the late 1400s, for example, a sheep boom in England severely depleted the nation's crucial woodlands. The Merino in particular had a devastating record, which included the ravaging of large tracts of vegetation in the Iberian peninsula and the Spanish holdings in Italy. Farmers were not even allowed to protect their crops or woodlots from passing flocks; if they did, they could be executed. In many places, countless trees were killed, and precious topsoil blew away after severe overgrazing.

No such disaster occurred in this region, one reason being that the sheep-owning craze peaked early. The Merino population hit its ceiling in the decades preceding the midpoint of the nineteenth century, continued to be an important regional economic force until about 1870, and then declined steadily. The causes of the collapse were several, including western competition, local economic depression, and tariff restrictions. The absence of a clear-cut environmental disaster notwithstanding, Merino mania contributed to the long-held perception that continued destruction of first-growth forests was justified. In traditional New England agriculture, the acreage devoted to domesticated animals was substantially greater than that used for crops. When many Yankees moved west after the opening of the Erie Canal, they often did so because they were running out of room for their livestock, not because they were getting hemmed in by other people. The role played by Merinos in the sustained clearing for pasturage was considerable. In 1840, their tally stood at two and one-quarter *million* in Vermont and New Hampshire alone, and in some towns, the sheep outnumbered Congregationalists by a factor of more than four to one.

At the north end of town are some large clumps of staghorn sumac, the same species you saw along the Bryant Farm Trail. This cashew-family plant takes its common name from its branching pattern, which resembles antlers. The wood has long been esteemed as kindling, and the plant was also used by early English settlers for dyeing and tanning. In addition, it could be dubbed the sugar maple of the shrubs, because it turns a stunning flame-orange in the fall. Farther north, about 2.0 miles from town, is a pleasant though intermittent waterfall on the left side of the road. As you might surmise, the best time to see it is during the spring freshet. An additional 0.9 mile brings you to the right-hand entrance to Wilgus State Park; turn in here. This small preserve, named for the Vermont engineer who designed New York City's Grand Central Station, is home to white pine, Canadian hemlock, American beech, red oak, and paper birch. Some human disrespect for the paper birch has resulted in the posting of a DO NOT PEEL THE BIRCHES sign.

Take a stroll to the water's edge. Somewhere in this vicinity in the year 1814, a Massachusetts farmer and contractor named Asa Sheldon forded the river with his horse, four yoke of oxen, and a load of grapeshot destined for Commodore Macdonough's forces at Vergennes, Vermont. Macdonough subsequently defeated a British squadron at Lake Champlain, and so became one of the U.S. Navy's War of 1812 heroes. Sheldon was not a man given to outbursts of poetic sentiment, but when he recounted his life half a century later, he clearly recalled the beauty of the scenery here. What impressed him most was the Connecticut's straight course and unvarying width. Had he ventured a little farther to the south, he would have seen the great stream make a sharp dogleg at Weathersfield Bow. (Incidentally, Sheldon's recently reprinted autobiography, *Yankee Drover,* is a good introduction to the no-nonsense, pennywise mentality of the preindustrial New Englander whom Thoreau knew and lambasted so well. It also shows how an enterprising person of that period had the mettle and ability to pursue more than one calling in a lifetime.)

Depart Wilgus State Park by turning right onto Route 5 North. At the stoplight 1.2 miles farther up, bear right again, onto Route 12 South, and cross into New Hampshire. As you travel over the bridge, take a guess: where does the boundary between the two states actually lie? It isn't down the middle of the river, as you'd expect, but at the mean low-water mark on the Vermont side. Virtually the whole waterway belongs to the Granite State. There is a persistent rumor, anecdotal but eminently plausible, that the high-minded Vermonters selflessly gave up their share of the river so that New Hampshire taxpayers could experience the joys of bridge repair and maintenance without distractions from outsiders. The boundary was a matter of dispute for a long time; the U.S. Supreme Court decided the issue in 1933.

As you reach the eastern bank, the tributary to your right is the Sugar River, which has its source less than 15 miles away, in gorgeous Sunapee Lake. Note the agricultural use of the floodplain; the only trees allowed to grow to maturity in this precious soil are the black locusts and maples that line the water's edge. At the first intersection, turn left onto Route 12A North. Directly across the way in Vermont is Mount Ascutney, which stands like the hub of the earth, dominating the landscape for miles upstream and down. If you glance at a topographic map of this area, you'll note an interesting phenomenon. The other hills in the region have a predominantly north-south trend, but Ascutney is almost perfectly circular in shape, as seen from above. In fact, it is another of New England's monadnocks, or detached high points that stand above a lower erosion surface. Ascutney is also significant in that it is a huge mass of rock known as syenite, which began some 200 million years ago as a concentration of upwelling magma. This and other disturbances in the region were a sideshow compared to the breakup of the supercontinent Pangaea farther to the east. As ancient as the mountain's rock is, it's only about half as old as the metamorphic formations in this area.

An 1850 rendering by the artist Nicolino Calyo, painted during the Merino boom times, shows the terrain around the mountain much more open than it appears today. Even part of Ascutney's upper slope appears to be cleared of trees. In modern times, agricultural activity is much less obvious, except here on the riverbank, where there is still a mixture of tillage, dairying, and nurseries. At 2.8 miles north of the last intersection, you'll come to the birthplace of Salmon P. Chase, the New Hampshire native who emigrated to Ohio and there became an anti-slavery advocate, a U.S. senator, and state governor before serving as Lincoln's secretary of the treasury. His distinguished career culminated with his appointment as Chief Justice of the Supreme Court. Today, his name is kept alive by a U.S. Coast Guard cutter and by one of the nation's largest banks.

After snaking through the Mill Brook ravine, the road passes the newly renovated Cornish-Windsor covered bridge. You'll return a little later to cross it, but continue on 12A for an additional 1.5 miles until you see the right-hand turnoff for Saint-Gaudens National Historic Site. Turn here and proceed 0.6 mile to the parking area. Notice the makeup of the forest along the access road: among the hardwoods are numerous examples of New England's two most prevalent softwoods, Canadian hemlock and white pine, doing just fine in their native state. The main attraction, however, is Aspet, the home of the great Irish-born American sculptor Augustus Saint-Gaudens, who lived here from 1885 to 1907. Saint-Gaudens was one of those especially lucky and talented souls whose efforts were widely acclaimed both during and after his lifetime. The Yankee composer Charles Ives based the first movement of his orchestral *Three Places in New England* on one of the artist's best-known works. The estate grounds contain much of horticultural interest and also offer an excellent view of Ascutney, if the mountain isn't hidden by clouds or by the frequent river fog.

At the entrance facing the parking lot, there is an ornamental barrier composed of the same two conifer species we saw in the woods a moment ago. It is quite a revelation to see them in this formal European guise, so severely clipped and hedged into geometry. To the diehard naturalist, the effect is much the same as seeing a timber wolf after it's been to the poodle parlor. However, no growth restrictions have been imposed on the massive honey locust that rises like a giant footman before the house entrance. This plant is a legume, in the same order with the diminutive clovers and trefoils of the roadside. Though its greenish flowers are much less distinctive, they too produce fruit in the form of seed pods. According to the park rangers here, the tree was planted during Saint-Gaudens' residency, so it must be near or even a bit beyond its one-hundredth birthday.

Another noteworthy sight for the tree lover lies behind the house and past the grape arbors, where ornamental birches were planted to line the paths. The

Ornamental birches define a path at the Saint-Gaudens National Historic Site.

The Cornish-Windsor Bridge. Mount Ascutney stands in the background.

effect is splendid. Also, a garden area features old-time herbaceous favorites that blend beautifully with the statuary. Note the annual bedding varieties, the biennial hollyhocks, and the selection of perennials—hostas, day lilies, garden peonies, and various types of phlox. One other perennial, deserving special attention, is a wonderful plant in the saxifrage family, known as astilbe. This plant is a dependable performer in partial shade. It comes in several good cultivars, thanks in large part to the work of a German hybridizer, Georg Arends. These have such Teutonically appropriate names as Deutschland and Ostfriesland; they vary in flower color from pure white to pink to stunning red. Japanese beetles can be a problem, but those pests are so catholic in their appetite that it's hardly a reason to resist using this otherwise terrific selection.

If you're in the mood for a hike, you might wish to take the adjoining Blow-Me-Down nature trail. When you're ready to leave Aspet, depart by the road you came in on, return to Route 12A, and head south. When you reach the Cornish-Windsor Bridge, take a right onto it and enjoy your transit of the longest surviving covered bridge in the United States. If you happen to be doing this tour in your four-in-hand or atop your favorite sorrel, watch out. As the sign over the entrance proclaims, WALK YOUR HORSES OR PAY TWO DOLLARS FINE. This impressive structure, a toll crossing until 1943, was originally raised in 1866 and consists of two spans of the Town Lattice design. Together these measure 460 feet. If you're an architecture or civil engineering aficionado, take a look inside at how the lattice is constructed. The crisscross pattern is formed not with the customary planking but with full squared timbers. This gives the interior an especially sturdy demeanor. (For more on covered bridge design, see Tour 30.)

On the Vermont side, where the folks refer to the Windsor-Cornish Bridge instead of the Cornish-Windsor Bridge, you enter the overtly historic town known to all good state officials and tour bus guides as the Birthplace of Vermont. Continue straight through the railroad overpass; at the stoplight intersection turn right, onto Route 5 North. This is the main street in Windsor. The first point of interest appears almost at once on your left—the Old South Church, designed by the great New England architect Asher Benjamin. Another 0.4 mile north, past the next stoplight, stands the Constitution House, where the Republic of Vermont came into being in 1777. Beyond this point, black locust trees are common and black willows fringe the drainage ditches. This native species can be distinguished from the common ornamental weeping willow by its less mop-headed habit. Additionally, on wet days before leaf break, the younger branches of the black willow are a basic yellow, while weeping willow stems are a golden brown. The willow genus as a whole is widely distributed around the globe, especially in the Northern Hemisphere. In England, it has served as the subject of one of the loveliest Elizabethan lute songs and is considered evocative of gentle streams and the pleasantries of the countryside.

At a point 2.1 miles north of the Constitution House, you come upon a striking hillside topped by white pine, a tree that requires full sunlight and well-drained soil. This area has sediments ranging from glacial lake-bottom clays to gravels characteristic of other Ice Age features called kame terraces. An abundance of sand and gravel has encouraged a good deal of quarrying, as is evident on this stretch of road. Just past the hill, Route 5 meets Interstate 91. Here bear right onto I-91 North.

As usual, the edge of the interstate displays more wild vegetation and fewer ornamental plant species than along the side roads. Familiar wildflowers appear in the summer: Queen Anne's lace, bird's-foot trefoil, common mullein, common St. Johnswort, and common milkweed. Central hardwoods that wouldn't be prevalent in the uplands to the east and west are doing fine here, in this gentler microclimate. You should spot red oak, for instance, mixed in with black locust and the red and white pines. If you're lucky, you may even catch sight of a pitch pine. Poplar or quaking aspen is also frequently seen, and its close relative, the bigtooth poplar, makes a showing here and there. In early spring, the latter tree is easy to find because its emerging leaves are covered with a distinctive silvery fuzz.

As you pass over the Ottauquechee River 5.8 miles north of your entry onto the highway, the Hartland Dam is momentarily visible high up on the left. Typical of this moisture-rich region, the vegetation gets a toehold even in the crevices of the rock outcrops. See if you can identify the woody pioneer plants precariously clinging to life in the road-cuts. Among their number are poplar, gray birch, and staghorn sumac. Another plant community thrives across from the rest area at 2.2 miles above the Ottauquechee. In a swale here, common cattail, a native monocot, and purple loosestrife, a dicot invader, compete in the miniature wetland setting.

Another 1.0 mile beyond this, and a mere 0.3 mile before you bear right onto Exit 10N for Interstate 89 North, the outcrops themselves put on a good show, featuring maroons, greens, yellows, and oranges produced by weathering. The massive units exposed here are part of the Bronson Hill zone, a distinct strip of metamorphosed volcanic rock that runs from southern Connecticut up to westernmost Maine. The Bronson Hill rocks appear to have begun their long career as an arc of islands lying a considerable distance off the coast of an older version of North America. When another great land mass approached from the east, this string of islands collided with our continent, causing the Taconian mountain-building event that produced the ancestral Green Mountains. Millions of years later, the eastern land mass—perhaps Africa, Europe, or parts thereof— came plowing in behind, with the Bronson Hill area sandwiched smack dab in the middle. This produced another episode of mountain building, the Acadian. In this geologic period, the Devonian, plants were still relative newcomers on the earth's surface, but they were developing and diversifying rapidly. If you could take a time machine back the necessary 380 million years, you might discover the earliest trees, as well as the humble forerunners of the ferns, horsetails, and club mosses of today's New England forests. If you insisted on going back further, to the time almost half a billion years ago when the volcanic islands first formed, your tour might reveal the earliest plants of all, ground-hugging types reminiscent of modern liverworts.

Once you're through the White River Junction interchange and onto I-89, say farewell to our long-time companion, the lovely Connecticut. Take a quick look at the White River valley and keep an eye out for Exit 1, some 3.1 miles down the road. Take that turnoff to Route 4. At the bottom of the ramp, bear left onto 4 West, toward your next stop, the Ottauquechee River's most scenic locale. Within 3.6 miles, you've arrived at Quechee Gorge State Park. Drive across the bridge and park just past it, at the picnic area on the right. Before inspecting the gorge itself, take a quick foray into the woods behind the clearing, where familiar conifer species can be examined at your leisure. First, check out the tall, narrow ornamental evergreens by the park sign. Their foliage, deployed in flay sprays that look as though they've been pressed with an iron, indicates these trees are what is sometimes called northern white cedar. Their other name, arbor vitae, is preferable, since the species is not a true cedar. Besides, how could any person of taste and erudition resist using the Latin tag when it means "the tree of life"? A classicist would pronounce it "*are*-bor *wee*-tie"; American botanists, who mangle Latin terms as a point of honor, say "*are*-bur *vie*-tee." The arbor vitae is widely employed in landscaping; to see it growing in the wild, it's best to travel to damp areas of the Green Mountains, the Northeast Kingdom, or Maine.

As you head into the forest proper, try to distinguish the wild conifers around you. Red pines have branches that curve up at the tips, giving the trees a shape somewhat resembling a candelabra. In addition, red pines have stout needles that are roughly 4 or 5 inches long and attached to the stem in fascicles, or bundles, of two. The bark often has a characteristic pinkish cast. This species grows to a respectable size, but is dwarfed by the Northeast's tallest tree, the white pine. White pine trunks like pillars rise high above the woodland floor. When deprived of sunlight, the lower branches die and fall away—a demonstration of unaesthetic

Quechee Gorge from the Route 4 bridge.

but effective natural pruning. The white pine has delicate needles about 3 inches long, attached five to a fascicle.

If you happen to visit this site in late April, you may also find a leafless shrub with fragrant, rose-pink, four-petaled flowers. This is daphne, an escaped European introduction. Also, spend a little time looking at the smaller trees in the understory. Significantly, they're not all baby pines, but rather Canadian hemlocks (dark green, with the classic Christmas-tree habit) and American beeches (broad-leaved with smooth, gunmetal-gray bark). As noted above, white pines need full sunlight to grow. Once the first generation has formed and the floor is relegated to shade, other plants more tolerant of low light levels sprout and await the chance to shoot up when a big pine falls. In this way the population of the forest gradually changes.

Now return to the road and take the bridge walkway for a magnificent view of Quechee Gorge. This narrow spillway was formed by a dramatic geologic event at the end of the Ice Age. As melting glaciers left the area, they deposited debris that blocked the flow of the Ottauquechee. As the river backed up, a sizable lake formed to the north. When the rubble dam finally gave way, the rushing waters scoured out this long channel in the tough metamorphic rock.

Amazingly, the steep walls of the gorge are being colonized by young hemlocks, which, despite their tenuous hold, seem to enjoy the moist air above the rapids. When you peer down to the bottom of the gorge, keep in mind that it's roughly 160 feet deep. That's quite a distance, of course, but imagine a tree that grows half again that high—say, from the riverbed far below to a point 90 feet over your head, as you stand on the bridge. Reliable sources, including geographer and former Yale president Timothy Dwight, reported that there were first-growth white pines that big, at an incredible 250 feet. In fact, there was even one behemoth in nearby Hanover, New Hampshire, that had been measured at 270 feet. Unfortunately, none of these awe-inspiring plants survive. According to the American Forestry Association, the tallest known modern white pine is 201 feet. To see it, alas, you'll have to travel to Marquette, Minnesota.

———————————— ❧ ————————————

After viewing the gorge, continue westward on Route 4 as it follows the Ottauquechee upstream. The town of Taftsville, less than 4 miles away, features a smart covered bridge of queenpost-and-arch design. It dates from 1836, which makes it a trifle older than everybody's favorite span over the Ottauquechee at Woodstock. That paragon of Americana, truly a handsome structure in its own right, was built in the Good Old Days when Americans were first going extra-vehicular on the lunar surface—in A.D. 1969.

A common lawn tree in Taftsville and other Vermont villages is the white poplar, also known, to fans of *New York Times* crosswords at least, as abele. Indigenous to Eurasia, it was the subject of a Greek myth. During a visit to the underworld, it was said, the hero Heracles made himself a crown of white poplar branches. The upper surfaces of the leaves remained an appropriately somber dark green, but the undersides turned a heavenly silver where they touched the demigod's damp brow. And so they are today; when a breeze stirs the foliage, the tree shimmers in a sparkling contrast of light and darkness. Unfortunately, many horticulturists these days consider this species a poor selection for yards because of its aggressive, pipe-seeking roots, and also because it naturalizes and gets weedy much too easily. As the nineteenth-century Vermonter George Perkins Marsh noted, "Wherever man has transported a plant from its native habitat to a new soil, he has introduced a new geographical force to act upon it, and this generally at the expense of some indigenous growth which the foreign vegetable has supplanted."

———————————— ❧ ————————————

Past Taftsville, the beautifully terraced valley of the Ottauquechee takes on an otherworldly look. Its low, sculpted hills make it difficult to resist thinking you've slipped into J. R. R. Tolkien's Middle Earth, with condos substituting quite aptly for hobbit holes. Some of the land is still cleared for pasturage, despite the siren call of real estate speculators.

308

Soon, you enter the downtown section of Woodstock. In the nineteenth century, this center of tourism was renowned for its spas—in the twentieth, for its ski slopes and upscale shops. Through it all, it has been a center of wealth and culture. One of its most famous natives, George Perkins Marsh (1801–1882), was the son of a congressman who was also a formidable Woodstock patrician. Despite bouts of serious eye trouble and nagging financial problems, George showed aptitude in such diverse fields as linguistics, the natural sciences, and politics. In 1864, he published the groundbreaking book *Man and Nature,* which described the human disruption of ecosystems and ultimately earned him the title of America's first widely influential environmental advocate. Here is Marsh's view of the future: "The earth is fast becoming an unfit home for its noblest inhabitant, and another era of equal human crime and human improvidence . . . would reduce it to such a condition of impoverished productiveness, of shattered surface, of climatic excess, as to threaten the depravation, barbarism, and perhaps even extinction of the species." Compare that resounding style to cute modern eco-speak as ladled out in television documentaries. In Marsh's day, nature had a Shakespearean tinge; now it's a subset of the family-entertainment industry.

The journey ends at the green, on Route 4 just west of the main shopping district and across from the imposing Woodstock Inn. Find a metered parking place in the green's vicinity. This is a delightful example of traditional New England town-commons design. The public parkland is ringed by august but modestly proportioned houses, most of which are excellent examples of the Federal style. The success of this open space is particularly evident on warm Indian summer days, when the local businesspeople pause to spend their lunch breaks amid the falling leaves.

The trees planted here include honey locusts, littleleaf lindens, and some good old New England stalwarts. Red oaks can be identified by their familiar, lobed leaves and vertical ski-track markings on the bark; red maples have three-lobed foliage, which turns scarlet or burgundy in the fall; and sugar maples have sap buckets, at least in March. For those seeking a lesson in Yankee frugality, travel no further. In Vermont, even the public trees get tapped.

If time and inclination permit, you may want to visit the Billings Farm and Museum, located a little more than half a mile up Route 12 from the main intersection. The farm features various fascinating displays of old-time agriculture.

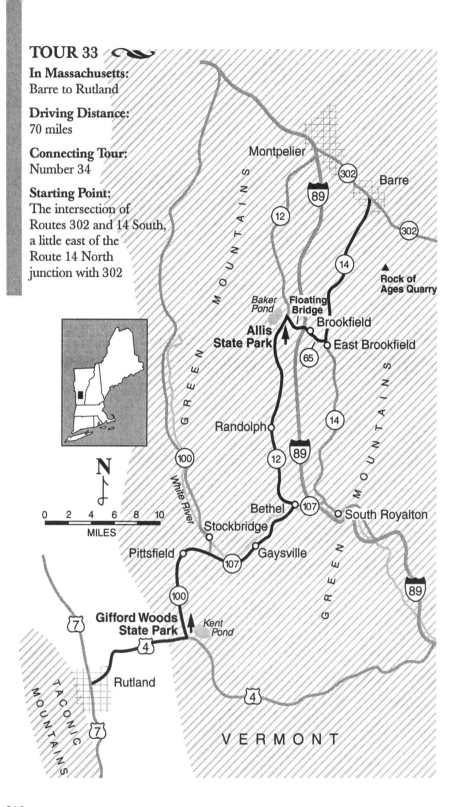

TOUR 33 ❧

In Massachusetts:
Barre to Rutland

Driving Distance:
70 miles

Connecting Tour:
Number 34

Starting Point:
The intersection of
Routes 302 and 14 South,
a little east of the
Route 14 North
junction with 302

Montpelier

302

Barre

89

302

14

▲
Rock of
Ages Quarry

Baker
Pond

Floating
Bridge

**Allis
State Park**

Brookfield

East Brookfield

65

Randolph

14

12   89

N

0 2 4 6 8 10
MILES

White River

Bethel   107   South Royalton

Stockbridge

Pittsfield   Gaysville

107

100

89

**Gifford Woods
State Park**   Kent
Pond

7   4

Rutland

4

7

G R E E N   M O U N T A I N S

T A C O N I C
M O U N T A I N S

V E R M O N T

Tour 33
The Green Mountain Leaf-Peeper Special

A Reccommended Fall Foliage Tour
HIGHLIGHTS
Barre and Its Quarry ❧ The Official Green Mountain Look ❧ A
Fall-Color Conifer ❧ At the Divide ❧ Dirt Roads and a Floating Bridge ❧
Hardwood-Taiga Transition ❧ Allis State Park ❧ Bryophytes and the Alternation
of Generations ❧ A Rare Fern and a Showplace of Wildflowers ❧ Brookies
and Bicyclists ❧ Why Leaves Turn Color ❧ Gifford Woods State Park ❧
First-Growth Trees and Where They Went ❧ Norway Spruce Offers
a Great Deal ❧ The Western Gateway ❧ Cars, not Cahs

*T*here is a good assortment of recommended fall foliage tours in this book, but if you're interested in seeing New England's autumn color show at its fullest glory, this tour, with its rustic scenery and maple-clad hillsides, is very likely your best bet. And there is plenty to see along this route at other times of the year, too. In spring and summer, the wildflower display is at its peak; and on the right day in winter, the snowy landscape and the intricate architecture of the bare trees definitely make the journey worthwhile. There are two things to keep in mind, however; in the cold season, access to the two state parks in this section is quite limited. In addition, the unpaved leg described below may be impassable in the ice and mud seasons.

We begin in Barre, the pleasant urban center southeast of Montpelier that is blessed with an ethnic diversity more suggestive of Hartford or Boston than of northern Vermont. Its name, botanically speaking, is pronounced something like the kind of fruit that has a thin skin, a fleshy interior, and usually more than one seed. From this starting point, go south on Route 14. Barre is renowned as a center of granite production; if you're at all interested in architecture, geology, or large holes in the ground, take a short detour to the famous Rock of Ages Quarry nearby. To do so, turn left at the intersection 1.8 miles south of the starting point and follow the signs. There is an indoor exhibit as well as a protected walkway to the edge of one pit. The granite here is a beautiful igneous rock type, white or pink with a uniform pattern of dark mica flecks. Unlike the famous sedimentary brownstone of southern New England's Central Lowland, this is tough stuff, which polishes admirably and lasts virtually forever. So it must be said that the people who named this operation were practicing truth-in-advertising.

Continuing from the main intersection mentioned above, proceed south on Route 14 past residential landscapes largely given over to Norway spruce and giant garage-mounted butterflies. In another 3.6 miles, you enter Williamstown, where there is a nice view that could serve as the Official Green Mountain Look. It is a miniature and almost self-contained universe, a chaste town nestled in a valley framed by low ridges. Compare this to the great trans-American clonescapes of endless fast-food joints and car dealerships. Here, the vegetation and underlying geology, rather than the bulldozer blade, remain the prime determinants of the terrain. With this idyllic setting, however, goes a sense of inescapable remoteness, even if the nearest drive-through restaurant is only fourteen minutes away. The all-encircling intimacy of the hills can be murderous to a transplanted flatlander, but in the course of generations, it has bred a race of people who, despite their feuds and squabbles, have transformed the necessity of local cooperation into an utter work of art.

Beyond the town, the wild woodland intergrades again and again with tended fields. Tamarack, the elegant deciduous conifer of wetter soils, shares low ground by Stevens Brook with common cattail, speckled alder, American elm, and spotted joe-pye-weed, to name a few of the species. Tamarack is the only native member of the pine family capable of putting on its own fall color show. Its needles turn an intense buttery yellow before they drop at the onset of the dormancy period.

On your right, approximately 3.2 miles south of Williamstown center, lies Cutter Pond and its associated wetland. Canadian hemlock dot the hillsides, along with basswood and yellow birch. Beyond this point, the road does something very significant. It crosses the drainage divide between the Winooski River to the north and the White River to the south. This is of more than academic interest because the Winooski flows westward to Lake Champlain, which in turn feeds the mighty St. Lawrence. The White, on the other hand, is a major tributary of the Connecticut River, a system that for all its beauty and geographical significance can't compare in scope with the great Canadian waterway. After all, the St. Lawrence also drains that great planetary reservoir of fresh water, the Great Lakes.

In places such as this, the continued erosion of a little land may one day result in a large-scale reshaping of the drainage pattern. According to one long-standing geological theory, headwater streams on different sides of a divide sometimes come so close that one "captures" the other, so to speak, and causes a reversal of flow on the captured side. Should you wish to stake a bet on which stream at this divide is more likely to end up stealing the other, it would be wise for you to put your money on the one with the higher gradient, the steeper slope. But you might have to wait tens of thousands of years for the results.

The valley south of the divide is that of the Second Branch of the White River. At 3.8 miles south of Cutter Pond, the valley floor is wide enough to accommodate corn fields and dairy pastures replete with holsteins, the type of cows that seem to be props from an old black-and-white movie. At the junction

0.6 mile farther along, turn right onto Route 65 West. This road is an excellent lesson in why most transportation and other cultural patterns in Vermont are oriented north-south. This sort of across-the-grain travel, with its repeated ups and downs, was not easily accomplished before the invention of railroads and the internal combustion engine. Notice how the land is cleared except at the ridgeline, where a narrow strip of trees sticks up like a Mohawk haircut.

At Brookfield town center, the pavement ends. In places, the dirt road, however well maintained, can be very washboardy. Keep in mind that in some instances a gravel surface provides even less braking control than icy or slushy pavement. In dry conditions, passenger cars have no trouble on this stretch, but do take it easy. If the road surface is wet and you don't have four-wheel drive, watch out. Vermont mud is a friendly substance and would gladly hug your vehicle up to its axles.

The chief attraction of Brookfield is its famous floating bridge, the latest in a long line of structures spanning Sunset Lake. The first was built in 1820. After you cross this current version, look for stands of arbor vitae. Misleadingly called northern white cedars, these chunky conifers prefer cool, moist places. In the same neighborhood, other trees indicate that this is a transition zone between the northern hardwoods (for example, sugar maple) and the montane version of the taiga regime (for example, red spruce and mountain ash). As you approach the 3.9-mile mark from your turn onto Route 65, look for the access road to Allis State Park, on the left. Take this uphill, and soon you'll be on pavement again, at least temporarily.

Allis State Park is one of the gems in the state's park system. When you've reached the parking lot, walk up to the summit of Bear Hill, and if you're not feeling vertiginous, climb up the fire tower, where the view may render you speechless, though not permanently. In the distance, Vermont's high mountains stand out: to the north, Jay Peak, almost on the Quebec border; to the left, Mount

Brookfield's floating bridge.

The observation tower at Allis State Park.

Mansfield, the state's loftiest mountain and the subject of Tour 37; and to the south, Mount Ascutney, the lord of the Connecticut River's Upper Valley. Far away in the east, New Hampshire's White Mountains are often visible. Below you, by the tower, are red spruce trees. These can be distinguished from balsam firs, also found in taiga communities, by their downward-hanging cones. In contrast, fir cones sit upright on the branch.

There are two nature paths, the quarter-mile Little Spruce Trail marked by yellow blazes, and the blue-blazed Bear Hill Trail, a bit over a mile. Both begin just to the south of the fire tower. Those willing to take the latter route are well repaid with botanical treasures. You'll find one of North America's most wide-spread bryophytes, haircap moss, in many damp and shady spots along the way. It grows quite happily on stone surfaces, for instance, where its taller companions couldn't survive. Bryophytes are described as nonvascular plants because they lack true foliage, true roots, and a highly developed internal circulation system. On top of that, they don't have cuticle, the waxy outer layer that helps more advanced plants resist water loss. In this wet climate, it is not a crucial factor, though, and many bryophytes thrive here.

If you look at the haircap moss closely, you may notice thin, capsule-bearing stalks rising from lower sections that resemble leafy stems. Actually, you're seeing

314

two different versions of the plant living at the same time. The lower portion is the first generation, which develops gametes that fuse and produce the younger spore-producing generation (the stalk and capsule). This pattern of alternating generations is common to all plants, though it often manifests itself in less obvious ways. In the bryophytes, the gamete-forming generation is the dominant one, but in vascular plants, it is the spore phase that is dominant. Indeed, it's the only one that people usually notice. For instance, many nature lovers see ferns all their lives, never realizing that the larger, frond-forming plants are only the spore phase, and only half of the full reproduction cycle. The gamete generation, consisting of a tiny structure called a prothallus, is almost impossible to catch sight of in the leaf litter of the forest floor. (Many ferns also reproduce asexually, especially by underground rhizomes. This method does not involve the generational issue at all.)

While we're on the subject of ferns, this trail presents a delightful treat. Along with such New England regulars as Christmas and sensitive ferns, a less common species occurs—Braun's holly fern. It does well in the deepest part of the woods, and is easily recognized by its stiff, rosette-forming fronds. These have coarse, chaffy stems and hairy, spine-tipped leaflet divisions. The plant prefers neutral soil, a condition none too common in New England but made possible in this locale by the presence of de-acidifying limestone.

Since there are several distinct habitats along this trail, ranging from deep forest gloom to sunlit thickets, the wildflower count is quite high. Among the more easily found species are meadowsweet, baneberry, blue cohosh, foam flower, herb Robert, tall lettuce, whorled aster, jack-in-the-pulpit, stinging nettle (careful!), and wild sarsaparilla. The roots of the sarsaparilla have been employed herbally as a cure for pulmonary problems and shingles. Among the woody plants are basswood, white ash, mountain ash, yellow birch, sugar maple, alternate-leaved and red osier dogwoods, and purple-flowering raspberry.

--------------------------- ∾ ---------------------------

When you depart Allis State Park, return to Route 65 and continue westward on it. The road quickly descends a steep slope into a new Green Mountain valley, where more tamaracks can be seen along the way. You reach the intersection of Route 65 and Route 12 at 1.3 miles from the park entrance; turn left onto 12 South. The rest of the tour is on paved roads. If you uttered a sigh of relief on regaining the blacktop, thank your friends, the bicyclists. It was the efforts of cycling lobbyists in the early 1900s, not the efforts of their automobile-owning counterparts, that first provoked the large-scale paving of American roads. You might also be interested to learn that here in Vermont, there was no state-sponsored maintenance or construction of roads until 1898. Instead, individual towns required their own able-bodied men to donate several days a year to the local road repair effort.

After turning onto 12 South, your first sight across the way is lovely Baker Pond, a good fishing hole for brook trout, or, as the initiated call them, brookies or squaretail. Whatever the name, this popular game fish species is restricted to the purest waters. As you enter the heart of Randolph, the valley of the White River's Third Branch opens up before you, revealing an idyllic landscape of well-kept farms. Besides possessing an overtly agricultural role, dairy farms are highly valued by regional planners because they provide scenic vistas for passing motorists. In a state as dependent on tourism as this one, that's no small matter.

The mill town of Bethel is next, and it is here that you meet the main branch of the White River. In Bethel center, Route 12 combines with Route 107 West. Follow the signs for the latter even after Route 12 splits away from it. This scenic stretch of mountain river along 107 is largely northern hardwoods territory, where fall foliage colors are often most splendid. For that reason, this is a suitable juncture to consider the biochemical magic that produces the effect.

When the deciduous trees of the New England forests begin to change color, they're doing so not to benefit the local motel industry, but to prepare for the prolonged period of dormancy that will allow them to survive the region's harsh winters. The trees could get the whole process over with quickly simply by dropping their leaves all at once while they're still green. To do so, though, would be to lose important chemical compounds that took a lot of effort to accumulate or make. Accordingly, the chlorophyll in each leaf is broken down, and as it disappears, other colors, previously masked by it, begin to appear. Species that turn predominantly red are likely to have anthocyanin pigments present; yellows and oranges are usually produced by carotenoids; and the duller browns of oaks and beeches are the result of tannins. When important substances such as sugars, amino acids, and magnesium ions have been withdrawn into the main body of the tree, a zone of weakness, caused by enzymes or a new type of cell growth, forms at the base of each petiole, or leaf stem. This zone of weakness permits the leaf to detach without damaging the remaining tissue beneath it. This process of leaf drop is called abscission. In the temperate regions of the Northern Hemisphere, abscission is largely triggered by the shortening daylight hours of autumn.

At the junction 10.0 miles below Bethel, Route 107 ends. Continue straight on Route 100 South. In a moment or two you enter Pittsfield. You've left the White River behind and are now in the valley of the Tweed. Our second stop, Gifford Woods State Park, is located 7.6 miles south of town. Go past its main campground entrance on the right, then turn left onto the Kent Pond fishing area access drive. Park where convenient at the end. In this spot there is a 4- or 5-acre stand of large trees alleged to be virgin, first-growth timber, with some plants more than four centuries old. If so, it is an extremely rare thing. As amazing as it seems, virtually all the original forest of New England, including its most remote portions, has been cut down. The reasons have been many. For one thing, the demand for firewood was once enormous. As Pehr Kalm noted in the mid-1700s, "An incredible amount of wood is really squandered in this country for fuel; day and night all winter, or for nearly half the year, in all rooms, a fire is kept going." But that factor, even when added to commercial lumbering, was smaller than the destruction caused by the clearing of land for livestock and crops.

All these activities had a deeper sanction. The European settlers carried with them memories of an Old World landscape of open fields and vastly diminished forests, almost wholly subdued to short-term human needs. When they crossed the Atlantic and faced the New England wilderness, their impulse was to reduce the complexities of the rampant vegetation to simple, agricultural order. Productivity, they thought, was nature's highest attribute. Early in the seventeenth century, Captain John Smith spoke for many people when he said that New England was a land of climatic extremes that would be dramatically improved by clearing and farming. Storms would lose their force; temperatures would moderate; this brooding land would be a gentle home at last. It wasn't true, but it did support some cherished prejudices. Generations of farmers and loggers would come and go before Vermonter George Perkins Marsh would sound the opening blast of the war for conservation:

> Man has too long forgotten that the earth was given to him for usufruct alone, not for consumption, still less for profligate waste. Nature has provided against the absolute destruction of any of her elementary matter, the raw material of her works. . . . But she has left it within the power of man irreparably to derange the combinations of inorganic matter and of organic life, which through the night of aeons she had been proportioning and balancing.

Take the trail along the water's edge on the right (south) to see the beautiful low falls of Kent Brook surging through the forest of beech, hemlock, yellow

Rock-strewn Kent Brook finds its way through a classic northern hardwood setting, at Gifford Woods State Park.

and paper birch, and sugar maple. As you wander among the undergrowth of sensitive fern, it is easy to imagine that this is the Vermont the Iroquois and Algonquins knew. It is not so easy elsewhere.

———————————— ❧ ————————————

To begin the last leg of the tour, turn left onto Route 100 South. Almost immediately, the road meets Route 4; merge right onto 4 West. This is a heavily traveled thoroughfare, much given over to ski resorts and their patrons. You quickly reach the high point, Sherburne Pass, and begin the long descent into the western lowland. Keep an eye out for plantations of Norway spruce at about 5.6 miles from your entry onto Route 4. This tall-growing European species with characteristically pendulous branches has been planted for lumber and ornament in many places in New England. Its timber, sometimes referred to as "white wood" or "deal," is harvested for paper, roofing materials, and many other uses.

On the final approach to Rutland, the impact of the modern urban sprawl is softened a bit by a few remaining roadside orchards, and by the silhouette of the northern end of the Taconic Range, directly ahead. Rutland is the Green Mountain State's second largest city, and is an important crossroads where the east-west flow of travelers merges with the great north-south corridor of the Champlain Lowland and Valley of Vermont. Our tour ends at the best place to see the crossroads, the junction of Routes 4 and 7. Should you stop in town and strike up a conversation with the local folks, you might be a bit surprised by their speech. This is the beginning of the vast Inland Northern dialect region. Instead of the clipped *r*'s and swallowed *t*'s of the eastern Vermonter, you're likely to hear those consonants fully pronounced along with flatter, nasal vowels; the singsong inflections are replaced by a Midwestern monotone. In this western gateway of New England, the cultural affinities point not backward toward the seat of Yankee heritage, but onward, toward the flat and windy reaches of the Great Lakes states.

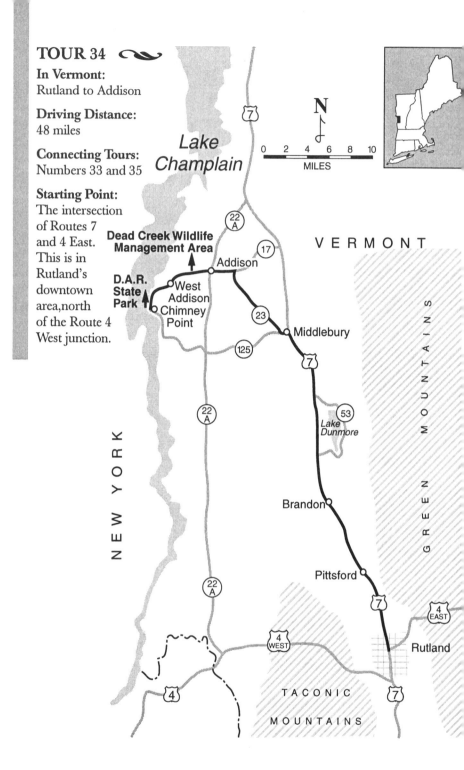

TOUR 34 ❧

In Vermont:
Rutland to Addison

Driving Distance:
48 miles

Connecting Tours:
Numbers 33 and 35

Starting Point:
The intersection
of Routes 7
and 4 East.
This is in
Rutland's
downtown
area, north
of the Route 4
West junction.

*Lake
Champlain*

**Dead Creek Wildlife
Management Area**

**D.A.R.
State
Park**

Addison

West
Addison

Chimney
Point

V E R M O N T

Middlebury

*Lake
Dunmore*

Brandon

Pittsford

Rutland

N E W   Y O R K

G R E E N   M O U N T A I N S

T A C O N I C

M O U N T A I N S

N

0 2 4 6 8 10
MILES

Tour 34
Northwestern Vermont–Part I:
A Flatlander's Delight

*T*ravel in northern New England involves so much up-and-down that a trip in just two dimensions seems an almost forbidden delight. In this tour, and the one that follows, you have the opportunity to experience the level ground of Vermont's western flank. If you complete both tours, consider yourself an Honorary Flatlander, even if you were born in a black-bear den in Green Mountain National Forest, or grew up with a bunch of wild turkeys in Smugglers' Notch. On the other hand, if the sight of a distant horizon upsets you, remember the two mountain ranges close by–you can always escape.

Head north on Route 7 from the Rutland starting point. This area is the uppermost portion of the Valley of Vermont, which extends in the opposite direction almost all the way down to the Massachusetts border. Both the valley and the analogous Champlain Lowland to the north are largely underlaid by limestone, dolostone, and their metamorphic equivalent, marble. This geologic fact has had a tremendous effect on the local plant communities. These soil-sweetening rocks neutralize acidity and thereby permit a much greater variety of species to flourish. If you've had the opportunity to see the staggeringly diverse fern population at Bartholomew's Cobble in southwestern Massachusetts, where a similar geologic regime exists, you'll understand why this narrow corridor of Vermont is more botanically complex than any other part of the state. (For more on Bartholomew's Cobble, see Tour 12.)

Once you're out of downtown Rutland, you'll be confronted with a mixture of wild and ornamental woody plants that may lead you to consult two or three different identification manuals simultaneously. Most of the plants are familiar species: the ever handsome sugar maple, white and red pines, black willow along the watercourses, and tamarack, the deciduous conifer that thrives in wet spots. You may also discern an unusual shrub by the name of smokebush, which is frequently planted in front of houses. This Eurasian introduction is a representative of the cashew family, and therefore a fairly close relative of the sumacs. Unless you're a trained systematic botanist, you'll probably have a difficult time proving the kinship. In Europe, this poisonous plant has been used for tanning and dyeing, but in the United States, its role has been exclusively ornamental. Its appeal is based not on its flowers as such, but on the intricate, hairy stems that support them. These are dense enough to resemble a cloud billowing over the foliage. If it's grouped at the back of a large perennial bed, smokebush can be very effective; but when it's used as a foundation planting, it may end up looking like a World War II destroyer frantically trying to lay down a smoke screen around its stricken flagship, the house. In recent times, purple-leaved cultivars of the smokebush have become quite popular, too. I happen to have a weakness for most dark-leaved plants, but some of these selections are just too much. In the wrong settings, they look like the plant kingdom's answer to the Rorschach test.

The center of Pittsford lies just 7.0 miles up the road from the starting point. This is a town much given to Norway spruce. Unlike native members of the spruce genus, this species has dangling branches that give mature specimens a brooding, swords-and-sorcerers look. North of town, take a final glance at the upper tip of the Taconic Range to the east. The Taconics are an intriguing feature because they are made of rock units that originated in an area of what is now the eastern side of the Green Mountains. In the course of a major mountain-building event almost half a billion years ago, this material was thrust westward to its present position and subsequently isolated from its source area by the forces of erosion. To see these mountains in their full glory, take Tour 30, which concludes with a visit to the summit of the highest Taconic of them all, Mount Equinox.

Dairying is the primary agricultural pursuit in these parts, as it is throughout the Green Mountain State. To the biologist, this industry turns the energy produced by thermonuclear fusion 93 million miles away into grasses and other herbaceous plants, and then into cows' milk—a source of energy that our nonphotosynthetic (and therefore intrinsically parasitic) bodies can use. Before the Civil War era, the process was more streamlined, biochemically speaking. The main agribusiness in those days was crop raising. The light from our bountiful parent star was transformed more directly into such human food sources as wheat, oats, rye, and barley. These grass-family staples were favored by farmers because they did not spoil readily—an important consideration in the days before railroads and other high-speed transportation could safely take perishable products to Boston, New York City, and other major population centers.

Here and there in the pastures and along the roadside stand large dead trees, or snags. For the most part, these are American elms. Normally, snags more than compensate for their unsightliness by providing shelter and food for a variety of animals. In this case, however, it would be extremely prudent to remove them as soon as possible and burn the wood, since they serve as breeding sites for the bark beetles that spread the dreaded Dutch elm fungus disease far and wide.

If by any chance you grew up on an elm-lined street and then saw the old neighborhood years later with the trees all gone and the soul of the place destroyed, you have a good understanding of what this plant plague has done. The DED (Dutch elm disease) pathogen has several strains of varying virulence; unfortunately, the most aggressive one is now present across most of North America. Normally, when a tree is attacked by a foreign organism, it responds much the way a warship does in battle: it compartmentalizes itself to localize the damage. In this way a tree can prevent the disease from spreading throughout its vascular system. One or two branches may die, but the main onslaught is checked or at any rate delayed. With DED, the fungus travels through the internal tissue so quickly that the elm shuts off the affected area too late—a classic example of shutting the barn door after the horse is gone. Human beings have actually helped the fungus along tremendously by using the American elm ornamentally in large single-species plantings known as monocultures. The elm-lined streets were beautiful, but their designers were asking for trouble. Instead of losing a few trees here and there to a particular malady, entire neighborhoods, lined with American elms and nothing else, were blighted.

The disease strikes in summer. The first indication of infection is the rapid dieback of a section of the plant's crown of branches. This symptom, called "flagging," is a heartbreaking sight for anyone in the tree-care business. Immediate removal of the dying stem might keep the problem from spreading, but often it's too late; the whole elm will die unless it's one of the lucky few that has enough natural resistance. Various remedies and preventive injection devices have been tried since DED appeared in the United States in the 1920s, and many claims about their success or utter worthlessness have been bandied about. There have also been various "resistant" American elm cultivars produced and distributed by a strange combination of capable researchers as well as charlatans with hidden agendas. The latter cloak shoddy science and clever marketing campaigns in the guise of philanthropic ventures to elevate their particular brand of American elm to the monoculture status that largely caused the mess in the first place. *Caveat emptor.* The future may bring us a truly reliable, DED-resistant American elm. If it does, the tree must be planted wisely, in the context of other species—as every professional arborist worth his or her salt will tell you. (For a good example of a modern-day street planting that avoids the monoculture time bomb, see Tour 39.)

———————————— ❧ ————————————

Your entry into Brandon is marked by the white cliffs of Hawk Hill on the left, and then by Jones Mill Pond on the right. This is the gateway to the Champlain Lowland. The Brandon area is renowned for an abundance of fern species, and for a unique deposit of low-grade coal, lignite. The lignite contains a number of plant fossils that give a good glimpse of the forest in this area some 30 or 40 million years ago. Among the remains are those of oak, holly, and tupelo species. These and the other plants and pollen discovered help demonstrate that the climate of the time was balmier than it is today—something resembling the conditions of modern Alabama. A few other plant-fossil localities exist in New England, but they represent very different parts of the geologic record. It's extremely fortunate that this Cenozoic era deposit is here, in a landscape so wholly given over to Paleozoic rocks about twenty times more ancient. When these marbles, limestones, and quartzites formed, not even the earliest vascular plants had appeared.

Brandon's downtown street plantings include several selections well known to urban forestry projects. Ash trees are here, for instance. When not bothered by borer insects, they perform well. Several cultivars are available; most are varieties of the green ash species. The plant with the feathery compound or doubly compound foliage is the honey locust. Its pods have a high sugar content, and in previous centuries they were used for animal fodder and as a source of sweeteners for human consumption. In common with many other woody legumes, this species is normally armed with vicious spines. These form not only on the branches but also on the lower trunk, and pose an unacceptable risk in an area of high pedestrian traffic. For that reason, a thornless variety has been widely substituted. In many places, honey locust is now threatened by a variety of pests and diseases. It was planted wholesale as a replacement for the American elm, and the result was predictable. It's just another example of how difficult it is to break the monoculture habit.

A common yard tree, the weeping willow, is not up to the rigors of the curbside environment. This dramatic ornamental is often relegated to cramped spaces, and the effect can be sloppy or downright dreadful. Many horticulturists and landscape architects have consigned it everlastingly to the eighteenth hole in hell, but it can be effective in the right place—for instance, when massed by a still body of water on the grounds of a hospital or corporate headquarters, where its somnolent grandeur is particularly fitting.

North of town the western scarp, or cliff face, of the Green Mountains rises to the right. If you wish, you can take the Route 53 Lake Dunmore loop at 9.4 miles above downtown Brandon, and then rejoin Route 7 a few miles up the road. The view of the cliffs from the lake is definitely worth the diversion. Continuing north on 7, another set of mountains is now evident on the left, beyond the low hills in the foreground. These are the mighty Adirondacks of northernmost New York.

In a few minutes, you enter the college town of Middlebury. At the junction of Route 125, bear left onto it; a mere 0.3 mile farther along, bear right onto Route 23 after you've crossed Otter Creek. Ash and honey locust street trees are joined by another common selection, the Norway maple, but the main attraction is the behemoth cottonwoods, on private property, which come into view a bit farther down the way. Once out of town, the road cuts across the broad-shouldered, open-faced terrain typical of the Champlain Lowland. This is Vermont's best farmland, enjoying fertile soils, a gentle climate, and unhindered light. When you reach the T-intersection of Route 17, go left on 17 West. After the turn, a feast for the eyes spreads out before you as a different world rises across the water.

The Adirondacks are a spectacle and a mystery. Utterly distinct from the hills of New England, they defy the stable nature of the Northeast by generating earthquakes and actually growing at the rate of about 1 foot per century. What are they doing there? What's going on under them? Strangely, they are new mountains rising on the trailing edge of a continent, where the forces of mountain building have otherwise slumbered for hundreds of millions of years. And as youthful as they are, they contain ancient Precambrian rocks, including anorthosite, a type common enough on the moon, but rare elsewhere on earth.

In recent times, the Adirondack upland has borne the brunt of acid rain caused by airborne sulfur and nitrogen compounds. The range has acted like a buffer, protecting northern Vermont from the worst. But in the process, it has suffered significant environmental harm. Some of its beautiful lakes have lost their entire fish populations.

———————— ∾ ————————

Continue on Route 17 through the Route 22A junction, then 3.4 miles later, stop at the right-hand turnoff where the road crosses a major stream. This is part of the Dead Creek Wildlife Management Area located in Addison. The wetlands are home to many waterfowl, muskrats, and other creatures of the marsh. The water flow is sluggish, which permits fragrant water lily and common cattail to become established. Whereas the water lily has adapted to its aquatic environment by producing flat, floating leaves, the cattail is, in ecology lingo, an emergent—it rises high above the water surface. Few are the souls who haven't noticed its unique sausage-on-a-stick flower clusters at one time or another. When they first form, these clusters each contain a zone of tiny male flowers, set just above the group of female flowers that eventually turn into the sausagelike structures. These contain a huge number of seeds.

As mentioned in other tours, the angiosperms, or flowering plants, are grouped into two taxonomic classes, the dicots and the monocots. The cattail is a monocot, as a careful examination of its flowers and seed structure under a dissecting microscope will reveal. Many monocots also have leaves with parallel veins running the length of the blade, rather than the more hierarchical arrangement featuring one main vein and branching subsidiaries. Look at the cattails here. Which pattern do they have?

———————— ∾ ————————

As you leave the Dead Creek pullout, resume your westward heading on Route 17. In another 3.9 miles, you reach our final destination, D.A.R. State Park, located on Addison's Lake Champlain frontage. As you turn left into the park, notice the zigzag fencing, known variously as the worm, snake, or Virginia rail

A traditional zigzag fence marks the front of D.A.R. State Park.

style. This type of barrier was commonly used on New England farms, especially before the heyday of stone fences. (For a fuller discussion of traditional fences and their significance, see Tour 19.)

As you head down to the lovely Lake Champlain overlook, you'll come upon the site of the cabin of John Strong, an early English-speaking settler who staked his claim here in 1765. Twenty-six years before he arrived, the botanical explorer Pehr Kalm passed this way on his trips to and from French Canada. One of his diary entries, penned a few miles south of here, paints a picture of this area quite different from our modern perception.

> A little while ago there was a crackling sound in the woods just as if something had walked or approached slowly in order to steal upon us. Almost everyone arose to see what was the matter, but we heard nothing more. It was said that we had just been talking about scalping and that we could suffer the same fate before we were aware of it. The long autumn nights are rather terrifying in these vast wildernesses. May God be with us!

Nowadays, it would be difficult to imagine a more placid spot. The woody plants that ring the picnic area include two alien species, Tatarian honeysuckle, and common buckthorn. They have staked their own claim here, very successfully. Both species are spread far and wide by birds that relish their fruit. In this country, the buckthorn especially is considered a noxious pest, and rightly so. In Europe, however, it has an appreciative following of herbalists who swear by its cathartic and laxative powers.

One other shrub species present here is a genuine native, the American prickly ash. Look for its pinnately compound leaves and its spiny leaf stems and branches. It isn't really an ash—in fact, it belongs to the rue family, which is best known for such tropical and subtropical species as the oranges and other members of genus *Citrus*. The Champlain littoral is probably the only place in New England you'll see the prickly ash, since its geographic range is mainly restricted to the North Central states. Having said that, though, it should also be noted that this shrub can be quite invasive. It is considered something of a pest by the locals. The Indians found it useful, however; when they chewed its bark, it helped to numb aching teeth and gums. To this day, its alternative common name is toothache tree.

Just as significant from an ecological standpoint is the red cedar, the narrow, upright juniper tree. It is a common sight in abandoned pastures of southern New England, but its presence here provides an important clue about the climate of Vermont's western reaches, compared to the rest of the Green Mountain State and New Hampshire. The lowland is a relatively warm corridor where a number of southern species have been able to extend their distribution far northward. The result is a crazy-quilt community in which species typical of Connecticut coexist with others characteristic of Quebec.

Red cedar mixes with its shrubby next-of-kin, common juniper. You may spot suspicious growths on some of the juniper stems. Depending on the time of year, they can look like offbeat, horned Christmas ornaments designed by an eccentric aunt who never left the 1970s, or like gelatinous globs of the worst grade-school pudding you'll ever encounter in your nightmares. These are galls, tumorlike swellings caused by the presence of a foreign organism. Galls occur on all sorts of plants, and are often triggered by mites or egg-laying wasps. In this case, though, the galls are the manifestation of a fungus, the dreaded cedar-apple rust. This disease has as its alternate host the cultivated apple tree.

If you wish, you can feel indignant that the fungus preys opportunistically on harmless plant tissue; but remember your own participation in the animal kingdom. In common with the fungi, we animals base our existence on the continual sacrifice of other organisms. This fact is only half-perceived by vegetarians, who swell in righteous outrage at the sight of a veal cutlet, but who gladly accept the wholesale slaughter of defenseless garden vegetables. Whether it's trendy or not, or morally elevated, the system of life has a tendency to gnaw on its own toes.

TOUR 35 〜

In Vermont:
Addison to Swanton

Driving Distance:
91 miles

Connecting Tours:
Numbers 34 and 36

Starting Point:
The intersection of
Routes 17 and 22A

QUEBEC   CANADA

UNITED STATES

Alburg Center

**Missisquoi National
Wildlife Refuge**

Swanton
**North Hero
State Park**

Isle La Motte

North Hero

**Jedidiah
Hyde
Cabin**

Grand Isle

South Hero

EXIT **17**

Lake

Champlain

Winooski

Winooski River

Burlington

Shelburne

Charlotte

**Vermont
Wildflower
Farm**

**Mount
▲ Philo**

Ferrisburg

VERMONT

Vergennes

Lamoille River

**N E W
Y O R K**

Addison

**D.A.R.
State
Park**

Chimney
Point

N

0   2   4   6   8   10
MILES

Tour 35
Northwestern Vermont–Part II:
The Land of Lake Champlain

HIGHLIGHTS

Freestanding Trees Develop Good Habits ∾ Vermont's State Whale ∾
The Vermont Wildflower Farm ∾ Free Market Botany ∾ A Debate over
Aliens ∾ Plant Material and the Shopping Mode ∾ Willow Whistles and Bean
Poles ∾ The Floating World of the Islands ∾ Jedidiah Hyde and Log Cabin
Chic ∾ North Hero State Park ∾ America's Deadliest Plant ∾ The World's
Oldest Reef ∾ A Federally Mandated Duck Crossing ∾ Missisquoi National
Wildlife Refuge ∾ A Plant to Make You Gag ∾ The Franco-American Waterfront

*T*he second and longer leg of our Champlain Lowland trek features not only the mainland shore of the big lake but also the waterborne world of its major islands. From this flat terrain the surrounding mountain scenery is superb. Closer at hand, the plant life varies from farm trees and meadow flowers to water weeds and lush swamp growth. This is a study in abundance.

The Addison starting point is situated approximately 7.3 miles east on Route 17 from the final stop of the previous tour, D.A.R. State Park. When you're ready to set out, head north on Route 22A. In this open farmland, the population of woody plants is small, in both terms of individuals and species, but you get the chance to see freestanding trees express their habit, or overall shape, much more fully. In contrast, you could cross the entire length and width of Green Mountain National Forest and see thousands of sugar maples, for instance, without realizing they'd form broad, rounded canopies if given enough elbowroom to do so.

In this stretch, the American elms lucky enough to escape Dutch elm disease demonstrate why their species was once so highly prized as a street and boulevard tree. Their habit is often termed "vaselike." In contrast to the European elms, their trunks are uncluttered by burls or clumps of water sprouts, and their branches arch gracefully from the top of the plant to produce a fountainlike effect.

Starting roughly 2.0 miles north of the 17-22A junction, the road runs along an ancient Champlain lakeshore, clearly visible on your left. The fertile valley soil once made this area a major wheat-producing center. Beneath the soil lie deposits of silt, sand, and boulders that long ago fell from floating blocks of ice. After the last glacial retreat had begun, a giant meltwater lake formed in the lowland and in the valleys of its major tributaries. Still later, when the ice had regressed farther into the Canadian interior, an arm of the Atlantic Ocean actually reached inland as far as this, bathing the area in salt water. The most dramatic proof of this marine episode is the whale skeleton that was found just north of here, in Charlotte. It wasn't a bad find for the one New England state currently without a seacoast.

Soon you enter Vergennes, the smallest city in the United States, with a little more than two thousand residents—a population about one-thirtieth that of Skokie, Illinois, the Chicago suburb that claims to be the world's largest village. Don't expect much in the way of backed-up expressways, smog clouds, or urban blight in Vergennes. The air quality and way of life haven't changed much since Commodore Thomas Macdonough readied his small freshwater fleet for an engagement with a British force during the War of 1812. To arm his gunners with grapeshot and other grim necessities of war, the commodore was forced to rely on supply lines that stretched across the Green Mountains and New Hampshire all the way to Boston. (For a description of the contribution of Asa Sheldon and his oxen, see Tour 32.) The vessels were no doubt constructed from local timber, a fact that reminds us that the Champlain Lowland once had extensive tracts of white pine, ideal for masts and other traditional shipbuilding uses. Macdonough won the battle, and is remembered as one of the Navy's more illustrious heroes from the age of sail.

———————— ∾ ————————

Stay on Route 22A through town and over Otter Creek until the road meets Route 7. Turn left onto 7 North and proceed through Ferrisburg into Charlotte. The name of the second town, incidentally, is pronounced "shar-*lot*." Many people assume that this is a reflection of the persistent influence of the French language in this area; actually, the pronunciation was used at the insistence of eighteenth-century royal governor Benning Wentworth, who wanted his English-speaking settlers to emulate the pronunciation then in vogue at the court of George III.

Just over the Charlotte town line is a right-hand turnoff to little Mount Philo and its state park. The side road leads to the summit, where there is an excellent view of the surroundings. Approximately 2.3 miles farther north on 7 is our first main stop, the Vermont Wildflower Farm. Turn left into the driveway entrance and park in the lot.

This facility is not government administered nor is it held in the public trust by a nonprofit organization. Rather, it is an experiment in what its owner, Ray Allen, calls ecotourism (other resounding terms, such as "free market botany" and "phytocapitalism," also come to mind). In partnership with his wife, Chy, Allen established the farm in 1981, certainly to promote the cause of wildflower appreciation and gardening, but also to make it a paying proposition.

The concept of ecotourism, if not the name, has been around in one form or another for a long time. Owners of popular natural attractions were charging

The Vermont Wildflower Farm is one of Charlotte's most appealing attractions. –Vermont Wildflower Farm

customers for viewing rights long before anyone considered the possibility of establishing national or state parks. To this day, some of the most beautiful spots in New England are privately owned business entities, such as Mount Washington and the Castle in the Clouds, in New Hampshire, and another Vermont attraction along Route 7, Mount Equinox. The more you visit such privately run places, the more you realize that they must be judged by the same basic criteria that apply to their public counterparts. For instance, does the place in question fulfill its potential or pander to the dog-and-pony-show mentality? Does it give people a better appreciation of nature and environmental awareness? Is the facility well cared for? And, last but definitely not least, is the visit worth the admission fee?

Happily, at the Vermont Wildflower Farm, the answer to all these questions is a consistent yes. The interaction between botany and the bottom line seems a pleasant one. The display the Allens have created, with the help of their staff and the silent complicity of thousands of plants, is a tribute to their enthusiasm. Beyond the gift shop and introductory sound-and-light show lies the real proof, in the 6-acre expanse of meadow and woodland. The Allens have succeeded where some of the best botanic gardens fail, by providing excellent plant identification tags and other educational information of interest to gardeners, herbalists, and even anthropologists and etymologists. It's obvious that a lot of loving dedication has gone into this well-documented mapping and signage. One wonders if the same dedication would have been manifest had this operation been run by a large corporation rather than a small family-business team.

One of the local factors that led to the establishment of the Vermont Wildflower Farm in Charlotte was the area's impressive floral diversity. Thanks to the relatively warm climate and the beneficial, soil-sweetening effects of the underlying limestone, there are roughly four or five times as many native species here as in the nearby uplands. The Allens have also kept or added a broad assortment of alien herbaceous plants to the meadow communities. These include everything from chicory and Queen Anne's lace, anathematized by most naturalists but eternal favorites with seed-buying customers, to such exotics as Indian blanket, Rocky Mountain columbine, and that scourge of Sunbelt median strips, the Cape marigold (also known as African daisy). This guilt-free blending of plants foreign and domestic points to a hot controversy that is just about as polarizing to modern botanists and ecologists as the true nature of Christ was to churchgoers of the sixth century.

The issue is centered on the question of whether the sale and use of alien plants should be encouraged or curtailed. Purists would prefer to have only certified native wildflower species promoted, lest the aggressive invaders outcompete them and threaten the traditional floral associations of the region. The problem is, some aliens have become so well established and have been ineradicable for so long that you have to wonder whether they haven't earned themselves a de facto citizenship. In one case, at least, this has already occurred: the imported red clover is so thoroughly accepted in the Green Mountain State that it's been designated the state flower.

As early as the 1670s, botanist John Josselyn noted the widespread presence in New England of such Old World baddies as dandelion, groundsel, sow thistle, shepherd's purse, stinging nettle, and common plantain. He also observed that the Indians called the last-named plant Englishman's foot, because it seemed to spring up wherever the new colonists stepped. Looking back even further, you might wonder what the natives-only advocates would have done eight or ten thousand years ago, when the species they now strive to protect were themselves busy invading the area from gene pools far to the south. In those days, the good guys were ruthless invaders, outcompeting the dwindling cold-weather flora that had lived in the postglacial tundra.

More far-sighted purists are well aware of the ebb and flow of plant communities over geologic time. Their point, that natural change by itself is not the culprit, while accelerated, human-induced change is, needs to be taken very seriously indeed. As with many other disputes, both sides have valid things to say, and it's important to realize that a synthesis of these seemingly exclusive views is possible, after all. For example, while the Allens do sell some aliens, they clearly state that the more aggressive species should be yanked out, not bought and sold. Visitors to the wildflower farm often wish to purchase seeds of the lovely plant that grows in ditches just up the road, off the Allen property. Instead of getting the seeds, however, they get a quick sermon on why the plant, purple loosestrife, is a highly destructive pest.

Make sure that your walk through the display areas in the back includes the forested zone. As advertised, the spring-flowering white trillium in full bloom is a sight worth a visit all on its own—as is the blazing summertime red of the cardinal flower. Also, if you visit in late spring or early summer, hunt for the white-blossomed wild leek. And take a look at—but do not touch, sniff, taste, or masticate—the native poisonous plants on display. Formerly, the deadliest of all, water hemlock, could be found here. Water hemlock can cause seizures and death to a human within thirty minutes of ingestion. Ray Allen wanted to educate his visitors about the danger by exhibiting an actual living specimen that was very clearly marked; but he had to remove it after a particular very special person ignored the warnings and tasted the water hemlock's seeds because they resembled those of dill. Certain unprintable thoughts about poetic justice come to mind; fortunately for the spaced-out taster, nothing serious resulted. Later in this tour, you may find water hemlock growing in the wild. If you do, don't be tempted to think the seeds, or any other part of it, are safe to try.

When you're ready to continue on, resume your northward heading on Route 7. The area becomes increasingly urban, or at least suburban, as you approach Burlington. As you pass strip malls and shopping centers, note how ornamental trees and shrubs, known to the hardened landscape architect as "plant material," are used to enforce the dictates of modern consumer-oriented society. More often than not, vegetation is used sparingly and symbolically, as high points of symmetry or color, to suggest a zone of ordered cheerfulness. There is nothing to fear. Every eventuality has already been predicted. Challenging forms and a lack of balance have lost all value because they have nothing to do with the human mind in the shopping mode. But like it or not, this global design ethic didn't happen by mere chance. The least ambitious storefront in the farthest corner of New England is the product of directed thought. In our particular use of plants, as in our use of architecture or speech, we reveal the essence of our age. What is your own opinion of it?

A bit of freeway driving begins at a point about 10.1 miles from the Vermont Wildflower Farm. Here bear right onto I-189; then, 1.4 miles down this spur, bear left onto Interstate 89. You are now passing through the heart of Burlington, Vermont's largest city. This fact might escape you, since the metropolis is largely invisible behind dense stands of white pine. This is a good place to see firsthand why the federal Census Bureau deems Vermont the most rural state in the nation. Even its chief center of population is home to only forty thousand souls—that is, approximately one one-hundredth the population of the metropolitan Boston area, and only about one four-hundredth the population of the greater New York City region. Vermonters, characteristically unimpressed by the sheer weight of numbers, tend to be rather proud of this fact.

Just 2.6 miles from your entry onto I-89, you pass over the Winooski River, where it makes its final winding approach to Lake Champlain. The name of this major Vermont waterway does not commemorate one of those Polish noblemen who took time out to fight in the American Revolution; rather, it's an Indian term that translates into something botanical—wild leek. (Compare this with Chicago, which is reputed to mean "little onion weed.") Wild leek—which you may have seen at the Vermont Wildflower Farm—is well suited to the damp riverbank woodlands nearby. It is said that Ethan and Ira Allen used it as survival fare when they settled here.

At Exit 17, depart I-89 by bearing right onto Route 2 North. After passing by some excellent road-cuts into Cambrian period rock units, you reach a large red pine plantation on the left, at 3.1 miles from the interstate exit. Red pines can often be identified by the pinkish gray highlights on their bark. Many group plantings of this valuable softwood lumber tree appear, from one end of New England to the other. Most date from the 1930s, when workers of the depression-era Civilian Conservation Corps set out row after row of seedlings on abandoned farm land. The admirable legacy of the CCC can also be noted in the sturdily constructed roads and public buildings in numerous state parks throughout the region and across the country.

Soon after the plantation, the road enters the first wetland of the tour, where the Lamoille River meets Lake Champlain. Cottonwood and black willow, two members of the same family, are dominant here. Both are extremely well adapted to waterlogged soils. In Thoreau's day, boys made whistles from the willow stems, probably to torment their parents and the local dogs. Also, willow whips were used as bean poles, though, as Thoreau noted, they often grew more than the beans themselves. This demonstrates the remarkable ability of willow twigs to sprout after they've broken from their parent trees. When the twigs fall in water, they float and eventually drift to a new location. This extremely effective form of asexual reproduction has made the willow genus a highly successful colonizer of streambanks and lakeshores.

In a moment or two, the road takes you across the main body of the lake to South Hero. This island and its northern counterpart are named in honor of Vermont's often outrageous role model, Ethan Allen, who died early in 1789 after crossing the lake ice from Grand Isle to Burlington. Although Allen's exploits, real or imagined, were many, he most certainly did not found a na-tionwide chain of furniture stores. I say this for the benefit of those who think Herbert Hoover built dams and vacuum cleaners.

The causeway en route to South Hero is lined with volunteer trees—cotton-woods and black willows again, as well as white ash. Purple loosestrife, the dreaded herbaceous invader mentioned earlier in this tour, grows nearby. It is not exactly visible from the road, but one serious aquatic interloper grows in the open water. Known as Eurasian milfoil, it now plagues Champlain and other Vermont lakes. Its rampant growth has been encouraged by runoff rich in farm fertilizers and domestic soap residues.

Many of the houses on South Hero Island were built with local rocks, and there is a succession of cornfields, willow thickets, pastures, and residential areas typical of the lovely Champlain isles. Elsewhere stand basswoods and American elms, and old cultivated apple trees that can be distinguished even in overgrown lots by their thick trunks and low, spreading crowns. As you enter the next town, Grand Isle, keep an eye out for the Jedidiah Hyde Cabin, on the right. This simple house of hand-hewn logs was built in 1783, when Vermont was an independent nation. Humble the cabin may be, but it has accomplished what few modern homes will—it has served as a family dwelling for 150 years. Old-time log cabins are rarely found elsewhere in New England. A cherished myth, first promulgated by presidential campaign promoters in the early 1800s, holds that the classic dwelling of the Pilgrims and other early settlers was the humble but intrinsically noble log cabin. In truth, this round-logged, saddle-notched structure was unknown to the Indians, the English, the French, and the Dutch. Apparently, it was first brought to these shores at a point far from New England, in the old Swedish settlements in Delaware. This fact is no doubt lost on purveyors of the Log Cabin Chic style prevalent across the region. People may think they're going back to their roots, but in fact, the style is no more authentically Yankee than Ingmar Bergman.

Incidentally, Jedidiah Hyde was a man who saw more than his share of Revolutionary War battles, including the one fought primarily against hunger, cold temperatures, and demoralization—Valley Forge. Another significant battle took place in the waters within view here. A most daring and resourceful American leader, every bit as good a sailor as a soldier, conducted a costly but cleverly executed delaying action against a superior and better trained British fleet. This was thirty-eight years before Thomas Macdonough produced his less equivocal results against the same foe. The Revolutionary commander, who had earlier distinguished himself by capturing Fort Ticonderoga in the company of Ethan Allen, was a Connecticut native by the name of Benedict Arnold. This remarkable man, whose frustrations with a treacherous Continental Congress caused him in the end to become the symbol of treachery itself, earlier came within an eyelash of conquering Canada, with the help of disgruntled French- and English-speaking Canadians. On top of that, Arnold played a decisive role at Saratoga as a dazzling, fearless, and insubordinate front-line leader.

───────────── ∿ ─────────────

As you continue northward and cross over to North Hero Island, look to your right across the lake. On a clear day, Mount Mansfield and other Green Mountains rise high above the valley floor. This was the same sight beheld and described by botanist-explorer Pehr Kalm on his peaceful expedition from British to French territory in 1749. As you reach the center of the town of North Hero, note your mileage; then, 3.3 miles beyond, go straight when Route 2 bears off to the left. This puts you on the shoreline access road to our next stop, North Hero State Park. The park entrance is 3.8 miles down the road, on the left. Once you're inside, proceed another 1.0 mile, through the darkling woods that enclose the group camping area, all the way to the beach and boat-launching facility on the island's northernmost tip.

When you've parked in the lot, walk out to the beach and head to the left (west). If the lake level is low enough, you'll be able to examine the plants of

The Lake Champlain shore at North Hero State Park. Note the progression of habitats, from lower left to upper right: *stony beach subject to frequent submergence, where the aquatic water plantain grows; the grass zone, dominated by freshwater cord grass; and the swamp forest, which includes maple species, swamp white oak, cardinal flower, and the lethal water hemlock.*

three environments: shallow water, shore, and the swamp just inland. In the submerged zone, you may spot water plantain, a monocot closely related to the more common arrowhead plant. In summer, water plantain puts up intricately branched inflorescences that appear to have been designed by a frustrated toy company engineer. It has gained notoriety as a herbal cure for everything from kidney stones and epilepsy to rattlesnake bites.

In the rocks and sand close at hand grow a variety of interesting dicot wildflowers: wild mint, water smartweed, swamp smartweed, and cut-leaved water horehound. Unlike its cultivated equivalents, wild mint is not a welcome sight to herb growers. It sometimes invades peppermint fields and adulterates the crop. Also, it has been said that when cattle eat it, they give noncurdling milk. If you twirl the stem of the plant between your fingers, you'll find it's square in cross section. The edges of the stem are composed of a strengthening tissue called collenchyma.

The monocots of this beach zone include freshwater cord grass and nut sedge. The first is the inland version of the salt-tolerant *Spartina* cord grasses of the

Atlantic tidal marshes. Not only does it inhabit lake edges but it also inhabits damp sections of prairie land, far to the west. The nut sedge has thin leaf blades that superficially resemble those of grass, but its flower structure is sufficiently different to warrant its classification in a separate family. Many sedges have stems that are triangular in cross section, rather than round. The old saw for this is "sedges have edges." The common name of this particular species refers to the small, edible tubers found on its roots. One variety of the plant, called chufa, is grown as a food crop in Africa and elsewhere.

If the swamp forest behind the beach is dry enough, take one of the paths through it, but definitely refrain from handling or picking any plants. By mid-summer, its most vivid bloomer, cardinal flower, can be seen blazing through the dense shade. Also present is the subject of an earlier passage in this tour, water hemlock. This member of the carrot family is the most poisonous plant in the United States. It bears no resemblance or relation to the coniferous tree known as Canadian hemlock. It resembles Queen Anne's lace and dill, though, so beware. In summer, white, flat-topped flower clusters are one clue to its identity; also look for compound foliage composed of pointed, toothed leaflets, and for purple splotches on its otherwise green stem.

The swamp's overstory community includes another tree species well adapted to flooded ground. This is the swamp white oak, long prized for its premium shipbuilding timber. True to its name, the leaves are similar in overall design to the regular white oak, except that they are more entire, with toothed, wavy margins instead of well-defined sinuses and lobes. Also present are maples, which often seem to combine the characters of the red and silver species. Hybrids between the two types are not unusual.

On leaving the park, return to the junction of the access road and Route 2. Make a hard right onto 2 North. This stretch features many black locusts, which proclaim their status as legumes by producing handsome clusters of fragrant, white flowers in early June, which, in the fullness of time, give way to dark, flat seed pods. About 2.0 miles from your reentry onto Route 2, bear left onto the bridge to Alburg. You can continue on 2 North or take a short detour on 129 West to lovely Isle la Motte, where a fossilized coral reef, the most ancient one known, stands exposed in places in the private pastures. If you decide to go reef-hunting, you're well advised to stop at the Isle la Motte town hall and get precise directions. Another attraction is a mixture of look-alike cypress-family conifers you'll encounter in the fields. See if you can distinguish arbor vitae, a plant with northern affinities, from red cedar, a plant well known in southern New England and surviving here thanks to the unusually moderate conditions of this far northern locale.

Route 2 North passes through Alburg center; soon thereafter, you reach its junction with Route 78. Turn right onto 78 East and enjoy the wetland view of the Mud Creek Waterfowl Area. Then take the bridge that leads across the narrows to Swanton and our final stop, Missisquoi National Wildlife Refuge.

Make sure you heed the official Duck Crossing signs. The refuge entrance is on the right-hand side, about 4.8 miles from the bridge. Park in the lot by the headquarters building and head down the path that leads from the lot, over the railroad tracks and into the land framed by Maquam and Black Creeks.

During a research visit here, I paused at the trail entrance and read the visitor log provided by the refuge staff. The most recent animal sighting recorded was "LIONS TIGERS & BEARS, O MY!"—which means that some budding naturalist had a better grasp of the history of the American cinema than of this site's inventory of mammalian carnivores. Depending on when you visit, however, you may run into the largest assortment of peeping and croaking amphibians you'll probably ever want to see.

The fauna aside, this is a splendid place to see all sorts of bottomland plants, starting with excellent specimens of the royal and sensitive ferns. Three flowering plants demand your immediate and sustained respect throughout the jaunt: poison ivy, poison sumac, and arrow-leaved tearthumb. Poison ivy has the three-leaflet pattern familiar to all good outdoorspeople. But the nastier poison sumac shrub is more difficult to identify because it has compound leaves quite like those of an ash tree. If you're susceptible to poison ivy—in fact, even if you aren't—beat a wide path around its more toxic relative here. The other plant to avoid, the aptly named arrow-leaved tearthumb, is a rambling herbaceous plant armed not with toxic chemicals but with small, vicious hooks.

As you'd expect, swamp white oak and red maple are some of the larger trees here, while the trailside understory includes speckled alder and withe rod viburnum. This is a good spot to examine the leaves of the red maple. If you see round, dark blotches on them, you've found tar spot fungus, a common though not very harmful foliage disease. While you're at it, also look for the herbaceous vine known as ground nut. Like the unrelated nut sedge mentioned above, this legume forms tubers that are quite edible. In his role as precursor to Euell Gibbons, our mentor Henry David Thoreau noted that ground nut has "a sweetish taste, much like that of a frostbitten potato." Also, for what it's worth, he said it's better boiled than roasted.

To balance all this edibility, another wildflower found here, Indian tobacco, shares the distinction with the showier cardinal flower of being a member of the horticulturally important genus *Lobelia*. Lobelias have had a long career as herbal remedies. The purported properties of Indian tobacco are best indicated by some of its other common names: gagroot, vomitwort, and even pukeweed. In other words, it was employed as an emetic and expectorant. If you do happen to come across a lion, tiger, or bear, offer it some of this and make good your escape before the choking stops.

———————— ∾ ————————

The last stretch of driving is a short one, from the refuge east on Route 78 into Swanton center and the Route 7 intersection. On the way, you parallel the bank of the Missisquoi River, Vermont's northernmost main tributary of the gigantic Champlain–St. Lawrence drainage system. As residential landscapes supplant the wetlands, you enter the very different world of lawn trees—catalpas, Norway maples, and weeping willows. You are now at a gateway of two cultures. A mere 7 miles up the road, as the duck flies, lies French-speaking North America. Not so very long ago, this spot, as well as lands much farther to the south, were claimed and colonized by forebears of the current Quebecois. In places this fact can still be detected in the long, narrow lots of Champlain lakefront property. This pattern was part of the old French land use system that permitted more settlers to get a portion, albeit a small one, of the premium water frontage.

TOUR 36 ❧

In Vermont:
Swanton to Westmore

Driving Distance:
77 miles

Connecting Tours:
Numbers 35 and 45

Starting Point:
The intersection of
Routes 7 and 78

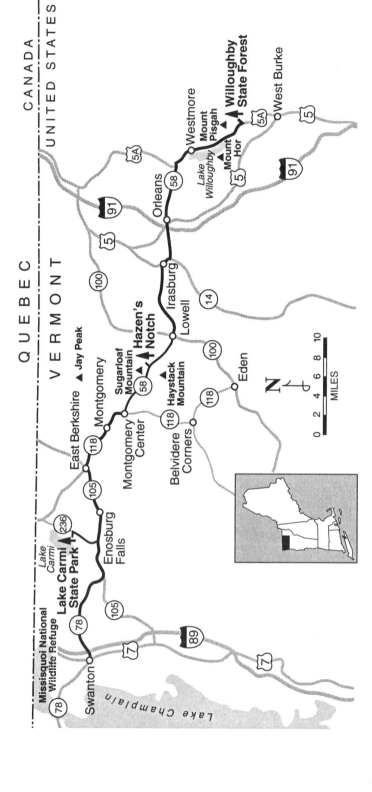

Tour 36
The High Points of Northern Vermont

A Recommended Fall Foliage Tour

HIGHLIGHTS

Dairy Country and a Fossil Landscape ❧ A Pocket Guide to Forest-Free
Farming ❧ Lake Carmi State Park ❧ A Flower for Horace ❧ A Beautiful
Bog or a Deadly Morass? ❧ A Window on the Past ❧ Fireweed and the
Virgin's Bower ❧ At Lake's Edge ❧ This Is Atlantic Ocean Country ❧
Hazen's Notch ❧ Canadian Yew ❧ Taiga Trees ❧ Discerning Spruces ❧ Lake
Willoughby and the Ah Factor ❧ Rare Arctic Hangers-On ❧ Willoughby State
Forest and an Ascent of Mount Pisgah ❧ A Cavalcade of Wildflowers ❧
Meditation on the Form Internal ❧ The Invisible Web Made Visible

*T*rips across New England have a way of involving rivers. Wherever
possible, the roads follow their lead, for the streams offer the best
passage through a hilly land. Our guide for the first half of this journey
through northernmost Vermont will be the comely Missisquoi River, the major
Lake Champlain feeder seen in its final reaches at the end of the previous tour.
Beyond that, we'll rely on other smaller watercourses, until we reach the most
breathtaking lake in the Green Mountain State.

Begin by heading east on Route 78. East of Swanton center is a mixture of
trees—poplar, white pine, red maple, and some cottonwood. Soon you're in the
kind of dairy country typical of this part of the state. Unlike the wooded expanses
of Green Mountain National Forest to the south, this terrain is still largely cleared.
In a certain sense it represents a fossil landscape, one that gives the observant
traveler the opportunity to see how the majority of New England looked in the
eighteenth and nineteenth centuries, when agriculture was the predominant way
of life.

This is not to suggest that the current mode of agriculture, dairy farming,
was the typical means of livelihood in those days. Indeed, dairying is largely an
offshoot of the Industrial Revolution and its improved transportation systems.
After all, milk in its basic liquid form cannot reach its most lucrative market,
the cities, unless it's taken there swiftly, before it spoils, in refrigerated trains or
trucks. Before milk production rose to prominence in the early 1900s, cheese
and butter were the dairy farmer's main products; but even these were not

especially well established before the Civil War years. Further back, in the days of Ethan Allen, you'd look in vain for the holstein-dotted hills that today stand as the irrefutable symbol of vintage Vermont. In colonial and Federal times, oxen were the most common type of cattle. These docile and enduring beasts provided transportation and aided the small landowner in a multitude of tasks, the most important being the clearing of land and plowing. One historian has suggested that the reason Indians did not alter the look of New England as much as the colonists was because they lacked the extra power potential cattle provided. For herd animals, most settlers preferred sheep to cows because they were less labor intensive. Periodic sheep shearing may be no picnic, but it isn't as wearying as hand milking and churning on a daily basis. (For a discussion of the Merino sheep movement in Vermont, see Tour 32.)

When its time did come, however, large-scale dairy farming proved to be a winning proposition. In recent years, the trend has been toward a smaller number of large farms and herds, interspersed with hobby farms kept by people with other professional demands. Whatever analysis is attached to this, the fact remains that more than three-quarters of the agricultural income in the Green Mountain State comes from dairying.

The ideal environment for milk cows includes a cool climate where rain is plentiful enough for pasturage and hayfields, as well as rocky land, cleared of course, but not so good as to be used for premium tillage crops. These conditions, all manifestly part of the Vermont picture, would have been insufficient without the proximity of Boston and other large population centers, with their steady demand for milk.

A common dairy farm feature, the silo, is a relatively recent addition to the scene, though it has become an intrinsic part of the rural Vermont landscape. The earliest models date from the late 1800s. Silos are used to store chopped-up fodder, or ensilage. When fodder is kept this way for some time, it ferments somewhat, but does not spoil.

———————— ∽ ————————

At a point 10.8 miles from the start, Route 78 ends at the junction with Route 105. Turn left onto 105 East. Despite the prevalence of cleared land in this area, openness is by no means the natural order of things. Wherever a lot is abandoned, the woodland reasserts itself with an implacable intentness. About 1.5 miles past the turn, the handsome profile of Jay Peak rises in front of you. Look to the right, across the Missisquoi to the far bank, where Canadian hemlock and hardwoods have reestablished themselves.

At this juncture, take a jump back in time. Imagine yourself as a member of an early settler family who must create farmland from this stretch of forest. Environmental concerns are not part of your value system, but the grim task of survival is. The trees you see here take up valuable land and threaten your ability to subsist at the most basic level. Furthermore, you've been told by your fellow settlers that the woods harbor wild predators that will attack your domesticated animals. Add to this the fact that the timber in front of you has considerable commercial value. You know you want to remove the forest, but you live in an era that has no chainsaws, bulldozers, or tractors. How do you clear the land?

To kill the trees outright, you could chop them down, of course, or just girdle them. Girdling is a simple but effective form of tree assassination. A ringed incision is made all the way around one section of the trunk. This destroys the continuity of the vital phloem tissue. Phloem transports food and other substances from the leaves to other parts of the plant; but when its way is thoroughly blocked, the tree slowly but surely starves to death. Large numbers of trees can be girdled by just one axe-wielding person, but then the leafless hulks are left standing. In the short run, you won't worry about this; instead, you'll plant your first crop of maize Indian-style, in small hills that do not require heavy plowing. Eventually, the snags will rot, allowing their easy removal at your leisure. Finally, when the land is clear enough, you'll resort to more conventional Old World tillage practices, with the corn set in rows, instead of mounds.

If you have a mind to, you can speed up the process by cutting down all the trees at once. You'd probably elect to do the felling late in the warm season, when the trees are still in full leaf. As every woodlot manager knows, leaves continue to pull moisture out of a tree even after it's severed from the roots. Hence, the wood dries out more quickly than it would if cut in the cold season. You can harvest the firewood you'll need during the winter, but next spring, you'll probably burn the remainder. One by-product of the burning, potash, will bring you extra income, because it is an important ingredient in the manufacture of soap and gunpowder.

―――――――――――― ∽ ――――――――――――

In a few moments, you reach the intersection of Routes 105 and 236. Bear left onto 236 North. This is a lovely road, with sweeping vistas. Before long, you'll see our first stop in the distance: a large body of water with a low, forested wetland on its southern end. This is Lake Carmi State Park, just a couple of miles south of the Canadian border. Continue on for 3.0 miles from the previous turn, and bear left onto the park access road. If you visit in late summer, you may spot a large, coarse-leaved herbaceous plant with yellow sunflower-like blooms growing on the forest edge. Known as elecampane, this species in the composite family is an alien that has been appreciated as a remedy and food source since ancient times. The Roman poet Horace, living at the time of Christ, touched on the plant's culinary uses in one of his satires; and over the course of two millennia, its roots and rhizomes have been employed to fight bronchitis, tuberculosis, and even skin problems in sheep.

Once through the park entrance, turn left again and head down the campsite road. Another alien composite, Tyrol knapweed, grows along the pavement. Farther down, the wetland begins to assert itself where common cattail and jewelweed signal the presence of waterlogged soil. You then come upon one of the region's most mysterious and rewarding plant environments, a beautiful sphagnum bog. Black spruce is everywhere. As usual, it looks pretty scraggly,

Spindly black spruce trees rise over the sphagnum bog at Lake Carmi State Park.

but it's in no danger of being monopolized by the Christmas tree industry. As noted in Tours 10 and 22, this remarkable conifer has developed a full suite of tricks to help it survive in the acid soup of the peatland water. It was prized by the Indians (who used its finer roots for sewing birchbark canoes together), by the early European settlers (who thought its wood was the best possible for rafters), and by the modern lumber industry (which uses its pulp for high-quality paper products). Interspersed here among the spruces are tamaracks, often draped with *Usnea* lichens.

If you choose to pull over and investigate the bog more closely, keep in mind that it is not a solid surface but a mat of sphagnum moss that undulates and slowly gives way underfoot. This is not to say the bog proper is impassable; but if you venture forth, wear ooze shoes, be prepared to get wet, and be very careful not to disturb this fragile community of highly specialized plants. It is true that the full wonder of the place is revealed only to the bog-trotter willing to slosh into its midst, where leatherleaf, Labrador tea, few-fruited sedge, and tan-tufted cotton grass hold sway.

The traditional Yankee view of bogs and other interior wetlands was often far from appreciative. Even Vermont's great nineteenth-century environmentalist George Perkins Marsh, himself the owner of a good wetland name, described these crucially important areas as "mischievous accumulations of moisture," "deadly morasses," and "stagnant waters." Ironically, it is the stagnant nature of bogs that makes them one of the best windows on the past. In New England, the successive layers of pollen and partly undecomposed plant matter preserved in them reveal the fascinating story of how the living landscape has changed since the Ice Age from tundra to taiga to temperate forest.

Next, go back down the same road toward the main entrance. The woods of the lakefront camping area to your left contain good examples of the northern conifer arbor vitae, an evergreen used extensively in landscaping in the United States. When you reach the park's entry area, continue down to the beach parking lot—which, translated into Parkspeak, is the "Day Use Area." An excellent short nature trail begins at the far end of the parking lot. Follow it down to the lake. On the way, you pass through a meadow with a profusion of shrubs and wildflowers: jewelweed, wineberry, self-heal, speckled alder, arrow-leaved tearthumb (watch out!), goldenrod species, common milkweed, black-eyed Susan, and spotted joe-pye-weed. Among these are three other noteworthy species: the striking magenta-flowered fireweed, so named because it springs up on fire-cleared land; hemp nettle, a herb in the mint family, with the hand-prickling properties of its name-sake; and the vining virgin's bower. The last of these is a native member of the large *Clematis* genus. Plant experts are just about evenly divided on how to pronounce the generic epithet. Take your pick between "*clem*-ah-tiss," "cluh-*may*-tiss," or "clem-*mat*-is." This charmer has eye-catching flowers with four white sepals apiece. By summer's end, they develop into fuzzy or plumed fruit resembling the hairdos of those plastic trolls sold in souvenir shops along the Ohio Turnpike.

At lake's edge, another plant community takes over. In front of a cluster of red maple stand common cattail, water parsnip, and blue vervain, all leading contented lives in the muck. A bit farther out, in the shallowest open water, float leathery bullhead lily pads. A close relation of spatterdock and fragrant water lily, this species has unusual, rounded yellow flowers that seem not to have opened fully. As you head back to the parking lot, you may wish to explore the longer trail spur leading away from the wooden bridge over the ditch. One word of friendly advice: curb your enthusiasm when you cross the bridge. If you step on it too sprightly, it may launch you over the Green Mountains into New Hampshire.

───────────── ∾ ─────────────

Depart Lake Carmi State Park by returning to Route 236 and retracing your track to its junction with Route 105. Turn left onto 105 East and resume your gentle ascent of the Missisquoi River valley. As you enter Enosburg Falls, follow the signs for Route 105 and take a gander at the northern fringe of the Green Mountains, ahead of you. Can you figure out the best way to cross them? This will be a matter of more direct interest in a little while.

The landscape close at hand may not be imposing, but it has its own story to tell. One of the most amazing events occurred just yesterday, geologically speaking, when this valley was nothing less than a narrow arm of the Atlantic Ocean. As noted in Tour 35, the Champlain Lowland to the west experienced more than one spate of massive flooding when the last Ice Age glaciers retreated a few thousand years ago. In one episode, the lowland and its tributaries were submerged in fresh meltwater; in another, the ocean actually reached far down the St. Lawrence into northern Vermont. Here, for example, geologists have mapped sand deposits of marine origin. It's difficult to imagine this peaceful inland setting bathed in salt water.

Continue on Route 105 East 5.5 miles past the center of Enosburg Falls to the intersection of Route 118 in East Berkshire. Turn right onto 118, cross the Missisquoi, and on the left note the beautiful river terrace on private farm property. You now ascend the smaller Trout River. This area is blessed with several covered bridges. Try to spot a common ornamental conifer with reddish bark and short needle foliage that has a distinctive glaucous, or waxy blue, cast. This is the Scots pine, an introduced species that is a favorite lawn planting in New England. It will even establish itself in abandoned lots, if given the chance. The Scots pine is also grown on a huge scale for the Christmas tree trade. For more on the history of this conifer, see Tour 41.

Looking straight ahead, you'll notice an indentation in the mountain front. That is Hazen's Notch. The following leg of the tour takes you on an unpaved road through this high passage in the hills. The road is well maintained, but "Road Closed" signs are put up in winter and during the mud season. During those times of year, it's probably best to take the alternative track of Route 118 South to Eden, and then Route 100 North to Lowell. However, if you're making this tour in dry conditions in late spring, summer, or the fall foliage season, you should have no problem proceeding through the notch. Try not to miss it.

When you reach Montgomery Center, proceed past the intersection of Routes 118 and 242 to the next junction, where Route 58 meets 118. Turn left onto 58 East. Soon, the 8.5-mile stretch of graded gravel begins, and you rise into a terrain unsullied by more than the occasional residence. This is an upland of remoteness and light. Almost at once, the great and lonely profile of Jay Peak comes into view on the left; and Hazen's Notch looms ahead, framed by Sugarloaf Mountain on the north and Haystack Mountain on the south. As you continue to ascend, you see red spruce of the montane taiga regime mixing with northern hardwoods species. The world gets ever lovelier as you make the final approach to the forested base of the Sugarloaf cliff face. Once there, carefully pull off to the side and take a moment to inspect the vegetation. The overstory is composed largely of sugar maple; beneath it grows hobblebush viburnum, with its large, rounded leaves arranged in pairs along stems. This opposite pattern of buds and branching is one of the main characters, or identification traits, of viburnums and other members of the honeysuckle family.

The Canadian yew may be more difficult to locate in summer, though its evergreen branches stand out well in the dormant season. It is the native equivalent of the Japanese and English yews typically used as hedges and foundation plantings. Ornamentally speaking, this species doesn't have much to commend it because it has a low, unpredictable, and often ratty habit. Still, it's reassuring to know that this sole representative of the unusual yew family is holding its own in New England woodlands. Botanists have had much fun debating the exact position of the yews in the complex record of plant evolution. Some experts are inclined to think that yews represent a line quite separate from pines, cypresses, cedars, and other conifers. These theorists suggest that the yew family came from a different ancestor altogether. If you look at its foliage, which closely resembles Canadian hemlock and other conifers, it may be difficult to see what all the fuss is about. But if you locate its distinctly unconelike seed structures, you'll see how unique the yews really are. These structures aren't woody containers holding a number of seeds, but fleshy red jackets, known as arils, each enclosing a single seed. (Yews are also discussed in Tour 50.) In the much less stratospheric realm of horticulture, this plant type has been the pretext for several dreadful landscaping puns, mostly of the it-had-to-be-yew and it's-yew-baby variety. Any attempt to castigate the author for mentioning these will simply be a case of shooting the messenger.

The lush growth of moisture-loving herbaceous plants, especially yellow and orange jewelweed and sensitive fern, indicates that this is a damp environment. Look also for blue cohosh, wild sarsaparilla, tall meadow rue, and the mat-forming haircap moss on top of rock ledges and boulders. Blue cohosh, a relative of the barberry shrubs, is a classic spring ephemeral, blooming in this location by late April before its leaves are fully deployed. Even this shy character has not escaped the eye of the herbalist; it has been applied in treatments for rheumatism, dropsy, and epilepsy.

A dense second-growth forest fringes the road as you descend the eastern side of the notch. When you reach Lowell, Route 58 returns to its previous paved state. Stay on it as it continues eastward. Beyond Lowell, the hillside tree communities demonstrate that you are far up-country indeed. Red spruces appear time and again, with such cold-hardy deciduous species as paper birches and poplars. The red spruce is a particularly striking tree, in a rather formal way. When it's predominant, it transforms the landscape into an almost abstract surface of spirelike projections. It can be distinguished from its close look-alike, the white spruce, by the more forward-pointing needles, and by its habitat. Red spruce is the more common species in northern New England and on the upper slopes of the mountains; white spruce prefers the harsher conditions of Maine's Down East seacoast (see Tours 53 and 54) and farther north, of the Canadian lowland taiga. Ironically, roughly ten thousand years ago white spruce covered much of New England.

Remain on Route 58 through Irasburg. Near Orleans, the road passes under Interstate 91. In this vicinity you enter Vermont's beautiful and sparsely populated Northeast Kingdom. On the eastern side of Orleans, as you head up the Willoughby River, another taiga denizen comes into view—the balsam poplar, the cold-climate equivalent of the cottonwood. It can be found here and there along the watercourses of the southern Green Mountains, but generally it's more commonly located closer to the Canadian border. (For more on this striking tree, see Tour 45.)

Approximately 7.1 miles beyond the I-91 overpass, Route 58 meets Route 5A at a T-intersection. Bear right onto 5A South. In another 1.2 miles, you reach the town of Westmore and come to one of the most stunning vistas in New England—the long glittering body of Lake Willoughby bracketed at its far end by the glacially sheared cliffs of Mount Pisgah on the left and Mount Hor on the right. This is a place that produces what I call the Ah Factor. This syndrome is restricted to situations of acute natural beauty, when the human nervous system is suddenly confronted with a massive dose of the sublime. The road runs the length of the lake on its eastern shore, under Pisgah's hulking mass. Arbor vitae grows amid the rushing cascades of the lower slopes; far above, out of reach of everyone except birds and rock climbers, is a rare community of arctic plants, the last children of the Ice Age. This tiny remnant population is priceless, one of a very few that have survived on the inaccessible Vermont mountain cliffs. Another such plant community, at Smugglers' Notch, is mentioned in the next tour.

If all this breathtaking scenery encourages you to get out on foot and encounter it at close quarters, use the tour's endpoint, the Mount Pisgah trailhead, as the beginning of your hike. This trail into Willoughby State Forest meets the left side of Route 5A about 0.5 mile below the lake's southern tip, just across from the big parking pull-off on the right. If you plan to climb all the way to the Pisgah summit, a one-way distance of 1.7 miles, keep in mind that it is one of the most challenging hikes in the book. There is a good trail all the way, but the slopes are steep in places. In wet conditions, ooze shoes are a necessity. As always, you'll find an abbreviated excursion more inspiring and more informative than none at all.

The footpath first heads across a wetland boardwalk. This must be nirvana for beaver and brook trout; but then it's nirvana for botanists, too. If you're visiting in summertime, look in this low area for the humorous blooms of turtlehead, a wildflower in the snapdragon family. If you see these white or pinkish blossoms, you'll understand the plant's common name (though John Josselyn thought they resembled serpent heads instead). Also, note the arbor vitae trees. They definitely are conifers, but their cones are minuscule by pine or spruce standards. On the lower slope beyond is a wonderful collection of wildflowers, many of them spring ephemerals. The roster includes goldthread, coltsfoot, blue cohosh, trillium species, plantain-leaved sedge, round-leaved yellow violet and its assorted relatives, trout lily, shinleaf pyrola, and that wan and lovely masterpiece of fragility, spring beauty.

Another fairly frequent sight is a miracle of miniaturization, the bunchberry. This ground-hugger is actually a herbaceous dogwood that delights the hiker in late spring or early summer with its white-bracted flower heads. Bunchberry is a good example of closely related species taking on drastically different forms. To the casual observer, this small plant has nothing to do with its high-growing shrub equivalents. To the botanist, however, the kinship is amply proven in the details. As Sir Walter Raleigh observed, "It is not the visible fashion and shape of plants, and of reasonable Creatures, that makes the difference, of working in the one, and of condition in the other; but the form internal."

No New England forest would be worth its salt if it didn't also contain a good assortment of pteridophytes, or free-sporing vascular plants. You won't be disappointed here. For instance, the tough evergreen fronds of Christmas fern are very much in evidence, and the creeping stems of stiff club moss in some places march straight across the path. Returning to the realm of seed plants—specifically, the overstory trees—there is a blend of Canadian hemlock, paper birch,

The view from Mount Pisgah's Pulpit Rock. Both Lake Willoughby and the sheared cliff of Mount Hor behind it owe their existence to the grinding and plucking forces of an Ice Age glacier.

the silvery-barked yellow birch, white ash, and American beech. Below them are some good examples of our recent acquaintance, Canadian yew. Near the cliff edge, in the vicinity of Pulpit Rock, you'll see a taller-growing shrub, mountain alder, that shows its kinship with the birches by producing flower clusters called catkins.

Extra care is required where the trail parallels the drop-off. In my opinion, Pulpit Rock provides one of the most spectacular views anywhere in the explored solar system. The lake and the opposing hillside, monuments of great time and silence, bear testimony to the scouring, plucking, and gouging powers unleashed in Vermont's glacial past. Still, it is a dangerous place for children and adults alike. Be very cautious.

The taiga regime, in the form of red spruce and even balsam fir, makes its appearance more fully on the upper slope. Though this is the most tiring segment of the ascent, the world of the summit is worth it. On top, lichen-encrusted stretches of rock alternate with dwarfed conifers exposed to the full blast of the wind; sheltered communities of sugar maple and scarlet elder cling to the lee of the saddles. The Northeast Kingdom terrain, visible at several viewpoints, seems almost totally unaffected by human forces, though its forests have been stripped by farmers and commercial loggers for at least two hundred years. Thanks to the incredible persistence of the plant kingdom, though, there is a pervading sense of things being intact. In their joint survival as a complex web of species, the plants, lichens, and other beings of this mountaintop have much to teach the receptive human soul. This was the understanding of George Perkins Marsh, one of the first Americans to seriously advocate the protection of forests. "All nature," he wrote, "is linked together by invisible bonds, and every organic creature, however low, however feeble, however dependent, is necessary to the well-being of some other among the myriad forms of life with which the Creator has peopled the earth."

349

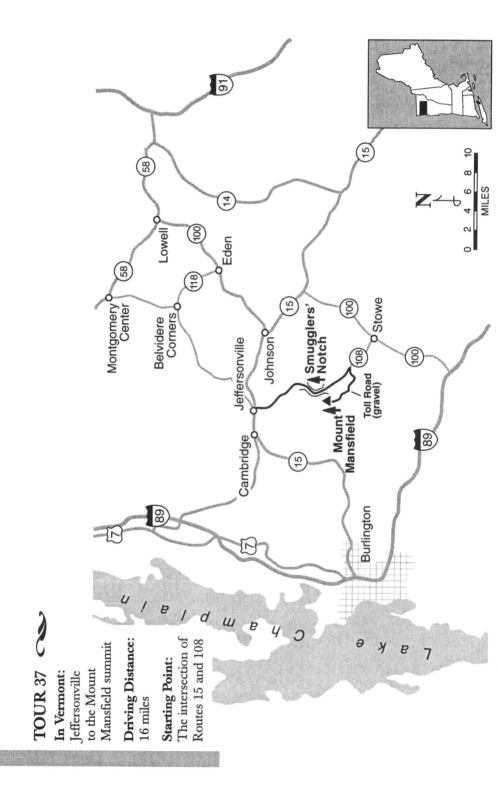

TOUR 37

In Vermont:
Jeffersonville
to the Mount
Mansfield summit

Driving Distance:
16 miles

Starting Point:
The intersection of
Routes 15 and 108

350

Tour 37
Smugglers' Notch to Mount Mansfield

A Recommended Fall Foliage Tour
HIGHLIGHTS

Mountains and Music ∾ Up the Brewster River ∾ Meadowsweet and
the New England Aster ∾ Smugglers' Notch State Park ∾ A Profitable War ∾
The Rare Plants of the High Cliffs ∾ Bush Honeysuckle and a Plant with
Hypodermic Needles ∾ A Boulder-Strewn Road ∾ Ski Bunnies and a Second-
Rate Switzerland ∾ The Mount Mansfield Toll Road ∾ Northern Hardwoods
Prevail ∾ All Sorts of Bloomers ∾ A Soverain Remedy for Bruises, and a
Dancing Sedge ∾ Entering the Taiga ∾ The Path Chinward ∾ Krummholz
and Lovely Blossoms ∾ Turf It Isn't ∾ The Gulf of Space and Time

O nce, while traveling through mountainous country, the great Russian-
born composer Igor Stravinsky was seen to be getting increasingly
irritated by something. At last the truth came out. When he and his
companions stopped high up the road, to take in the glorious view, he announced
that he couldn't stand mountains; they didn't tell him anything. Several significant
composers before and since have taken inspiration from the high places, but no
matter; the maestro had no time to waste on them.

It may be a matter of contention in the field of music, but few botanists would
ever say that mountains fail to tell them a great deal. The simple fact that the
earth's surface can rise to higher elevations is in itself a tale of tremendous
significance. That tale involves environments and plant communities that are
distinctly different from those a few miles away in the lowlands. As Vermont's
loftiest prominence, Mount Mansfield has an especially gripping story line. Reaching
up to just about the 4,400-foot mark, this peak is one of only two in the Green
Mountains where true alpine tundra conditions exist. In effect, its summit is a
tiny island of northern Canada set in a lower sea of forest greenery.

Before your ascent of Mount Mansfield begins, you have a rendezvous with
its dramatic northern gateway, Smugglers' Notch. (Note that both the notch and
the Mansfield auto road are closed in winter.) Begin this tour at the Jeffersonville
starting point by driving south on Route 108. The road heads up the valley of
the Brewster River, a tributary of the Lamoille. Before long, Mount Mansfield
lends its majesty to the scene.

The southern approach to Smugglers' Notch.

Along this stretch of road, Canadian hemlock is the dominant conifer. Also present are poplar, American elm, black locust, and red maple. A mere 1.9 miles from the start, you enter an area of pastures and tillage farming. In this cleared stretch, white pines have found the full sunlight they need to establish themselves. See if you can identify white ash and sugar maple, as well. In summer, old wildflower favorites line the shoulders to cheer your uphill progress. Their number includes common mullein, ox-eye daisy, Queen Anne's lace, and various goldenrods.

In another 4.2 miles, the composition of the forest community changes to reflect the climate of the higher elevation. Now American beech and striped maple are intergrading with the rest, and there is some mountain ash, too. About 0.6 mile beyond is an excellent view of the notch. From late July on, this is a good spot to see the native spiraea known as meadowsweet, as well as the incomparable New England aster. Higher up, ferns and other low vegetation clamber over the roadside rocks. When you reach Smugglers' Notch State Park just a moment or two later, carefully pull into one of the turnoffs on either side of the road. Congratulations. You're in the groove, so to speak, at an altitude of about 2,200 feet—or almost exactly halfway between sea level and the top of Mansfield.

The name of this breathtaking landmark derives from the fact that it was frequented by pro-British smugglers during the War of 1812. While Commodore Macdonough and friends were busy beating up the Royal Navy contingent in nearby Lake Champlain, many Vermonters and other New Englanders preferred to make love, or at least a profit, and not war. Here, in a region so close to Canada, more attention was paid to the bottom line than to President Madison's unpopular embargo; and the temptation to transact a little treasonable business often proved irresistible.

On the cliffs high above the road exists an extremely rare arctic flora similar to the one on Mount Pisgah at Willoughby State Forest (see Tour 36). These plants are experts in dealing with the harshest weather. Interestingly, their inaccessible habitat is quite different from the tundra environment you'll soon see on the summit of Mount Mansfield. In some ways, it's even more demanding, since a protective snow cover can't form to any extent on the vertical cliff faces, the way it does on the more horizontal Mansfield ridgeline.

Down at the level of the roadway, however, the plants need not face the full brunt of ice and wind, and the forest is brimming over with familiar faces. Sugar maple is the dominant overstory tree, and there is some yellow birch as well. Make sure you pay heed to the multiple-stemmed woody plants beneath them. As a group, the shrubs are often unjustly overlooked. Here, among the clumps of elder and hobblebush, you should be able to locate a species with leaves that are opposite, minutely toothed, and taper-pointed—the bush honeysuckle, also known by its genus name, *Diervilla*.

Much of the lush look here is imparted by the herbaceous plants. Yellow jewelweed is prevalent, but in other places, particularly on the fringes of the parking areas, a more formidable species predominates. This is the famous stinging nettle, an Old World alien that loves nothing better than rich, disturbed soil. In case you aren't already aware of its antisocial tendencies—resist any urge to touch it. Were you to examine its leaves under a dissecting microscope, you'd see they are equipped with one of nature's meaner tricks: hairs redesigned as tiny hypodermic needles, each filled with a dose of formic acid.

When a poor, unwary animal grasps the nettle, or just happens to brush against its foliage, the needles pierce its skin and inject their contents. If you get stung over a large enough area, it can be a very bothersome experience. Take it from someone who once waded unsuspectingly through a patch of stinging nettle on an island in the Mississippi River. My first thought afterward was that I must be in water moccasin country. A few minutes later, I wondered whether I'd ever see the splendors of mainland Illinois again.

According to some accounts, this prickling experience was actually appreciated by Roman soldiers serving at the outposts of their empire, in windswept northern Britain. There, the story goes, they collected leaves of a closely related nettle species and rubbed them into their skin—a rather masochistic substitute for the Mediterranean warmth they'd left far behind. Despite its wickedly effective system of defense, the plant has been used for a variety of things, from a herbal remedy, said to control gout and internal bleeding, to a sought-after vegetable dish and pudding ingredient. English countryfolk have even brewed a nettle beer. Its more elegant relative, slender nettle, grows on the lower slopes of Mount Mansfield.

Leave Smugglers' Notch by continuing south on Route 108. Mind your helm: the road descends steeply, winding obligingly around great fallen boulders. Soon you pass into an area largely given over to the industry that caters to people who strap rails on their feet and slide down snowy mountainsides. An anthropologist from some far distant time would have a field day here and farther down, in Stowe. Not only does the area exhibit all the elaborate social rituals peculiar to the Ski Bunny Cult but it also sports one of the world's finest collections of pseudo-Swiss architecture. An anthropologist-to-be might wonder why the people of our era were intent on transforming the Green Mountain State into a second-rate version of the Alps, when it already had the most appealing image possible. Gingerbread and cuckoo-clock decor makes about as much sense in this land of divine sparseness as would the ruins of Angkor, or the institutional excesses of the Spanish Baroque.

As you approach the mark 3.7 miles from the notch, keep an eye out for the right-hand entrance to the Mount Mansfield toll road. The road, approximately 4.4 miles long from bottom to top, is mostly unpaved. It's traveled by thousands of cars each summer, but be careful, especially in the hairpin turns, where the surface tends to be a little mushy.

The first half of the journey, from the toll gate to the Octagon House, is principally the domain of the northern hardwoods. Sugar maple, American beech, and yellow and paper birches form the canopy, while striped maple, hobblebush, and blackberry and elder species occupy the understory. Along the roadside are interrupted and hay-scented ferns and a host of wildflowers, including devil's paintbrush, ox-eye daisy, fireweed, goldenrods, joe-pye-weed, turtlehead, wild sarsaparilla, common milkweed, common St. Johnswort, New England aster, coltsfoot, pearly everlasting, and yarrow. Yarrow is a lacy-leaved Old World herb traditionally used against colds and fevers.

In damper, shady locations yellow jewelweed, a wild member of the horticulturally important *Impatiens* genus, forms substantial communities. In the first field guide to New England plants, seventeenth-century botanical explorer John Josselyn wrote this appreciation of the species:

> This Plant the Humming Bird feedeth upon, it groweth . . . in wet grounds, and is not at its full growth till July, and then it is two Cubits high and better. . . . It spreads into many branches . . . garnished on top with many hollow dangling flowers of a bright yellow Colour, speckled with a deeper yellow as it were shadowed. . . . The Indians make use of it for Aches, being bruised between two stones . . . there is not a more soveraign remedy for bruises of what kind soever; and for Aches upon Stroaks.

Here it easily achieves Josselyn's stated height of two cubits—that is, between 3 and 4 feet. Another herbaceous plant worth looking for is the fringed sedge, a monocot that develops dangling, tassel-like flower clusters. The languid elegance of these plants makes up for their lack of insect-attracting colors; even the faint slipstream of a passing car may set them gently dancing.

When you reach Octagon House, you may wish to stop for a moment at the gift shop and purchase a copy of the tour booklet for the summit's Tundra

Trail. The upper half of the road is a transition zone, where montane taiga plants come to the fore. Chief among these are mountain ash, mountain maple, and red spruce. Paper birch is here in a more stunted form—this is the heart-leaved upland variety.

In the final half-mile, you will have a splendid view of the summit ridge—with its profile resembling a horizontal human face—and the other Green Mountains. It's a fitting home for balsam fir, a hardy pine-family conifer distinguished from the spruces by its blunt-tipped needles and upright cones. Interestingly, the herbaceous American white hellebore, normally an inhabitant of lowland swamps and streamsides, is well established here, just as it is on the slopes of Mount Greylock, some 200 miles to the south (see Tour 29).

Park in the lot at road's end. You are now under Mansfield's nose. The walking trail to the real summit, the chin, is 1.4 miles one way. Before you set out, ignore the fashion statements of your fellow visitors, and make sure your footwear is appropriate for the rocky path ahead. Spike heels and flipflops simply won't do; sneakers or other high-grip shoes are a much wiser choice. In addition, you are about to enter one of the most fragile plant communities within public reach in all of New England. *Please stay on the rock-surfaced paths and boardwalks. Do not tread on any plants, even if they appear to be lawn turf.*

As you wend your way along the ridgeline, you encounter dwarfed versions of mountain ash, paper birch, and balsam fir. When tree growth is distorted by high winds and other stressful factors, the result is krummholz, a German term

The windy summit ridgeline of Mount Mansfield. Note the balsam fir krummholz.

meaning crooked wood. Most of the species here are intrinsically prostrate, or nearly so. These miniature alpine shrubs include crowberry, lowbush blueberry, bog bilberry, mountain cranberry, creeping snowberry, and Labrador tea. All except the first are ericads, or plants in the heath family, and all produce beautiful flowers, either subtle or dramatic, in summer. They are joined by two herbaceous dicots of note, diapensia and mountain sandwort.

Ask yourself why the low growth habit of these plants would be a distinct advantage in this blustery place, where freezing temperatures normally occur nine months a year. Would it be helpful or detrimental for these plants to be totally buried in the snow? If you're a New England gardener, you should be able to figure this out easily enough. (For the answer, see Tour 47.)

The final approach to the chin is pure tundra, unblemished by even the most gnarled little krummholz tree. What seems to be carefully laid-out turf is actually not grass, but Bigelow's sedge, close kin to the fringed sedge noted before. You are walking through a precious duplicate of a plant regime that covers the earth 1,000 miles to the north. This is one of a handful of New England mountain fastnesses where the old ways of life, the ways of everlasting cold and glacier ice, still triumph, however narrowly, over the world of warmth below. In looking across the gulf of space to Lake Champlain, you also look across ten thousand years of time, onto fields and forests that owe their existence to the gentler winters of modernity. The experts say the ice may come again. If so, the Bigelow sedge and its tough companions will spread anew. Life adapts.

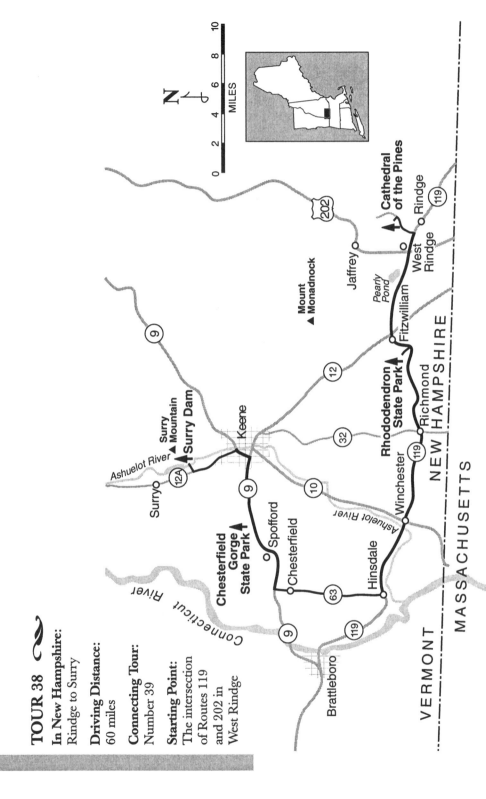

TOUR 38 ❧

In New Hampshire:
Rindge to Surry

Driving Distance:
60 miles

Connecting Tour:
Number 39

Starting Point:
The intersection
of Routes 119
and 202 in
West Rindge

Cathedral
of the Pines

Rindge

West
Rindge

119

Jaffrey

Pearly
Pond

Fitzwilliam

Mount
▲ Monadnock

202

9

Surry
Mountain
▲ Surry Dam

Ashuelot River

Surry

12A

Keene

Rhododendron
State Park ▲

Richmond

119

Winchester

NEW HAMPSHIRE

12

32

9

10

Chesterfield
Gorge
State Park ▲

Spofford

Chesterfield

Ashuelot River

Hinsdale

63

Connecticut River

9

119

Brattleboro

VERMONT

MASSACHUSETTS

N

MILES

0   2   4   6   8   10

Tour 38
A Swing Through Southwestern New Hampshire

HIGHLIGHTS

The Cathedral of the Pines ∿ New England's Tallest Tree: Its Products and Higher Uses ∿ A Gathering of Elders ∿ Monadnock's Bald Pate ∿ A Second Finland ∿ Rhododendron State Park ∿ Rosebay Glory and a Yankee Jungle ∿ Avoiding the Intricacies of Sex ∿ The Ashuelot River and a Covered Bridge ∿ The Slough of Despond and Fiddlehead Cuisine ∿ Chesterfield Gorge State Park ∿ Witch Hazels Will Be Different ∿ Surry Dam and One Wet April ∿ The Interrupted Fern ∿ A Hunt for the Bitternut Hickory

*T*hough it may not top the list of favorite New England tourist attractions, the lower left-hand corner of the state of New Hampshire presents the plant lover with many fascinating points of interest. In terms of ecology, climate, and culture, it has strong links with leafy, rural Massachusetts just to the south; at the same time, it is a distinct part of the ruggedly independent Granite State. This tour is an excellent demonstration of how the second-growth forest has reclaimed open land. It also features one of the region's great natural flower shows. This is woodland botany in some of its best locales.

We begin in the lake-dotted town of Rindge, in easternmost Cheshire County. The first stop is the Cathedral of the Pines, a nondenominational open-air chapel that welcomes visitors from May through October. In contrast to the vast majority of public and private attractions, the cathedral charges no mandatory entrance fee, though it is supported by freewill donations. To reach it from the starting point, proceed 1.5 miles east on Route 119, then turn left (north) onto Cathedral Road. The entrance to the facility is located on the left another 1.5 miles up the road.

The cathedral site is indeed an inspirational setting. The chapel area is situated in a grove dominated by New England's tallest tree, the white pine. It would be virtually impossible to tell the history of New Hampshire without mentioning the preeminent role this stately conifer has played. Besides being the focus of an intense political dispute between early British settlers and their mother country (see Tour 50), this famous tree has been utilized for centuries in a number of ways. In southern New England, Indians used its trunks for dugout canoes.

It does not furnish especially good firewood because of its low heat output, but it has long been a common type of kindling. More important is its continuing role as a multipurpose building material, suitable for shingles, clapboards, wainscoting, and flooring. In addition, it's been used in lighter pieces of furniture. If you're an antiques buff familiar with the term "pumpkin pine," you know it doesn't refer to another tree species, but to old white pine wood that has mellowed over the years to a rich orange-brown tone. Specialists sometimes call the well-seasoned wood "apple pine" instead.

Timber is by no means the only product the white pine provides. The tree has also been a major source of tar, pitch, resin, tanbark, lampblack, and turpentine. At one time, both the natives and the colonists ate its inner bark, sometimes candied, and drank a tea made from its needles. This infusion, rich in vitamin C, helped prevent scurvy. According to John Josselyn, even the earliest white settlers used the pine's turpentine medicinally, to clean cuts and aid in the healing of wounds.

The cathedral is the perfect place to understand that the white pine has a value transcending its economic worth. One nineteenth-century Yankee, an expert woodworker and builder who incidentally knew this neighborhood well, put his finger on the crucial point:

> I have been into the lumber-yard, and the carpenter's shop, and the tannery, and the lampblack factory, and the turpentine clearing; but when at length I saw the tops of the pines waving and reflecting the light at a distance high over all the rest of the forest, I realized that the former were not the highest use of the pine. It is not their bones or hide or tallow that I love most. It is the living spirit of the tree, not its spirit of turpentine, with which I sympathize, and which heals my cuts. It is as immortal as I am, and perchance will go to as high a heaven, there to tower above me still.

This is the sly and ringing prose of Henry David Thoreau. As usual, he noted what few people of his own time, or ours, are able to see.

As you walk about the chapel, look at the periphery of the seating area, where scarlet elder, a shrub with bright red fruit, grows in the thicket. In common parlance this fruit is a berry, though it's really a drupe. Drupes differ from berries in having their seeds surrounded by a hard covering. Good examples of this are the cherry and the peach: their pits are seeds enclosed in a stony outer layer.

The scarlet elder, yet another member of the horticulturally important honeysuckle family, can be told apart from common elder by the brown pith of its stems and its much larger buds. Additionally, common elder fruit is much darker, practically black, at maturity. The Indians depended on both shrubs as reliable sources of summer food; but common elder was probably the more appreciated of the two, since it blooms late and ripens at a time when little else does. The use of elder fruit as an ingredient in Grandma's homemade wine is now an indelible part of the American legend. Or at least it should be.

The elder's common name comes from the Anglo-Saxon *aeld,* or "fire." This alludes to the fact that the hollowed stems of a European elder species were used as blow-tubes to kindle blazes. In the Old World, the elder was sometimes held to be a symbol of ill tidings. Some superstitious gardeners refused to prune the plants for fear they'd have bad luck. In one part of England, it was thought that a young child disciplined with an elder whip would permanently stop growing. Nevertheless, the elder was often planted near homes to prevent witches from entering them.

The view northwestward from the altar is dominated by the silhouette of illustrious Mount Monadnock, that is, "muh-*ned*-nock." In these parts, the best way to get yourself branded as one of those expletive-deleted flatlanders is to pronounce it "mah-*nod*-nock." You might gather, reasonably enough, that the mountain's bald summit indicates that it's above the natural tree line. After all, it stands about 2,000 feet above the surrounding terrain. In actuality, it used to be cloaked in woods all the way to the top. In the year 1800, fire killed the spruce tree forest there; the result was a maze of dead and fallen trunks that supposedly harbored marauding wolves. Two decades later, local farmers used the wolf story justification for setting the tangle ablaze. With all the remaining plants finally destroyed, the soil was no longer anchored in place by roots, and it was quickly blown away. The result was the bare rock surface we see today, which suggests that the most wolfish creatures to hit Mount Monadnock weren't the wolves.

On your way out of the main seating area, note the red pines near the chapel entrance, and also take a look at the woody vine plants established on the exterior of the Hilltop House, on the far side of the bell tower. They belong to the Virginia creeper species, a veteran of New England forests and many ornamental settings as well. Its leaves are composed of five leaflets each, in a pattern botanists describe as palmately compound. Like its next of kin, Boston ivy, Virginia creeper turns a striking red in the fall. (If you'd like to see how Boston ivy compares, see Tour 39.)

———————————— ∾ ————————————

When you're ready to press on, return along the same route to the intersection of Routes 119 and 202. As you leave the cathedral grounds, check out the big red oaks that have been "set out"—that's New England for "planted"—in the open area in front. These handsome specimens have ample growing room, and have developed their species' full, spreading habit. That's quite a contrast to their equivalents in the woods across the road, where elbowroom can't be bought for love or money.

On reaching the Route 119-202 intersection, continue straight on 119 West. You are passing through Rindge, a town that in modern times has become the Little Finland of New Hampshire. Presumably, this was an excellent place for settlers from that northern land to become established in the New World. The earliest European immigrants here, the English, left the cleared fields and milder weather of their homeland and found themselves confronted with something quite different. The Finns, on the other hand, have the advantage of a culture firmly rooted in glacial lakes and coniferous forests. It's not a bad fit.

Just 0.8 mile past the junction, tamarack, the American larch tree, is seen on the right, in its accustomed damp-ground habitat. Next, the campus of Franklin Pierce College and the charmingly named Pearly Pond come into view. Almost 1.0 mile farther on, a large wetland appears on the left, where pickerel weed and other attractive aquatics thrive. In drier sections along this stretch, the woodland environment is a direct extension of northern Massachusetts' second-growth woodland. The prevalence of both red and white oaks indicates that this is still prime central hardwoods territory. Local power company contractors can chop away all they want at the swelling vegetation that continually threatens to engulf the roadside power lines. It will be back next year, or the year after that.

Route 119 meets Route 12 in Fitzwilliam. Once again, continue straight on 119 West through the town center. Here stands the lovely Fitzwilliam Town House, the near-duplicate of the Templeton, Massachusetts, Federated Church mentioned in Tour 24. As you pass the Blake House and the next building, note the tamarack and white poplar specimen trees, both on the right. One mile past the Route 119-12 junction, turn right onto the road leading to our next stop, Rhododendron State Park. Proceed an additional 2.0 miles through densely forested land—typical of modern New Hampshire—until you see the park entrance on the right. Turn here and drive past the house and into the parking area.

As its name suggests, the park's main attraction is a huge stand of rosebay rhododendrons; here, they are near the far northern edge of their natural range. These grand shrubs are renowned for their summertime floral display, which attracts appreciative human beings and ecstatic bees from miles around. The sign near the parking lot states that peak bloom is in mid-July, though my own experience, admittedly in a time of unusually mild winters, is that the local folks are right in saying the plants like to bloom somewhat earlier, to celebrate the signing of the Declaration of Independence. Interestingly, this conforms with one of Thoreau's journal entries from 1853, in which he writes that the Fitzwilliam rhodos are "in perfection" around the Fourth.

Rosebay rhododendrons in full flower at Rhododendron State Park.

To inspect these giants among shrubs, enter the woods through the upright posts and take the Rhododendron Loop Trail, an easy hike of 0.6 mile. You'll find all sorts of things to enjoy along the path in any season. Most of the way, Canadian hemlocks rule the roost, but in addition, there's a mixture of central and northern hardwoods, as well as mountain laurel and white pine. It doesn't take much sleuthing to discover a smattering of red spruce—a taiga regime indicator clearly thrown in to confound any attempt at ecological generalizing.

Shade-loving wildflowers are also present in force—partridgeberry, sharp-lobed hepatica, the splendid poppy-family delegate known as bloodroot, wild sarsaparilla, Canada Mayflower, checkerberry, goldthread, starflower, Indian cucumber root, and the two most aristocratic species here, painted trillium and trailing arbutus. The last is a special-delivery gift from heaven, even though its herbal application as a cure for bladder problems is a little less than ethereal. The humble goldthread, a plant in the buttercup family, yielded a traditional colonial cure for "thrush," a fungal mouth disease once commonly contracted by children.

Rosebay rhododendrons are acid-soil plants that are most at home in low, swampy areas such as this. When the trail brings you into their domain, you may think you've entered a Florida mangrove thicket or a Tarzan set; one almost expects exotic bird calls and spear-carrying Hollywood extras running through the underbrush. When these big, broad-leaved evergreens are in full flower, the air is filled with the low-pitched hum of countless bees, all too intent on the pollen orgy to bother anyone.

The main reason the rhododendrons have prospered and formed such a dense, monotypic community is that they reproduce willy-nilly by layering. Whenever a branch on a mature plant touches the ground, it forms roots and new upright stems. Before long, there's a thicket of asexually-produced shrubs. If you prefer, you can think of this assemblage of clones as one super-individual, since all the layered plants have a genetic makeup identical with the original organism. It's as though you were an exact duplicate of one of your parents, who for the sake of raising a family as quickly as possible, had chosen to avoid the intricacies of sex. As every high-school biology student knows, asexual reproduction is normally not a hot idea, at least in the long run. After all, it doesn't permit the reshuffling of genes that allows species to keep pace with a changing world. Layering, rhizomes, and other asexual methods are, nonetheless, a good way for a species to establish itself over a large area in a minimum amount of time. This reproductive strategy isn't restricted to swampy ground in the northeastern woodlands, either. The ubiquitous creosote bush of Southwestern deserts owes much of its amazing success to the same overall approach. (For more on the interesting adaptations of the rosebay rhododendron, see Tour 13.)

As imposing as the rosebay community is, there are additional attractions here. Two herbaceous plants—dark green bulrush and fringed sedge—grow next to the first boardwalk; they are both members of the same monocot family but distinguishable from one another by their flower structure and other traits. Naturalists often can tell the two types apart, simply enough, by feeling their stems. Bulrushes are round in cross section, while sedges tend to be tri-angular; which prompts the shopworn mnemonic, noted elsewhere in this book, that "sedges have edges." Farther down the path are examples of mountain holly, a shrub with long-stemmed fruit and simple, entire leaves, each equipped with a tiny point, or mucro, at the tip. On your way back to the parking lot, take the short, connecting Wildflower Trail, where some of the inhabitants of the forest floor, such as the coy white-flowered dewdrop, are labeled for easy identification.

When you depart Rhododendron State Park, retrace your track to the intersection of the forested road and Route 119. Turn right onto 119 West. This is still central hardwoods country, with much white pine and Canadian hemlock as well. Along the shoulders in midsummer, you may catch sight of tall meadow rue, one of the loftiest native herbaceous plants of the buttercup family. Sometimes reaching a height of 6 feet or more, it has intricate, fernlike foliage and unusual white flowers that owe their attractiveness not to showy petals or sepals but to their prominent stamens.

Continue straight on Route 119 until it meets Route 32 in Richmond. The terrain now becomes somewhat hillier, and American beech is visible in some of the higher spots. As you approach the junction with Route 10 in Winchester, you enter the valley of the Ashuelot River. Carefully follow Route 119 as it takes a jog to the left through town for half a mile, and then resumes its westward course by crossing the Ashuelot. Look at the vegetation on the riverbanks. It's composed of classic bottomland tree species: American elm, silver maple, red maple, and white oak. About 2.5 miles downstream, you pass a double-spanned, Town Lattice–design covered bridge, on the left. The 160-foot bridge dates from 1864, the year America's most truthful general, William Tecumseh Sherman, took Atlanta and thereby revitalized Abraham Lincoln's flagging reelection campaign. Common elder shrubs are common indeed in this area. The best time to see them from a moving vehicle is late June or early July, when they finally get around to blooming. Flat-topped clusters of white flowers stand out against a backdrop of dark green leaves. Scarlet elder, the species described earlier in this tour, blooms in spring. It too has white flower clusters, but they are more elongated. Scarlet elder fruit is already set by the time its tardy relative blossoms. (Here's a good opportunity for you to test your retention of botanical jargon. What is the correct term for the type of fruit elders have? It's mentioned in the Cathedral of the Pines description.)

You reach the junction of Routes 119 and 63 in Hinsdale approximately 3.5 miles past the covered bridge. Here, depart the lovely Ashuelot Valley by turning right onto 63 North. You'll revisit this important Connecticut River tributary farther upstream, but for now head into the forested seclusion of the Pisgah highlands. During the French and Indian Wars, this was a perfectly safe place to be, providing you were either French or Indian. The land was claimed by Massachusetts, but the forests harbored their share of enemy guerrilla fighters. The Indians, by the way, were not one monolithic anti-British force in this era. By and large the Algonquins sided with the French, but their enemies, the formidable Iroquois, were British allies.

In her memoirs, Susanna Johnson, one of the early colonists of western New Hampshire, described this landscape in lurid, Bunyanesque terms, full of Hills of Difficulty and Sloughs of Despond. Mrs. Johnson also noted, however, that by post-Revolutionary times, this section of the state was full of "well-till'd farms" that occupied "each rod of ground"–the clear result of more peaceful times and a burgeoning population. As you approach the center of Chesterfield, the well-till'd farms, bordering the mighty Connecticut and framed in the distance by Vermont hills, are still in evidence. This is lovely countryside.

In this area, in the great river's bottomland, local folks seek out New England's special springtime delicacy, fiddleheads, the newly emerging fronds of the tall-growing ostrich fern. It takes an experienced eye to differentiate them from the fiddleheads of other sprouting fern species, some of which are distinctly toxic. Fortunately, the baby ostrich fern fronds are now sold (when in season) in local supermarkets. If you'd like to try your hand at pteridophyte cuisine, I highly recommend you use the commercial source. To cook fiddleheads, rinse them well and briefly dowse them in boiling water; then rinse them again and steam them until they're tender but not mushy. They have a taste and consistency that reminds many people of asparagus. Some cooks prepare fiddleheads by frying them in batter.

When you arrive at the Route 9 intersection, turn right onto 9 East. A substantial wetland soon appears on the right. Many wetlands are relatively long-lived features associated with lakes or river systems, but others are more recent, the result of a process called paludification. Basically, this involves a rise in the water level to the point where the water is at or above the soil surface. It can be caused by a number of factors, but often it's the result of damming. Two sorts of creatures, both mammals, are compulsive paludifiers. One of these, a biped, likes to invest in real estate, watch football games, and drive around in large, beeping, yellow machines. Can you guess the identity of the other species? Here's a hint: given a choice between a medium-rare T-bone steak and a juicy poplar sapling, it will invariably take the sapling.

At a point 3.2 miles from your turn onto Route 9, carefully bear left into the parking area for Chesterfield Gorge State Park. (It's also referred to as a State Wayside.) This facility—not to be confused with the Massachusetts Trustees Reservation of the same name, described in Tour 27—features a short trail through a small, creek-formed canyon. Head down that trail. At first, Canadian hemlocks cast a dense shade, and consequently, little exists in the way of understory. The

At Chesterfield Gorge State Park, modern forest trees grow atop rock that once formed an arc of volcanic islands off ancient North America.

rock units exposed in the gorge have quite a story to tell. They are part of the Bronson Hill zone, which hundreds of millions of years ago was an arc of volcanic islands lying quite some distance off the coast of ancestral North America. In the fullness of time, the forces of plate tectonics pushed the island arc into the continent proper. Later still, a much larger land mass came rolling in behind, to sandwich it into place.

As you reach the bottom of the gorge, follow the path downstream—it is marked by blue blazes. This is where most of the wildflowers are, including quite a few of the species listed at Rhododendron State Park. The low-growing Canadian yew crouches near the stream, and without too much trouble you should also be able to find a higher, deciduous shrub with asymmetrical, wavy-edged leaves. This is witch hazel, a fascinating species belonging to a group of primitive flowering plants. Whatever its pedigree, its habits of reproduction are all its own. In fall, when most broad-leaved species are preparing for their next period of dormancy, witch hazel is happily putting forth its zany yellow flowers. Thoreau's journal entry for October 9, 1851, comments on it:

> The witch-hazel here is in full blossom on this magical hillside, while its broad yellow leaves are falling. Some bushes are completely bare of leaves, and leather-colored they strew the ground. It is an extremely interesting plant,—October and November's child, and yet reminds me of the very earliest spring. Its blossoms smell like the spring, like the willow catkins; by their color as well as their fragrance they belong to the saffron dawn of the year, suggesting amid all these signs of autumn, falling leaves and frost, that the life of Nature, by which she eternally flourishes, is untouched.

Witch hazel's unique flowering cycle is matched by its extremely effective method of seed dispersal. When its fruit capsules dry out, they split open violently, discharging their contents a distance of 40 feet or more. For centuries, witch hazel has been used medicinally, particularly as the base for a popular, soothing lotion. In addition, early European explorer Martin Pring reported that it was highly prized as a source of premium-grade soap ashes. Its common name refers not to witchery or the black arts, but an Old English word, allied to *switch,* referring to the plant's pliant stems.

———————— ∾ ————————

Leave Chesterfield Gorge State Park by resuming your eastward trek on Route 9. This road is the main thoroughfare across southern New Hampshire and Vermont. The laws of the Green Mountain State ban off-premises advertising signs, so this route is mostly free of eyesores. In the Granite State, however, things are predictably more laissez-faire, and as you can see in this vicinity, the landscape is somewhat marred by billboards. Regrettably, these have a long history in New England. Some of the earliest of their kind adorned covered bridges, as old-timers will attest. Approximately 1.9 miles from the last stop is a dramatic road-cut at the crest of a hill. The exposure is granodiorite, a granitelike rock that came into being almost half a billion years ago, as a molten mass injected deep into the volcanic material of the Bronson Hill zone. Today, birch seedlings sprout on

the exposed surfaces and valiantly cling to life long enough to set seed and launch the next generation on its way.

In another 3.2 miles, you've descended onto the floor of ancient Lake Ashuelot. As with Vermont's Champlain Lowland and the Connecticut River valley, this wide bottomland was flooded with meltwater after the most recent retreat of the continental ice sheet. The natural dam of glacial debris that created the lake was located between downtown Keene and where you first met the Ashuelot River, earlier in this tour. The exposed lake bottom is an essay in the changing patterns of land use; once a renowned potato-growing center, it is now a mixture of wetlands and cornfields, which are giving way more and more to the encroachment of car dealerships and shopping centers. The contest between conservation and our mania for consumerism, and the battle between long-term and short-term interests, does not always have a happy ending. That should come as no surprise to anyone who has lived through the 1980s and early 1990s.

When you reach the three-way stoplight junction of Routes 9, 10, and 12, turn left and follow the signs for Route 12 for 2.5 miles. Then bear right onto the Maple Avenue exit. At the base of the ramp, turn right on Route 12A. Soon you're passing through one of New England's many red pine plantations, most of which are the legacy of the depression-era Civilian Conservation Corps. See if you can detect the pinkish cast often indicative of this conifer's bark.

At the next stoplight, take note of your odometer reading and continue straight on 12A North. You're now passing through one of Keene's outlying residential neighborhoods. The Ashuelot River is to your right; beyond it rises the impressive ridge of Surry Mountain. Keep an eye out for a fork in the road at 1.9 miles north of the last stoplight, and bear right off Route 12A onto the Surry Dam access road. The final stop of this tour lies just ahead.

When you reach the facility's entrance, stop for a moment to inspect the impressive engineering feat. Surry Dam was completed in 1941; the agency that built it, the U.S. Army Corps of Engineers, still maintains it and administers its associated public recreational area. As impressive as it is all by itself, the dam is but one part of a much larger flood-control network that involves much of the Connecticut River drainage system.

Make sure you see the spillway, which seems to have been cut with a single, giant knife stroke through the tough metamorphic rock. So far, this artificial river channel has seen action only once, in the freshet of April 1987, when the area was subjected to heavy rains and rapid snowmelt. In that short but unforgettable episode, the world seemed to be in danger of a second drowning. Sections of local roads were carried away by sheets of water appearing where no streams had been before. The customary grimness of mud season gave way to something much more unsettling. The torrent that passed down this narrow groove in the rock relieved the immense pressure on the dam. It also served as a reminder that nature still has the force, infrequently applied but unimagined in its potential, to dwarf all human works.

Cross the dam, either on foot or in your car. The view of Surry Mountain and the submerged river valley seems to refer more to the time before human settlement than it does to the great structure that makes it possible. At the far end, look at the woody plants that form a thicket on the edge of the parking circle. Among them is red osier dogwood, a shrub with beautiful ruddy stems somewhat reminiscent of ornamental cherry trees. The bark has been used as a malaria cure, as a treatment for typhoid fever, and, when ground up, as a tooth powder. Take a few moments to hike up the forest path that leads north from the circle. In spring, wood anemone puts forth its cheerful and uncomplicated white blossoms. A bit later, in the warm season, the alien self-heal becomes one of the most obvious members of the trailside community. Its name demonstrates its reputation as a herbal remedy for sore throats and practically everything else.

This is also one of the best places to see the interrupted fern. Many ferns have their spore-producing structures borne on the undersides of their leaves, or on separate modified fronds that almost look like flower stems. This species, however, has its spore structures placed in one fertile area partway up its otherwise normal fronds. The resulting gap in the pattern of leaflets gives the plant its common name.

Examine the overstory trees. You are still in the central hardwoods regime—a fact of special interest when you consider that there are sizable lowland tracts of taiga conifers in Acworth, a mere 15 miles or so to the north. This is a terrain of rapidly changing microclimates. Here, though, it's not a matter of red spruce but of red oak, with basswood and hefty white pine specimens present as well. Try to find the hickory trees, and if you can, try to pin down their exact species. You should be able to find the type known as bitternut. Its leaves could be mistaken for those of the shagbark, but they generally have more leaflets—seven or nine, instead of the shag's customary five. An even more obvious indicator of the bitternut's true identity is its furrowed but unpeeling trunk.

TOUR 39 ～

In New Hampshire:
Keene

Connecting Tour:
Number 38

Starting Point:
In downtown Keene,
along Main Street near
Central Square

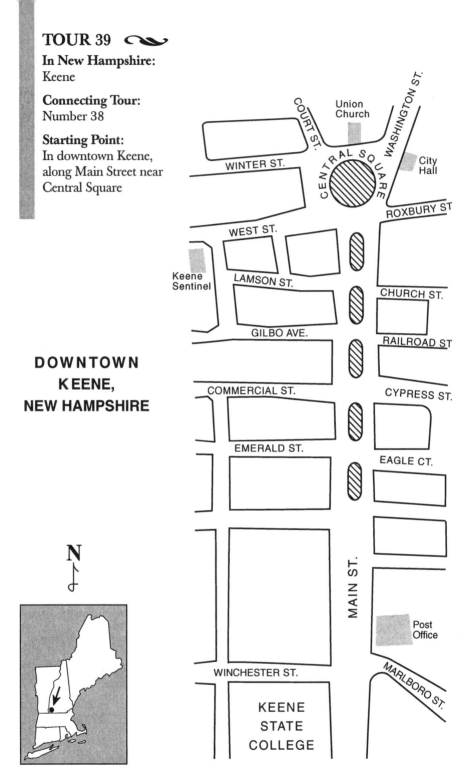

**DOWNTOWN
KEENE,
NEW HAMPSHIRE**

N

Tour 39
A Stroll Through the Street Trees

When John Donne, the Jacobean poet and divine, considered humanity's place in the scheme of things, which was pretty often, he usually came away from the experience with hope for the human soul. He reasoned each of us is a little world, a microcosm of a greater order. But he recognized the chafing limit of our short existence: "What a Minute is Mans life in respect of the Sunnes, or of a Tree!"

Donne was hardly the first person to use the tree as a symbol of permanence. Another writer, the Japanese poet Basho, wondered

> *How many priests*
> *How many morning glories*
> *Have perished under the pine*
> *Eternal as law?*

Indeed, this commonly held view of arboreal longevity is often justified. In direct contrast to animals, trees have the potential for continued growth throughout their entire lives. They have no equivalent, as far as we can tell, to the automatic time bomb of the aging process we know all too well. With good luck and the right conditions, our leafy friends can measure their lives not in decades but in centuries. Be that as it may, there is one type of tree, the kind planted on city streets, that has a much shorter average life-span than our own. Why is this the case? What can be done to better the odds?

This tour, taken entirely on foot in the pleasant surroundings of downtown Keene, New Hampshire, will give you some answers to those questions. It will also introduce you to a number of tree types you won't find in the wilds of New England. As the hub of southwestern New Hampshire, downtown Keene is large enough to have an urban feel to it; in addition, it boasts a relatively new, well-labeled collection of plants that exemplifies the latest generation of street trees in use throughout the Northeast, and is an ideal place to acquaint oneself with the subject.

<hr />

Before we commence, take a moment and put yourself in the shoes, or rather the soil, of an urban street tree. Unlike your relatives in the forest, you require periodic human care if you're going to survive. More often than not, that care won't arrive. The government agency or civic organization that installed you probably had just enough money budgeted for planting, with nothing left over for sustained maintenance. Unlike your other relatives, the ones planted on private property, you belong to no one person or family; no homeowner appreciates your role in increasing property values and adding aesthetic appeal.

As though that weren't enough, your streetside environment is much more stressful than a forest or lawn. The most important proof of this lies in the pit that is your inescapable home. Most New England soils are somewhat acidic to neutral, but it's likely yours is distinctly alkaline, thanks to your enclosure's cement lining and other factors. On top of that, the soil may be badly trampled and compacted into impervious hardpan. As a result, you have trouble growing new roots. There is too little vital oxygen and too much toxic carbon dioxide. You're starved for nutrients, and chances are you're either dying of thirst or stuck with too much water that can't drain out the bottom of the pit. In this icy climate especially, you're subjected to massive overdoses of poisonous rock salt, not to mention the occasional splash of car oil. Dog walkers, who rationalize that their pets' wastes help fertilize the ground you live in, are actually adding to the unsightliness and toxicity of the immediate environment.

Add to the list of woes the fact that most street trees occupy heat islands, where conditions are challenging even on a seemingly mild day. Forest trees don't have to worry about thermal energy reflected off the pavement, sidewalks, and nearby buildings; street trees do. In winter, when woodland trees enjoy the comfort of numbers and the lessened exposure to freezing winds, city trees must bear the full brunt, because they stand quite literally alone. Sometimes winter winds are actually intensified by the layout of the streets themselves. When all these things are added together, the result is an inordinate amount of physiologic stress. Trees in this situation are acutely prone to attack by pests and diseases. It's the classic domino effect: one set of problems engenders another, which in turn triggers another.

Despite these harrowing factors, some street trees manage to beat the odds and live to a ripe old age. It's these, no doubt, that encourage people to plant still more. If things work out, the benefits are enormous. Not only do the trees provide shade, they also moderate temperature extremes at ground level, help cleanse the air, and mitigate air and noise pollution. They also have a tremendously positive psychological effect on people, who consciously or unconsciously prefer leafy surroundings to the monotony of unadorned building fronts.

It's difficult to say when the movement to add greenery to city thoroughfares first got under way, but we know the idea was alive and well in colonial North America. When botanist Pehr Kalm visited Philadelphia in the middle of the eighteenth century, he found native sycamores and black locusts, often planted alternately, along the main streets. Even earlier, some New England towns passed laws discouraging damage to trees on their public boulevards; this was the case, for instance, in Groton, Connecticut, in 1665. By the early 1800s, the American elm, an ornamentally superb species tolerant of oxygen-starved soil, had become all the rage and had been planted in many parts of the country in monocultural stands. For well over a century, this presented no problem; but then Dutch elm disease (often abbreviated DED) arrived from Europe. So began one of the most disastrous plant plagues of modern times.

The densely packed elm monocultures ensured that this largely untreatable fungus would spread like wildfire. Most American elm street trees are now long gone, and because more virulent strains of the pathogen have recently come to the fore, the disease continues to ravage the few survivors, as well as elms growing in the wild. One still hears certain people espouse a return to the monocultures, using "resistant" American elms. Even if these varieties ultimately prove to be resistant to current forms of the fungus, there's no guarantee they will be a match for the next strain down the line. Even now, American elms are seriously affected by other diseases unrelated to DED. But there is a stronger point to be made here: massed plantings of any kind are the horticultural equivalent of Russian roulette. There are various other examples of monoculture fiascos, from the anthracnose-stricken London plane trees of New York City to the lethal yellowing disease of coconut palms and related species in Jamaica and Florida. The only safe strategy, therefore, is to plant a mixture of different species in one area, to lessen the chances of all trees succumbing at once.

After this grim overview, you're ready to set forth. If you're arriving in Keene by car, park in the vicinity of Central Square, located at the head of Main Street several blocks north of Keene State College. Many metered spaces are available on Main Street and the adjoining side streets. Proceed to the corner where Emerald Street meets the west side of Main Street. The tall, narrow tree along the sidewalk on the north side of Emerald is a columnar cultivar, or cultivated variety, of sugar maple. This is a tree type that has been selected and bred specifically for its constricted shape. On the Main Street traffic island nearby stand examples of the ginkgo, the famous "living fossil" that has generally proved to be one of the most problem-free of street trees. In this vicinity, unfortunately, at least one of the ginkgos has already succumbed—perhaps because here the

species is at the northern limit of its cold-hardiness. (See Tour 18 for a fuller discussion of the traits of this unique species.)

From an ornamental aspect, the ginkgo is a bit of a wild card, in that its habit can become gawky or downright humorous as it matures. It's not unusual for a tree to behave quite properly for a decade or so, then throw out one long, lateral branch that disturbs its simple symmetry. Personally, I think this can be a character-building experience for a city street, but many landscape architects, hardened to the dictates of corporate boredom, insist on the regularity of lollipop trees. Since the ginkgo species is the sole survivor of an ancient order of the gymnosperms, it is not a flowering plant. Don't expect nice puffy blossoms in spring or any other time. However, its deciduous leaves, intrinsically interesting due to their strange fan shape and medium green color, are all the more eye-catching in autumn, when they turn a glowing golden yellow.

Take the crosswalk just above the columnar sugar maple to the other side of Main. On the eastern side of the street, head north toward the square. At the corner just this side of Cypress Street stands a Japanese zelkova, an elm-family

The bark of one of Keene's young Japanese zelkova street trees. Note the horizontal rows of lenticels. As the tree matures, its bark will take on another pattern, resembling brown-and-orange patchwork.

species that has been touted as a replacement for the American elm. Zelkovas are indeed handsome trees, and their presence here does the town credit. But their habit is slightly stiffer and less graceful than that of their illustrious predecessor. Fortunately, this species, when taken for its own merits, is highly commendable for other reasons. Its dark green foliage, when not nibbled too badly by elm leaf beetles, is very elegant in shape and overall texture, and the bark is extremely attractive. In its youthful stage, it is often covered with a striking pattern of bumps, called lenticels. These structures, also common in cherry trees and many other woody plants, are a sort of auxiliary breathing apparatus that supplies the inner stem tissues with gases they need from the surrounding atmosphere.

The block north of Cypress contains still more street tree types. First, though, look at the zelkova here and notice its small individual pit, and the gravel and grating that have been put over the soil proper. This design somewhat lessens the chances of compaction, but it can be a real chore to remove the grating and check the soil later on. The next tree, also in an isolated pit, is a Callery pear. The identification tags along the way say "Bradford Pear" instead. "Callery" refers

A raised turf island on the west side of Central Square. Such continuous strips offer better growing conditions for street trees than small, separated pits do.

to its species, and the epithet "Bradford" is the name of this particular cultivar. Callery pear, alias *Pyrus calleryana,* is now offered in several varieties besides this one: Chanticleer, Aristocrat, and Redspire, among others.

Visitors from the Sunbelt might easily mistake the tree for the look-alike evergreen pear, *Pyrus kawakamii.* Both originally hail from the Orient, and both flower splendidly, but only the Callery is consistently deciduous enough to put on an excellent predormancy color display. This can be a real treat, with the leaves turning scarlet or a rich burgundy. In addition, this species is resistant to the dreaded bacterial fire blight disease that has hobbled the American pear-growing industry. In my experience, the evergreen pear is considerably more susceptible. (In the Phoenix area, I've seen large parking lot plantings blossom beautifully, then suddenly suffer massive branch dieback.) In the eyes of the landscape architect, the Callery pear fills an important gap in street tree design because it rarely grows to an unmanageable size. Alas, all these attributes tempt reckless or inexperienced planners to use it as the latest monoculture tree. Fortunately, that hasn't been done here in Keene.

The next trees of note on this block should count their blessings, because they've been placed not in single holes but in consolidated turf islands. Recent research shows that continuous plantings such as these create microenvironments that are much preferable to the ones the stand-alone specimens have. The species here include pin oak, an old curbside standard; another offering from the elm family, hackberry; and two green ash cultivars, Summit and Marshall's Seedless. Pin oak has been widely used as a street tree, but as noted in Tour 27, it sometimes has its problems. The hackberry, seen in one of its wild settings in Tour 32, is an intriguing new choice for urban street use. It is a tough plant, but it does have the potential for one particularly disfiguring ailment. In New York's Central Park, and many other locations, it is afflicted with "witches' brooms," clusters of small twigs on the ends of the larger branches. This is caused, in some way not yet fully understood, when two different invading organisms, a type of mite and a powdery mildew fungus, are present.

Meanwhile, the green ash represents one of the most handsome and dependable shade trees in general use. If you glance at the median strip, you'll see they've been used there, too. Naturally, every living thing has some Achilles' heel or other, and ashes can be particularly susceptible to boring insects. Despite their name, these pests do not kill the trees by talking to them endlessly about economics, the movies, or other dull subjects, but by chewing larval galleries through the trunk tissue. From an ornamental standpoint, the green ash is a tree of understatement. Its flowers do not win a poet's praise, nor does its autumn yellow rival the ginkgo's. But to the connoisseur of geometric patterns, ash bark is a work of art. The matrix of ridges and furrows forms a pattern of narrow, upright diamonds.

Continue up the eastern side of Main and cross Church Street. Here you'll spot more hackberry and green ash. Once you're beyond Roxbury Street, take the crosswalk over to the fountain island. Despite the swirling traffic, this is a pleasant spot. In summer, bank tellers, appliance salespeople, and bookstore owners pause to chat with the gardeners watering beds of annuals under the great red oaks. It is wonderful to see one town center not yet depopulated by malls and shopping centers far from the true heart of things.

The fountain area contains two other tree types worth mentioning. The first, thornless honey locust, grows to an impressive size but has light, feathery foliage; the other, littleleaf linden, is a European equivalent of the basswood native to New England forests. Like the basswood, this linden has deliciously fragrant cream yellow blossoms that usually open here in late June. When it's not the victim of mitigating circumstances (as we'll see in a few minutes), the littleleaf linden develops a symmetrical, broadly pyramidal habit that makes it an especially beautiful shade tree.

Take the crossing on the northwest side of the island to the group of striking zelkovas that stands to the right of the junction of Court Street. This gives you the chance to take another look at this appealing introduction from the Far East. Next, head south on the western side of the square by crossing Court. When you reach West Street, cross it, then make a quick detour a couple of blocks to your right (west) to the headquarters of the award-winning local newspaper, the *Keene Sentinel*. Besides containing some of the nicest people in Cheshire County, this unpretentious building is covered, at least partially, by Boston ivy. This woody vine is yet another Asian import, despite its Yankee common name. It sports glossy, three-lobed leaves that flush wine red in the fall. The softening nature of this plant, coupled with the dark, evergreen foliage of the yews and rhododendrons in front, makes a statement that is solid, simple, and effective—in direct contrast to the badly overplanted insurance company grounds across the street. Also notice how the crowns of the streetside lindens on this block have been sheared and damaged by passing trucks—an all-too-common problem for street trees.

When you return to the corner of West Street and Main, turn right and head south, down the near side of Main Street. Once again, the wise placement of trees is obvious: ginkgo is blended with green ash, pin oak, and zelkova. The one flaw in the overall design can be spotted when you reach Gilbo Avenue and see the littleleaf lindens squashed into a cramped location along the side street. As these potentially hefty trees get larger, they will seem even more awkward and inappropriate. This points out the need to think ahead. The rule is simple but often forgotten: Never plant a tree based on its current size!

It would be unfair to end this tour through the street trees on a critical note. Few other New England towns of Keene's size have done such an attractive or canny job of enhancing their boulevard plantings. If they're kept in decent shape, these trees will fill in well, provide appealing shelter, and give Keene the dignified focal point this regional center of commerce and history deserves.

(If you're in the mood to stroll a bit farther, head south along Main Street to the Keene State College Campus. Some large American elms survive in this neighborhood, and you can catch a glimpse of how this street looked in the glorious era before Dutch elm disease.)

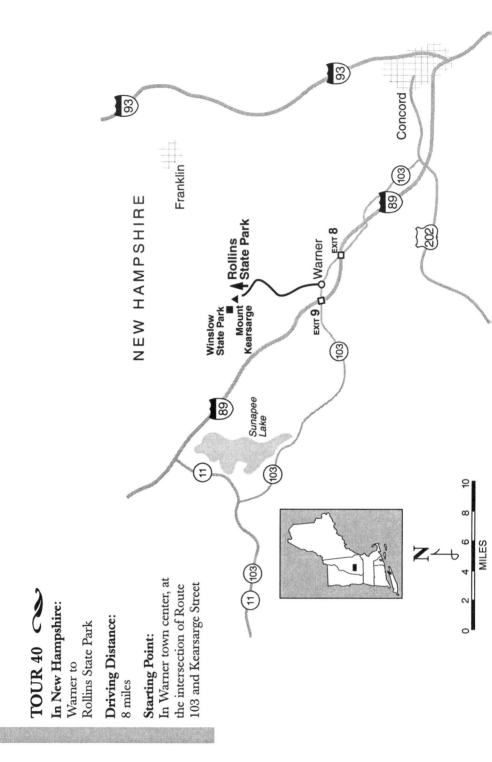

TOUR 40 ~

In New Hampshire:
Warner to
Rollins State Park

Driving Distance:
8 miles

Starting Point:
In Warner town center, at
the intersection of Route
103 and Kearsarge Street

NEW HAMPSHIRE

Franklin

Winslow
State Park

**Rollins
State Park**

Mount
Kearsarge

Warner

EXIT **8**

EXIT **9**

Concord

89

103

202

89

Sunapee
Lake

11

103

103

11–103

93

93

N

MILES

0   2   4   6   8   10

Tour 40
Mount Kearsarge

A Recommemded Fall Foliage Tour

HIGHLIGHTS

The Undiscovered Country ∿ A Town That Fits Its Surroundings ∿ Of Barns
and Trees ∿ Rollins State Park ∿ A Leitmotiv of Birches ∿ Beginning with the
Gray Birch ∿ Woodland Wildflowers ∿ A Miracle of Horsetail Architecture ∿
Paper Birch, Mountain Style ∿ The Yellow Birch ∿ A Mini-Peatland ∿ New
England's Sordid Jezebel ∿ The Lethal Sheep Laurel ∿ A Dwarfish Sort of
Taiga ∿ Crustose Lichen and a Base for British Soldiers

*A*t times, visiting New England's natural wonders can be a bittersweet
experience. The gorgeous scenery is relatively easy to get to, but it's
often viewed at the price of considerable distraction from crowds of
fellow tourists. In some cases, getting away to the great outdoors resembles getting
away to rush-hour Manhattan.

That's the thing that makes the area around Mount Kearsarge and the town
of Warner, New Hampshire, so appealing. Even though it's situated midway
between popular Sunapee Lake and the state capital, and is served by an interstate
highway carrying considerable traffic from Boston to Montreal, its remote and
sparsely populated soul manages to survive intact. As any passing motorist can
see, the landscape imparts the perfect North Country image of mountains, lakes,
and apparently limitless forest. One would have to look very carefully to find
another unspoiled expanse as beautiful as this. On this tour, you'll probably see
a few other people along the way, but by and large, they'll be local folks who
regard Rollins State Park as their own special place, a well-kept secret best left
that way.

Begin your ascent of the mountain in Warner, one of the most unobtrusively
beautiful towns anywhere in America. Notice how perfectly the spirit of the place
matches the character of the surrounding terrain. When you reach the intersection
that marks the tour's starting point, turn north on Kearsarge Street. Soon the
name of this picturesque country byway changes to Troy Hill Road; continue
straight ahead. This is so quintessentially the Rural New England Look that you

381

may have to pinch yourself to make sure you're not dreaming. It has a timeless, legendary quality. Sugar maples stand guard next to old barns, whose simple lines match the architectural merit of many a European cathedral.

This is a land of the northern hardwoods. Besides the big maples, there are American beeches with their smooth, gunmetal bark, and basswoods, with leaves sometimes the size of flapjacks. Nevertheless, red oaks of the central hardwoods regime find the local environment, especially the warmer south-facing hillsides, suitable to their own tastes. The entrance to Rollins State Park is 4.7 miles up the road from the start. If by chance you find the gate closed, take the gravel driveway that runs behind the park house.

The comparison of three native birches will be one of the focal points of this tour. Just up on the right, you'll find a stand of the first type we'll consider, the white-barked gray birch. One of the more traditional common names for this species is poverty birch, no doubt conferred because the tree sprouted so readily on abandoned farmland lost, as it often was, through debt. Ecologists can frequently determine the type of preexisting land use by simply noting which plants have replaced the fields. For instance, gray birch often colonizes areas originally under tillage, while common juniper indicates old pastures. Both kinds of plants can be found in this vicinity—which suggests that the land once served both livestock and crops. That certainly would be no surprise.

Given the similarity in stem markings and color, it's easy to mistake this for the more beloved species, paper birch. A little scrutiny clears up the confusion. Gray birch leaves are smaller, and have a more triangular shape as well as a longer, tapered tip. The bark of paper birch exfoliates by peeling into curly strips; gray birch bark stays put. Although one lucky specimen in Somers, Connecticut, has managed to reach a height of almost 80 feet, gray birch as a rule does not have a good record for longevity and does not achieve the other's loftiness and girth.

Red baneberry blooms in the woods, on the lower slope of Mount Kearsarge.

It may have a scrubbier look, but it plays an important role as a nurse tree for tender white pine seedlings. Without the light shade provided by gray birches, the pines are susceptible to attack from sunlight-loving weevils. In addition, the species has been put to all sorts of household uses. The Indians found it soothed burns, bruises, and even the pains of childbirth; English settlers made brooms from its pliant stems. And our favorite marker of metes and bounds, Mr. H. D. Thoreau, reported that this and the other white-stemmed birches made excellent surveying rods, easy to sight on.

The road to the Kearsarge summit trail is paved all the way to the parking lot. The lower slope has been thoroughly reclaimed by the forest. If you pause here and there along the road, you'll discover a thriving wildflower community with such notables as Canada Mayflower, baneberry, corn lily, wild strawberry, starflower, goldthread, bellwort, wood rush, and wood aster. There are also plenty of pteridophytes. The most obvious of these is hay-scented fern, an aggressive plant that forms large colonies of yellow-green fronds rising singly from the ground, rather than from clusters or rosettes. However, the most beautiful free-sporing plant is the elegant woodland horsetail. It's a miracle of fine branching and tracery in three dimensions, which prefers to raise its lacy spires in damp spots, including the roadside drainage ditch.

The upper portion of the road, beginning roughly 0.8 mile from the park entrance, features a transition to a new tree population. Below, white pine and red oak blended in with the northern hardwoods species; now, red spruce signals the onset of the montane version of the taiga regime. In some places, another woody plant with a much different habit appears on the shoulders. This ground-hugger is lowbush blueberry, an expert survivor in harsh environments. At 2.2 miles from the entrance, you'll come to an especially good scenic turnoff to the right. The view from here, especially at the height of the fall color show, is stunning. Above this point, mountain maple and mountain ash trees increase in number, and two birch species begin to take on stunted forms in response to the challenging conditions.

The first of these is the paper birch, here represented by its highland variety. If you check its leaves, you'll see that their bases are rounded, giving the foliage a somewhat heart-shaped form. The lowland variety of the same species has leaves with tapered bases. As everyone knows from kindergarten on, the Indians of the northern reaches of New England used paper birch bark for their canoes, though tribes in the southern part of the region relied on dugout craft formed from such straight-trunked species as white pine or tulip tree. The natives harvested the bark during the trees' winter dormancy when the temperature rose above freezing, knowing that was when the cambium layer underneath was least likely to split away. The tree's outer layer has seen additional duty as a source of perfume and a good yellow dye. But in modern times, it is the wood of paper birch that is most sought after. It is an important source of pulp for paper and has also been fashioned into toothpicks, bobbins, spools, and veneer material.

The other species present is yellow birch. It poses few identification problems, since its shredding silvery bark sets it apart from all other northeastern tree types.

Its wood is prized for handles, doors, floors, railroad ties, and particularly for cabinets and furniture. As with some other types of birch, it can be stained to look like maple, cherry, and even walnut; therefore, it often does duty as a low-budget substitute for them. Its leaves generally turn a clear yellow, the best autumn color of any northern birch. Yellow birch is normally the most cold-hardy of all the northern hardwoods and in mountain locations such as this, is usually the last to disappear from the scene as you travel upslope.

When you reach the parking lot, pause a moment to look at the mountain ash and pin cherry plants nearby. Both are rosaceous—members of the rose family. True to its name, mountain ash usually sticks to the high places, at least until you approach the Canadian border; but the pin cherry can be found in the lowlands of this area as well. It is most common in habitats recently cleared by fire. Its seeds have the ability to remain inert for years—fire triggers their sprouting. In common with the gray birch and the junipers, it prefers full sunlight. It too benefited tremendously when white settlers turned thousands of square miles of woodland into open farmland.

The path to the summit fire tower is quite stony in places, but otherwise not too formidable. The bedrock here is part of the Devonian period Littleton formation, an assemblage of supertough metamorphic rocks that also make up Mounts Washington and Monadnock. Before long, you reach a miniature peatland containing genuine sphagnum moss, no less, and the small but showy shrub known as rhodora, one of New England's wild azaleas. In case you didn't know it, azaleas are card-carrying ericads, now included in the genus *Rhododendron*. In spring, rhodora puts on quite a show, in classic painted-lady azalea fashion, with lurid magenta blossoms that pop open before its new crop of leaves are fully out. To see this splash of color when the rest of the plant world is still trying to stumble awake is a bit like hearing an orchestra's timpanist let loose three beats too early. In a land as persistently Calvinistic as this, it's a wonder this shameless Jezebel hasn't been banned for unseemly behavior. Still, most early-season hikers welcome the rhodora's outrageousness after a long and dreary winter.

Keep an eye out for other rosaceous plants: three-toothed cinquefoil, a prostrate alpine species; and the shrubs or small trees grouped together in the genus *Amelanchier* (pronounced "am-ell-*ank*-ee-er"). The latter go by the common names of shadbush, shadblow, juneberry, and serviceberry. Their white flowers are one of the most pleasing sights on the mountainside in May.

A short, woody ericad is here, too. It is sheep laurel, one of the three New England species in the genus *Kalmia*, named by Linnaeus to honor his friend and protégé Pehr Kalm. This species is not as impressive as the hefty mountain laurel, which ironically prefers lower elevations in New England. The sheep laurel is formidable in one respect, though: it produces toxins that can be deadly to livestock. Farmers, well aware of its properties, bitterly nicknamed it "lambkill." Occasionally, domesticated animals nibbled the sheep laurel, survived the poisoning, and actually became addicted. When this happened, their owners would

The view from Mount Kearsarge's upper reaches. Note the krummholz in the foreground.

follow them out to pasture. The beasts were sure to locate the offending plants, which were then destroyed.

As you climb the last leg of the trail, you'll pass through a community of dwarfed red spruce and balsam fir. Look for those other hardy organisms, the lichens, inhabiting Kearsarge's upper reaches. The most attractive of these strange forms is known as British soldiers, because of the bright red knoblike structures on its tips. Lichenologists call these projections "apothecia"; they are one of the ways these fascinating forms of life reproduce. As you reach the fire spotter's tower at the top and take in the wonderful view, pay your respects to another sort of lichen, the crustose type. It endures the wind-lashed cold of the summit by assuming the flattest habit possible. In fact, it's difficult to imagine that this is living tissue at all. Rather, it seems to be a thin crust of paint, dabbed onto the weathered face of the rock by some prehistoric hand.

TOUR 41 ～

In New Hampshire:
Seabrook to Portsmouth

Driving Distance:
20 miles

Connecting Tour:
Number 50

Starting Point:
The intersection
of Routes 1A and 286

MAINE

Piscataqua River

Portsmouth

○ Kittery

Moffatt-Ladd House

NEW

HAMPSHIRE

Odiorne Point State Park

○ Rye

Fuller Gardens

N

0 1 2 3 4 5
MILES

Atlantic

Ocean

Hampton

Hampton Beach State Park

Seabrook

286

1A

Newburyport

Parker River
National Wildlife
Refuge

MASSACHUSETTS

Tour 41
The New Hampshire Seacoast

Highlights

A Land of Dunes and Bayberry Candles ∿ Hampton Beach State Park ∿
How to Build a Seacoast ∿ Nature's Own Cradle ∿ Plant Commandos Hit
the Beach and Take It ∿ Suntan Land and Shotgun Horticulture ∿ The
"Scotch Fir" ∿ Welcome to *Rosa*'s Place ∿ Fuller Gardens ∿ Tulipomania ∿
New England's Darling Lilac ∿ Odiorne Point State Park ∿ The Rampant
Buckthorn ∿ A Drowned Forest and Our Great Transgression ∿ A Study in
Green, Red, and Brown ∿ The World's Longest Organism ∿ The Aesthetics of
Beer Suds ∿ Portsmouth's Moffatt-Ladd House and the Georgian Imperative

C ompared with its Connecticut River frontage or even with the shoreline
of Winnipesaukee, its largest lake, New Hampshire has a seacoast of un-
impressive dimensions. On the map, it's nothing more than a brief plug
in the breach between Massachusetts and Maine. Notwithstanding its small size,
it offers the plant explorer a well-mixed sampler of ecology, botany, horticulture,
and marine biology, garnished with trimmings of geology, history, architecture,
and beautiful scenery. And all these are located within a few hours' drive.

The starting point lies just north of the Massachusetts border. Take Route
1A North. This section of the New Hampshire littoral is a highly built-up area,
but natural features persist in the form of salt marshes located inshore to the
left, and sand dunes visible to seaward on the right. Soon more dunes appear,
on the left; they resemble hummocks and are fixed, or anchored in place, by
colonies of bayberry, a common coastal shrub. The fruit from this much-loved
plant yields "wax" for exquisitely fragrant candles. It thrives in the poorest
seacoast soils because its roots are infected with beneficial mycorrhizal fungi. (See
Tour 13 for more on this important symbiotic relationship.) The narrow, upright
forms of red cedar, actually a species of juniper, can also be seen time and again.
Farther in the distance looms one of New England's most controversial human
artifacts, the Seabrook nuclear power plant. Continue over the bridge spanning
the mouth of the Hampton River. At its far end, turn right into the entrance
of the tour's first stop, Hampton Beach State Park. This is but one segment of
a much larger beachfront that is obviously the area's main tourist attraction.

In the broadest sense, New England has two kinds of seacoast. One is the erosional type, where for various reasons the ocean confronts a dwindling beach or a rocky shore and leaves no sediments behind. Down East Maine has many examples of this. Then there's the type of seacoast that is depositional: water-borne sediments such as sand or gravel are deposited, and the beaches are built up and out. From lower Maine southward, a good portion of the shoreline of eastern North America is the depositional type. And wherever beaches form, there is the potential for a burgeoning tourist industry. After all, sand is a tremendously attractive substance to people, firm to the step but also yielding, a sort of halfway point between true earth and the nearby sea. As you'll certainly see here, it invites the human body to come into full contact with it. It's nature's own cradle, a mattress of tiny rocks. Ever influenced by the forces that delivered it in the first place, it forms its own distinctive terrain that you can observe as you head seaward.

From the parking lot, walk past the beach house and over the dunes. This barrier of low sand hills, set a good way in from the high-water mark, is the first main component of the beachscape. Notice the plant that's growing on and around the dune crests. You can probably recognize it as a member of the grass family. How can it possibly survive in this environment? It's growing in pure sand, virtually devoid of organic matter and nutrients. The winds and salt spray are powerful desiccants. Further, whatever fresh water falls here is quickly lost; sand has no absorptive capacity. The ever shifting surface is also a real threat, because blowing sand can bury even the hardiest plants.

It so happens that this species, beach grass, has developed the almost unbelievable ability to deal with each of these problems in turn. In doing so, it has an immensely beneficial impact on its surroundings. Like most grasses, it sends out underground stems, rhizomes, that form colonies and help stabilize the dunes. Eventually, other hardy plants, such as beach pea and seaside gold-enrod, take root in the places pioneered by the grass. You should be able to spot these other plants here as well.

As you head past the dune front, you come onto the surface known as the backshore, a zone that often includes one or more terracelike surfaces called berms. The backshore slopes gently down to the foreshore, which is the part of the beach alternately covered and exposed by the motion of the tides. In these places, even the toughest members of the plant kingdom cannot thrive in any numbers; the combination of ocean blast and countless thousands of trampling feet forbid. As you reach the undulating edge of the continent, you must put your hopes in another system of chlorophyll-based life, the Protists. These organisms, the marine algae, are sufficiently plantlike to be called the seaweeds.

Generally speaking, though, smooth sandy beaches are not the best place to find even these organisms, because they mostly require a firm surface on which to anchor themselves. However, if you walk to the stone breakwater at the southern end of the beach, you'll find wispy green forms of *Enteromorpha,* as well as a few other algae. Later in this tour, you'll visit a much more spectacular showplace for the kingdom Protista.

⁓

Depart the state park by continuing north on Route 1A. This is the heart of Yankee Suntan Land, and the businesses here are not particularly renowned for the subtlety of their allurements. The first English-speaking settlers here had other uses for this real estate—the abundant cord grasses of the tidal marshes made it premium cattle country. One can't help wondering what those stolid, God-fearing folk would have made of the video arcades or the motels with hourly rates. They probably would have condemned the mammon wholesale, then examined the books carefully to see if the good Lord offered a reasonable profit.

On the north end of the public beach area, the road is lined with vacation homes and fairly new condominium complexes. The grounds of the latter demonstrate the modern predilection for shotgun horticulture: when it comes to ornamental trees and shrubs, the contractor has generally tried one of everything, in the off chance that something might fit. One plant that often does, though not by anyone's conscious design, is the Scots pine, known in colonial times by the inaccurate name of Scotch fir. The upper portion of its trunk usually has a distinct reddish cast, but an even better identification trait is its blue-green, or glaucous, needle foliage.

This native of northern Europe and Russia provided many masts for the Royal Navy's ships of the line, but once the New England lumber trade developed, it was recognized that the New World equivalent, the white pine, was far superior. Not only did the white pine provide wood that remained intact a good twenty years longer, it also produced larger trunks, which could be used singly as entire mast segments. Scots pine, initially somewhat stronger pound for pound, usually had its structural superiority nullified by the fact it had to be spliced together for the bigger masts. Ironically, the white pine never came close to ousting the other species from its predominant role. Baltic lumber merchants charged more for the less preferable tree, but the higher cost of transporting white pines all the way across the Atlantic proved to be a major drawback.

⁓

Some 5.6 miles north of Hampton Beach State Park, a mixed stand of pitch pine, red maple, and black cherry appears on the left, on the edge of a salt marsh. Pitch pine, very much a native of this region, was not a suitable mast tree, but for many years it provided a well-known item of the naval stores trade— tar. A bit farther up, clumps of rugosa rose, a resilient, salt-tolerant Oriental import, grow along the roadside to seaward. If you can safely pull over for a moment, do so and examine this species; it serves as a good introduction to the great genus *Rosa*. In a few minutes, you'll be looking at an assortment of its more refined relatives.

If you catch the rugosa rose during its summer flowering period, first inspect the blossoms. These have five petals apiece (some cultivated varieties have "doubled" flowers with more), and there are numerous stamens and pistils. As every botany student learns, rose stamens are perigynous—that is, they are attached around the ovary, rather than below it, as is the case, for instance, with buttercup flowers.

The ovaries eventually develop into the kind of fruit known as hips. In this species, the hips are large and turn a bright red, which makes them one of the plant's best ornamental attributes.

Now look at the foliage. It has classic rose features: the leaves are compound and have at the base of their stems structures called stipules. Some plants—the black locust, for instance, and the mesquite trees of the Southwest—have stipules in the form of paired spines; but in the roses, they take the form of winglike or wispy appendages. Another salient rose characteristic is the one most frequently mentioned by poets—the thorns. Technically, they aren't thorns at all, but prickles, since they're composed of bark tissue instead of being modified stems or stipules.

Continue your northward track on Route 1A. Once you've entered the town of North Hampton, keep an eye out for the junction of Route 101D. Stay on 1A through that intersection, then turn left onto Willow Avenue. In another 0.1 mile you'll see the entrance to Fuller Gardens on the left. Turn in here. This showcase of ornamental outdoor plants, administered by the Fuller Foundation, is the first of two horticultural focal points on this tour.

Fuller Gardens is a small but fascinating place to see formal hedge plantings and perennial beds. If you visit in May, you'll have an excellent opportunity to witness skilled professionals turning traditional bulb plantings, and particularly tulips, into a thing of artistry. Tulips are monocots in the lily family and have been highly prized members of Western Hemisphere gardens since they were first introduced from Turkey in the 1500s. In the following century, they even became the basis of one of humankind's most dramatic departures from common sense, in an episode of zany financial speculation known as Tulipomania. During the craze, a single tulip bulb would fetch thousands of Dutch guilders. Certificates resembling the junk bonds of recent memory were issued, and in the end, many eager but unwise investors lost their fortunes. Meanwhile, the inhabitants of Paris found a new way to express fashion's triumph over reality: women were simply nobody if they failed to decorate their décolletage with cut tulip flowers.

Any garden plant with a history stretching back to the Turkish sultanate and the Netherlands of the Reformation must command our respect, but here you'll see another section of the garden devoted to a plant that has been in continuous cultivation and development for well over two millennia. This plant is the subject of our recent scrutiny—the rose.

Fuller Gardens is a rosarian's delight, and it is also a good place for the naturalist who prefers vegetation in the wild state and enjoys pondering the interaction between our own species and the world of plants. Look at the way the rose bushes have been set out. This is their standard display format. To use the unforgettable prose of one highly publicized American, they are an off-the-shelf, stand-alone entity—refined and coddled specimens, isolated and exempt from the complexities and interconnections of perennial-border design. They simply don't refer to anything else in the world of shrubs or flowers, and certainly not to their rampant relative, the rugosa rose.

Many authorities believe the cultivated rose originated in ancient Persia. From there it spread to Asia Minor and the Mediterranean Basin, where it was extolled from one generation to the next by the likes of Homer, Sappho, and Horace. Pliny the Elder, that double-dealer of pure humbug and laudable scientific inquiry, devoted a portion of his endless scribblings to a description of how the soil should be prepared for rose plantings. True to their hedonistic reputation, Romans of the later empire refuted the stern legacy of their ancestors by strewing rose petals practically everywhere: on the floors of their banquet halls, in their hair, and in their spiced Falernian wine. Even the prows of Roman warships were decorated with rose garlands. (Nineteen centuries later, the Imperial Japanese Navy would have a similar idea, but it would use the symbolic chrysanthemum instead.) One of the rose types in use in ancient Italy was the so-called Alba; much later it would become the emblem of England's House of York.

Since medieval times, the rose has also been highly regarded and widely employed by Islamic cultures. The Sufi poet Omar Khayyam had the good fortune to have his tomb covered in rose petals scattered by the north wind. One of the great horticultural productions of the Arabic-speaking world was the famous damask rose, which, thanks to the mayhem of the Crusades, was imported into Western Europe. This type is the source of attar, a fragrant oil distilled from its petals. Another pleasing floral by-product is rose water, a perfume solution that has seen many uses over the years. A less well known application of garden roses in general was supplied by the English herbalist, Mrs. M. Grieve:

> Put a layer of red rose-petals in the bottom of a jar or covered dish, put in 4 oz. of fresh butter wrapped in wax paper. Cover with a thick layer of rose-petals. Cover closely and leave in a cool place overnight. The more fragrant the roses, the finer the flavour imparted. Cut bread in thin strips or circles, spread each with the perfumed butter and place several petals from fresh red roses between the slices, allowing edges to show.

In more recent times, garden roses have been grouped into several grand categories. Among these are the hybrid perpetuals, favorites on both sides of the Atlantic until they were finally superseded in the late 1800s and early 1900s by the hybrid tea varieties. The hybrid teas, well represented here at Fuller Gardens, owe their name to the fact that their leaves smell somewhat like tea when they are crushed. (True tea comes from a woody plant that is a close relative of the camellia used so extensively in southern climes.) According to *Hortus Third,* there are some twenty thousand extant rose cultivars, and hundreds of new varieties appear each year. Just keeping track of them all is a staggering task, as the management of the American Rose Society would no doubt gladly tell you.

Nowadays, hybrid teas come in a spectrum of flower colors that would have astounded Pliny or Omar Khayyam. Not only can you choose such old standards as white, pink, and red but also peach, orange, yellow, lavender, purple, and a strange tone that apparently wants to be khaki or maybe café au lait. And there are plenty of selections that offer striking combinations of the above. If you ever have the chance to visit one of the world's great collections, such as the splendid ensemble at the Brooklyn Botanic Garden, you'll go away convinced that rose development is fast approaching the state of terminal complexity. The rose hasn't exhausted it possibilities, but can the poor human mind keep up?

─────────────── ❧ ───────────────

When you're done at Fuller Gardens, return to Route 1A, and once again head north on it. Some of the best ocean views occur in this stretch. In the foreground, at low tide, huge communities of brown algae lie exposed in the sun, and beyond them roll the deeper stretches of salt water. The Atlantic seems so elemental it's difficult to grasp that it was born in fairly recent times, geologically speaking. The ancestors of the conifers that now adorn our parks and forests were growing in this region before it had an ocean washing its eastern edge. In those days, this was the heartland of the supercontinent Pangaea, the most unified land mass in the history of the earth. It stretched from one pole practically to the other.

Roughly 2.0 miles north of Fuller Gardens, Eel Pond appears on the left. You are now in the town of Rye. In another 1.3 miles, salt-marsh areas display large colonies of cord grasses and smaller groups of red-tinged glasswort plants. The latter have fleshy, water-retaining succulent stems, which help them deal with their saline environment. Homes along the road are fronted with common lilac, the tall foundation shrub that has been the unassailable darling of the Yankee soul since pre-Revolutionary times. The original species comes from southern Europe, but hundreds of cultivars have been produced since the sixteenth century. Many lilacs spend their lives standing guard at the entryways of houses, a fact noted by Walt Whitman in his poem "When Lilacs Last in the Dooryard Bloom'd." Though you couldn't tell by this plant's simple, broad, almost heart-shaped leaves, the lilac genus is in the olive family. Hence, it is a dicot, sharing a common ancestry with ash trees, privets, jasmines, and those masters of spring-time garishness, the forsythias. Common lilac was once used medicinally as a quinine substitute for treating malaria, and it also figured in a treatment to expel intestinal parasites.

─────────────── ❧ ───────────────

Our next stop, Odiorne Point State Park, is located another 4.4 miles north along Route 1A. This facility marks the spot where New Hampshire's first white settler, a Scots fisherman, had his dwelling. Turn right at the entrance and park in the lot. Before you proceed on foot, take a look at the shrubs bordering the parking area. Among them is another Old World introduction, common buck-thorn. As a rule, all plants, even forsythias, have their redeeming features; but in the case of the buckthorn genus, they're difficult to locate. Buckthorns have about as much ornamental value as windblown trash. Bird-watchers probably love them, since they're a copious food source for you-know-whom. And therein lies the problem, for the birds spread the seeds far and wide. In some places— for instance, in the excellent forest preserve system that rings Chicago—buck-thorns have reached nightmare status, and are busily crowding out the native species in what seems to be a program of calculated genocide.

On your way to the shore, you'll see more Scots pine specimens, and also tree-sized privets that demonstrate what they can do when not hedged six times each summer. However, the main attractions of this stop are situated below the high-tide mark, in a nearby cove. Watch for the sign that will lead you to the Drowned Forest. This refers to an ancient stand of trees—white pines and perhaps other species as well—that once grew where now the water rules. If you're very lucky and you hit the right low tide, you may actually spot their stumps. Their presence points out an interesting fact confirmed in other parts of the New England seacoast: in the past several thousand years, the ocean has been gradually rising. Geologists call this process transgression, though they don't use the term in its biblical sense. Fluctuations of sea level along this shoreline are nothing new. During the Ice Age, the immense weight of the glaciers actually pushed the continent down into the underlying mantle. It's been estimated that each 3,000-foot thickness of ice caused the land to sink about 1,000 feet. When the ice melted, the land recoiled slowly, and regression, or the retreat of the ocean, was the rule of the day. These days, a new cycle of submergence, probably caused by the continued melting of ice in polar regions, is well under way. It should be pointed out that this current spate of slow-motion flooding started long before the theorized global warming could have had any effect.

The other notable thing about the cove is its bountiful crop of seaweeds. At low tide, the dedicated beachcomber, shod in old sneakers or ooze shoes, can

Hefty representatives of the brown algae: kelps of the species Laminaria saccharina, *stranded at low tide at Odiorne Point. Normally, kelps do not wear baseball caps; this one was supplied by the author to provide scale.*

have tremendous fun identifying the strange forms of these important organisms. Three basic groups, once included in the plant kingdom but now considered part of the kingdom Protista, are represented here. These are the divisions Rhodophyta (the red algae), Phaeophyta (the brown algae), and Chlorophyta (the green algae).

The green algae, thought by many paleobotanists to be the group that gave rise to the true plants, have been residents of the earth for a long, long time. Primitive one-celled forms were thriving in the Precambrian time, almost one billion years ago; the more complex forms, such as those you may see here, had their beginnings in the Cambrian period, roughly half a billion years before present. Their bright green color is caused by chlorophyll, the primary photosynthetic pigment. Here you may find the broad, flat blades of sea lettuce, which is sometimes actually eaten in salads, and the ghostly, strandlike shapes of the genus *Enteromorpha*.

The red algae, which also probably originated in the Cambrian, contain another class of pigments, phycobilins, as well as chlorophyll. These accessory compounds allow their organisms to absorb a broader band of the light spectrum, which is a distinct advantage in turbid or deep water. Members of this division are most abundant in warmer climes, but there are some important kinds this far north. For example, in this cove it isn't difficult to locate the crisp-edged Irish moss, used for food and industrial applications by some maritime nations, and the species *Polysiphonia lanosa*. It's one of the most unusual kinds you'll see, since it's epiphytic—that is, it grows on other algae, the same way such plant epiphytes as Spanish moss grow on tree branches.

The origin of the brown algae is harder to pin down, because they do not leave calcareous remains that are preserved in the fossil record. They are best adapted to cooler parts of the ocean. Their accessory pigment, fucoxanthin, often imparts an olive color to their tough, rubbery tissue. Here, they appear in great numbers, particularly the knotted rockweed, with its bulblike terminal projections, and the species of wrack, which are many-branched algae that somewhat resemble miniature shrubs. The most bizarre alga you're likely to find looks as though it has been riddled with buckshot. This is the highly perforated *Agarum cribosum*. Finally, if you come across a long, entire, bannerlike blade, you've found kelp, a source of fertilizer, iodine, and the substance algin. It's hardly a household word, but algin has had quite a record of service, especially as a food ingredient and as a thickener and suspension agent in a variety of industrial processes. Mr. Joe Six-Pack has reason to be thankful for this seaweed derivative, because it's added to beer to give it a better head of smaller, more persistent bubbles. (Apparently, aesthetics extend into all sorts of human activities.) Kelp can grow to an astounding size; one specimen measured 710 feet, well over two football fields in length. In all likelihood, it was the world's longest known organism.

———————— ❧ ————————

The last leg of the tour takes you out of the timeless domain of the seaweeds and into the urban heart of New Hampshire's sole seaport, at the mouth of the Piscataqua River. Leave Odiorne Point by resuming your track on Route 1A

North. This leads you into the outskirts of Portsmouth. Even in colonial times, this small and handsome city was an important center of overseas commerce. One special export to Britain was the great pine tree masts cut in the Concord and Winnipesaukee regions of New Hampshire, and also in nearby Maine. At the peak of that trade, in 1742, five hundred masts were shipped from this port alone. (For more on the white pine's role in what was a particularly contentious episode of colonial history, see Tour 50.)

At the intersection of Routes 1A and 1 (otherwise known as Middle Street), bear right into the small and lovely downtown area. Park in one of the public lots and proceed the few remaining blocks on foot to the Moffatt-Ladd House, at 154 Market Street. Portsmouth first went by the good botanical name of Strawbery Banke, because its earliest white inhabitants discovered an abundance of wild strawberry plants growing on the site of their new settlement. That name is now used to designate the main district of historic houses in the vicinity of Court and Washington Streets. In contrast, the Moffatt-Ladd House stands quite alone, in a delightful spot almost within a doubloon's toss of modern freighters loading their cargoes.

The main justification for this visit is the pleasant 2.5-acre garden out back, but you probably will want to begin by touring the interior of the house. As

The front exterior of Portsmouth's Moffatt-Ladd House, done in finest Georgian style. Note the common lilac shrubs flanking the doorway, in accordance with time-honored New England custom.

the scroll pediments over the second-story windows indicate, the structure is an example of the Georgian style, and a very good one at that. Georgian architecture reached its zenith in the mid-eighteenth century; and sure enough, this building dates from 1763.

When you reemerge into the sunlight, you'll get some idea of the role horticulture played in the lives of the colonial well-to-do, even though the current plantings largely reflect the tastes of an owner who followed about a century later. In the Georgian era, gardens were meant to provide a pleasing diversion as well as a lighthearted form of good order. Here, in their composite form, the grounds also have an underlying theme of agriculture and utility, as evidenced in the herb plantings, grape arbor, and beehives. In the straight Georgian style, greenery simply was not allowed to obscure or seclude the grandeur of a building. You can judge for yourself whether the later interpretation has changed that. In any case, the brand-new Moffatt-Ladd House must have been a far cry from the rude farms that fronted the forest a few miles inland. Had you stood in this garden in the 1760s, in one sense you would have been much closer to the London of George III than to the culture that embraced sedition from the British crown, just one decade later.

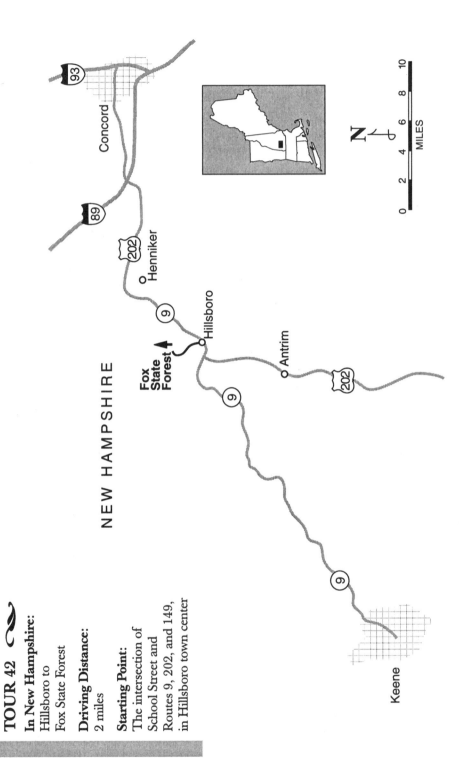

TOUR 42

In New Hampshire:
Hillsboro to
Fox State Forest

Driving Distance:
2 miles

Starting Point:
The intersection of
School Street and
Routes 9, 202, and 149,
in Hillsboro town center

NEW HAMPSHIRE

Fox
State
Forest

Concord

Henniker

Hillsboro

Antrim

Keene

93

89

202

202

9

9

9

N

MILES

0   2   4   6   8   10

Tour 42
The Wonders of Fox State Forest

Highlights

A Town with Two Spellings ∿ Farm Country and Sugar Maples ∿
Escaped Day Lilies and Antimalarial Mullein ∿ Say Hello to Doug Fir ∿ The
Butternut ∿ Forest Mushrooms ∿ Big Leaves and Little Leaves ∿ Circe's
Wildflower ∿ God Smiles on Stone Fences ∿ A Deathly White Marvel ∿
Buds ∿ First-Growth Trees ∿ A Spine-Tingling White Ash ∿ The Mud Pond
Boardwalk ∿ A Kettle and a Fen ∿ Peatland Shrubs and Trees ∿ The Little
Guys Take Over ∿ A Low-Budget Triffid and Two Other Killers

"The cheapest way to travel, and the way to travel the furthest in the shortest distance, is to go afoot." These are the words of our leading woodland pundit, Henry David Thoreau. This tour demonstrates once again that Mr. Thoreau knew what he was talking about. The 2-mile round-trip hike featured here takes you through a succession of settings, from forest brooks fringed with wildflowers to rare first-growth trees and an outstanding peatland habitat. Botanically speaking, this could be one of the most information-rich journeys you'll ever take.

Since a good portion of the trek through Fox State Forest involves damp woods and wetlands, be prepared. Ooze shoes are a must. Also, take along your own preferred form of protection against insects. The folks up Canada way scoff at the scrawny southern New Hampshire mosquitoes, which rarely have a wingspan greater than the Andean condor's, but their number and sincerity make up the difference. And if you visit in black fly season, don't set out until you've really convinced yourself that everything good in life involves a little pain. Or a great deal. (See John Josselyn's assessment of the charming little black fly, quoted in the Mounting Your Expedition section.)

The tour begins in the center of a town with a name routinely spelled two ways, Hillsboro and Hillsborough. Starting at the intersection, turn north onto School Street. This takes you past the post office and the Middle School into beautiful farmland lined with stone fences and sugar maples. Just 1.9 miles from the intersection, you will see the headquarters for the state forest on your right; turn into the parking lot.

There are several worthwhile trails to take. They are marked with a system of blazes, but play it safe and pick up a trail guide booklet in the dispenser located in front of the headquarters house. Then proceed to the north end of the parking lot and let the exploring begin. In more open or disturbed areas, you may catch sight of two well-established alien species: escaped day lilies, the descendants of old garden plantings, and the tall, spirelike forms of the yellow-flowered common mullein. Day lilies love to hang out along stone fences, and it's the perfect setting for them. What crass and heartless soul would ever want to rid the landscape of these splendid monocots? They should be given Honorary Native status. As far as the mullein goes, Pehr Kalm wrote that Swedish settlers in the Central Atlantic states wrapped its felty leaves on their feet and arms to counteract malaria—which constitutes just about the strangest substitute for quinine on record.

As you head down the main trail, you may wish to increase your dendrological IQ rating by taking the short tree-identification circuit. Besides showing you labeled specimens of common New England trees, this path gives you the opportunity to see a few examples of rarer species and introduced types. One of these is the mighty Douglas fir, the great lumber conifer of the American West. Naturalists and other compulsive name-droppers often call it "Doug fir" instead, prompting listeners who aren't familiar with the tree to think it's someone running for Congress. See if you can find any of Doug's cones lying around. They look as though they were designed by a Chinese artisan; their scales are ornamented with flamboyant, toothed bracts.

Also look for the butternut specimen. This indigenous broad-leaved tree is a member of the walnut genus. Its sap has occasionally been used for sugar production; butternut fruit husks yield a dark brown dye. The wood reputedly saws and carves better, and is even more attractive, than black walnut, and that's no mean feat. Butternut was used extensively for furniture making in the eighteenth century.

If you happen to be intrigued by the world of forest mushrooms, another nearby trail deals with these elegant and sometimes colorful members of the kingdom Fungi. When you're ready to continue on with the main route, head downhill on the wide trail marked by white, rectangular blazes. Your ultimate destination is Mud Pond. Follow the signs for it, and keep in mind that the type of blazes you should follow will change more than once.

Soon the Mud Pond trail bears off to the left (here, for the time being, you'll be guided by the white mushroom markings). One of the attractions in this neighborhood is the tall-growing interrupted fern. You might well ask why it and so many other forest plants produce large leaves, and why so many meadow or pasture plants, such as the grasses, have narrower, less substantial foliage. The answer lies in the amount of sunlight each environment receives. In the twilit gloom of the woodland, plants need to present as much photosynthetic surface as possible, to make the best of a bad situation. More leaf area also means the plant loses water more quickly; but usually this is not a crucial factor in such a damp and shady locale. On the other hand, plants that live in the sunlight's full glare must cut down on leaf area to minimize water loss and prevent wilting; their smaller surfaces suffice in the great energy glut of direct sunlight.

When you reach the brook, where red rectangular blazes take over, keep an eye out for a small but engaging wildflower called smaller enchanter's night-shade. This is one of the least showy members of the evening primrose family. Its flowers are borne on a single stem and are constructed on a scale that would please a microchip designer. If you have a hand lens, you can see the fine points of this miracle of minimalism, where practically everything is expressed in pairs: two stamens and also two styles, fused into one threadlike tube, rising above just two deeply notched petals. Beneath them all are two backward-curved sepals. Even the leaves are opposite—that is, arranged in twos. The plant's genus name, *Circaea,* derives from the legendary sorceress Circe, who turned Odysseus' crew into swine. Perhaps it was her so potent art that created this plant's unusual game of pairs, and kept it hidden from the swiftly passing eye.

The path continues through a stand of Canadian hemlocks and mixed hardwoods. Note the stone fences that were painstakingly built from local field rocks. These enduring reminders of New England's agricultural era bear witness to the fact that this land was, believe it or not, cleared for farming. The reason the fences have withstood the onslaught of the years is that they were very ably constructed. Those that weren't have long since fallen away. First, the farmer dug a long trench in the soil, to a level below the frost line, which in this region is several feet down. (If you've ever had the jarring experience of hand-digging this deeply in rocky glacial till, you'll wonder how New England's politicians manage to get to hell when they die.) Then a bed of sand, gravel, or small stones was put in, and on top of that, a foundation of larger stones. Once these had been given a chance to settle in, smaller rocks, plucked from nearby fields and often carefully chosen for their ornamental value, were put in place as super-structure. The process of fencing one family farm could take a generation or more to complete.

To many European visitors, the Yankee obsession with boundary barriers seemed incomprehensible. Even some Americans wondered at the practice. In one journal entry, Thoreau, himself a highly esteemed surveyor, wrote, "I am amused to see from my window here how busily man has divided and staked off his domain. God must smile at his puny fences running hither and thither everywhere over the land." Yet these low walls of stone are also the hallmark of local craftsmanship in its highest form. Today, they constitute one of the most characteristic aspects of the landscape. No tourist-conscious New England chamber of commerce is likely to push for their removal.

The look of your surroundings soon changes. On the right, a new plant community composed of willows and other water-loving species signals the onset of swampy conditions. Wildflowers in this area include forget-me-not, bunchberry, the evergreen goldthread, Indian cucumber root, and the most ghostly plant of all, Indian pipe. This bizarre species, a deathly white, has given up its birthright to chlorophyll. It can afford to do so because it is a parasite, deriving its sustenance from the roots of other plants.

This section of the trail is also premium-grade pteridophyte territory. Royal fern, one of the most obvious and one of the comeliest of the group, is a truly global species. It is native to six continents, and its remains have even been found on the frigid seventh, Antarctica. Keep a sharp lookout for woodland horsetail too, which without doubt is one of the finest examples of highly embellished delicacy in the natural world. In addition, pay some attention to the woody flowering plants. The offbeat witch hazel, described in Tour 38, is here, as is the round-leaved viburnum known as hobblebush. Its chaste flowers are among the earliest delights of spring, the serene white answer to the yellow outrage of countless household forsythias.

In winter, hobblebush buds are one of the most intriguing features of the forest. Buds are a plant's embryonic shoots; most are equipped with external scales to prevent desiccation, restrict oxygen intake, and help insulate the inner tissues from heat loss. The intriguing thing about hobblebush buds is that they lack protective scales. Instead, the primordial leaf and flower structures curl and huddle naked in the cold. It doesn't seem to make much sense, but the plant survives just fine—more proof that nature is a practicing exponent of the do-whatever-works philosophy.

The trail gradually angles away from the wetland. At the Y-intersection, pause and take a good look around. This small area is reputed to be virgin first-growth woodland. One powerful bit of evidence supporting the contention is the pillow-and-cradle terrain found around here. This is characteristic of old forests. When trees fall, their root systems wrench soil out of the ground and so leave cradles, or hollows; the hummocks between them are the pillows. Had this stretch ever been plowed, its surface would be more level.

Notice the open, parklike feeling of the grove. This too suggests the antiquity of the community. As the hemlock canopy matured, less and less light reached the floor, and the understory growth dwindled, leaving a gallery of uncluttered columnar trunks. This cathedral-like effect was a common feature of New England in early colonial times, and it took a deadly combination of thoughtlessness and systematic greed to obliterate all but a few acres of it. Thoreau, who lived during a peak period of forest destruction, watched the process with dismay, and gave a bitter sort of thanks that the woodchoppers couldn't cut down the clouds.

At the Y, take the trail on the right, marked with white dot blazes. You come to another wet spot, lorded over by a magnificent beast of a white ash. Perhaps it too is first growth. If you don't feel a quickening of your heart and a slight shiver down your spine at the sight of this magnificent organism, you'd better check your pulse. This is a place of maximum contrasts. In the damp at the base of the great tree grow masses of a small, often-overlooked herbaceous plant. This is water carpet, a well-named species of the saxifrage family. The differences seem so fundamental, yet the tiny herb and the giant ash share a common ancestry as dicots. Together, they demonstrate what a broad range of forms, sizes, and survival strategies the flowering plants have developed in their reign of close to 100 million years.

Next, the trail rises into dry upland, where paper birch and some American beech largely supplant the Canadian hemlock. Cross the gravel road and head down the slope to the Mud Pond boardwalk. You are now entering a world of plants so distinctly different from the preceding that it appears to have been the subject of a special creation, one in which all the normal rules of form and behavior have been rewritten by an evil, or at least prankish, genius. Before we consider the local plant wonders, though, a brief discussion of the origin of the pond itself is in order.

Geologically speaking, Mud Pond is a "kettle," a common New England landscape feature that is the result of a rather complicated chain of events. More than ten thousand years ago, when the most recent Ice Age glaciers began their slow retreat northward, they left behind vast deposits of debris: large boulders, cobbles, gravel, and much finer sediments. All of this made up the widespread

A superb peatland example: the Mud Pond fen. This seemingly solid surface is actually a yielding mat of sphagnum moss.

mantle of material geologists call "till." In places, the till buried or surrounded detached blocks of old ice. Some of these ice blocks were truly huge. Eventually, the hidden ice melted away, causing the overlying surface of sediments to fall inward and form a large depression. As with Thoreau's favorite kettle, Walden Pond, this depression permanently filled with a body of water.

While earth scientists apply one term to Mud Pond, ecologists use another. To them, this is a fen, a type of peatland in which the water flow is partially impeded, though not completely as in a real bog. Fens are probably the most prevalent kind of peatland in North America. Peatlands by definition are wetlands with organic soil; in New England, the organic matter is largely supplied by the remains of higher plants and one of the world's most important nonvascular plant types, sphagnum moss. You may not realize it at first, but the trees and shrubs growing along the boardwalk are largely rooted in sphagnum. Solid ground stops long before the untrained eye notices it.

Distinct changes in the vegetation are clearly visible as you proceed from the pond edge out onto the floating mat. To begin with, there are large woody plants: shadblow, highbush blueberry, mountain holly, speckled alder, and black chokeberry, as well as two *Kalmia* species, sheep laurel and the shorter bog laurel. As you continue, you come upon black spruce and lichen-festooned tamarack. Then the little guys take over; against a lovely red-and-green sphagnum backdrop, you should discover the clambering swamp loosestrife, a wetland regular also known by its genus title, *Decodon*. Joining it are the prostrate small cranberry and two distinctive monocots, arrow arum and white-fringed orchis. (For more on the last two, see Tour 22.)

The majority of human visitors to Mud Pond are most dazzled by the strangest residents, the carnivorous plants. The first of these, spatulate-leaved sundew, often blends imperceptibly into its surroundings. Using the minute sticky hairs on its leaves, it traps insects, which apparently provide it with nitrogen otherwise lacking in this nutrient-poor environment. Even more eye-catching is the green or purplish pitcher plant, which science-fiction fans could liken to a scaled-down triffid. This is what seventeenth-century naturalist John Josselyn had to say of it, in the run-on sentence style typical of his day:

> Upon the top standeth one fantastical Flower, the Leaves grow close from the root, in the shape like a Tankard, hollow, tough, and alwayes full of Water, the Root is made up of many small strings, growing only in the moss, and not in the Earth, the whole Plant comes to its perfection in August, and then it has Leaves, Stalks, and Flowers as red as blood, excepting the Flower which hath some yellow admixt.

When insects venture into the plant's gaping trap, they often fail to get out again, because the inside of the "tankard" has several devices that cleverly inhibit escape. Those creatures too tired to fight on fall into the water, die, and are eventually digested by the enzymes the pitcher plant secretes. Indians knew this species well, and used it to treat smallpox.

The third plant carnivore, the horned bladderwort, is best spotted during its midsummer flowering cycle. Its method of entrapment is the most improbable of all. It doesn't have real roots; moreover, its leaves and stems are located under the peat surface. Mounted on the leaves are small bladderlike traps which suck in and digest floating insect larvae and other tiny organisms. This process is normally hidden from view, but the plant's bright yellow, spurred blossoms stand high aloft the mat and provide one of the most delightful sights of the whole strange scene.

When you return to the parking lot, you can take a left onto the gravel road and then a right farther down; or you can return the same way you came. I recommend the latter. Every nature trail reveals a good deal more on the second pass; and in the ever fascinating Fox State Forest, this is especially true.

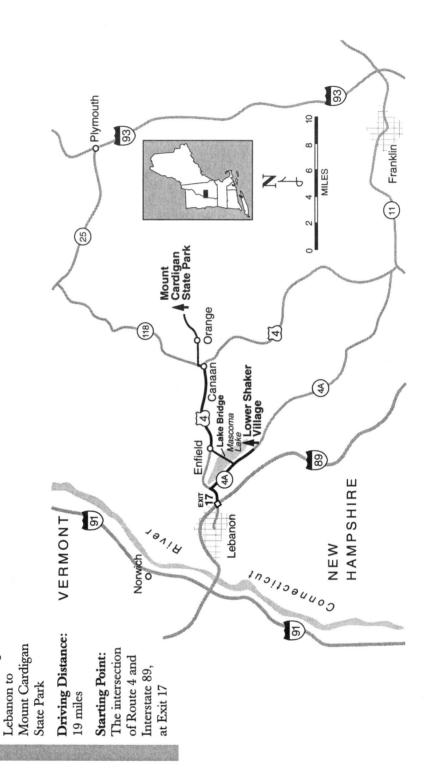

TOUR 43

In New Hampshire:
Lebanon to
Mount Cardigan
State Park

Driving Distance:
19 miles

Starting Point:
The intersection
of Route 4 and
Interstate 89,
at Exit 17

Tour 43
A Lakeside Shaker Paradise and the Alpine Islands of Mount Cardigan

A Recommended Fall Foliage Tour

HIGHLIGHTS

A Drive Down Mascoma Lake ∾ The Lower Shaker Village ∾ Heavenly
Fields and Handsome Buildings ∾ Mother Ann and the Millennial Church ∾
Famous Shaker Seeds ∾ The Elder Explains His Love of Trees ∾ A Sunny
Garden for Herbs ∾ Plant Profligacy ∾ Crossing the Waters ∾ Mount
Cardigan State Park ∾ Bryophytes and Their Wildflower Companions ∾
Walking on a Roof of Giant Crystals ∾ Discrete and Perfect Worlds ∾ Alpine
Plants by Special Arrangement ∾ A Second Kind of Heaven

*T*his tour could have been entitled "Looking in on Alternative Ways of
Life," because it features two communities—one of humans, the other
of plants—where the inhabitants have lived by different rules and have
occupied a kind of earthly heaven. Both communities, each in their own way,
have been a fascinating part of the New England story, and in their uniqueness
they've pointed out how much diversity this corner of the country contains.

The starting point of this journey is located just up the Mascoma River from
downtown Lebanon. Head east on Route 4. This is country dominated by white
ash, Canadian hemlock, poplar, and paper birch. Just 1.5 miles from the beginning, you'll reach the intersection of Routes 4 and 4A; bear right onto 4A. This
leads you along the southern edge of one of New Hampshire's prettiest bodies
of water, Mascoma Lake. Summer homes and cottages line the shore, and you'll
notice the predictable mixture of wild and cultivated trees so common in the
residential stretches of New England. Among the species present are basswood,
red oak, and black locust.

Within 2.3 miles of the last turn, you will see a low bridge on the left. A
bit later, you'll be taking it across the lake, but for now continue straight on 4A
past the rather arresting piece of property dotted with religious statuary. This
is the Shrine of Our Lady of La Salette. Just beyond it, on the left, is your initial
destination, the Lower Shaker Village. Go down the driveway and park in the
lot located between the imposing stonework inn and the museum.

Now that you've arrived on Shaker ground, welcome to your first paradise of the day, and also to the first of the two alternative ways of life. To grasp why this lovely setting qualifies for heavenly status, it's essential to have some understanding of the people who built and once lived in this complex of fields and unadorned buildings. Their view of life was radically different from that of the American mainstream. For instance, in contrast to most Christians, they believed not in a heavenly reward deferred to the end of history, but in a paradise of their own making, in the here and now. To them, salvation was not the hope eternal, but an obvious fact of their everyday life. The key to happiness, so they contended, was the renunciation of worldly cravings.

The Lower Shaker Village, established in 1782 and active until 1923, was one of nine settlements in New England created by the people who called themselves Believers in Christ's Second Appearing. The other Granite State site, the one better known to most tourists, is located in the town of Canterbury, north of Concord. The Shaker movement, also known as the Millennial Church, reached its peak in the nineteenth century. Its adherents once numbered in the thousands, but the sect has virtually disappeared, no doubt as a result of the movement's emphasis on celibacy.

One of the most important points of reference in the history of the faith was the life of its first great leader, English-born Ann Lee. Known to her followers as Mother Ann, this amazing woman had one of the most torturous yet productive spiritual journeys imaginable. Her youth was spent in England, working in the textile mills; later, she was forced to marry against her will. She suffered the early deaths of all four of her children. Subsequently, she suffered from what today would probably be diagnosed as schizophrenia. She had visions and became morbidly obsessed with the avoidance of carnality. At the far end of this harrowing ordeal, she emerged convinced that she was Christ's bride and successor. Her zeal attracted the attention of other seekers, and eventually, Ann and her followers obtained the wherewithal to emigrate to America, the place a vision had told them would be their promised land. They arrived in New York in 1774.

In stark contrast to conventional Christian teaching, Shakers believed that the deity combined male and female attributes in equal proportion. In its manifestation as Heavenly Father, the godhead was the creative force; as Mother Wisdom, its essence of order. The Shakers' obvious dislike of normal family values, coupled with their strange ritual dancing and songs—some of which were purported to be in the language of the Moon People—soon brought them a full dose of ridicule and scandal.

What the outside world never doubted was the high quality work of Shaker artisans. Today, furniture and other handiwork made in the spartan style of the Believers are all the rage. At the height of their movement, however, the Shakers were better known for their contribution to American horticulture. In those days, their vegetable and flower seeds were marketed far and wide, and were extolled by countless satisfied gardeners who had little or no knowledge of the unusual doctrines of the Millennial Church.

The success of their seed business was just one aspect of the Shakers' plant-centered way of life. One of the most enduring images in their art is based on a vision beheld by James Whittaker, the man who succeeded Ann Lee as head of the movement. He saw the Millennial Church revealed in the form of a great tree that bore leaves so bright they seemed to be aflame. In the everyday running of their villages, orchard work received special attention. One visitor, Hepworth Dixon, left this account detailing the Shaker view of the interaction between people and plants:

> This morning I have spent an hour with Frederick in the new orchard, listening to the story of how he planted it, as to a tale by some Arabian poet. "A tree has its wants and wishes," said the Elder, "and a man should study them as a teacher watches a child, to see what he can do. If you love the plant, and take heed of what it likes, you will be well repaid by it. I don't know if a tree ever comes to know you; and I think it may; but I am sure it feels when you care for it and tend it, as a child does, as a woman does. Now, when we planted this orchard, we first got the very best cuttings in our reach; we then built a house for every plant to live in, that is to say, we dug a deep hole for each; we drained it well; we laid down tiles and rubble, and then filled in a bed of suitable manure and mold; we put the plant into its nest gently, and pressed up the earth about it, and protected the infant tree by this metal fence."
>
> "You take a world of pains," I said.
>
> "Ah, Brother Hepworth," he rejoined, "thee sees we love our garden."

Try extracting that attitude from the second sub-assistant of a modern-day landscaping contractor.

<hr />

Nowadays, the horticultural highlight of this settlement is not fruit trees but a large herb garden located along the Route 4A roadway. This can be found to the north of the handsomely austere inn and the incongruous neobaroque church that flanks it. As you walk over to the garden, note the excellent sugar maple and arbor vitae specimens in the lawn area. Also pay heed to the hill-framed view across the lake. The Shakers certainly knew a good parcel of land when they saw it.

The herb planting, for the most part, is well labeled. It will show you examples of many interesting species not found in most home garden plots. As mentioned in Tour 19, the term *herb* in its broadest sense refers to any plant that lacks woody tissue; hence the widely used botanical adjective *herbaceous*. In general horticultural parlance, however, a herb is a plant with a distinct role in cookery, in other household matters, or in medicine—orthodox or otherwise. For centuries, the most accomplished botanists were physicians, since plants formed the basis of the vast majority of remedies. It is interesting to note that even in this age of mass-produced synthetic drugs, approximately one-quarter of all doctors' prescriptions rely on plant products.

The herb garden at Lower Shaker Village. Make sure you also visit the cleverly designed barn, to the left in the background.

This bright and tidy setting is a good place to consider some of the biological processes governing these plants and all others. One surprising statistic is that our chlorophyll-bearing friends, for all their amazing adaptations, can generally make use of only 1 percent or so of the sunlight that falls on them—not the sort of efficiency rating a solar energy expert will envy or emulate.

Nor does the wastefulness end there. Pound for pound, plants need much more water than we members of kingdom Animalia do. Animals have an impressive pump-driven circulation system involving blood plasma and other fluids. But much of the water entering a plant's roots is on a rather useless one-way ride up the stem and out through the stomata, or leaf pores. Some of the water gets used en route by cells, either for photosynthesis or other life-sustaining processes. But once again, plants are not high on the list for efficiency awards. (With so many saints standing around in this vicinity, mention should be made of one of botany's own elect, the eighteenth-century researcher Stephen Hales. His experiment, demonstrating the fact that a sunflower loses seventeen times as much water as a human being, is a classic. It's a good thing, though, that he didn't use a cactus or some other specialized water-retaining plant instead.)

There is one other measure of plant prodigality, which can be seen with an electron microscope. For some reason, plant cells have up to one hundred times as much genetic material, or "junk DNA," as human beings. If you'd like to make a big name for yourself in the field of microbiology, figure out why.

———————————— ∾ ————————————

Depart the Lower Shaker Village by returning via Route 4A to the bridge mentioned above. Turn right, onto it. Note your odometer reading, so you won't miss the next turn. It isn't often one can cross the main body of a New England lake in one's car, so enjoy the experience. On the far side, you enter Enfield town center. Continue straight through town. About 1.3 miles from your turn onto the bridge, bear right at the intersection onto Route 4 East. Note that there may not be a sign for it here.

Farther down, you cross another bridge, over the Mascoma River. Take a good look at the tree population in this area. There is a great deal of red maple, poplar, and white pine. The last two are pioneer species—they indicate that this was once cleared land. Another 2.5 miles down the road, a substantial amount of taiga regime red spruce mixes with the northern hardwoods. You'll also see some surviving hayfields. Within a moment or two, the bald summit of our second objective, Mount Cardigan, looms into view.

When you reach the good town of Canaan, bear left at the Y-intersection onto Route 118 North. In an additional 0.6 mile, turn right at the sign for Mount Cardigan State Park and proceed past the fairground. In roughly another mile, you've entered Orange; eventually, the road turns to gravel. It can be quite bumpy, so drive carefully. There is still some red spruce lurking about, but the northern hardwoods, particularly yellow birch and American beech, rule the scene in conjunction with paper birch. The trailhead parking lot is about 3.0 miles from the Orange line. When you reach it, pause a moment before you start up the mountain to observe the increased proportion of red spruce. Clearly, you are in a transition zone between the taiga and northern hardwood regimes.

The Mount Cardigan summit trail has a one-way length of 1.3 miles. I once saw a very athletic man run all the way to the top in the company of a faithful but weary golden retriever. Generally, even the most casual hikers find their arrival at the summit a well-earned and timely event. If you're not the decathlon type, the secret of success lies in taking your time and pampering your oxygen debt whenever you feel the need. In fact, you'll see a great deal more botany that way.

The trail can be divided into two parts: the shaded woodland phase and the final approach across an open rock slope. The forested section features an overstory of Canadian hemlock, paper birch, and some bigtooth poplar; beneath grows striped maple, lowbush blueberry, sheep laurel, and hobblebush, which is a lovely sight in May, when it blooms. The peak wildflower show is in the spring, when the lurid beauty of red trillium is offset by cheerful yellow corn lilies and the splotchy-leaved trout lily. The attractive foliage of the trout lily has been used in a treatment for skin problems. If you see these plants in bloom, examine the flowers—their parts are grouped in threes or multiples thereof, which is a good indication that they're monocots. Dicots usually have flower parts in fours or fives.

The woodland stretch is also prime territory for the humble but endlessly intriguing nonvascular bryophytes. The ubiquitous haircap moss, once used to combat dropsy and kidney stones, is only one of several species to be found on wet banks and rotting tree stumps. At one point, where a brook creates a miniature wetland, there is even some sphagnum. It too is considered a true moss, though clearly its ancestors diverged from the main types a long time ago. Bryophytes, in common with the lichens, are often indicators of good air quality, because they tolerate very little in the way of pollution.

As the trail gains altitude, montane plant species such as mountain ash and the lowly three-toothed cinquefoil become more prevalent. When you reach the clean bedrock surface, you come face-to-face with one of the most beautiful rock types in all the Northeast. This is a distinctive granite called the Kinsman Quartz Monzonite. It is a product of those wild days in the Devonian period when the Acadian mountains, roughly the same scale as the Alps, reared up in these parts. Back then, the monzonite was definitely not exposed to the air, nor was Cardigan itself a part of the upland heights. Rather, this great granite mass was a pluton, or body of molten material, far underground. To the climber, the rock is an excellent surface because it contains elongated feldspar crystals that act like deck treads. To the geologist, the feldspar pieces, also known as phenocrysts, are a sure sign that the monzonite took a very long time to change from the liquid to the solid state. Other igneous rocks, which cool much more quickly, do not form such large crystals.

When this monzonite was slowly but surely hardening in the sub-basement of that long-vanished mountain range, the planet's earliest plants were busily adapting themselves to a multitude of new environments on the earth's dry surface. In one sense, they were confronted by a scene similar to the one you see here: expanses of bare ledge, with not so much as a pinch of good soil anywhere in sight.

As you scuttle up toward the summit, you suddenly come upon a truly remarkable tableau. Alpine plants have established themselves in small, island communities. They represent species with needs and capabilities that are fine-tuned to this meaner place. You have come to the tour's alternative way of life. These specimens are the rare remnants of a large assemblage of plants that thrived near the retreating glaciers ten thousand years ago. You'd expect to see them on the slopes of Mount Washington, which is twice as high as Cardigan's 3,121 feet. That they've survived here is a testament to the remoteness and climatic harshness of this place. By all means scrutinize these island habitats, but *do not harm them in any way.* Their very existence is fragile and miraculous—nature's answer to fine art.

One of the most impressive things about these windswept clusters of vegetation is the way they're zoned by plant type. Around the edge of each zone, only the prostrate crowberry and mountain cranberry can inch their way out onto bare rock. A little farther in, Labrador tea, a lovely mini-shrub with white flowers and woolly, orange lower leaf surfaces, finds a foothold next to a fellow ericad, sheep laurel. Only in the centermost portion of each island is there enough shelter for such relatively upright plants as mountain ash and dwarfed red spruce. Each ensemble reminds you of an embattled platoon, huddling under heavy fire. May these warriors survive for centuries to come.

The view from the fire spotter's tower at the summit of Mount Cardigan is undoubtedly one of the best in New Hampshire's lower half. It even beats the vista from that hiker's equivalent of O'Hare Airport, Mount Monadnock. It is here, on Cardigan's crystal-ribbed dome, that the human mind and soul get a priceless lesson in the true scale of things. In the distance, the hillside forests contend with height and climate in a way that can be read like a road map. In some areas, northern hardwoods find the protection they need to outcompete their neighbors and survive; elsewhere, montane spruces proclaim the onset of harsher conditions. These are patterns that establish themselves again and again without a discernible will, in ways quite alien to what we call consciousness, yet we can use our consciousness to comprehend it as best we can. However difficult you found the climb, you'll probably agree that this is the trip's second heaven.

One of Mount Cardigan's island communities of alpine plants.

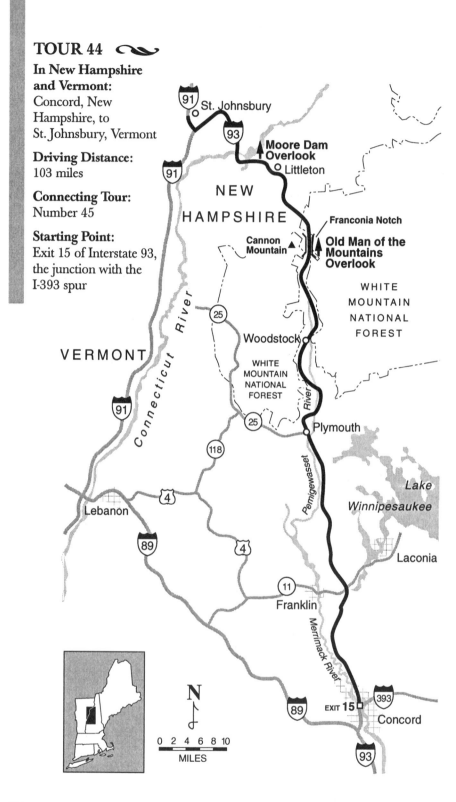

TOUR 44

In New Hampshire and Vermont:
Concord, New Hampshire, to St. Johnsbury, Vermont

Driving Distance:
103 miles

Connecting Tour:
Number 45

Starting Point:
Exit 15 of Interstate 93, the junction with the I-393 spur

St. Johnsbury

Moore Dam Overlook

Littleton

NEW

HAMPSHIRE

Franconia Notch

Cannon Mountain

Old Man of the Mountains Overlook

WHITE MOUNTAIN NATIONAL FOREST

VERMONT

Connecticut River

Woodstock

WHITE MOUNTAIN NATIONAL FOREST

Plymouth

Pemigewasset River

Lake Winnipesaukee

Lebanon

Laconia

Franklin

Merrimack River

N

EXIT 15

393

Concord

89

0 2 4 6 8 10
MILES

93

414

Tour 44
A Roadside Vegetation Tour:
North by Northwest

A Recommended Fall Foliage Tour

HIGHLIGHTS

An Appreciation of Interstates ∿ Monitoring the Merrimack ∿ The Living
River ∿ Pitch Pines and Mini-Wetlands ∿ Rapids by Another Name ∿ Coping
with Mowing, or the Strange Growth of Grass ∿ The Elusive Dandelion ∿
A Farewell to Oaks ∿ Franconia Notch State Park ∿ The Old Man of the
Mountains ∿ A Showplace of the Northern Hardwoods ∿ The Descent to
Lovely Littleton ∿ The Realm of Green Rocks ∿ Salmon and the Great
River Dammed ∿ The Wildflowers of Vermont's Northeast Kingdom

*T*his is one of several trips in the book that give you the opportunity
to participate in an exciting new field of science, known, at least by
me, as High Speed Botany. Though it lacks official academic accredi-
tation—perhaps because of that—it can be both worthwhile and fun. As you'll
see, it's surprising how much information can be gleaned about the plant world
in the course of a drive down a stretch of the interstate highway system.

Interstates are frequently maligned; they are characterized as nothing more
than sterile zones plowing their way through a landscape without harmonizing
with it in any way. In actuality, they're often a precious gift, not only to travelers
who need to reach their distant destinations quickly, but also to observant
naturalists. The plant explorer soon finds that a trip on two-lane roads leading
from one town to another is largely a blurred and inconclusive picture of wild
species mixing and alternating with cultivated ones. Besides, many side roads
are edged with second-growth woods too close for the eye to focus on. In contrast,
the interstates were built to avoid residential areas wherever possible. Feel detached and alienated in using them, if you must, but don't miss the benefits they
offer. Some of the best unspoiled scenery in New England is beheld only from
this controversial ribbon of concrete.

───────────── ∿ ─────────────

Our piece of the concrete ribbon begins in New Hampshire's capital, Concord.
The starting point lies within view of the glittering gilded dome of the State House.

Head north on Interstate 93. For the next 65 miles or so, the road follows the Granite State's second largest river, the Merrimack, upstream to its very source. At 1.4 miles from the beginning, you have a good opportunity to grasp its size, where the roadway crosses it.

Anyone familiar with Civil War history will recognize the waterway's name, since it was the one supposedly borne by the most famous ironclad in the Confederate States Navy. Originally, the ship was indeed the *Merrimack,* but when she was captured in port by Southern forces and radically rebuilt, she was rechristened CSS *Virginia.* Be that as it may, the Northern press that reported her standoff battle with USS *Monitor* clung to her old name.

In this locale, the shifting channel of the waterway has resulted in the creation of several oxbows, or curved lakes. Originally these oxbows were meanders, or bends in the river, but eventually the stream found a more direct route, which left them isolated as pools of standing water. Considering the river's capacity for change, both majestically slow and terrifyingly sudden, it is tempting to give this great and capricious form of power full status in the system of life. Interestingly, many of the terms geologists use to describe rivers are unconsciously organismic: streams are "youthful" or they have attained "old age"; or, in some cases, they have even been "rejuvenated."

To the plant ecologist, New England's north-south river valleys are of great interest because they constitute corridors with warmer microclimates than the surrounding uplands. As a consequence, species more typical of southern areas can compete quite successfully, as long as they stick to the low ground. A good example of this is the red oak, a main indicator of the central hardwoods regime. It's visible here and far upstream, all the way to the threshold of the White Mountains.

On this leg of the tour, you should have no problem spotting New England's tallest tree, white pine, as well as such deciduous species as poplar, red maple, white ash, and staghorn sumac. In the past few years, the gypsy moth infestation has been severe, and consequently, this area has been struggling with massive summertime defoliations. (For more on this worrisome insect problem, see Tour 23.)

On the right, 3.6 miles from the previous river crossing, is a stand of tamarack. This hardy conifer, unusual because it sheds its needles each fall the way many broad-leaved trees do, is one of the world's most cold-tolerant trees. Its natural range extends up to northernmost Labrador, across the shore of Hudson Bay and on to the Yukon and the Alaskan interior. The wood of this North American larch is a challenge to split, but it resists rot, and is therefore used for house sills, railroad ties, and fence posts. Indians used tamarack to treat bronchitis and other afflictions, and its roots made good cordage. Josselyn, writing in the late 1600s, reports that early colonial herbalists sought out agaric, a fungus growing on its trunk, and used it as a purgative.

Not long after the stand of tamaracks—beginning about 1.8 miles north of them—you'll notice the road embankments become distinctly sandy. To the south, in the vicinity of the confluence of the Merrimack and the Soucook Rivers in Pembroke, the look is similar. That downstream location is the Concord Pine

Barrens, home of rare southern plant species that include the New Jersey tea shrub. (To visit another New Hampshire Pine Barrens, see Tour 48.)

Pitch pine, a fire-resistant conifer most often associated with Cape Cod, is quite common around here. In the next few miles, you get a good look at drainage-ditch ecology, where two very different wetland plant types, the herbaceous monocot common cattail and the woody dicot speckled alder, have found a suitable home. Additionally, see if you can spot the pin oaks that have been set out on the median strip. While this is a bit north of their normal distribution in the wild, these specimens have survived so far. From a distance, they can be distinguished from red oaks by their flared trunk bases and drooping lower branches.

Exit 20 to Tilton is the turnoff to Laconia and Lake Winnipesaukee to the east, and to the town of Franklin to the west. At Franklin, the Merrimack suffers a name change and becomes the Pemigewasset. Though it's difficult for an English speaker to note the connection, both names stem from Algonquin terms to indicate the existence of rapids. Here, as in many other parts of the United States, American Indian vocabulary, however modified, survives and enriches our own. According to one interpretation, even that most basic Anglo term, *Yankee,* is a slight corruption of the Indian *Yengees,* which was the Algonquin pronunciation of the word *English.*

———————————— ∾ ————————————

Between Exits 20 and 25, the interstate first swings quite far to the east of the river, then returns to follow alongside. This is a good locale in which to examine the most overlooked roadside plants—the grasses growing on the shoulders. These highly specialized monocots have wind-pollinated flowers, are adept at asexual reproduction by means of spreading rhizomes, and possess an unusual manner of growth that permits them to survive the periodic sweeps of the mowing crews. In the vast majority of plants, new upward growth takes place only at the tips of the stems. That's why a sign tacked onto a tree trunk does not move upward in the passage of time, despite what most people think. The tree has a second kind of tissue formation, which accounts for the thickening of its trunk and branches, but its increase in height is due only to the elongation of its stem tips.

In grasses, however, elongation is centered near the base of the plant, so the stems are actually pushed up from below. One result of this is known to every suburban lawn owner: if a blade of grass gets clipped halfway down its length, it keeps inching its way upward. The growth center is unaffected and the plant doesn't have to go through the fuss of creating active side buds. The herbaceous dicots along the shoulder have no such advantage, so they either sprout, flower, and fruit quickly, or develop a prostrate habit that protects their vital parts from the mower's blade. Some common weeds, including the alien herb self-heal, have actually developed lower-growing races in response to human grass-cutting practices.

One of the best-adapted dicots of all, as every lawn owner knows, is the dandelion, a plant in the composite family that received its current taxonomic

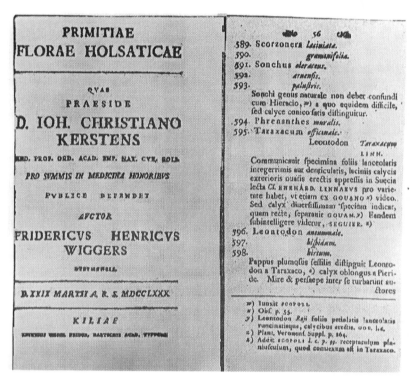

From F. H. Wiggers' 1780 book, Primitiae Florae Holsaticae. *The title page* (left) *and the page describing* Taraxacum officinale—*still the correct taxonomic name for the dreaded dandelion.* —Marlin Roos

classification in the late eighteenth century. According to some modern texts, including *Hortus Third,* the first person to use the name in print was a Dane, Friedrich Heinrich Wiggers (1746–1811). Both a physician and a botanist, Wiggers' magnum opus was an important book on the plant life of northern Europe. Published in 1780 and entitled *Primitiae Florae Holsaticae,* the work does indeed mention the dandelion by the new name *Taraxacum officinale.*

Much of the dandelion's notorious reputation for indestructibility is based on the fact it has contractile roots—that is, roots that pull the stem a little deeper into the ground each growing season, so that only the flat-lying leaves and rapidly developing flower stalks project above the surface. The dandelion may be Public Enemy Number One in today's lawn-obsessed culture, but it was a crop plant in New England in the nineteenth century. Believe it or not, its foliage provided a favorite salad green. (For pointers on making a rather dubious-sounding dandelion sandwich, see Tour 1.)

The section of the road just before Plymouth is a good spot for pine watching. Can you tell the two wild species apart? White pine branches have a flatter, finer, almost feathery texture; those of the red pine are somewhat coarser,

418

with tips often curving upward. In addition, red pine bark is frequently tinted pink. Just after Exit 25, the highway crosses the Pemigewasset for the first of several times, as the stream rises into the White Mountains. Check your odometer reading here, since it will serve as a benchmark for a couple of the sights ahead.

Another interesting plant appears on the roadside 1.3 miles north of the benchmark bridge. This is sweet fern, a low shrub in the bayberry family with curious, frondlike leaves. The species loves nothing better than disturbed ground and lots of sunlight. It is one of a number of plants that has a symbiotic association with bacteria that obtain vital nitrogen otherwise unavailable in the nutrient-starved soil. You'll also notice a change in the tree community in this area. Red oak has finally yielded its prominent role in the valley to its more cold-hardy relation, the American beech.

At 16.1 miles above the benchmark bridge, a magnificent view of Franconia Notch and its glacially scoured cliffs opens up ahead. In a few minutes, you'll be up there. In this zone, the northern hardwoods regime is well established, with sugar maple and yellow birch much in evidence, along with that happy participant in several regimes, the paper birch. As you enter Franconia Notch State Park, keep in mind that there are a number of interesting turnoffs to take, if you're so inclined. The stop described here is on the right, at 5.7 miles beyond the park entrance.

The sense of grandeur in the notch is reminiscent of the Rockies. On your left stands Cannon Mountain, where the famous Old Man of the Mountains

Tree roots have the ability to slowly crack and dislodge rock. This yellow birch grows perched in a crevice at Franconia Notch.

resides. You are almost at the headwaters of the Pemigewasset. See if you can locate the landslide scars and the slopes of loose rock, or talus, where vegetation attempts to gain a foothold. Your turnoff is the viewing area across the road from Profile Lake. It is here that you can best see the craggy countenance of the Old Man jutting out into the void. The trees on the parking lot periphery are textbook examples of the northern hardwoods: sugar maples, American beeches, and yellow birches.

Take an especially good look at the yellow birches. Their silver, shredding, horizontally marked bark is a foolproof identification trait. Under the hardwood canopy, meanwhile, stands the forest shrub known as hobblebush. Its white flower clusters are a delight to see in the early spring. Depending on when you visit, you may also see meadowsweet, a low rose-family shrub of the genus *Spiraea,* bearing bouquets of tiny white or pinkish flowers. The species usually blooms in this area by early July.

———————————— ∾ ————————————

When you're finished at the viewing station, resume your northward direction on the highway. The ski-slope development marks the approximate point of the drainage divide. You are leaving the Merrimack basin and heading into the larger and longer Connecticut River system. Northern hardwoods mix with white pines here, but at a point approximately 9.0 miles from your last stop, you'll notice a fairly sudden shift to the red spruce of the taiga regime. In July, this locale is a great place to spot black-eyed Susans nodding in the breeze. Like the dandelion, they are composites—each striking orange-and-brown flower head contains many miniaturized blossoms that fit together with the precision of a Swiss watch.

You soon arrive at the crossroads of Littleton, one of the nicest and most scenic towns in New Hampshire. If you look back to the east and the sky is clear, you will see the greatest mountains in the Northeast, the Presidential Range, including Mount Washington. As you approach Exit 42, you'll cross Littleton's own Ammonoosuc River, a lovely stream that for many years was used for log drives (see Tour 31). At 1.5 miles beyond it, there is a road-cut of beautiful green-tinted rocks that are almost half a billion years old. These are part of the metamorphosed volcanic units of the Bronson Hill zone, a narrow corridor that runs up from southern Connecticut, across the northern part of the Granite State, and just a bit into Maine. For a fuller description of this curious geologic terrane, see Tour 32.

This is a portion of the state where each point of scenic interest stands cheek-by-jowl with every other. The most magical thing about the Littleton vicinity is how the looming mass of the White Mountains and the mighty Connecticut River, otherwise separate entities, have agreed to share a single landscape. In a mere 4.7 miles from your crossing of the Ammonoosuc, you come upon a huge reservoir created by the Samuel C. Moore Dam and hydroelectric station. There are a number of similar facilities up and down the waterway, except in the state of Connecticut, where tidal effects prevent their large-scale use. The impact of all this dam construction has been ecologically significant, particularly on the fish population. For example, salmon were blocked from their breeding grounds in northern Vermont and New Hampshire as early as the opening decades of the nineteenth century. In recent times, they've gradually been reintroduced with the help of fish ladders.

A moment later, you're gliding across the Connecticut. This splendid, high bridge is your gateway to the Green Mountain State's Northeast Kingdom. At once, as if by plan, new types of trees, more typical of the far north, can be seen. In the vicinity of the visitor center, look for the evergreen arbor vitae, a member of the cypress family, and for balsam poplar. (For more on balsam poplar, see Tour 45.) If you are making the trip in midsummer, you'll probably also see a variety of roadside wildflowers: Queen Anne's lace, yellow hawkweeds, rabbit's-foot clover, purple loosestrife, and early goldenrods.

The tour ends at the junction of Interstates 93 and 91 in St. Johnsbury. While much of the Northeast Kingdom is a sparsely populated expanse of spruce and tamarack and soundless lakes, this section is largely cleared for dairying. The pastures are composed of a variety of alien forage species. Among them are timothy, red clover, white clover, and orchard grass. (Another common pasture species, Kentucky bluegrass, prefers the sweeter soil of limestone regions. Ironically, it's not a native of Kentucky, but of Europe and perhaps the northern parts our own continent as well.) One glance at this bucolic view of placid cows and proudly maintained farms makes you feel that you're back home at last—even if this is your very first time in Vermont.

North-country forests and green-tinted rock outcrops provide beautiful scenery along Interstate 93 in the vicinity of Littleton.

In Vermont and New Hampshire:
St. Johnsbury, Vermont, to Gorham, New Hampshire

Driving Distance:
197 miles

Connecting Tours:
Numbers 36, 44, 47, and 49

Starting Point:
The intersection of Routes 2 East and 5

Third Connecticut Lake

Second Connecticut Lake

First Connecticut Lake

3

MAINE

Pittsburg

Lake Francis

QUEBEC

Clarksville

3

145

Monadnock Mountain ▲

Colebrook

Dixville Notch State Park

114

Columbia

26

105

Island Pond

Bloomfield

16

105

North Stratford

Errol

Brighton State Park

3

26

East Haven

Connecticut

16

5

114

River

Burke Mountain ▲

NEW

Milan

91

East Burke

Androscoggin River

Lyndonville

HAMPSHIRE

VERMONT

Berlin

2

5

Maple Grove Museum

Passumpsic River

St. Johnsbury

2

Gorham

93

Littleton

3

16

Mount Washington ▲

N

Connecticut River

91

93

0 2 4 6 8 10
MILES

422

Tour 45
The Northeast Kingdom and a Journey to the Source

A Recommended Fall Foliage Tour

HIGHLIGHTS

The Maple Grove Museum and Factory ❧ Making Maple Sugar ❧ Brighton
State Park ❧ Acid Rain ❧ Birches Battle Nasty Nibblers ❧ A Kame Moraine
Terrain ❧ Monadnock Mountain ❧ Far North Fertility ❧ The Glorious History
of the Indian Stream Republic ❧ Podzol ❧ Connecticut Lake Country ❧ The
Balsam Poplar ❧ The Border and the Source ❧ Hills as Thick as Mole-Hills ❧
An Intervale Ascent ❧ Dixville Notch State Park ❧ A Comparison of Maples ❧
A Stream Divided ❧ Berlin and Its Atmosphere ❧ The Clutter Subservient

*I*t took all my willpower not to name this excursion The Spruce 'n'
Moose Tour, even though the same concept has popped into the mind
of every motel owner and campground director from Maine to British
Columbia. Taiga conifers and large mammalian herbivores are just two of the many
items on this trip's substantial agenda. Just short of a full 200 miles, this itinerary
is one of the book's longest. So, if you plan to complete it in one day—and you
certainly can, in six to nine hours, depending on how long you linger at the
stops—plan to start early. You might wish to call your first destination, the Maple
Grove Museum, to confirm their opening time; then set your pace by that.

The starting point is located in downtown St. Johnsbury, where Routes 2
East and 5 are otherwise known as Portland and Railroad Streets, respectively.
Take 2 East across the Passumpsic River for 1.3 miles. Turn into the Maple Grove
Museum visitors' parking lot on the left. As you'll see, the facility is located just
across the road. The museum is very much a commercial enterprise. The Maple
Grove company was founded in 1915 by two local women who first set up shop
in their own home. The public relations end of the current operation includes
a small exhibit building where traditional sugarhouse equipment is displayed, and
a large, highly emphasized gift shop. The real attraction is the factory, where
for three shiny quarters you can take a quick guided tour of the production floors
where Vermont's most addictive product is turned into candy and other con-
fections. The strong, sweet smell of the premises may send you floating up to
the ceiling.

Today, the Green Mountain State is the largest producer of maple sugar, with New York a close second. When you compare Vermont's tiny land area to the Empire State's huge expanse, you see how seriously the sugar trade is taken here. Other front-runners in this most saccharine industry are Ohio, Wisconsin, Michigan, New Hampshire, and Pennsylvania; surprisingly, the other New England states lag somewhat behind them. This list, restricted to the Northeast and the Great Lakes region, points out the importance climate plays. Sugar maples growing farther south often do not yield sap in sufficient quantities for large-scale business interests.

As with many aspects of Yankee agriculture, maple sugar production was first developed by the Indians. They recognized that the sugar maple has the best sap, but also used silver maple, box elder, and even black cherry trees. Sugaring was a factor in colonial farm life by no later than the eighteenth century, though most of it was for the fairly modest needs of home consumption. Not until the twentieth century did Vermont really push the commercial potential of this instantly appealing commodity. But when it finally did, it quickly succeeded in marketing the basic ingredient as syrup and candy, as breakfast spreads, and even as a sweetener for chewing tobacco. The syrup, which nowadays is expensive enough to qualify as an upscale, Fifth Avenue sort of gift, is rated as Fancy, with a light amber color; Grade A, medium amber; and Grade B, dark amber. The slippery (or rather, sticky) slope down into maple syrup dependence takes many forms; but my own preferred transgression is Grade A poured straight over top-of-the-line vanilla ice cream.

Of course, the starting point of the whole industry is the sap of the sugar maples. The process of extracting the sap from the trees can be tricky. A forest stand that includes suitable trees at least 10 inches in trunk diameter is known as a sugarbush. Most of the tapping is done between late February and late March, when the plants, still leafless, are transporting the sugar-rich food supply they manufactured the year before up the sapwood part of their xylem tissue. Holes are normally drilled from 2 to 3 feet above ground level, and about 2.5 inches into the stem. The real experts incline them upward a little bit for better flow. Then the taps, which are metal spigots, are inserted—only one in each small tree, but up to three or even four in a really large specimen. One obvious sign of a sloppy, inexperienced tapping job is a streak of darkened sap running down the bark under the insertion point.

Weather conditions are of paramount importance. Collection is worthwhile only on days preceded by freezing nighttime temperatures. During the day, the thermometer must reach 40 degrees Fahrenheit or higher, and it's best if it's sunny and windless as well. The traditional and much more picturesque method of sap collection is with individual buckets, but nowadays many people are using disposable milk jugs or plastic tubing that connects a number of taps to one large reservoir. The tubing is convenient but has its drawbacks—if one tree along the line is producing inferior sap, it gets mixed automatically with the better stuff; and occasionally, a sweet-toothed squirrel will take a nip out of the plastic to get a free pick-me-up, and so cause a wasteful leak in the line.

The sap flow of each tree varies, but one common estimate holds that the average sugar maple provides roughly 12 gallons a year. Since it takes from 35 to 40 gallons of high-quality sap to make a single gallon of syrup, the typical tree ends up providing only one quart or a little more of the finished product. To be profitable on a full-fledged commercial basis, a producer has to have well over a thousand taps. Sugarbushes decline with the passing of the years, but many remain viable for more than a century.

The driving portion of this tour gets under way in earnest when you retrace your route to the starting point in St. Johnsbury. Turn right onto Route 5 North. Watch for the point 0.4 mile up the road where the Passumpsic cuts through an esker. This snakelike ridge was formed by the deposition of sediments in a stream tunnel located inside a melting Ice Age glacier. You'll follow both it and the river quite a way to the north. As you will note in this vicinity, eskers are often quarried for their clean gravel and sand. St. Johnsbury's lovely Main Street runs along the top of this geologic marvel.

Not surprisingly, white pines are common esker residents. They do especially well in the dry, sandy soil. In the Passumpsic bottomland nearby, black locust and two flood-tolerant species, American elm and black willow, are present. Willows, in common with many other woody wetland plants, tend to reproduce asexually as well as sexually, to produce a dense thicket community. Thoreau described their vigor: "They multiply and clan together. . . . May I ever be in as good spirits as a willow! How tenacious of life! How withy! How soon it gets over its hurts! They never despair."

A Yankee wildflower that once honored the goddess Artemis. The familiar ox-eye daisy of New England roadsides is actually a European import with a long history of ceremonial and medicinal use.
–Vermont
Wildflower Farm

At the point 9.3 miles north of your previous turn onto Route 5, bear right at Lyndonville, onto Route 114. Here the esker mimics the river by splitting into two branches. Route 114 follows the eastern arms of both. After another 0.6 mile, you have a good view of Burke Mountain, one of the Northeast Kingdom's major peaks, directly ahead. The esker tags along on your left. In June, this locale is home to many white-rayed, yellow-centered ox-eye daisies. Known since ancient times, this Old World introduction was dedicated to the goddess Artemis, and in more recent ages has served as a remedy for asthma and bronchitis. As practically every popular botany book notes, the term *daisy* comes from the earlier English version of the name, *day's eye*. Strictly speaking, though, the ox-eye is not a true daisy, but rather is a wild version of the well-known genus *Chrysanthemum*.

Pastures and hayfields cluster in the vicinity of the Burke Mountain Resort entrance. Above them, on the high ground, is a community of mixed evergreen conifers. Even without any particularly distinctive landscape features, this sort of open land in the foreground, framed by woodland in the background, is a simple but effective formula for good scenery—as the works of Frederick Law Olmsted and other great urban park designers testify. In the East Haven area, the esker finally peters out. To the north, in Newark, the presence of low, wet ground is indicated by tamaracks, red maples, and speckled alder shrubs. In case you're wondering whether red maples can be tapped for sugar, the answer is yes, they can; however, their less concentrated sap yields only half as much finished syrup, and as a result, few people bother with them.

Another plant in this neighborhood is the June-blooming American mountain ash. Farther south in New England it is generally restricted to mountain slopes. In the next township, Brighton, you come to the source of the Passumpsic's East Branch, as well as the Route 114 junction with Route 105. Bear right onto the combined 114 North-105 East and proceed into the Northeast Kingdom's loveliest settlement, Island Pond. It is located on a body of water by the same name. The town is an old railroad center that owes its origin to its strategic location, halfway between Montreal and Portland, Maine. When Route 114 continues its way northward, follow 105 East, to the right, instead. Go approximately 2.8 miles to the right-hand access road to Brighton State Park and turn down it. In less than 0.5 mile you'll see the park, fronting Spectacle Pond.

The beach setting makes a pleasant place to get out, stretch, and enjoy the refreshing waterside breeze. Given all the talk in modern times about acid rain, you might be curious whether this pristine-looking lake, and all the other glittering waterholes in the general vicinity, have been afflicted with that serious environmental threat. The answer lies in how acid rain comes into being in the first place.

It all begins when two types of human-produced chemical compounds, sulfur dioxide and oxides of nitrogen, are wafted aloft, where they interact with atmospheric moisture. Sulfuric and nitric acids form and create acid rainwater. Aquatic ecosystems are perhaps the most vulnerable of all to the lowering pH level. Few fish can live and breed, happily or even unhappily, in water approaching the acidity of tart orange juice. Land vegetation is also at risk. Mycorrhizal fungi, which help many plant species living in poor soils to obtain crucial nitrogen, are easily killed by acid rain. Having said that, it so happens that this area has fared better than points to the east or west. The high Adirondacks and White Mountains apparently act as barriers, protecting Vermont from the brunt of the problem. Nevertheless, the fact that these ponds are better off than the dead lakes of upstate New York doesn't mean they won't be affected in the future. Vigilance and a healthy suspicion are most definitely called for.

Before you jump back into your car, take a look at the arbor vitae and paper birch specimens on the park grounds. In earlier times, our neighbors to the near north—the ones who have a sugar maple leaf on their flag—used newly cut arbor vitae branches as brooms. The flattened sprays of foliage might not have been as efficient as a vacuum cleaner, but they did do one thing better, by imparting their pleasantly bracing, camphorlike smell to the places they swept. The paper birches, not to be outdone, are a good example of how plants muster chemical defense systems to protect themselves against bothersome animals. The birches may not be quite as overt as poison sumac or stinging nettle, but they know how to deal with marauding herbivores. Snowshoe hares have the same love for tasty young birch shoots that the cartoon character Wimpy has for hamburgers. Hares get a free lunch, so to speak, the first time they munch on a birch; however, the tree responds by growing new stems that contain higher levels of phenols and resins. This is as appetizing to the hares as a juicy, medium-rare burger would be to Wimpy, provided it was first cooked in liquid shoe polish.

Return up the park access road to the junction of Route 105. Turn left onto 105 East. Before long, you will see large stands of one particular conifer, red pine. These plantations are a common sight throughout the entire New England region. Just past the settlement of East Brighton, the road begins to parallel the small Nulhegan River. This is lovely, remote up-country, and as you would expect, tamaracks are an important tree species in the low, wet spots. Generally, this is a transition zone between the northern hardwoods and the taiga spruces and firs. The terrain itself is a classic kame moraine—a gently rolling surface characterized by relatively poor stream drainage, a smattering of kettle ponds, and plenty of wetland. This sparsely populated area is the very heart and soul of the Northeast Kingdom.

In a matter of minutes, however, you and the Nulhegan have descended the remaining distance to the parent river, the Connecticut. Take the bridge across it to North Stratford, New Hampshire, and there turn left onto Route 3 North. The valley here is none too broad, but what bottomland there is remains largely devoted to agriculture. Poplar, black cherry, and American elm must be content to occupy a narrow strip between the fields and water's edge.

Within 9.1 miles of your turn onto Route 3, you will see rising majestically before you the most prominent landmark of the last reaches of the Vermont–New Hampshire border. This is Monadnock Mountain, not to be confused with the more frequented Mount Monadnock far to the south. This impressive high point is formed of igneous rock known as syenite, and it shares a common origin

Monadnock Mountain, in far northern Vermont—as seen from Route 3 in New Hampshire, across the Connecticut River.

with Mount Ascutney, roughly 100 miles downstream (see Tour 32). Next, you enter Colebrook town center, where you are presented with a sweeping view of hills and pastoral countryside. This is quite far north for such fertility. Through this enchanted scene winds the Connecticut, so little now that it's difficult to believe early colonists called it the Great River.

You are on your way to the very source of the Connecticut River. When you reach Stewartstown, you are just a few minutes' drive from the Quebec line. This placid little bridge town is the perfect place to test your global perspective. With Canada so close at hand, just how far do you think you are from the North Pole? How far from the equatorial rain forests of South America? In another 0.6 mile, a road marker gives you the answer: you are now on the parallel that is 45 degrees north latitude, exactly halfway between the pole and equator. If you are familiar with the climate in this locale, you might have a little trouble believing that the land of everlasting snow and ice is still really that far away. But it is.

— ∾ —

Route 3 begins its final leg through wilder ground toward the uppermost part of New Hampshire. Here, the Connecticut River is a rocky, shallow stream, and it merely marks the boundary between two wilderness towns. The line between the United States and Canada, finally agreed to in 1842 after decades of contention, is formed by the western tributary, Hall's Stream. Just before you arrive in Pittsburg town center, you pass the covered bridge connecting it with Clarksville. This single-span structure, built in the Paddleford design, requires a total length of only 91 feet to span the Connecticut. Downriver, in the southernmost third of the state, the great Cornish-Windsor covered bridge requires two spans and 460 feet to accomplish the same task.

Although it is not a widely advertised fact of North American history, Pittsburg has something to share with Vermont and Texas, since it too was once an independent country. Named the Indian Stream Republic, it came into being in 1832 when its inhabitants, sick of the squabbles and uncertainty that went with the overlapping territorial claims of Canada and the United States, decided to secede from both. This made the republic one of the least populated nations in human memory, since it numbered fifty-nine residents—hardly the raw material of an enduring civilization. Still, this outpost of backwoods orneriness managed to resist the dictates of its titanic neighbors until 1840, when that cruel instrument of imperial expansion, the New Hampshire state militia, finally stopped by for a visit and squelched this experiment in woodland autonomy. Had the secessionist Indian Stream Republic survived another two decades, one has to wonder whether it would have allied itself with President Jefferson Davis and tried to march on Gettysburg from the other direction. In this neck of the woods, one gathers, anything is possible.

The center of town is located on Lake Francis, a body of water conspicuously absent from older maps since it is the result of Murphy Dam, constructed in 1940. Heading farther north, you enter the mythic landscape of the Connecticut Lakes region. This federally designated route looks increasingly like an extended private driveway that might just lose its sense of purpose altogether. My favorite time to come here is also the least comfortable—frost-heave season, February through April, when the pavement feels rockier than the river itself. It is then that you can see the taiga conifers uncloaked by their broad-leaved, deciduous counterparts.

The spirelike forms of red spruce give the terrain a primordial, yet abstract, look. Some New England Indian tribes associated spruce trees with the disembodied souls of their ancestors. In recent years, a rather mysterious decline has hit the red spruce in particular. It appears to be linked with increasing air pollution. Nevertheless, the tree remains a major paper pulp source, and its wood is still used for the sounding boards of violins and other string instruments. When the taiga regime is predominant, the layer of fallen spruce and fir needles, known as duff, produces a sterile soil type called podzol. As rainwater percolates through the duff, it becomes acidic and removes the organic content of the underlying material. These particular conifers can tolerate this impoverished soil, but few hardwood or wildflower species can.

Keep in mind that all sorts of wildlife, from ambling moose to barreling logging-truck drivers, lay claim to this road. So proceed carefully. A pleasant turnoff by the First Connecticut Lake dam is 7.5 miles beyond Pittsburg. Between that and Second Connecticut Lake, take a good look at the community of woody plants. There are very few truly large trees. In this heavily logged area, the wild growth is harvested systematically and relentlessly in the race to keep the insatiable demand for wood products supplied. Once again, you are in a transitional

A moose ambles along the Route 3 roadside, between the First and Second Connecticut Lakes.

area where northern hardwoods–sugar maple and American beech accompanied by black cherry and red maple–mix with such taiga residents as red spruce, balsam fir, American mountain ash, and balsam poplar.

The last-named species can be found quite a distance to the south, but is more typical of northernmost New England. Its taxonomic classification is somewhat blurred. According to some botany texts, there are two distinct species: one called balsam poplar, as listed above, or tacamahac; and another called balm of Gilead. The distinction between the two has been based primarily on the details of the leaves. However, the great nineteenth-century American botanist Asa Gray considered balm of Gilead only a variety of balsam poplar, and the more modern sources tend to support his view. For what it's worth, the trees I've examined in this locale most often seem to fit the official classification of *Populus balsamifera* var. *subcordata*. In the following discussion, I'll stick with the common name, balsam poplar.

Whatever its variations, this tree has stout twigs, furrowed mature bark, and buds that are a dead giveaway–they're coated with a shiny, dark brown resin scented like balsam. Indians and white settlers alike used the resin as the main constituent of a soothing ointment. By late spring, after the buds have broken and the new leaves are unfurled, the foliage is often stained brown or a rusty color by the resin. This makes an excellent identification tool, for no other poplars in the region have this highly visible tint.

The dam for the Second Connecticut Lake is a good place for lichen spotters. *Usnea* species grow rampantly here, perched on the branches of the conifers. Now you're getting very close to the cartographic high point of the tour: 3.7 miles up from the dam, you have one last glimpse on your left of the brook that grows to be New England's greatest waterway. Here one not-so-ambitious beaver could seriously block its flow, yet it runs 410 miles, descends 1,600 feet in elevation to the sea, and drains four states and a bit of Quebec as well. Its long and narrow basin occupies more than 11,000 square miles.

Just so you can say you reached the border, drive past Third Connecticut Lake, swing around at the sleepy U.S. Customs outpost, and retrace your way on Route 3 South to the first fishing-boat access ramp. This is a gravel driveway on your right, where three balsam poplars stand. Here you have a splendid look at the Third Lake. A tiny Fourth Lake lies above it, but in essence this is the river's source. It's a beautiful spot in summer, but to witness the place in the last stage of a lingering winter, when the snow is falling soundlessly here but

Balsam poplar trees at the boat landing, Third Connecticut Lake.

nowhere else in New Hampshire, is to confront a serious thing and a very inhuman land. Over and over, the forests of this place have been scraped and erased and replanted, but the sanctity is indelible. Stand at the water's rim and consider where you really are.

There seems to be an unwritten law that the more breathtaking a place is, the more botanic detail it contains. Perhaps the eye is inspired to become more active. A short hike along the roadway, both above and below the turnoff, yields all sorts of treasures in spring and summer. Wildflowers abound, especially on the far side of the pavement. Even here dandelions flourish, coexisting with buttercup species, wild strawberries, bluets, tall meadow rue, and one of the best yellow bloomers anywhere, golden Alexanders. An unexpected Old World herb, caraway, can also be found. Since antiquity, it has been a favorite for cooking, candies, and even liqueurs, and has seen duty as a traditional earache cure.

The most wonderful wildflower of all is a plant in the rose family, purple avens. As with other members of the genus *Geum,* it has compound leaves that look as though they were designed by the proverbial committee. On the lower foliage, the terminal leaflet is large and lobed, while the other ranks of leaflets each have their own shape and size. Some are downright tiny. The flowers, which open here by early June, are true to the plant's common name; they're a handsome purple with a hint of brown. One of the species' other nicknames is cure-all, which indicates its high standing with herbalists. It was particularly favored for treating upset stomach. In addition, Pehr Kalm reported that both colonists and Indians used powder from its roots, mixed with cold water or brandy, to counteract the ague—malaria.

The woody plants, too, are well worth your scrutiny. On the lake side of the road look for beaked (or Bebb) willow, a good North Country species, and mountain maple. Kalm noted that the people of Quebec called the latter *bois d'orignal,* or elk wood, because large herbivores, probably the wapiti, like to nibble its stems in winter. Farther south, this shrubby tree is mostly confined to upper elevations, but here the climate is challenging enough to suit its tastes, even on low ground. Both the mountain maple and the beaked willow bloom in this area in late spring, or just over the line into summer proper.

———————————— ❧ ————————————

When you're ready to press on, return down Route 3 to the intersection of Route 145 in Pittsburg center. Turn left onto 145 South, cross the fledgling Connecticut, and head toward Clarksville and Stewartstown on this agreeable interior road. Here fields alternate with woodlands and can sometimes be caught in the act of reverting to them. This terrain still bears testimony to its role in Revolutionary and Federal times, when it was, along with the valleys of the White Mountains, one of New England's greatest wheat- and rye-producing centers. In early colonial times, the first-growth forest ruled all. In the 1670s, John Josselyn wrote of this inland area beyond the great peaks with no little awe: "The Country . . . is daunting terrible, being full of rocky Hills, as thick as Mole-hills in a Meadow, and cloathed with infinite thick Woods."

This pleasing stretch lasts 12.9 miles, until you reach Colebrook and Route 3 once more. Turn left onto 3 South, but only for a moment. At the town's main intersection, about 0.4 mile down, turn left again onto Route 26 East. Now you ascend a typical White Mountain intervale—a good Yankee term for the valley floor between the heights. Intervales are one of the chief scenic delights of Coos County, the Granite State's northernmost main subdivision. It is pronounced "*coe-oss*," definitely two syllables; it's an Indian word meaning pine tree. The entire New Hampshire portion of this tour takes place within its borders.

The stream that carved this valley is the Mohawk River, which obviously has no connection with New York State's great waterway. More balsam poplar can be seen along the road. As in Pittsburg, the local folks like to use it as a lawn tree. Things get more dramatic as you rise eastward toward your next destination, Dixville Notch and its state park. At the top of the notch, just past the stunning setting of the famous The Balsams resort, you may wish to pull off at a safe and convenient spot and examine the cliffside flora, but watch out for passing traffic. In this imposing pass, some genuine alpine plants are nudged in among the metamorphic rock cliffs of the Ordovician period. Among those more easily found are mountain cranberry and Labrador tea. As usual, when the going gets tough, the plants get to be ericads.

Half a mile down, you'll find a nice picnic area on the left. Pull in and investigate. When the flying insects aren't busy conducting a major blood drive, this is the perfect place for a picnic-with-waterfall backdrop. The northern hardwoods are firmly in control of the situation here, and there are various examples of American beech and yellow birch, as well as Canadian hemlock and a characteristic understory associate, hobblebush. This, however, is really a showcase for the maples. In one small area, you can find the sugar, red, striped, and mountain species. Compare their leaf forms; it makes an interesting study of subtle variation in characters within a single genus.

The herbaceous plants also beckon. Bracken is here in large numbers with its usual bag of toxic tricks (see Tour 27); so are the coarse, light green fronds of sensitive fern. This species is so called not because it has trouble dealing with criticism, but because it swoons away at the first frost. In more open spots along the driveway or trail, you might also catch sight of a fine little angiosperm by the name of large-leaved avens. This is a close relative of the somber-hued purple avens described earlier. Its foliage is very similar, but the flowers are a cheerful buttercup yellow.

———————————— ॐ ————————————

And so begins the final leg of the tour. Before leaving the picnic area, note your odometer reading. Follow Route 26 East down the valley of Clear Stream, through Millsfield to Errol. There, at the junction with Route 16, turn right onto 16 South. This is the valley of the Androscoggin, a broad river whose very name reeks with a sense of wilderness, even though it really just means *fish-curing place*. If roads could be given titles that refer to animals instead of generals and political nabobs, this one would be christened Moose Alley. One fine May morning, when

433

I was driving nearby, I saw three adults of this long-legged species in the space of ten minutes. Should you spot one of these majestic beasts on the pavement or feeding in the drainage ditch alongside it, be extremely careful. Hitting an animal of that mass with your vehicle makes an especially gruesome demonstration of the laws of physics. Many people have lost their lives this way, accomplishing nothing more than the destruction of one of the world's most impressive creatures. I refer to the moose.

The woody vegetation along this shaggy-edged river is a mixture of shadblow, red maple, and yellow birch, with speckled alder and hazel shrubs on the banks. Alders also form thickets on the midchannel islands. At the Androscoggin State Wayside, 15.0 miles from the last Dixville Notch stop, there is a particularly good view of the surroundings. In another 7.0 miles, the loftiest ramparts of the White Mountains, the Presidential Range, loom ahead. Within a few minutes, you've come abreast a wetland and the dam that caused it. A little farther downstream, the Androscoggin plays the part of a real mountain stream, laden with boulders.

As you enter the northern end of Milan, regularly spaced structures begin to appear in the water. These are old piers for booms that separated the river into two channels for log drives. On this side of the state, the drives remained a regular annual feature all the way up to 1963. The cut tree trunks were marked upstream according to their final destination, either the International Paper Company or the Brown Paper Company. The piers continue to march downstream to the great mills at Berlin, the area's lumber center. There is more than a glimmer of irony in the fact that some of these structures, which once herded thousands of dead and floating trees, are now slowly being pried apart by the roots of living plants that have sprouted atop them. As a Henrik Ibsen character noted, "The forest avenges itself."

―――――――――― ∾ ――――――――――

Downtown Berlin provides quite a contrast to the hours you've recently spent in the wild. Its name is spoken "*ber*-lin," leaning on the first syllable. The story goes that this practice was instituted during World War I, perhaps in the hope that the offbeat pronunciation would deal a stunning psychological blow to the Kaiser. Old residents of this classic lumber town have told me that its "air quality," or rather its lack of it, used to manifest its unique chemical properties by ruining automobile paint jobs.

Whether this vivid image is anecdotal or not, it says something about the local priorities, and also about the priorities of our civilization as a whole. When New Hampshire lumber baron J. E. Henry remarked that he never saw a tree on a hillside that looked as good as one hitting the saw, he made the kind of high-profile, Bambi-kicking bully who is the dream target of environmental activists everywhere. For all his flamboyance, he was only echoing a time-honored human attitude: the sum of a tree or a forest is its material usefulness. In this mill town and others like it in Maine and the Pacific Northwest, Thoreau could preach his higher uses of nature until the earth's last red spruce and Douglas fir are turned into candy wrappers. And yet it is in all the towns at the other end of the supply stream, full of people who buy the candy wrappers and then throw them into the gutter, where the ancient and magnificent first-growth forests really meet their end.

Route 16, south of Berlin, with its car dealerships and other less-than-subtle artifacts, is your deprogramming phase. You've been to the source and have crossed rivers and mountains where this clutter has no meaning. But this is the modern world, and the clutter pertains. On your way to the end point at Gorham's Route 2 intersection, isn't it a little reassuring that there are still hardwood forests on the slopes above the waving flags and blinking lights? Mount Washington stands directly ahead of you. The modern world sits at its feet like an adoring puppy. This sense of scale still has its ring of truth, and all who see it know it.

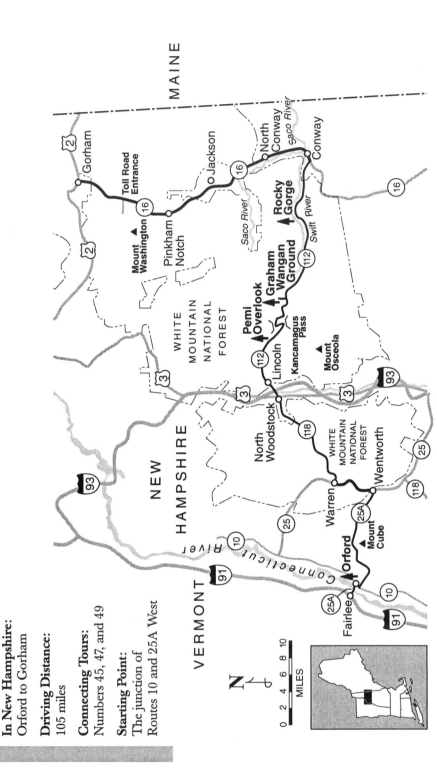

TOUR 46 ∾

In New Hampshire:
Orford to Gorham

Driving Distance:
105 miles

Connecting Tours:
Numbers 45, 47, and 49

Starting Point:
The junction of
Routes 10 and 25A West

436

Tour 46
A Grand Tour of the White Mountains

A Recommended Fall Foliage Tour

HIGHLIGHTS

Elegant Orford ❧ America's Own Style ❧ Into the Uplands ❧ The Popular
Popple ❧ Do Pines Point the Way? ❧ A Town of Many Fragrances ❧ A Land
of Many Uses ❧ Hawkweed's Giant Genus ❧ Golden Alexanders ❧ The
Kancamagus Highway ❧ Pemi Overlook and the Pass ❧ A Brave Ballerina ❧
Face to Face with the Balsam Fir ❧ The Montane Taiga ❧ A Meditation on
Deciduousness ❧ Rocky Gorge ❧ Sweet Gale, Shadblow, and Mountain Alder ❧
The Way of the Covered Bridges ❧ What Hath Consumerism Wrought? ❧
Is Mount Washington Hollow? ❧ Praise for the Common Milkweed

*T*here's irony in the fact that this traverse across some of New England's wildest terrain begins in a place that is the perfect opposite of wilderness. It is a place where, in the context of architecture, the region's human environment reaches one of its high-water marks. The place is Orford, New Hampshire, the Connecticut River town second to none in its air of old-money refinement and reserve. In a moment, you'll be heading into an ocean of trees, but first take a moment to consider what an earlier generation of Americans could contribute to the art of building things—an art that had its raw materials in the nearby forest.

Has any river town ever had a better site? Just across the water stand the Palisades, the sheer granite cliffs of Morey Hill, in Fairlee, Vermont. The trip's starting point is at Orford town center, close by the Fairlee bridge. Without moving an inch, you are within sight of the first point of interest, an ensemble of superb Federal-style houses that stands like a squadron of ships in line abreast. The setting is so perfect you might suspect that one master plan and several bulldozers and graders were responsible for the effect. But it was the river that did most of the work, carving and sculpting a well-defined terrace known, simply enough, as the Ridge. The group of houses is sometimes called Bulfinch Row, a name that derives from the long-standing belief that the preeminent American architect Charles Bulfinch (1763–1844) had a hand in designing one of the houses. Now the prevailing opinion is that all these lovely structures were the result of local talent, builders who relied on the handbooks of Bulfinch's associate, Asher

The Ridge at Orford—elegant Federal architecture backed by a wooded hillside.

Benjamin, and other disseminators of the burgeoning Federal school. (To see a certified example of Benjamin's own handiwork, see the reference in Tour 31 to Old South Church in Windsor, Vermont.)

If you have taken Tour 41 up the New Hampshire seacoast and seen the Moffatt-Ladd House in Portsmouth, you're acquainted with the Georgian style that preceded the Federal. The Georgian look was very much an English import. At its best, it was stately and formal; at its worst, it was stuffy or downright pompous. The Federal architects, active from 1780 to 1820, were influenced by contemporary British ideas, but nevertheless succeeded in creating the first authentically Anglo-American style—one that combined the time-honored vocabulary of Greek and Roman forms with a distinctly Yankee sense of restraint and good proportion. It is a measure of their success that these masterpieces of unheroic and cheerful elegance blend so perfectly with the rugged landscape they occupy. It's also fitting that they most certainly need a row of big sugar maples in front of them.

As you proceed southward down the Ridge on Route 10-25A, take heed of some of the trademarks of Federal-style architecture. The main doorway is usually flanked by sidelights (narrow vertical windows) and topped by a semi-elliptical wooden fan or window fanlight. Georgian homes typically have a triangular or elaborate scroll pediment instead. Another common Federal element, found in Georgian designs also, is the Palladian window, composed of one large arched opening framed by two small rectangular ones. You'll see these features over and over again as you travel throughout New England. The fascinating thing about them is that their variations and combinations are endless.

─────────────── ∾ ───────────────

At the point 0.4 mile down from the starting junction, turn left with Route 25A East. You quickly rise out of the Connecticut Valley, but its bottomland tree species—cottonwood, white ash, and American elm—faithfully follow you up the track of the nearby brook. In early summer, you will surely spot two old English wildflowers, yarrow and ox-eye daisy, that were established in this country long before the Daughters of the American Revolution had any revolution to be distinguished by. And, as usual, ferns thrive along the roadsides.

Another 5.4 miles from your last turn, you'll have a good view of the uplands, with Mount Cube directly ahead. White pine and gray birch are evident, as is red oak, a central hardwood that does fine this far north when snuggled into the valleys. The presence of American elm snags probably indicates that Dutch elm disease is no stranger to the area. After 4.9 more miles, you reach Lower Baker Pond, on the right. The red oak remains, but now there are plenty of northern hardwoods: wild sugar maple and yellow birch, as well as white ash, striped maple, and arrowwood.

At the intersection of Routes 25A and 25, turn left onto 25 West. You are now in the western reaches of the White Mountains. Notice how the branches of the white pines extend out quite a distance over the roadway. Thoreau observed that the Indians determined which way was south by the size of the pine limbs: supposedly, the longest pointed in that direction, where maximum sunlight would result in maximum growth. This rule is not as applicable in modern times, at least with roadside or yard trees, because the longest branches grow in whichever direction is least obstructed. Naturally, white pine stems growing over the pavement, even when it lies due north of the plant, have more encouragement to elongate than those facing inward, toward dense vegetation.

Another tree in this vicinity is the relatively small, short-lived pioneer species variously known as poplar, popple, or quaking aspen. If you're a gardener or professional horticulturist, you might be familiar with one common use of poplar wood, as tie logs for edging or containing flower beds. In earlier times, its bark was used as a quinine substitute in controlling malaria. We are not the only animal to make use of poplar—ruffed grouse and other wildlife species depend on it for protective cover. And, predictably, it is a favorite food source and construction material for the beaver.

In a short time, you're in the town of Warren. In the 1800s, this mountain settlement was very likely one of the best-smelling places in the Granite State, because it was a center for the production of essences—oils or other extracts from plants distilled into concentrated form. All sorts of herbs were used in the essence trade, and public demand for them was considerable. Among the more popular offerings were peppermint, spearmint, wintergreen, and common tansy; coniferous trees with pleasant-scented resins, such as spruce and Canadian hemlock, also accounted for part of the output.

On reaching the junction of Route 118 East, near town center, turn right onto that road. On your left, by the stream, you should be able to identify black locust and balsam poplar. In contrast to regular poplar, the balsam species has distinctly furrowed bark. Another 2.9 miles up the way, you enter the vast expanse of White Mountain National Forest. Its entrance sign sports the motto affixed to national forests throughout the country, stating that this is a "Land of Many

Uses." Does anyone else find that a little unsettling? At any rate, it's a good unconscious reference to the long-standing belief that it isn't enough that forests exist—they have to be utilized for something. However, the Forest Service rangers who manage this great tract of land deserve high praise for the splendid work they do maintaining the trails and facilities enjoyed, and sometimes abused, by many thousands of people each year. However controversial latter-day federal conservation policies may be, the fact remains that what you're doing today, studying the glories of nature, is one of those "many uses."

As the road steadily gains elevation, the last vestiges of bottomland plant community fall far behind. This is classic northern hardwoods territory, but there are telltale signs of other regimes as well. Look for yellow and paper birch, American beech, Canadian hemlock, red oak, striped maple, and red spruce. About 4.9 miles from the entrance, you'll have excellent views from the heights; yellow birch is prevalent. Among the summer wildflowers are hawkweeds, composites of the gigantic genus *Hieracium*, which contains hundreds of species. The one that seems most native here is actually an Old World import: the orange-blossomed devil's paintbrush. You can find it growing nearby, along with its bright yellow relatives.

Another beautiful herbaceous dicot you may see is golden Alexanders. This fetching native has doubly compound foliage and broad, flat-topped inflorescences that proclaim its membership in the carrot family. Unlike its more common relative, Queen Anne's lace, golden Alexanders blooms mellow-yellow, as its common name indicates. Other herbaceous plants here include tall meadow rue, hay-scented fern, and, in the neighborhood of the Moose Crossing sign, fringed sedge, with its pendulous, bobbing green flower tassels.

The road descends to meet Route 112. At the three-way junction, merge with 112 East by going straight. On your right is Moosilauke Brook, which soon will meet its end in the waters of the Pemigewasset River, the northern extension of the Merrimack. Continue east on Route 112 through North Woodstock and the Route 3 intersection. After you cross the Pemigewasset, the route takes on the additional name of the Kancamagus Highway.

You are now entering a zone of truly great mountains, descriptively named: Scar and Breadtray Ridges, Mounts Whiteface and Tripyramid, the Sleepers, and the Fool Killer. Before the road rises again, you pass through Lincoln's version of the Never-Never Land of the Ski Bunnies. It's a good time to remember Orford fondly. The contrast between the self-contained architecture and these sprawling resort clones is a bridgeless gulf. The high points nearby could aptly be renamed Speculation Hill, Mount Megaprofit, and the Heights of Invasiveness, but no matter. This wonderful boost to the economy, sealed with the devil's handshake, is soon far behind you.

And now comes what many people consider the best scenery in New England. Beauty is practically ensured wherever paper birch dominates, and here its effect is nothing short of magical. Mountain ash and mountain maple are also evident by the Otter Rocks Rest Area. The most dramatic part of the climb takes place in a series of switchbacks. As you can see, the summits of even the highest peaks here are not above the tree line, which in this area is roughly 5,000 feet above sea level. Gradually, the montane taiga trees, red spruce and the even hardier balsam fir, become more apparent. In a moment, you'll stop at Kancamagus Pass, but if you first want to admire the stunning view at the Pemi Overlook, pull over and enjoy the panorama. On the high slopes, landslide scars stand out vividly where cascading rock debris has effortlessly bulldozed its way through dense tree growth. The tallest mountain visible to the south is Mount Osceola, at 4,326 feet. There's something of interest in the foreground—if you examine the paper birches, you'll find their leaves are the cordate, or heart-shaped, kind, characteristic of the highland variety of the species.

The main stop, however, lies 0.5 mile farther down the road, past the striking road-cut into Mesozoic-era Conway granite. This igneous rock was intruded into other formations at least twice as old when very possibly the region was passing over a heat plume originating deep in the earth's mantle. If the heat-plume theory is correct, it may explain why the supercontinent Pangaea split apart approximately 50 miles to the southeast. During that period in Earth history, about 200 million years ago, New Hampshire would have been a geological madhouse, with numerous volcanoes burying the landscape in a blanket of ash thousands of feet thick. (For more on this dramatic episode of prehistory, see Tour 48.)

The here-and-now of Kancamagus Pass is infinitely more peaceful, especially if you have the good fortune to see it in the stillness of an early summer morning. Pull into the left-hand entrance of the Graham Wangan Ground picnic area. One of the great delights in being able to identify different kinds of plants is the unpredictability of what you'll find. Occasionally, species show up where common sense says they don't belong. Two cases in point appear nearby. As you stroll over to the walkway entrance by the information sign, you'll come upon a conifer that is neither a spruce nor a fir. A closer look reveals it has short needle foliage arranged in bundles of two, and cones that tend to be curved. This is jack pine, the most cold-hardy and northern growing of its genus. It was purposely planted here. Even though it is native to the uppermost tier of states, it is not really a montane species, preferring instead sandier or rockier soils, such as those found at Schoodic Point in Maine (see Tour 53), in parts of the Saranac intramontane basin of the Adirondacks, or on upper Wisconsin's Lake Michigan shore. One relict jack pine community even manages to survive as far south as the Indiana Dunes National Lakeshore, to the east of Chicago. It is a fragile reminder of the colder climate of several thousand years ago.

The other surprise plant grows nearby. It is a broad-leaved shrub drawing attention to itself in late June by producing globular white flower clusters. This, believe it or not, is the European snowball viburnum, a showy cultivar of a familiar garden species. It too must have been planted here. It's a strange sight amid the grizzled stands of fir and spruce. When it blossoms, it looks like a ballerina dancing across an active battlefield. Still, as Emerson wrote, "Nature suffers nothing to remain in her kingdoms which cannot help itself." This brave

The characteristic upward-facing cones of a balsam fir, as seen at Kancamagus Pass.

plant has somehow managed to eke out an existence in a challenging and supremely incongruous setting.

Before you forge on, walk out to the parking lot entrance and take in the view of the near hillsides to the east. The montane taiga conifers intergrade with hardwoods in a beautifully subtle pattern. Close at hand, by the turn-in, you can brush up on the characters of the balsam fir. In contrast to the red spruce, the tips of its needles are rounded, and its cones sit upright on their stems. The balsam fir has come up with an excellent adaptation, given its preferred environment: its branches are extremely supple and bend, rather than break, under the weight of snow and ice. Indians sought this species for its sap, with which they sealed the seams of their birchbark canoes.

———————— ❧ ————————

Continue east down Route 112 into the valley of the Swift River, a stream that is almost wholly contained within the confines of White Mountain National Forest. The relatively modest size of the trees along this and every other part of the Kancamagus Highway reminds you that even this terrain has been heavily logged. There is a wetland 8.3 miles down from the last stop, where tamarack

442

can be easily spotted. If you happen to be in a puzzle-solving frame of mind, you can speculate on why this particular conifer species has opted to be deciduous, when almost all of its relatives are evergreen. It might be a way of preventing wintertime water loss or tissue damage in bitterly cold climates; but then why doesn't the black spruce, at least as cold-resistant, do the same thing? On the other hand, perhaps the tamarack's complete annual leaf drop has something to do with its wet habitat. Can you support that idea by thinking of any other deciduous conifer that inhabits wetlands elsewhere? If you can't recall any, check Tour 2 and you'll discover the bald cypress, the knobby-rooted glory of our Southern swamps. Yet a monkey wrench can be thrown into that theory, too: our old friend the black spruce also happens to be the perfect bog plant, and its soggy surroundings don't force it to change its evergreen ways.

It would be ideal if we could trace the history of deciduousness all the way back through the story of plant evolution, and see when and why it first came into being. Unfortunately, the fossil record, however helpful, will probably always be too incomplete to answer such a question definitively. In the absence of certainty, scientists do what the rest of us do, they make the best assumptions they can. At this point, one prevailing assumption is that the deciduous broad-leaved flowering plants first developed in warm areas beset by periodic droughts. When the earth's climate began its Cenozoic era cooling trend about 33 million years ago, their ability to shed their leaves at stressful times of the year proved to be an immense advantage, and soon they dominated the scene. Alas, this may not help us figure out why the tamarack, a gymnosperm, developed the same ability. Incidentally, the ginkgo is another gymnosperm, though a much different kind, that drops its leaves each fall. Its history extends back before the first sign of flowering plants, and long before the Cenozoic global cooling began.

———————————— ᙡ ————————————

When you've gone 4.7 miles beyond the tamaracks, turn left into the entrance to Rocky Gorge Scenic Area and proceed to the main parking lot. This is a refreshing and photogenic stop. The Swift River tumbles through a granite terrain and creates large, still pools that are totally irresistible to youngsters. On the walkway from the parking lot to the footbridge, you can find an excellent collection of woody plants that include both lowland and upland types. Red oak mixes with sugar maple, cordate paper birch, and mountain maple; and, if you visit in May, you'll have no trouble spotting the profuse white blossoms of shadblow, a thicket-forming plant in the rose family, with gray, vertically striped bark and long, gracefully tapered buds. Indians used it as a sort of natural timer. They knew that the plant blossomed just as the shad run was beginning in New England rivers, and they would fish accordingly. Later, in the warm season, they collected shadblow fruit and dried it for a backup food supply.

The showcase of trees and shrubs continues down at the water's edge. Mountain alder, one of the lesser-known members of its genus, is well established here, as is red maple, sheep laurel, and witch hazel. The last of these plants is almost as plentiful as the NO SWIMMING signs. In addition, there is one extra-

The Swift River at Rocky Gorge.

special old-time favorite, the native bayberry called sweet gale. The Indians used this plant as a mosquito repellent, and even as a form of magical protection against snakes. Early colonists placed its leaves in their bedding to ward off fleas. The species has a wide distribution in the northern parts of our continent and Eurasia. It is highly regarded in Scotland and has been made the badge of Clan Campbell. Somewhat closer to home, in French Canada, it has been used to make a yellow dye for woolens, as well as a flavoring for broth.

Depart Rocky Gorge by continuing east on Route 112. A distance of 3.8 miles farther down you will see a pretty covered bridge on your left. It is a Paddleford Truss design, though some sources prefer to call it a Long Truss with added arches. It is 136 feet long and dates from 1858. One of the main features of the Paddleford design is the pair of interior-arch trusses, which add a sense of shiplike grace as well as structural support. If you stop and take a look, you'll find the Forest Service folks have replaced some of the old timber flooring with steel. In another 6.1 miles, you come to the end of your Kancamagus Highway traverse, at the intersection of Routes 112 and 16. Turn left onto 16 North, in the town of Conway. In less than a mile, you cross the Saco River just below its confluence with the Swift, where two more Paddleford bridges have been built, one over the Swift and the other over its parent stream.

In this neighborhood, first impressions are not deceiving. It looks as though all the minions of world tourism have descended to ply their wares and services along the narrow corridor of the poor besieged Saco. After the pristine mountain heights, consumer culture avenges itself so thoroughly that you can only watch the spectacle go by, as though it were a brand-new type of forest to observe. Here is your chance to be the modern Linnaeus, by devising a taxonomic classification system of kitsch architecture. At the bottom of my own list, somewhere in the Precambrian ooze, are the fern bars and frimpy little restaurants with inescapably cute names. At times, though, the American genius for overstatement stumbles its way into pure poetry. A bit farther up the road, for instance, stands the Luv Tubs Presidential Waterbed Motel, the very sight of which neutralizes the effects of heavy traffic.

------------------------------ ∽ ------------------------------

Between Conway and North Conway stand some big white pines, and at 8.0 miles up from your turn onto Route 16, you'll have a good panoramic view of the White Mountains, at the roadside turnoff on the left. The road goes on through Bartlett to Jackson, where another Paddleford bridge can be seen on the right. The Saco has by this point angled off westward toward its source in the breathtaking environs of Crawford Notch. The stream here is the Ellis River. In this vicinity, the conifer most likely to catch your eye is not a wild species but a lawn tree common throughout the northern United States, the Colorado blue spruce. Practically every homeowner succumbs to its striking foliage color in the nursery, but when it's taken home and planted in the front yard, it tends to upstage everything around it. Besides, enough is enough. Hula hoops, Nehru jackets, and Baby-on-Board signs have all had their day; isn't it time the Colorado blue became a treasured part of our national past?

Wilderness returns when you reenter White Mountain National Forest. Ahead and to the right is the loftiest of all northeastern mountains, Mount Washington, standing in company with the other Presidentials. As you rise to Pinkham Notch, you have a clear view up Tuckerman Ravine, the range's most dramatic cirque, or glacially formed amphitheater. Seventeenth-century naturalist John Josselyn, whose combination of naïveté and canny observation makes for delightful reading, believed these heights were hollow, because of the sound they made when the rain fell on them. He also theorized that they had been raised by earthquakes. The earthquake hypothesis is a good one for many mountain-building events, but it probably doesn't pertain in this case. (For more on the current theory on how these heights formed, see Tour 47.) Here is Josselyn's description of the natural wonder spread out before you:

> A Ridge of Mountains run Northwest and Northeast an hundred leagues, known by the name of the White Mountains, upon which lieth Snow all the year, and is a Land-mark twenty miles off at sea. It is rising ground from the Sea shore to these Hills, and they are inaccessible but by the Gullies which the dissolved Snow hath made; in these Gullies grow Saven [i.e., juniper] bushes, which being taken hold of are a good help to the climbing Discoverer; upon the top of the highest of these Mountains is a large Level or Plain of a days journey over, whereon nothing grows but Moss.

Once again, the details may be wrong in places, but the overall picture is compelling and basically correct. From your vantage point, you can almost see

Josselyn's ever inquisitive but accident-prone self dangling from a vegetation hold (which more likely would have been a bough of red spruce or balsam fir than real juniper).

The entrance to the Mount Washington Toll Road, and the beginning of Tour 47, is located on the left-hand side, just 0.4 mile beyond the notch. The remaining 7.8-mile stretch to Gorham's Route 2 junction, the endpoint of this tour, features a new tree, the white oak, growing along the near bank of the Peabody River. This top-of-the-line timber species was the apple of the Yankee shipbuilder's eye. In addition, its bark was used to tan cowhides; and the Indians found its acorns, lower in tannic acid than the red oak's, a premium food source. It shares this heavenly landscape with paper birch, red spruce, poplar, white pine, and mountain ash. Among the wildflowers of summer is the stout-leaved, white-sapped common milkweed. Of this widespread native Pehr Kalm wrote, "Its flowers are very fragrant, and, when in season, they fill the woods with their sweet exhalations." On top of that, they signal the onset of autumn by launching their silky-tufted seeds across this highway and into the mountain air.

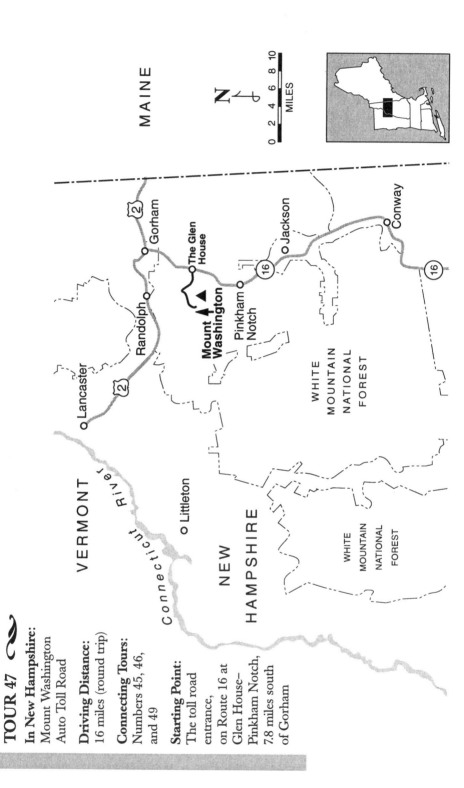

TOUR 47

In New Hampshire:
Mount Washington
Auto Toll Road

Driving Distance:
16 miles (round trip)

Connecting Tours:
Numbers 45, 46,
and 49

Starting Point:
The toll road
entrance,
on Route 16 at
Glen House–
Pinkham Notch,
7.8 miles south
of Gorham

MAINE

N

MILES
0 2 4 6 8 10

Gorham

The Glen
House

Jackson

Conway

2

16

16

Randolph

**Mount
Washington**

Pinkham
Notch

Lancaster

2

WHITE
MOUNTAIN
NATIONAL
FOREST

Connecticut River

VERMONT

Littleton

NEW
HAMPSHIRE

WHITE
MOUNTAIN
NATIONAL
FOREST

Tour 47
Mount Washington and a Rendezvous with the Ice Age

A Recommended Fall Foliage Tour

HIGHLIGHTS

Automotive Rites of Passage ∾ The Worst Weather in the World? ∾ Choosing the Right Day to Ascend ∾ The Lower Slope's Hardwood Forest ∾ Shaking Hands with the Taiga ∾ Flag Trees, Krummholz, and Lapdogs ∾ The Tree Line ∾ The Alpine Tundra Begins ∾ A Narrow Window of Opportunity ∾ Survival by Burial ∾ Water, Water Everywhere ∾ Thermal Jump-Starts ∾ The Presidential Upland and a Close Approach to Mars ∾ Midget Willows, a Tundra Pteridophyte, and a Host of Lichens ∾ The Summit ∾ Engulfed in Ice

You don't have to linger in New Hampshire very long to see one of its most popular automotive status symbols, the bumper sticker that says THIS CAR CLIMBED MOUNT WASHINGTON. Anyone can go to the local auto supply store and get a shotgun rack, flying-duck decals, or a tailgate net for the pickup; but to obtain the Mount Washington sticker, you have to be there, at the big mountain, and pay the not-so-insignificant fee to go up the summit road. The fact that native New Hampshire people display the sign at least as proudly as flatlanders do suggests that they consider the trip something of a rite of passage. Indeed, a day spent winding up and down the unpaved, 8-mile stretch of 12 percent grade is not the same as a drive to the corner convenience store.

To demonstrate the sanctity of the bumper-sticker cult, I offer this brief case history. Once I took my own hard-earned sticker and, with a little creative scissors work, changed the legend to read MOUNT WASHINGTON CLIMBED THIS CAR, which was supposed to be a droll reference to the down-at-heel condition of my long-suffering Chevy. However, when I put it on, right next to the other stickers declaring my eternal allegiance to the New England Wild Flower Society and the Arnold Arboretum, I started getting bucketfuls of dirty looks at red lights. In retrospect, I see I was being disrespectful. I might as well have toasted the memory of William Tecumseh Sherman at a crowded bar in rural Georgia.

One reason everyone holds Mount Washington in such awe, and considers the ascent such an accomplishment, even on the gentlest summer day, is that the mountain is touted as having the worst weather in the world. It's not the worst in any single sense—other places on the globe have higher records for cold, windiness, and total precipitation. What makes the claim stick, though, is the fact that the summit arguably sustains the nastiest of these factors in combination. At the top, 6,288 feet above sea level and some 4,700 feet above the toll road entrance, the average temperature is 27 degrees Fahrenheit, the average snowfall is 20 feet, and the wind reaches hurricane force an average of 104 days a year. The highest measured wind velocity, an astounding 231 miles per hour, was measured in 1934, close to the middle of one of the region's worst bad-weather decades.

But for all the statistics and the legendary status, Mount Washington is a safe and enjoyable place for human beings, provided they display a little common sense. The first rule is to pay close attention to the meteorological forecast, or check with the local office of the National Weather Service, before you set out. The special weather radios sold at those franchised gizmo stores are extremely helpful, and not expensive, even by Yankee standards. It can be a perfectly splendid day down in Gorham or Conway when it's total murk up at the summit, which is obscured, truth to tell, about 60 percent of the time. Absolutely cloudless days are best, and by all means avoid thunderstorm weather. As you'll see, a substantial part of the tour is across terrain with cover for nothing larger than a red-backed mouse.

The second rule is to scrupulously follow the printed driving tips you'll get from the tollhouse person. Do not emulate the extremely self-confident style of the professional shuttle van drivers. They know every turn and pebble by heart; you don't. The road is traveled by thousands of people each year with no problem; but it is narrow, unpaved, and edged in places by deep ditches and drop-offs. The scenery is beyond the reach of adjectives, but don't let it kill you. Save your sense of wonder for the roadside pull-offs. There are a number of these, well spaced along the way.

———————————— ❧ ————————————

When you've reached the starting point and have settled your account with the keeper of the cash box, your glorious climb begins. Note that all points of interest on this tour will be given in total elapsed mileage from the tollhouse. If you have taken other big-mountain tours in this book, you are already familiar with the transition that takes place from the lower hardwoods to the upper montane taiga community. You may have also had the chance to see an alpine tundra regime surmount the taiga on higher peaks. The difference with Mount Washington is in the matter of scale. There seems to be, and often is, much more of everything. The alpines best exemplify this. Here, they are not on duty at an isolated, embattled outpost, but rather, appear to have a whole world to themselves. Which is not to say they should be trampled on.

The first regime you'll see here on Mount Washington is indeed the northern hardwoods, with its familiar triad of American beech, sugar maple, and yellow birch. These are joined by Canadian hemlock and paper birch. Understory plants include hobblebush, striped maple, and luxuriant masses of ferns. One frequently cited rule of thumb is that the environment changes in the following way: each

450

American mountain ash on Mount Washington, blooming at about the 1.3-mile mark, in the third week of June.

1,000-foot gain in elevation is equivalent to moving another 250 miles northward. In other words, you are quickly rising into the characteristic conditions of Quebec, and it shouldn't be long before you hit a fair approximation of the taiga found there. The first person to describe the relationship between increasing elevation and increasing latitude was the great nineteenth-century German traveler and polymath Alexander von Humboldt. Humboldt, who inspired countless other scientists, had a major ocean current and various rivers and mountain ranges named after him.

At 1.3 miles (roughly 2,000 feet in elevation), the presence of mountain maple and mountain ash signals that a change is under way. By 2.7 miles (about 3,000 feet), red spruce and balsam fir are already dominant. In the corresponding part of Canada, the white and black spruce take the place of the red. There is a good pull-off here, should you wish to get out and shake hands with the taiga.

Before long, even these tough conifers are showing the effects of climatic stress. At 4.0 miles (4,000 feet), balsam fir, the hardier of the two, is predominant, but even it has had to assume smaller, more ragged shapes. Flag trees, with branch growth contorted in the direction of the prevailing airflow, are a common sight here and elsewhere on higher mountains. From this point onward, the reduction in fir size continues until the prevailing form is krummholz—the German word for "crooked wood." The severe dwarfing and horizontal habit characteristic of krummholz is a woody plant's way of responding to the deadly cold and wind. A conifer of this kind isn't sick or warped or pathetic; instead, it's successfully responding to the local conditions, and in effect it's extending the range of its own species. Some of the krummholz specimens you'll see today are more than a century old, even though they're no larger than bonsai plants. As a matter of fact, both bonsai and krummholz trees are admirable reminders that size alone is no indicator of a plant's age.

If it seems strange to you that a single species, balsam fir, can have widely differing identities and be either a tall, stately tree or a gnarled midget, just remember Great Danes and Chihuahuas, both members of *Canis familiaris,* or Clydesdales and Shetland ponies, which represent the long and the short of *Equus caballus.* This analogy is of course a limited one, since the differences in dogs and horses is largely the result of human breeding. Modern science confirms that lapdogs are not krummholz equivalents, and didn't originate in the stormy mountain heights. Krummholz firs are individual plants responding to a specific environment, not a distinct variety of their species. If they were given gentler conditions, their progeny would assume the normal tree shape. In contrast, Chihuahuas are a distinct race of dog. They don't revert into German shepherds when the weather changes.

A little farther on, at the 4.5-mile mark (4,200 feet), you reach the timberline, and the edge of the tundra. Stop at the pullout here and take a good look around, but please do not tread on any plants or lichens. These organisms are amazingly tough in one sense, but they're not accustomed to being used as Astroturf. One careless step may cause the destruction of decades of growth.

This stop gives you a terrific introduction to Mount Washington's alpine species. In this tundra environment, the growing season is only three months, or even less. Accordingly, the crucial activities of growth, food-making, and flowering take place within a narrow window of opportunity. Peak bloom usually occurs in the latter part of June. The more conspicuous members of the plant community here are bog laurel, mountain cranberry, bog bilberry, Labrador tea, three-toothed cinquefoil, and the early blossoming diapensia. All but the last two are ericads, or members of the heath family. In general, ericads are plants with precise requirements for survival—requirements, however, that would be the death of many other plant types. They love acidic soil—no minor attribute, since acidity normally prevents roots from using the available nutrients and water supply. Ericads are one of the plant groups that have joined forces with certain fungi to produce mycorrhizae (literally, "fungus roots"). The relationship permits vegetation to extract desperately needed nitrogen directly from its greatest source, the atmosphere.

These woody ericads, in common with alpine plants of other families, are faced with other big challenges as well. Ironically, water retention is difficult despite the huge amount of rain that falls here. Besides the acidity problem, the runoff from the rocky surface is very rapid, and the plants have to conserve whatever water they can grab hold of. In terms of water loss, this cloudbound slope has more in common with the sunbaked alkali flats of Clark County, Nevada, than it does with the valley forests below. If you examine the foliage of these plants, you will find a whole arsenal of interesting adaptations designed to meet the challenge; these include fleshy or leathery leaves, which preserve internal moisture, and leaves that are densely huddled together to reduce air circulation and maintain a higher level of humidity around themselves. Labrador tea has yet another approach—the undersides of its leaf blades are densely coated with an orange-tinted protective wool.

The plants here must also contend with the plunging temperatures and desiccating winds of winter. Look at the low habits of these plants and you'll get an inkling of how they survive. Their secret lies in their being buried alive, deep in the snow, where it's not only windless but also much warmer than out in the open. Any alpine tundra plant foolish enough to adopt an upright habit would quickly suffer the condition known as frost drought. Another species found here, the dwarf birch, demonstrates how even a genus known for its tree forms has developed a less vulnerable, miniaturized species that is more likely to be blanketed by the snow.

One more feature found in some of these species is a red or reddish brown tinting of stems and leaves. This coloration allows plants to absorb more energy at the infrared end of the spectrum as they finally emerge from the melting snow. In effect, it's a self-generated warm-up mechanism, which gets the current year's growth cycle off to a quick start. In the short spring and summer of the tundra, this kind of thermal jump-start is a distinct advantage.

———————————— ∽ ————————————

As you continue up the road past the 5.0-mile point (4,600 feet), you head into a rough-and-tumble dreamscape that would remind a planetary geologist, or a science-fiction buff, of the boulder-strewn Viking landing sites on Mars. To hikers, this zone is sheer heaven after traversing the lower, steeper slopes. Swing into the pullout at 5.7 miles (5,000 feet). I admit that this is my favorite spot. It is truly high ground by anyone's standards, and the lay of the land is fascinating. Above you rises the cone of the mountain summit; the surface under it is a rather even expanse, which is part of the Presidential Upland. It's a feature that has provided generations of geologists with the opportunity to theorize and countertheorize about its origin. One persistent and appealing interpretation amid the conceptual give-and-take is that the upland is nothing less than a fossil lowland, one that was pushed in two or more main pulses from its original position near sea level, to this much higher station. If true, this means that before about 20 million years ago, the summit cone was a fairly small hill projecting above the otherwise low terrain. Estimates indicate it will take another 30 million years for Mount Washington and its surroundings to be eroded back down to their flat, almost level starting point.

At first glance, the environs may seem barren, but take a closer look. Amid the jagged boulders are at least a few prostrate shrubs, including bearberry willow. It's difficult to believe, but taxonomically and genetically, this carpet-forming plant stands right next door to the flamboyant weeping willow of lawn and golf course fame. Herbaceous species are also present: the dicot mountain sandwort, the monocot highland rush, and perhaps most surprisingly, the fir club moss, a lowly pteridophyte representing a line that once included the great coal swamp plants

Abundant life above the timberline. Lichens are admirably adapted to the harsh conditions of the upper slope.

of the Carboniferous period. It is a very long way, in time and temperature and elevation, from that ancient wetland environment, where dragonflies with 2-foot wingspans skittered among the trees. In those lush and humid days, eastern North America was perched on the equator.

Another form of life, closer to the nature of the stone itself, grows abundantly here. It is actually a composite organism, part sheltering fungus, part food-producing green algae or blue-green bacteria. It is the lichen, a classic example of symbiosis, which, strictly speaking, is just a sustained interaction between two different types of organisms. It would be nice to think that the fungi and their photosynthetic associates, like the mycorrhizal partners mentioned earlier, are also a demonstration of mutualism, wherein the two contracted parties benefit more or less equally. There is evidence, though, that at least some fungal members actually indulge in controlled parasitism. In a few cases, they even end up killing the other half of the team.

Whatever the exact nature of their working relationship, no one could doubt the lichens' success. They inhabit an amazing range of environments, including some underwater habitats and others within 4 degrees of the South Pole. They are known for their numbingly slow growth, but they take advantage of every opportunity to snap out of dormancy. The lichen way of doing business is to soak up water—from three to thirty-five times their own weight's worth—to get the photosynthesis process going. That is why lichens on a rainy day are softer, spongier, and usually much greener than they are in their brittle, dry state. Most botanists still classify lichens by the usual taxonomic rules, with each combined form considered a species within a particular genus, family, and order. The species here include dark brown rock tripe, pale yellow snow lichen, ashen curled lichen, and the relatively lanky, brown-tipped Iceland moss. Remember that true mosses are nonvascular members of the plant kingdom, with no evolutionary connection to the lichens.

The remaining drive to the summit is spectacular. The terminus parking lot is situated just below the complex of structures that includes the modern Sherman Adams Summit Building and its meteorological observatory. The human hubbub comes as a bit of a shock after the intrinsic remoteness of the tundra, but the view from the glassed-in walkway is terrific. On your way up the steps, note the crustose lichens that have coated much of the boulder surface. One of the most common species of this type, distinguished by its yellow or chartreuse color, is map lichen.

If you snoop about the summit, you might find one or two colonies of alpine wildflowers, as well as other herbaceous plants. Still, at this highest point of the northeastern United States, the emphasis must be on the breathtaking panorama and the altered sense of scale it produces. Around you, the surface is littered with detached blocks of Littleton schist, the same formation that underpins so many of New Hampshire's mountains. These blocks have been broken loose over the centuries by the patient force of the frost. Below are the sweeping surfaces, or lawns, of the Presidential Upland. Beyond them stretch the forests of the lower slopes and valleys. As this tour's title promised, this is a rendezvous with the Ice Age. Here, on what seems to be the top of everything, you can find telltale grooves, or striations, as well as rocks that came from different outcrops miles to the north. These and other facts confirm that even lofty Mount Washington was overridden by at least one of the massive ice sheets that moved across the continent at the dawn of our own species' existence. How this could possibly happen, and how the plants and lichens of the tundra could thrive in this unending wake of the Ice Age, is open to enlightening scientific description. But all the explanations in the world cannot erase the spine-tingling wonder of just standing here, windswept in the high-altitude light.

The view from Mount Washington summit.

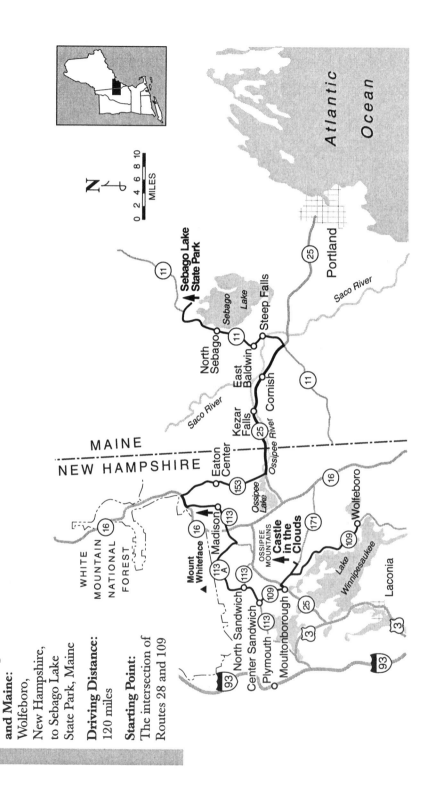

TOUR 48 ～

In New Hampshire and Maine:
Wolfeboro,
New Hampshire,
to Sebago Lake
State Park, Maine

Driving Distance:
120 miles

Starting Point:
The intersection of
Routes 28 and 109

456

Tour 48
A Northern New England Lakes Tour

A Recommended Fall Foliage Tour

HIGHLIGHTS

The Birth of Lakes ∾ The Great Spirit Smiles? ∾ Wolfeboro's Charm and
the Norway Maple ∾ Gypsy Moths and Woodland Musicians ∾ Into the
Ossipee Mountains ∾ The Castle in the Clouds ∾ A Ring Dike and the
Age of Fire ∾ The Vines of the Fantasy House ∾ Dog Lichen and Volcanic
Rocks ∾ The Antisocial Black Walnut ∾ The Rustic Look and Whiteface
Revealed ∾ The Ossipee Pine Barrens ∾ The Madison Boulder: Not Your
Average Erratic ∾ Lichen Anatomy 101 ∾ A Yankee Armenia? ∾ A Country
Rocky and Mountainous ∾ Inland Maine ∾ The Great Sebago Lake

*I*t has been said that all of New Hampshire's extant lakes owe their
origins to the work of Ice Age glaciers. This may not give the tireless
activities of beavers or the U.S. Army Corps of Engineers their full
due, but the state's pond-dotted surface is one of the major legacies of recent
glaciation. A notable example of how ice transformed the landscape is the state's
largest body of fresh water, Lake Winnipesaukee. It sits on an extensive low area
of not particularly resistant granite. When the ice sheet moved across the rock,
it hollowed out a great depression, then dumped a barrier of glacial till and other
debris, which impeded drainage at the southern end.

The same basic process of natural lake construction can be seen over and
over again in many places in New England, in upstate New York, and especially
in the interior freshwater seas of the Great Lakes. Such places provide us with
striking scenery and some of the most pleasant vacation spots in the northern
part of the country. They're one of the planet's most short-lived features, though,
often filling up or draining away within a few thousand years. We're witnesses
who managed to arrive on the scene at exactly the right time.

This lake-framed tour begins at Winnipesaukee's southeastern corner.
Wolfeboro lays claim to being the oldest resort town in America, and its appeal
certainly hasn't waned in modern times. It is now part of a constellation of towns
fronting the water, all competing for tourists' attention. The area's summer
population is huge. The attractions are many, and not always natural. From the
teen-flocking rituals of Laconia's Weirs Beach to aquatic drive-through restaurants

for boaters, the shoreline is almost wholly given over to the frantic pursuit of leisure. The modern rendering of the Indian term *Winnipesaukee* is the rather neutral "Lake Land," but Thoreau thought it translated into something much nicer, "The Smile of the Great Spirit." Nowadays, you have to suspect the Great Spirit isn't smiling quite as much as the New Hampshire Chamber of Commerce; especially on hot August afternoons when the lake's surface is so littered with speedboats and other craft that it looks like a full-scale reenactment of the Battle of Midway.

Through it all, Wolfeboro has managed to retain its charm. As you proceed north on Route 109 from the starting point, you will note that the town is heavily planted in Norway maple. In the eastern United States, this species is a widely used—and overused—ornamental, but it is a forest tree in its own right throughout much of Europe. To the untrained eye, it resembles the sugar maple; but in fact, a number of distinguishing traits separate the two types. For one thing, the Norway maple usually has a more dependable (or more boring) lollipop-shaped habit. Especially prevalent here is one of the tree's purple-leaved cultivars. It's probably the older Schwedler variety, since the purple foliage fades to a rubbery sort of olive drab, not unlike brown-algae seaweed, as the summer wears on. The plants that maintain the full somber hue throughout the warm months belong to Crimson King or a similar cultivar.

North of the town center, oaks, black locust, and white pine assert themselves in the less built-up areas. By 3.9 miles north of the starting point, you are in Tuftonboro, with Mirror Lake on the right and Winter Harbor on the left. The waterfront is very much a private item in these parts. Summertime road hazards include platoons of distracted joggers who run up, down, and diagonally across the pavement in interesting and unexpected trajectories. In another 2.3 miles, the view across the water stretches all the way to our trip's initial objective, the Ossipee Mountains. This imposing cluster of hills has a dramatic tale to tell; but more on that anon.

One fact that may or may not be apparent at first glance is the impact that the gypsy moth has recently had here. The infestations have been severe. In 1990 and 1991, for instance, oaks and other species were defoliated by the moth's voracious larvae. If you were to see a picture of this vicinity taken in the summer, immediately after the damage was done, you might think it was early November instead.

The besieged trees have the ability to develop a second set of leaves each year, but if they must divert their resources to the costly task of regeneration for more than one or two warm seasons in a row, they definitely begin to suffer. In modern times, scientists have learned that plants subjected to this onslaught are not simply passive victims that helplessly stand by while the insect tormentors have their way. Very often an affected tree's second crop of leaves will be more saturated with phenols and other chemicals that discourage insect feeding. Ironically, this impressive system of defense can go awry. In the specific case of the gypsy moth, for instance, there is some indication that the production of special compounds may backfire, and harm a beneficial larva-killing virus more than it hurts the larva itself. Fortunately, the gypsy moth outbreaks usually wane for a number of years before the next peak occurs. (For more on the history of this insect threat, see Tour 23.)

As destructive and loathsome as it is, the imported gypsy moth is only the latest in a long record of New England pest epidemics reaching back to colonial times and no doubt before. Early English settlers, so dependent on plant products, were quick to note surges in insect populations. For example, there were bad tree-nibbling caterpillar problems in 1646, 1649, and 1668; in 1666, orchards were hit for the first time by the formidable cankerworm. On one occasion, a swarm of six-legged creatures turned out to be less harmful or frightening than it first seemed. Seventeenth-century Plymouth Colony governor William Bradford wrote this account of the phenomenon: "The spring before [1634], especially all the month of May, there was such a quantity of a great sort of flies like for bigness to wasps or bumblebees, which came out of holes in the ground and replenished all the woods, and ate the green things, and made such a constant yelling noise as made all the woods ring of them, and ready to deaf the hearers."

This was the famous periodic cicada, whose unearthly drone is one of the chief glories and most distinctive features of the forests of eastern America. In northern areas, individual cicadas follow a seventeen-year development cycle (a thirteen-year race occurs in the South). Actually, Bradford, otherwise accurate in his description, was wrong about the cicadas eating the vegetation. If the adults do feed at all, they suck a little sap from tree twigs without damaging them. In laying their eggs, the females may trigger some branch dieback, but even that is not a particularly worrisome condition. (To offer a little perspective: the Indians had mixed impressions about this strangely ravishing music maker. Some native peoples considered its appearance an ill omen—indeed, Bradford's Indian friends warned him that sickness would follow the cicadas—but others looked on their sudden arrival as nothing less than a wonderful culinary opportunity. In New York, for instance, the Onondagas still sauté them in butter, as a sort of all-natural insect snack food.)

In the more wooded sections of this area, mineral-green foliose lichens, growing like bull's-eyes on the trunks of oaks and other broad-leaved trees, are a common sight. At 4.7 miles beyond the settlement of Melvin Village, a clearing on the right reveals the impressive ensemble of the Ossipee Mountains, now much closer at hand. Another 0.6 mile up the road, you reach the junction with Route 171. Turn right onto 171 East. Proceed 2.1 miles to the left-hand entrance of the tour's first stop, the Castle in the Clouds. This privately owned facility is a magnet for sojourning tourists, most of whom come to see the castle itself or the other man-made attractions on the grounds, without ever learning that they're visiting one of the most notable geological locales in the eastern portion of the country.

After paying the entrance fee at the gate, take the one-way road that ascends through a dense woodland containing Canadian hemlock, American beech, white pine, and red and striped maples. At 1.4 miles from the entrance, a pullout to

Lucknow, the Castle in the Clouds.

the left beckons, shaded by a grove of white oaks. By all means stop and take in the splendid view. From this vantage point, looking westward, the world seems half water, half mantling forest. Winnipesaukee, as primeval as it appears this far from the fray, is an infant feature, more ephemeral than the surface that holds it, and younger even than our own johnny-come-lately species. However transitory it and other lakes and ponds are, and however subject to change the surrounding green tide of vegetation is, it is impossible to come away from this overlook without feeling a sense of vast order and stability. Near or far, everything stands in its proper place.

No such vision would have fit the scene 200 to 100 million years ago, during the Mesozoic era, when all hell was breaking loose. In that eventful time, this region may well have been riding over a plume, or hot spot, emanating from deep in the earth's interior. As this land slowly drifted north and west, the plume triggered the upwelling of massive amounts of molten rock, called magma. Much of the magma hardened before it reached the surface, but a great deal succeeded in causing volcanic flows and eruptions. As astounding as it seems, this outpouring buried ancient New Hampshire in a blanket of lava and ash well over a mile deep. In those superheated, sulfurous days, the landscape rose and erupted and collapsed into a jumble of forms we can still partly decipher today. The scene would have outdone the wildest imaginings of Dante or Hieronymus Bosch. After all, things were falling apart in a big way. The globe-spanning supercontinent Pangaea was splitting asunder just a few miles to the east, and a new saltwater sea, the Atlantic, was greedily filling the breach.

One of the main reasons we know of these events is that scientists have found a good record of the geologic mayhem right here, in the Ossipee Mountains. These hills constitute the best example of what is known as a ring dike complex, a circular feature that is as eye-catching on a bedrock map as a bullet hole in a car windshield. One of the most interesting aspects of this ring complex is that there are remnants of the volcanic ash and lava, which in most other places have been completely eroded away. There are several other ring dike areas; together, they form a more or less linear feature extending from Hudson Bay southward to the ocean. By extrapolating this track, some specialists have concluded that the plume causing the disturbance so long ago is still active, and is now located off the western coast of Africa, where it is busy constructing the volcanic Cape Verde Islands. Since continents move and hot spots don't, it can be deduced that New England was passing over that point, not far from the equator, when the supercontinent breakup began.

Geologists have plenty of important clues about what the earth's crust was doing in Mesozoic time, but paleobotanists are left without much hope of constructing a correspondingly vivid picture of the local plant life. In certain places, and for certain periods, the fossil record has been abundant, but the frustrating fact is that vegetational remains usually don't get preserved in upland areas where erosion predominates. Even if fossils manage to form, they are erased along with their host rocks as the highlands are inexorably worn down to near sea level. As a result, we know many fine things about the ancient flora of coastal swamps, river deltas, and other low-lying environments—but next to nothing about the elevated parts of the continental interiors. It's as though a paleobotanist from the far future had to frame opinions about our White Mountain forests based solely on what grows in the salt marshes miles to seaward. Such a scientist might be lucky, and find the remains of a battered balsam fir trunk that had miraculously bumped its way downstream to the coastal zone, but it wouldn't be much help in accurately describing the larger community of plants in which it lived.

The main parking lot lies just up the road from the turnoff. Park here to see Lucknow, the mansion better known as the Castle in the Clouds. The house, a laissez-faire mishmash of pseudomedieval stonework and tile roofing, was millionaire Thomas Plant's defiant answer to reality. It's impossible not to love it at first glance. It would make the perfect setting for the second coming of Tinkerbell. You have the option of taking the auto train up to the house, or being one of the few nonconformists who actually make the climb, a grueling three minutes or so, on foot. When you reach the building, abandon all hope of architectural unity and concentrate instead on the interesting details. Sturdy oak beams jut out beneath the roof; inside, elm wood has been used for decorative paneling. On the exterior, at least the last time I was there, bittersweet and Virginia creeper vines have become well established. The latter species, a widespread North American native, is the one with five leaflets attached at a central point. It was introduced to the Old World in the 1620s, and its berries, considered an important medicinal resource, were used to treat victims of the Great Plague of

London. This fruit and other parts of the Virginia creeper were reputed to kill everything—from lice to microbes to headaches.

The grounds surrounding the house feature staghorn sumac and naturalizing common lilac and forsythia shrubs. A nice stand of orange-barked Scots pine is in front—the tree once incorrectly called Scotch fir. In Europe, it has been a main source of pitch and tar, and has even been used as a cough remedy. (It is described in greater detail in Tour 41.) Specimens of poplar, ginkgo, and white ash are also nearby.

If time permits, stroll down the paved access road, where you may see another escaped ornamental, a woody vine called Dutchman's pipe, twining about the trees. The plant's big, heart-shaped leaves are its chief asset. Here and there in residential neighborhoods you'll see a front porch pretty well swallowed up by this distinctive foliage. It can transform a prim frame house into something straight out of Mother Goose. A little beyond, hugging the soil and moss by the rock exposure, is a colony of white and light brown dog lichen. This foliose species was one of several lichens once used herbally. It supposedly soothed liver complaints, but was even more highly esteemed as a cure for rabies (probably the reason for its common name). Also look at the rock itself, since it is the solidified form of the Mesozoic lava and ash mentioned above. When you touch it, you are touching a different New England, one full of dinosaurs, tropical volcanoes, and lakes of liquid fire.

Before leaving the parking lot, pause to examine the rhododendrons planted in the adjacent walkway area. Another shrub here, by no means as widely used, is bristly locust, a close relative of the black locust tree. It produces striking, rose-purple flowers, and being a legume, it actually enriches the quality of the soil.

When you depart the parking lot, the road will take you back to Route 171 at a point some distance up from the entrance. Turn right onto 171 West and return to its junction with Route 109. Go straight on 109 North for 2.9 miles to the T-intersection with Route 25. Remain on 109 North by bearing left, then turning right soon thereafter.

As you pass through the town of Sandwich, see if you can recognize the black walnuts in some of the yards. These splendid trees have long, pinnately compound leaves that often sport two terminal leaflets instead of just one. In the wild, the species grows northward only as far as Massachusetts; but in this and other areas of New Hampshire, it is employed successfully as a shade tree. Its fruit is edible, though not as choice as the Persian (or common) walnut used by commercial growers. Its best product, in fact, is its wood. When freshly cut, it has an almost purplish tone that subsequently ages to a handsome dark brown. This is perhaps the most prized material for cabinets, other furniture, waterwheels, and gunstocks. There have even been reports of black market operations in which ornamental black walnut trees have been stolen for their timber content.

The black walnut may not be equipped to fight back against human hijackers, but it is quite adept at waging chemical warfare against competing plants. Its roots secrete juglone, a substance toxic to a wide range of vegetation, from garden flowers and vegetables to apple trees, rhododendrons, and azaleas (and, I fervently hope, panicle hydrangeas as well). Woe be unto the unsuspecting soul who sets out tomato plants under its spreading boughs! Were it not for the mainstay of letters beginning, "I simply can't get anything to grow under my lovely walnut tree," the writers of advice columns in horticultural magazines would probably spend most of their time in the bread line.

————————— ∽ —————————

The next turn takes place at Center Sandwich, where Route 109 meets Route 113. Go right on 113 East. This is primal B & B (bed and breakfast) territory, with just about every vital component of the New England Rustic Look, including cedar-shingled barns, stone fences, and naturalized day lilies, converging in one locale. Approximately 3.2 miles farther down, head straight along the even prettier Route 113A. On this loop, you get a pleasant respite from the more heavily traveled thoroughfares, as well as a good taste of White Mountain scenery—especially the prominent bare cliff of Mount Whiteface, rising up ahead. Canadian hemlock and paper birch are the predominant trees, though there are also some tamarack and red spruce. The 113A circuit lasts about 13.3 miles, after which you return to Route 113 proper. Turn left onto 113 East. In this stretch, bracken fern, one of the world's most cosmopolitan plant species, takes plentiful advantage of its preferred roadside environment.

On reaching Madison town center, continue on Route 113 by bearing left. Now you're in one of New Hampshire's most fascinating natural communities, known to plant geographers as the Ossipee Pine Barrens. As usual, botany is intimately linked with geology: the growing conditions here are dictated largely by the sandy and gravelly substrate of the glacial outwash plain. True to their love of dry soils, pines abound here. In addition to the usual white and red species, there is the fire-resistant pitch pine, the rugged hero of Cape Cod (its identification traits are discussed in Tour 16).

One of the most ecologically significant features of the pine barrens environment is its highly acidic, flammable layer of duff that lies on the surface of the ground. It is made up of discarded conifer foliage and other nondecomposed organic remains, and it creates an impoverished soil beneath; many of the herbaceous and woody species common just a few miles away cannot survive in it. Duff is also a typical component of the taiga spruce-fir regime.

In another 2.7 miles, you'll see the sign for Madison Boulder State Wayside. Turn left onto the access road, and at the fork 0.9 mile farther down, bear right onto the gravel roadway. The boulder, the next attraction, is just 0.4 mile beyond. Take a look at the vegetation as you go; along with pitch pine, you'll see scrub oak, another indicator of poor soils. This rather humble and scraggly shrub surprises many people, who think oaks by definition are always hefty trees.

Madison Boulder is one of the area's major geological wonders, but it also provides the perfect habitat for an important type of organism. Pull off at the end of the road and proceed on foot the short remaining distance. In early spring,

One corner of the massive Madison Boulder.

you may catch sight of trailing arbutus blooming along the path, though later it is quite thoroughly cloaked by poison ivy and other greenery. Elsewhere, you can spot interrupted fern, and above it, red maple and Canadian hemlock. Despite the moronic graffiti, the big rock is a breathtaking sight—a massive block of granite 87 feet long, 37 feet high, 23 feet wide, and, at an estimated 4,662 tons, the equivalent in mass of two World War II navy destroyers. This is one of the largest erratics, or glacially transported boulders, in the world. Experts believe it was hefted by the ice sheet from a source in Albany, 2 miles to the north.

As you walk around the great mass of stone, you'll see that its hard, unyielding surface, seemingly so sterile, is actually well colonized by lichens. Many of these are the kind unflatteringly called rock tripe. Some sources state that these composite organisms were a backup food supply for both the Indians and the French Canadians; others report that they're indigestible and cause severe gastric distress. In any case, native peoples used lichens as a source of dye.

At first glance, rock tripe seems little more than an undifferentiated brown disk, but its thallus, or body, is actually quite complex. The uppermost fungal layer, the cortex, is underlain by a layer of photosynthetic green algae, a crucial component, since the fungus cannot manufacture its own food. Beneath the algae lies the medulla, composed of fungal strands, and the cortex's lower surface. Many lichens affix themselves to rock or other surfaces by means of small, rootlike rhizines. Rock tripe, an example of the umbilicate type, also has a central cordlike point of attachment.

Return to Route 113 and turn left onto it. As the local commercial quarrying operations testify, this area is a particularly sandy part of the barrens. Pitch pine and scrub oak are still prevalent. Another 2.5 miles to the north lies the junction with Route 16. Turn right onto 16 North toward downtown Conway and then, 2.9 miles later, bear right again onto Route 153 South.

Soon you come upon Crystal Lake and Eaton Center, together forming the very image of peacefulness. It's difficult to believe that this, the Ossipee region, is New England's most earthquake-prone area. Lest I get tarred and feathered by the Carroll County Committee for the Encouragement of Trade and Tourism, let me point out that this should not deter anyone from staying ten minutes or an entire lifetime in this locale. You could spend years trying to find the one native who loses sleep worrying about the situation, and for good reason. The quake-generating stresses are bled off harmlessly, a bit at a time, and the threat of a major event after a slow buildup of forces is insubstantial. In short, this in not the Yankee equivalent of San Francisco or Armenia. (I should add that this juicy little fact about earthquake frequency came from a source extolling the scenic wonders of Connecticut. I hope I haven't been duped into participating in a fiendishly clever disinformation campaign.)

In another 9.6 miles, you come upon the Route 25 intersection at Effingham Falls, just after you cross the Ossipee River. Turn left onto 25 East. This, too, is prime barrens terrain. There's plenty of opportunity to impress yourself and any co-explorers with your ability to differentiate white, pitch, and red pines from one another. One thing to keep in mind is that the prevalence of pines here and elsewhere in New England is not just a function of soil quality. In the 1800s, these conifers were often spared from cutting, while hardwoods were logged ruthlessly—because the pines' pitch content rendered them unusable for the region's large charcoal industry.

In a few moments, you have entered the expansive state of Maine. The first botany writer on the scene, John Josselyn, described it this way: "The Country generally is Rocky and Mountanous, and extremely overgrown with wood, yet here and there beautified with large rich Valleys, wherein are Lakes ten, twenty, yea sixty miles in compass, out of which our great Rivers have their Beginnings." Where there is slack water in this river, you should be able to recognize two beautiful aquatic monocots, pickerel weed and fragrant water lily. At Kezar Falls, 6.5 miles farther on, the road crosses the Ossipee at a dam. This stream joins the Saco River soon thereafter, east of Cornish town center. Continue on 25 East through Cornish for an additional 7.8 miles, then turn left onto Route 11 North.

The land looks much the same as it does across the Granite State border. The soil is still dry, sandy, and nutrient poor, so the pines are in the forefront, even though there is some red maple and a quantity of poplar occupying its accustomed habitat at the forest fringe. For what it's worth, about 25 miles north of here stands the largest pitch pine on record. Located in the town of Poland, it is 101 feet tall and 137 inches in trunk circumference.

About 3.3 miles to the north, you cross the Saco (pronounced, by the way, "*sock*-oh") at Steep Falls, a favorite testing place for kamikaze canoeists. At the next T-intersection, stay on 11 North by bearing left. From now on, keep a careful watch on the direction signs for Route 11. Wetlands appear in this vicinity, and there is a good deal of truck farming as well.

The final destination of this tour, magnificent Sebago Lake, first reveals itself on the right 7.8 miles from the Saco crossing. In size and the underlying granite base, this is a suitable match for Winnipesaukee. It is Maine's second largest body of fresh water; Moosehead Lake, far to the northeast, is number one. The local Indians were impressed by its size, but it didn't drive them to poetic rapture; *sebago* simply means "big lake." Here, on a breezy summer's day, the air has a bracing pine scent, and it's not difficult to see why the residents of nearby Portland so often repair to this inland retreat.

In another 5.5 miles, turn right onto the Sebago Lake State Park access road and continue 2.4 miles, past the camping facility and over two bridges to the so-called Day Area. Turn right into the entry drive, pay the admission fee, and follow the signs for Songo Beach. Before you reach it, however, you may wish to stop for a moment at the Pine Grove Picnic Area, where you are likely to see northern hardwoods species, wild sarsaparilla, tree lichens, sphagnum moss in the wet woodland depressions, and the boats of fisherman drifting soundlessly across the glassy surface. When you get to the beach and its pretty stand of white pines, you'll have the perfect opportunity to pull out a deck chair, soak your toes in the cool, lapping water, and ponder the awe-inspiring products of ice and sand and fire you encountered today.

The tour's pleasant and leafy ending, at Songo Beach in Sebago Lake State Park.

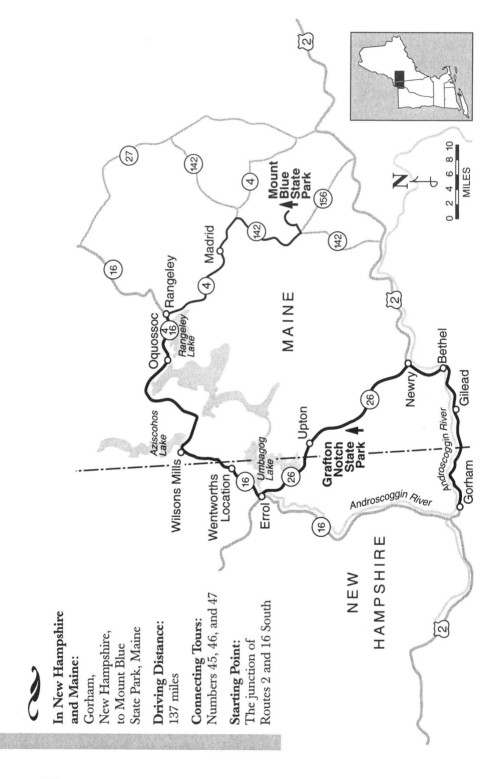

In New Hampshire
and Maine:
Gorham,
New Hampshire,
to Mount Blue
State Park, Maine

Driving Distance:
137 miles

Connecting Tours:
Numbers 45, 46, and 47

Starting Point:
The junction of
Routes 2 and 16 South

Tour 49
Two Great State Parks

"Wherever a man goes," wrote Henry David Thoreau, "men will pursue and paw him with their dirty institutions, and, if they can, constrain him to belong to their desperate odd-fellow society." If you're of a mind to break free from those institutions, if only for a day, this route, which winds through the wilds on either side of the Maine–New Hampshire border, will fit the bill. The tour has its share of crossroads towns and public facilities, but there is something unique and frontierlike about this inland region. Perhaps it has to do with the fact that human control has lessened here, not strengthened, in the past two hundred years. Once vibrant with busy farms, this land has reverted to solitude and its own nonhuman kind of productivity.

Head east on Route 2 from the tour's Gorham starting point toward the Maine border. The road parallels the beautiful Androscoggin River. Soon you pass groves almost entirely composed of paper birch. Somehow, the effect is both dramatic and refined. No other tree species looks this splendid when taken straight. The bark of the paper birch was used by northern Algonquin peoples not only for their canoes but also as protective covering for their dwellings. Horticulturally speaking, I venture the minor heresy that this species is superior in most cases to the widely used, nonpeeling European white birch. Members of the birch genus as a whole cannot compete in size or longevity with the real giants of the forest, but you might be interested to know that the largest paper birch on file with the American Forestry Association is located just a few miles downstream, in Hartford, Maine. At last reckoning, it was 93 feet tall and 217 inches in trunk circumference. That makes it a big tree by any standards, even if the giant white pines of colonial times were a staggering two and a half times that height.

You enter Maine, New England's largest state, 9.2 miles from Gorham. White pine and Canadian hemlock blend with the northern hardwoods, but in the somewhat gentler bottomland microclimate, a central hardwood, red oak, is also well established, along with the usual thickets of black locust. This tree community as a whole might strike you as a good example of the forest primeval, but in fact, each type currently found here is a relative newcomer. Estimates indicate there have been approximately one hundred generations of trees since the most recent Ice Age glaciers melted away between ten and fourteen thousand years ago. The first tall-growing woody species to reestablish itself here was the taiga-forming white spruce, a conifer now largely relegated to Canada and Maine's Down East seacoast. Pollen studies indicate that approximately nine thousand years ago, the climate had warmed sufficiently to permit the return of white pine. A millennium after that, more or less, the maples arrived, and at the end of yet another thousand years, hemlock and American beech were present as well.

In the last stretch before Bethel, the railroad tracks along the road provide the sort of disturbed, unstabilized environment where alien wildflowers such as Queen Anne's lace and rabbit's-foot clover thrive. Another beneficiary of the human activity is poplar, or quaking aspen as it is also known, a native pioneer tree that is an exemplar of what evolutionary biologists refer to as an r environment species—one that exploits an ecological vacuum. Most r environment plants emphasize quantity rather than quality. This means they depend on fast growth, prolific reproduction, relatively small size, and a high turnover rate. In effect, they are trying to grab as much territory as they can while the getting is good. At the other end of this ecological continuum is the K environment, a stable, densely populated habitat where competition is intense. One example of the K environment is the Amazon rain forest, where a 2.5-acre parcel of ground may contain seven hundred tree species, as many as are found in all of the United States and Canada. In such a place, the chance of any new seedling making it to maturity is poor, so each species must ensure that its few mature individuals survive a very long time. Hence, the name of the K game is a relatively low reproductive rate, large size, and increased longevity.

---------- ∽ ----------

On reaching the junction of Routes 2, 5, and 26 in Bethel, bear left with Route 2 as it crosses the Androscoggin and then follows it on a northward leg; 3.8 miles farther along, there's a rest area on the right; turn in for a moment. The river's floodplain is especially broad and fertile here; not surprisingly, it is still under tillage. The banks are lined with silver or soft maple, a key indicator of sustained-flooding areas. The tree has a shallow, spreading root system that occupies the uppermost and best aerated section of the otherwise waterlogged and oxygen-starved soil. Its weak wood will never compare with that of oaks, hickories, hornbeams, or sugar maple, but it has been used extensively for furniture, tool parts, and charcoal.

Next, continue up the road 2.6 miles and bear left with Route 26 North where it separates from the other two. For the time being, you are leaving the land of the Androscoggin, but you'll see it once again, near its source, later in the tour. You are now traveling up the much smaller Bear River, a stream that rises at our next destination, Grafton Notch. This is one of the region's prettiest intervales, or intermontane valleys. Before the onset of the Industrial Revolution, the White Mountain intervales were famous for their wheat and rye farms, but eventually, the burgeoning cereal centers to the west, linked to the new railroad lines and the Erie Canal, ended New England's role as the nation's granary.

The vista is splendid at about 3.0 miles from the previous turn. Soon thereafter, red pine makes an appearance among the more common white pine. Also, there is some black cherry. This tree, a member of the rose family, has been a favorite of furniture makers for centuries. When polished, its wood has a rich and dusky red tone unmatched by any other native timber. In addition, its bark has been used as a sedative and tonic.

Within a few minutes, you've reached the entrance to Grafton Notch State Park. The northern hardwoods predominate at first, but as the roadway climbs the montane, taiga conifers become more and more prevalent. In common with New Hampshire's own notch parks, this excellent public site has a number of stops of great geological significance—among them are Screw Auger Falls and Mother Walker Falls. (In case you haven't noticed already, Maine has the best place names of any eastern state. Many reach the high artistic standards set by Truth or Consequences, New Mexico, and Why, Arizona.)

Two stopping points described here are situated farther up the way. The first is Moose Cave, located 2.8 miles from the park entrance. Turn right into its parking lot, and set off on foot along the delightful 0.4-mile nature trail. Any plant or lichen fancier who fails to fall in love with this spot needs a comprehensive attitude exam. The woody species range from an overstory composed of sugar and red maples, paper birch, red spruce, white ash, and Canadian hemlock, to an understory containing hobblebush, mountain ash, fly honeysuckle, and mountain holly. The last-named plant is the sole representative of genus *Nemopanthus*. Its specific epithet, *mucronatus,* refers to the tiny point, or mucro, on the tips of its leaves. The best find of all, though, is a prostrate little charmer named creeping snowberry. Its flowers are tucked away underneath the foliage, but in season its distinctive, pure white fruit stands out well. It is joined by another low-growing plant, mountain cranberry, which is a member of the heath family.

Moose Cave is a stream channel that has become partially enclosed by large blocks of talus, or detached rock. This sort of misty, protected habitat is sheer fern heaven. The dependably humorous polypody perches in weird tufts on the tops of boulders, and you will also find examples of upland lady fern and long beech fern. The smallest of the bunch, long beech fern, can be identified by its bottom leaflets, which usually point down and away from the rest of the frond.

Continuing on, you come to another community of free-sporing plants, the bryophytes, here including haircap moss and some of its relations. Mosses normally do not have much in the way of commercial value, but herbalists have used the

The rock-loving polypody fern, at Moose Cave, Grafton Notch State Park.

haircap species as a diuretic and dropsy remedy, and as a cure for kidney stones. (For more on the bryophytes, see Tour 33.) Just above and beyond the moss display area is the real showstopper, a big colony of the fruticose ground-dwelling lichen called, quite misleadingly, reindeer moss. Lichens are not plants, but symbiotic associations of fungi and either green algae or blue-green bacteria. True mosses, on the other hand, are real plants, even though their growth habit is as humble as the lichens'. It's just another case of a common name that is based on supposed but unfounded similarities. The descriptive sign at this spot gives a very good rundown on this lichen's ecological significance. Most botanists now assign reindeer moss to the genus *Cladina,* and not *Cladonia,* as the sign states.

The second stop at Grafton Notch State Park, Spruce Meadow Picnic Area, lies 1.7 miles farther up the road. On your way, note the glacially scoured cliffs above. The montane taiga conifers definitely hold sway at this high elevation. The Spruce Meadow lot is on the left. When you've parked, take a good look at the trees and shrubs close by. Along with red spruce and balsam fir, you'll see some tamarack, withe rod viburnum, and lowbush blueberry. The roots of the withe rod once furnished the Indians with cordage; the blueberry supplied native practitioners with nothing less than a treatment for insanity. Its fruit has the reputation of being sweeter than that produced by the highbush blueberry; and, as seen in Tour 54, the lowbush type is used in commercial blueberry production.

If you'd like to investigate the low area that sits between the picnic area proper and nearby Old Speck Mountain, descend the brushy slope carefully. It doesn't take a great deal of scrutiny to see the depression is a type of wetland. As you get closer, you'll discover that it is topped with a monotypic stand of speckled alder. This shrub grows on a yielding, soggy mat of sphagnum moss. Ecologists call this environment a carr. It is often the result of beaver damming. If you're especially lucky, you'll spot the ethereal narrow-leaved gentian, with its powder blue flowers. Traditionally, gentians have been used herbally for tonics, and in the Middle Ages they were even considered an effective antidote to poisons. They are rare and fragile members of the wildflower community, so please don't dig them up and try to plant them in the home garden.

A number of other wildflower species also inhabit the carr or its periphery. In the trickling brook that feeds the peatland, there are clumps of water carpet, otherwise dubbed golden saxifrage. Don't expect the plant to put on a fancy floral display; its tiny green flowers are difficult to recognize. Less demure are the wood sorrel, wild iris, and trout lily specimens in the same vicinity.

On leaving Spruce Meadow, depart the park by continuing up Route 26 North. Almost at once, you're at the top of the notch, surrounded by the silence and grandeur of spruce, balsam fir, and tamarack. These gymnosperms, along with their dutiful angiosperm camp-followers, poplar and fireweed, have developed marvelous mechanisms to deal with the plunging temperatures of the North Country winter. The main challenge the woody plants in particular face in the most frigid months is frost drought. This condition involves the freezing of the xylem tissue and the blocking of life-sustaining water flow to upper stems and evergreen leaves. In the spring thaw, the ice in the xylem melts, but forms deadly air bubbles. Angiosperms, usually considered to be equipped with a more advanced and efficient kind of xylem, are actually more vulnerable to the bubble formation. To circumvent the problem, they must divert their energy to grow new, unclogged xylem cells each year. The gymnosperms, in contrast, profit from their "primitive" xylem cell structure, which localizes the bubbles more successfully. This gives us one plausible reason why conifers outcompete most hardwoods in the taiga zone.

There is another survival trick that both evergreen and deciduous plants have developed. To prevent the formation of damaging ice crystals in their foliage and buds, the plants supercool their internal water—the water is allowed to chill rapidly enough to prevent the crystallization process from really taking hold. White spruce and tamarack have the exceptional ability to resist freezing at a teeth-chattering minus 50 degrees Fahrenheit, but most supercooled species succumb about 10 degrees above that point.

When you are within a mile or two of recrossing the New Hampshire line, try to spot the abandoned apple trees, which indicate this upland area was once cleared. One pernicious alien perennial, Japanese knotweed, has managed to establish itself here, far from its more accustomed turf in city parks and suburbs. Soon, on the right, you'll see Umbagog Lake, the source of the Androscoggin River, close by the state line. At 7.3 miles west of the boundary, after you cross the Androscoggin, Route 26 meets Route 16 in Errol. Turn right onto 16 North.

As noted in Tour 46, this neck of the woods is Moose Alley. Keep a sharp lookout for this noble beast. One of its favorite hangouts is the roadside drainage ditches. Before long, the road leaves the uppermost section of the Androscoggin and tags alongside the Magalloway River instead. The moose find their own version of haute cuisine, fragrant water lily, floating in the slack water here. This aquatic dicot has met the challenge of the wet, oxygen-poor environment by developing aerenchyma, a specialized tissue containing air chambers, which help transfer atmospheric oxygen to where it is needed most, the roots. The air chambers also contribute to the plant's buoyancy.

At 8.9 miles above the Errol intersection, your second stretch of northern New Hampshire is over. Soon, Maine's Aziscohos Lake greets you on the left. This, too, is an area of pleasing intervale scenery, where white pine, poplar, red spruce, and speckled alder thickets share the spotlight. Within 11.6 miles, Canadian hemlock and the northern hardwoods have reasserted themselves, though none of the trees are particularly massive, thanks to the persistent activities of the logging industry. Pearly everlasting, a late summer wildflower near and dear to the heart of every northern New Englander, blossoms along this roadside. The name stems from its long-lasting, frosty white flower heads. It has traditionally been used in preparing a poultice for sprains and bruises.

The concluding section of the road trip, to Rangeley and Mount Blue State Park, crosses territory characteristic of the wilds of Maine's western interior. A glance at a topographic map reveals a crazy-quilt texture of irregular hills and lakes. At the nearest tip of Rangeley Lake, about 26.3 miles from the state line, Route 16 bears left and combines with Route 4. Follow both eastward. On reaching the center of the pretty resort town of Rangeley, bid farewell to Route 16 and follow Route 4 as it splits off to the right.

Alternating populations of red spruce and hardwoods seesaw back and forth along the way. Sometimes the conifer descends into the lower areas; at other times it stages a strategic retreat upslope to the cooler hilltops. Approximately 15.4 miles south of the Rangeley turn, you enter the small settlement of Madrid; 5.0 miles beyond that you come to the crossroads of Routes 4 and 142. Go right on 142 South. In another 9.7 miles, you'll see the first sign for Mount Blue State Park. This is the Webb Lake turnoff. Continue straight past it for 2.3 miles, into Weld town center. At its main four-way intersection, turn left onto Center Hill Road; 0.4 mile later, bear left at the fork onto the paved park access road. This passes a lovely line of sugar maples and a breathtaking view of Jackson Mountain and other surrounding heights. For the fall foliage devotee, this is an earthly paradise.

The roadway soon becomes a dirt surface, offering you three options. Depending on your time and inclinations, you can pull into the Center Hill site, which features a short nature path, a picnic area, and terrific scenery; or you can continue on to the Mount Blue summit trail and make the two- to three-hour hike up and down the mountain; or you can do both. The right-hand turnoff to Center Hill is located 2.2 miles from the fork at the park access road. From

One of the best views in New England, from Mount Blue State Park's Center Hill.

the parking area, you can take the trail uphill over tough Paleozoic metamorphic rocks into a woodsy spot containing everything from red spruce to red oak. The lichenologist who takes this brief stroll will be well rewarded. It's also worthwhile to walk back down the roadway and examine the low outcrops where mosses and more lichens are busy exploiting unused ground with a zeal that would befit a 1980s-era real estate developer. Some of the lichens here, unlike the reindeer moss seen earlier at Grafton Notch, are true members of the genus *Cladonia*. They sport squat cups tipped with scarlet bumps called apothecia. Collectors of invertebrate fossils might be surprised at their basic resemblance to the horn corals.

A little farther down, in a drainage swale, woodland horsetail deploys its lacy superstructure. Horsetails are primitive vascular plants, which lack the lignin that provides structural support for many upright plants. Instead, they rely on the silica they incorporate into their tissues. The silica gives the plant stems a gritty texture. Consequently, Indians used them to polish their arrowheads and bows, and early white settlers put them to service scouring pots and pans. The abrasive and apparently toxic nature of horsetails makes them indigestible to many herbivores, but Linnaeus observed that this woodland species was used as horse fodder in Sweden.

To get to the Mount Blue trailhead, drive down the dirt road an additional 3.3 miles past the Center Hill turnoff. The road becomes a one-lane affair, so be alert for oncoming cars. The dense forest it traverses is crisscrossed with stone fences, proof that this land was once under cultivation or used for pasturage. This is a good place to identify a variety of ferns, including the interrupted, ostrich, Christmas, sensitive, and hay-scented species. Near the parking lot, more naturalized apple trees contend with the thicket growth around them. Once planted to serve human needs, these faithful providers now cast their fruit for whatever wild creatures will take their bounty.

The hike to the top of Mount Blue begins at the trail sign, visible from the parking area. The ascent is uniformly (some would say relentlessly) steep the entire way; it favors the unhurried soul. Fortunately, botany offers the perfect pretext for the lingering, unathletic approach. As a matter of fact, if you're not careful, you could spend an hour identifying plants in the first 100 yards, so dense and varied is the mantle of vegetation. For much of the hike, the overstory includes white pine, sugar maple, red oak, arbor vitae, Canadian hemlock, red spruce, white ash, and basswood. Let's focus for a moment on the last two.

Here and elsewhere in the book, my references to white ash may in some cases include a similar species, green ash. Hybrids between different members of the ash genus are common and often confounding. If you take time to hunt for the green type, for example, you might find a tree that keys out well–except for one or two nagging characters that are straight out of the description for the white species. Be of good cheer, and accept the tree's disregard of taxonomic propriety as gracefully as you can. After all, white ash (or whatever) is a splendid tree with many redeeming attributes.

To find the ashes here, look for compound, pinnate foliage and bark with diagonal ridges that form a repeating diamond pattern. As every sports fan knows, ash wood, carefully cured for seven years, is the preferred material for baseball bats, because it combines elasticity with good strength. This, supposedly, is a function of the tree's growth rate–new wood cells are formed in an explosive spate each spring, and the resulting tissue is unusually resilient. For the same reason, ash wood has been used extensively for oars, tool handles, snowshoes, rake heads, and archery bows. In this country, ash trees have also served herbal ends. To cite but one application, their leaves and bark were made into a home-remedy laxative. In Europe, the native ash species were thought to possess supernatural powers, as well. Besides frightening away sickness and evil spirits, the trees are the answer to a snake-hater's prayers. According to time-honored legend, the mere shadow of an ash tree kills serpents of all descriptions. On top of that, a pin jabbed into an unsightly wart, then stuck into an ash, will make the former disappear. And in the sphere of weather forecasting, ash trees indicate whether the coming year will be dry or rainy: if they bloom before the oaks, get your umbrella ready, but if they bloom after, prepare for drought.

The other tree highlighted here, the basswood, presents no taxonomic problems whatsoever. Its toothed, uneven-based leaves, sometimes 8 inches or even longer, are utterly distinct. Ash tree flowers are nothing showy, but a basswood's cream-colored, fragrant blossoms are the delight of naturalists and bees alike. (Emerson noted that his friend Thoreau had a special basswood tree he always visited when it bloomed in early summer.) One of North America's contributions to the linden genus, basswood has had its offbeat common name for a long time. It derives from "bast wood," because its phloem tissue yields fibers used in making bast for cordage and cloth. The tree's wood is soft and much favored by carvers, and since it is extremely light when it dries out, it is perfect for baskets and boxes. It has also been fashioned into piano keys, beehives, and coffins. Native Americans used its bark for cordage and fish nets. In Europe, the flowers of related linden species were once an ingredient in a bath treatment for hysteria.

As engaging as the game of tree identification is throughout the climb, you shouldn't ignore the woody understory and the herbaceous plants. Among the striped maple and American hazel are two additional shrubs of note, chokecherry and bush honeysuckle (or *Diervilla*). Chokecherry has broader leaves than the tree-sized black cherry, and usually blooms in the spring a week or two before the other. Bush honeysuckle has oppositely arranged leaves with long, tapered tips. In the wildflower department, there are monocots such as bellwort and corn lily, and a broad range of dicots: wild sarsaparilla, stout goldenrod, rough hawkweed, gall-of-the-earth, and tall lettuce.

When you make your well-deserved final approach to the summit, you will find that balsam fir largely takes over from the red spruce. Dwarfed, heart-leaved paper birch, a familiar sight on most New England mountaintops, is present and accounted for, as are pin cherry and lowbush blueberry. Another plant you may come across is bristly sarsaparilla, a close relative of the wild sarsaparilla. Both are native members of the *Aralia* genus.

In recent times, the fire tower on the summit has reached a dilapidated state, but eager climbers still scuttle up it for the 360-degree view. In all directions, the terrain speaks of endless woodlands and blank-faced lakes. It is one of America's most ancient landscapes, but it is also intensely alive. Despite the seemingly insurmountable interference of continental glaciers and Thoreau's dirty human institutions, the ever changing forest triumphs intact.

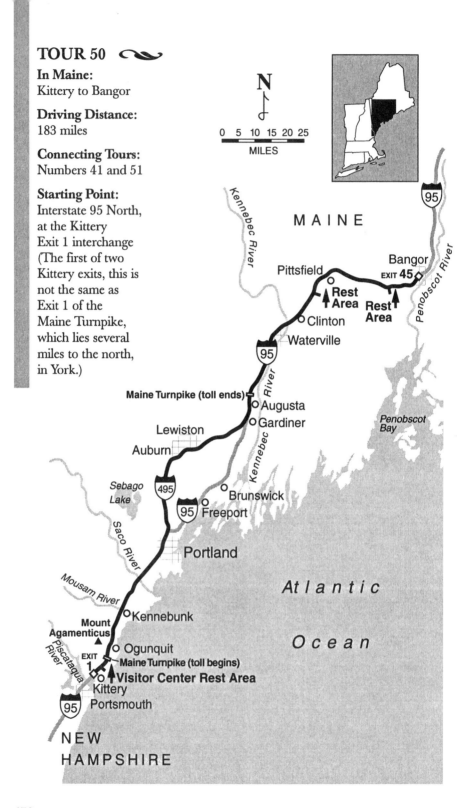

TOUR 50 ∿

In Maine:
Kittery to Bangor

Driving Distance:
183 miles

Connecting Tours:
Numbers 41 and 51

Starting Point:
Interstate 95 North,
at the Kittery
Exit 1 interchange
(The first of two
Kittery exits, this is
not the same as
Exit 1 of the
Maine Turnpike,
which lies several
miles to the north,
in York.)

N

0 5 10 15 20 25
MILES

MAINE

Kennebec River

Bangor
EXIT 45

Pittsfield
Rest Area
Rest Area
Clinton
Waterville

95

Penobscot River

Maine Turnpike (toll ends)
Augusta
Gardiner

Lewiston
Auburn

Kennebec River

Penobscot Bay

Sebago Lake

495

95
Brunswick
Freeport

Saco River

Portland

Atlantic

Mousam River

Kennebunk

Ocean

Mount Agamenticus

Ogunquit

Piscataqua River

EXIT 1

Maine Turnpike (toll begins)
Visitor Center Rest Area
Kittery
95
Portsmouth

NEW
HAMPSHIRE

Tour 50
A Roadside Vegetation Tour:
Kittery to Bangor

*T*his excursion deep into the great state of Maine is one of several tours
designed for travelers availing themselves of New England's extensive
interstate system. High Speed Botany is the perfect antidote for long-
distance automotive boredom. This is especially true here. The region you will
traverse is rich in plant-related sights and subjects, from the complex patterns of
shifting tree populations to their interactions with disease and insect invaders. And
there is the dramatic story of how, in colonial times, one native conifer you'll see
along this route became the focus of the Yankee resistance to overseas authority.

The starting point is located at a reasonable approximation of southernmost
Maine, just across the Piscataqua River from the handsome and historic seaport
of Portsmouth, New Hampshire. Head north on Interstate 95. Almost at once,
you're beyond the most built-up section of Kittery. One of the first sights to greet
you is an extensive community of what is potentially the largest indigenous plant
of the American Northeast, the white pine. The presence of this majestic gymnosperm at the start of the tour is fitting; after all, we're at the gateway of the
Pine Tree State. As we'll see a little later, this single tree species has played a
prominent role here and in neighboring areas.

Within 3.5 miles of the beginning, a pleasant rest area lies off to the right.
Get ready to pull in—it's the perfect place to do a little preparatory woody-plant
identification. Before you reach it, though, there are some other species to note:
red pine, black willow, clump-forming staghorn sumac, and pin oak planted along
the roadside. In winter or spring you may notice that the pines have brown needles

on branches facing the roadway. This most likely is a symptom of root damage caused by dissolved ice-clearing rock salt.

When you reach the rest area parking lot, take a good look at the white pines on the grounds. Pine foliage is attached to its stems in bundles called fascicles. A good way to distinguish one species from another is to carefully count how many needle leaves are contained in each individual fascicle. Fortunately, white pine is a dead giveaway, because it's the only native species with five needles per bundle. In most cases, its leaves range from 3 to 6 inches long. Even though pines are evergreen, they do eventually shed their foliage. In the case of the white pine, needles usually last for two or three years before they drop. What distinguishes evergreens from deciduous plants, therefore, is that they don't drop all their leaves in one short period, at the end of a single growing season.

There are other characters, or identification traits, that set white pine apart from its relatives. Its cones are quite distinctive. When mature, they are long and narrow (6 to 8 inches by 2 inches). The cones of most other pines are closer to a spherical shape, especially after they've opened. From a distance, the white pine is often unmistakable in appearance, with long horizontal branches stretching over the roadway. The slender needles impart a fine, almost feathery texture to the crown, in contrast to the coarser look of the red, pitch, Scots, and Austrian pines.

With this introduction to the fine art of white pine spotting, you should be able to see for yourself the plant's prevalence and ecological importance as you ascend the eastern side of the state. But the rest area presents additional botanical points of interest. Besides the ornamental juniper shrubs very much in evidence, two Asian introductions are important components of the New England horticultural scene. The first of these, and perhaps the more difficult to locate, is the evergreen shrub known as Japanese yew. You'll find it on the side and at the back of the visitor center. In modern times, in the United States at least, this species is generally considered superior to English yew, though the latter is by no means rare. It isn't always easy to tell the two apart. If you're lucky, the English yew leaves will have a small, brownish tip, while the Japanese ones will possess a more pronounced cusp, or elongated point. There is also a native yew, the Canadian (see Tours 36 and 38), but it is a low, scraggly survivor of the dim forest floor, and not much suited to life as an ornamental. All yews are highly toxic; still, a West Coast species has been in the news recently because it contains a chemical compound proven to be effective against ovarian cancer.

In Britain, a traditional use of yew wood was in the manufacture of longbows. Unlike their shorter equivalents, these deadly, high-velocity weapons could shoot arrows that penetrated even the best body armor. This fact was gruesomely demonstrated in A.D. 1415 at the battle of Agincourt, where the English yeoman archers unexpectedly neutralized the enemy heavy cavalry by literally nailing the French knights to their horses. So much for class privilege.

Whatever their medical or military applications, members of the yew family are of great interest, and of great bafflement, to students of plant evolution. If you happen to see the bright red "berries" on these plants at the rest area, you'll have a good clue to the yew's special place in the class of gymnosperms. As you'll recall, gymnosperms are the relatively primitive, naked-seed plants that bear neither flowers nor fruit. Most modern representatives of the class are conifers—pines, spruces, firs, and so on—and, as their name indicates, they bear cones instead of fruit. What makes the yews so distinctive is their seeds, which are sheathed singly in arils, or fleshy red jackets. Many experts now believe that the yew group doesn't belong to the conifers but occupies its own distinct gymnosperm order. How far back in the history of life its uniqueness extends is still very much a mystery.

The other introduced shrub is that old dependable denizen of New England highways, the winged euonymus. It is shown at best advantage in a pretty triangular ensemble in front of the visitor center entrance. This deciduous angiosperm of the bittersweet family has stems with corky ridges and oppositely arranged, pointed leaves, widest above the middle. One of the chief attractions of this Asian species is its stunning fall foliage, which ranges in color from salmon pink to the more usual throbbing red. This has prompted growers and marketers, always given to hype, to label it with splashy common names, such as burning bush. But it's just too far from Mount Sinai for that one.

Before you depart the rest area, note how the visitor center building, done very much in a latter-day style, nevertheless sports the customary shingled exterior. Shingles are most often made from the rot-resistant wood of the Atlantic white cedar or the arbor vitae; but over the centuries, oak, white pine, and American chestnut have been widely used for the same purpose.

———————————— ∾ ————————————

As you continue north on I-95, you cross the York River at a location where arbor vitae trees have been planted along the way. The red oaks here are a reminder that the central hardwoods regime extends up the seaboard all the way from the southern part of New England roughly to Cape Elizabeth. The oaks are a component of the Maine woodlands even beyond that, as you'll see later in this tour. At 2.1 miles north of the river, the toll plaza marks the beginning of the Maine turnpike. This is a good place to mention that our track takes us along the Turnpike all the way to its terminus in Augusta. So plan to take I-495 via Auburn and Lewiston when you arrive in the Portland area.

About 1.4 miles above the toll plaza, a road-cut reveals Mesozoic era granite or granitelike rock. It is part of a circular bedrock structure akin to various igneous units that are scattered across New Hampshire. Not far to the west, running roughly parallel to the turnpike, is the Norumbega Fault Zone. This is the putative boundary between an ancient edge of North America and Avalonia, the suspect terrane (possibly a mini-continent) that crashed into the mainland about 400 million years ago. Roughly 2.5 miles farther on, you come abreast of Mount Agamenticus, which lies a little farther inland. This hill, one of the few heights on Maine's southern coastal plain, has long been a welcome landmark for mariners.

The next stretch, from Ogunquit to the Mousam River in Kennebunk, is a good place to review the fascinating tale that combines sea power, imperial politics, and the noble white pine. The story begins centuries before the European settlement of New England, when human beings first carried their conflicts to sea in wooden-hulled warships. It is difficult for us, in our present cultural context of steel, plastic, and microchips, to conceive of the immense environmental effect shipbuilding once had. By the late Middle Ages and the Renaissance, for instance, Venice and the other naval powers of the Mediterranean Basin had denuded most of their coastal forests. Precious topsoil was lost, severe erosion set in, and busy harbors silted up. In fact, the deforestation problem in southern Europe became acute enough to force *La Serenissima,* the Venetian Republic, to enforce the first tree-conservation program in history, in a less than serene way. To build just one large war galley, fifty beech trees were required for the oars, three hundred conifers for the planking and spars, and another three hundred oaks for the hull. About a quarter of a million mature trees were needed to construct the Muslim and Christian fleets that fought the great battle of Lepanto in 1571. Later, in the seventeenth century, it was not unusual for two thousand trees to be used in the completion of a single Northern European man-of-war.

When Captain John Smith and other early transatlantic explorers reported that New England was covered with first-rate lumber trees, Britain's Royal Navy realized that an important resource lay within its grasp. England, traditionally a nautical country, was well on its way to becoming the greatest sea power the world had ever seen. But to do so, the small island nation had to devote up to one-third of its total expenditures on its fleet. The ecological impact was predictable. By the eras of Shakespeare and Oliver Cromwell, English woodlands had been severely depleted, and the Royal Navy was forced to turn to less predictable timber sources in the Baltic and Scandinavia.

It didn't take the mother country long to turn to its new North American colonies for wood products. The first consignment of naval timber left New England only fourteen years after the founding of Plymouth Colony. As time went on, Maine and New Hampshire emerged as centers of the lumber trade; it was there that the prime item, the mighty white pine, was biggest and most plentiful. This lofty, straight-growing tree provided excellent masts, even for the largest ships of the line. Colonial woodsmen carefully cleared rocks and stumps away from the vicinity of a chosen pine, felled it, and then hauled it as gingerly as possible, often with teams numbering forty or fifty oxen, to the nearest river. Generally, the felling was done in winter, so the hulk could be sledded rather than dragged; but occasionally, a giant trunk would be rolled onto a wheeled undercarriage instead. Once it reached the waterway, the mast-to-be was floated down to the coast, where it was loaded onto a specially designed cargo vessel and transported across the Atlantic.

As it turned out, the New England source never did beat out the Old World suppliers. By the mid-1600s, naval stores, including pitch, tar, turpentine, and oak timbers, were being sent to British shipbuilders—who promptly declared them inferior to the European equivalents. Additionally, the British Admiralty continued to favor Baltic trees (often called Riga or Scotch firs, but in fact Scots pines) for all but the largest masts. They were more expensive, but they were closer at hand, and barring war or embargoes, they presented fewer transportation problems. Ironically, this did not bother Yankee entrepreneurs, who cheerfully carried on a brisk illicit trade with the dastardly Popish lands of Spain and Portugal and with their holdings in the Azores and Canary Islands.

Despite all these developments, His Britannic Majesty's government continued to act as though the forests of this region were worth their weight in platinum, all the way up to the American Revolution. In what constitutes one of history's longest-running performances of imperial ineffectiveness, the British issued one unenforceable law after another, "protecting" white pines from commercial exploitation and "reserving" them for official Royal Navy use. The first of these was the Charter of 1691, which prohibited the unlicensed cutting of pines 24 inches or greater in diameter, as measured a foot above the ground. Anyone caught breaking the law was fined 100 pounds per tree; but if anyone was caught, he must have been a complete idiot, since there was little or no enforcement going on.

This slight oversight on the part of the authorities was addressed somewhat by the follow-on Naval Stores Act of 1705. It created the post of Surveyor-General of the Woods, a position that soon earned for its owner the additional responsibilities as Most Hated Man in New England. The beleaguered surveyor-general, sometimes zealous, sometimes corrupt, had to patrol a vast trackless region inhabited by some of the meanest characters in the English-speaking world. One

The King's Broad Arrow. –Karen Lommler

of his jobs was to mark reserved pines with the King's Broad Arrow, made with three hatchet strokes. It didn't take clever backwoodsmen long to turn this identification system to meet their own needs. They too would mark their own favored trees with the Broad Arrow, to discourage their competitors from stealing them before they could get around to it themselves. The main problem was that local sawmills paid the lumberjacks as much as the government would; so there was no material incentive for the colonials to toe the line imposed by the restrictions.

Still other acts, issued subsequently, did little to change the situation, and at one point in the 1730s, a surveyor-general by the name of David Dunbar even managed to precipitate a riot in Exeter, New Hampshire–not all that far from here. One statistic from that period, out of Northampton, Massachusetts, summarizes the Yankee defiance: of 363 pines marked with the Broad Arrow, only 37 were not surreptitiously taken for illegal use. Perhaps the height of abuse, however, occurred when New Hampshire governor Benning Wentworth managed to grab the surveyor-general post as an added sinecure, after Dunbar's departure. He did protect the pines–but more for himself and his landowner cronies than for the Crown. When he was finally dismissed from the post, dishonored but rich, his nephew John Wentworth took over. He played the game straight and followed the letter of the laws, thus fomenting so much anger that New Hampshire eagerly embraced the outbreak of the Revolution.

———————— ❧ ————————

When you reach the vicinity of the highway's Mousam River crossing, you come upon sandy road-cuts. This is an area of extensive glaciomarine deposits, sediments dumped by glacial meltwater into the ocean when it extended this far inland. About 0.5 mile beyond the river is another rest area, planted with pin oaks and purple-leaved cultivars of Norway maple. Beyond, in wild sections, note the young trees growing under the white pines. Most are not pines, but broad-leaved trees capable of withstanding partial shade as they grow through their juvenile phase. White pines, in contrast, need a fair measure of sunlight; for that reason, they often give way to the next generation of more shade-tolerant species. When a white pine grove does survive in the long term, its success is usually due to the dry, sandy, or gravelly soil, where broad-leaved trees cannot compete with their accustomed vigor.

A familiar but unwanted architect: the eastern tent caterpillar and its highly visible housing complex.

Maine is nothing if not a land of rivers. The next large waterway the interstate spans is the Saco River, one of New England's great streams. It rises in the supremely beautiful setting of Crawford Notch, in the White Mountains of New Hampshire. Here, it is just several miles from its end in the Atlantic. In the late 1600s, the nearby seacoast was home to the naturalist John Josselyn; all in all, he spent about ten years of his life residing in Maine.

If you make this journey in late summer, you may spot a dirty gray webbing in the branches of the black cherry trees nearby. This is the work of the eastern tent caterpillar, a native defoliator that was a part of New England's natural scheme of things long before it was first reported by white settlers in 1646. As with many plant pests, the size of its population varies considerably from year to year, though it tends to reach epidemic proportions in roughly ten-year cycles. At 2.3 miles north of the Saco, you can see evidence of another insect trouble-maker, the white pine weevil. This pest feeds on the upper growth of its namesake.

It doesn't kill the tree, but it deforms it by destroying the leader, or central uppermost stem. The result is a cabbage pine, as seen here—a plant with a broad, low, leaderless crown. This affliction is most common in old pastures, where the sun-loving weevil can feed on wholly exposed saplings. To counteract the problem, foresters planting new white pines will shade them just a little with young gray birch nurse trees.

As you approach the western outskirts of the ice-free harbor of Portland, keep an eye out for the continuation of the Maine Turnpike, which is designated Interstate 495 all the way to Augusta. Stay on it. You will rejoin I-95 later on. North of the city, the highway arches due north and inland, east of Sebago Lake. In this stretch to Auburn and Lewiston, see if you can spot yellowish reindeer moss lichen growing on the embankments, as well as the following woody species: a plantation of red pine, the bottomland-dwelling American elm, a nice colony of common juniper, and pitch pine. What do the last two tell you about their environments? The broad, shrubby common juniper might be termed a low-competition species, because it does best in open areas where larger plants can't shoulder it out. Luckily for the juniper, it is for the most part unpalatable to cattle, and therefore is one of the few large plants that survives in an active pasture. If the pasture is ultimately abandoned, other pioneer plants rapidly take over; before long, the juniper gets shaded out.

The pitch pine is a low-competition type as well, since it colonizes extremely poor soils such as sand plains, which most hardwoods cannot handle. In addition, the pitch pine has an effective way of dealing with less finicky competitors, its pine relatives. The majority of conifers are extremely vulnerable to fires, and will not resprout when their main stems are killed. But the pitch pine has thick, fireproof bark and can send up new stems from its base. Furthermore, its cones are serotinous—they delay their opening. This is possible because they are sealed by resin, even after their seeds are mature. These cones wait for the heat of the flames to open them. Accordingly, the species is the first to reseed the practically competition-free environmental vacuum created by burning.

───────────── ∽ ─────────────

The next great river you cross is the Androscoggin, an exceptionally beautiful waterway upstream, as you already know if you've taken Tours 45 and 49. Here, it divides Lewiston, a center of French Canadian settlement, from Auburn, to the west. In this area, American elms are numerous, though a number have succumbed, presumably to Dutch elm disease. It behooves the local powers-that-be, or the appropriate landowners, to remove the dead trees as soon as possible because they act as breeding centers for the elm bark beetles carrying the lethal fungus.

On the leg to Gardiner, the fine-textured forms of tamarack make an appearance in low areas, and red maple, basswood, and poplar are also seen. When I-495 and I-95 rejoin and merge into the latter, the turnpike enters the territory of the Kennebec River, which runs to the right of the highway for the next 30 miles or so. Had you been here in 1775, you would have seen Benedict Arnold leading his Yankee army northward up this waterway. As daring as ever, Arnold proposed to take Quebec from the British and extend the Revolution into the heart of Canada. After a terrible ordeal in the wilderness, his force reached its destination and laid siege to the city on the St. Lawrence, as planned. However, the subsequent assault was a failure and Arnold was badly wounded. What is not so well known, though, is that a separate American force succeeded in taking Montreal for a while.

In the environs of Augusta, the state capital, two naturalizing introductions, Scots pine and black locust, are frequent sights along the road. The first of these can be distinguished from native pines by its reddish bark and glaucous, or waxy blue, foliage. North of the capital, you enter a wonderful melting pot of forest regimes—everything from central hardwood oaks to taiga conifers. The latter predictably include red spruce and balsam fir. As you speed along, you may be able to distinguish the firs, with their upward-pointing cones, from the spruces. Meanwhile, herbaceous plants abound. In the wet drainage channels, common cattail and the invasive purple loosestrife are well established; on drier ground, Queen Anne's lace and goldenrods provide a pleasing mixture of white and yellow in middle to late summer.

The curious combinations continue in and about the Waterville area. The roadside wildflowers, such as rabbit's-foot clover and black-eyed Susan, are predictable; but behind them, the white pine of light soils stands cheek-by-jowl with the black willow of damp conditions. Beyond the Kennebec crossing, arbor vitae, the portly northlands evergreen, marches along the median strip. There are some speckled alder thickets, but the conifers more often than not rule the roost. Look for Canadian hemlock, white and Scots pines, and the spruce-fir team.

If you would like to take a closer look at some of the plant species that have been so conspicuous for most of your journey, pull in at the rest stop a few miles before the town of Pittsfield. You'll be able to examine a grove of red oaks, arbor vitae trees, Canadian hemlocks, and balsam firs. The next rest stop after that, on the approach to Bangor, is a good place to see native and introduced spruces, as well as bristly locust, a shrub used to fix embankments.

Past Pittsfield and the northward-swinging Sebasticook River, the highway traverses some of the best taiga scenery in inland Maine. The vertical dignity of balsam fir, tamarack, Canadian hemlock, and arbor vitae is softened somewhat by the shimmering foliage of their little angiosperm associate, poplar. The somber dark green of the conifers is balanced by the pleasant colors of two classic up-country wildflowers, magenta-blossomed fireweed and the icy-white pearly everlasting. You've finally come to the great North Woods, a well-earned sight after a trek of some 180 miles. But as you approach Exit 44 and the sheltered valley of the Penobscot River, the playful unpredictability of the forest reasserts itself. Conifers are still in evidence, but they share the landscape with black locust and red oak, main constituents of the woodlands of far-distant Massachusetts and Connecticut. As you reach the tour's end, at Exit 45 in Bangor, you've learned that the natural world of the Pine Tree State refuses to reduce itself to simple images.

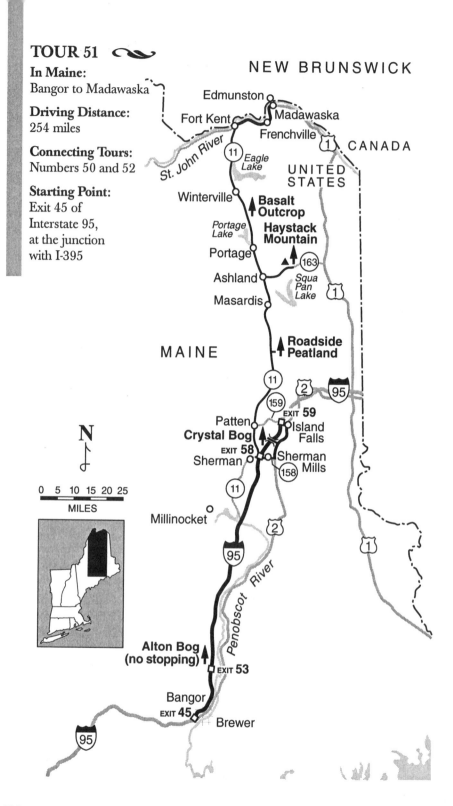

TOUR 51 ∽

In Maine:
Bangor to Madawaska

Driving Distance:
254 miles

Connecting Tours:
Numbers 50 and 52

Starting Point:
Exit 45 of
Interstate 95,
at the junction
with I-395

NEW BRUNSWICK

Edmunston

Fort Kent

Madawaska

Frenchville

CANADA

St. John River

11

Eagle
Lake

UNITED
STATES

Winterville

**Basalt
Outcrop**

Portage
Lake

**Haystack
Mountain**

Portage

163

Ashland

Squa
Pan
Lake

Masardis

1

MAINE

**Roadside
Peatland**

11

2

95

159

EXIT **59**

Patten

Island
Falls

Crystal Bog

EXIT **58**

Sherman

Sherman
Mills

158

11

Millinocket

2

95

Penobscot River

1

N

0 5 10 15 20 25
MILES

**Alton Bog
(no stopping)**

EXIT **53**

Bangor

EXIT **45**

Brewer

95

490

Tour 51
The Narrow Road to the Deep North

A Recommended Fall Foliage Tour

HIGHLIGHTS

Preparing for Our Longest Trip ∿ A Drive Through a Raised Bog ∿ The
Paper Birch Ah Factor ∿ The Gymnosperms' Tale ∿ Frontier Maine ∿
Beholding Katahdin ∿ Thoreau Meets Unloving Nature ∿ Convergent
Evolution and the Ferns That Weren't ∿ Husband-and-Wife Trees ∿ Crystal
Bog ∿ Welcome to Scenic T8 R5 WELS ∿ Sawtimber and Pulp ∿ Haystack
Mountain, the Rhyolite Hill ∿ Pipsissewa ∿ The Taiga, a Botanical Desert ∿
Highlands of Basalt ∿ Plants and Pillow Lavas ∿ The Surging St. John ∿
Life at the Top ∿ Red Maple Islands ∿ A Land of Two Languages

T he title for this tour comes from a wonderful travel book by the Japanese
poet Matsuo Basho. He was one of many writers and explorers who
have felt the irresistible pull of the north. Here in the New World, there
is a mystical quality about any northern woodland, and Maine's contribution
to the North Country mystique, from the eras of Josselyn and Thoreau to our
own, has been considerable.

Part of the appeal lies in the sheer size of the wilderness. In total land area,
Maine is roughly the equal of the other five New England states combined. And
scanning a road map, you'll see just how vast its northern hinterland is. Which
leads to this tour's most important factor: it is, by a wide margin, the book's
longest route. Plan accordingly. Choose a day when the weather will be good
across the whole area. If you intend to do this tour along with the next in the
course of a single day, begin at the crack of dawn. Bear in mind that the distance
between service stations can be considerable; keep an eye on your gas gauge.
Also, equip yourself with a good spare tire, a safety kit, and the willingness to
encounter roads that may be bumpier than usual.

One more word before you begin. One of the stops along the way, the
magnificent Crystal Bog, is largely owned by The Nature Conservancy, one of
the nation's finest nonprofit environmental organizations. I've included directions
into the bog and a description of some of its sights, but I recommend you contact
the Conservancy's Maine Chapter, at 20 Federal Street, Brunswick ME 04011.
By doing so, you can get more information on this priceless natural treasure,

and you can check on current access to the bog. First and foremost, yield to the Conservancy's wishes in this matter.

When you are ready to get under way, head north on Interstate 95 through Bangor. In a moment, you cross Kenduskeag Stream, a tributary of the Penobscot River that divides the city in two. Before long, residences, businesses, and mallscapes disappear from view. Beyond the university town of Orono and just past Exit 53 lies the first natural wonder of the tour. Surprisingly, the superhighway passes right through it. It's a 2-mile expanse of low vegetation, dotted here and there by the spindly forms of black spruce. This is Alton Bog, an excellent example of peatland found in only a few places in the United States besides coastal and up-country Maine.

Known as a raised bog, this unusual environment is somewhat domed in cross section, with the central portion higher than the edges. This is the result of the gradual accumulation of peat in the middle—a process that can only occur in extremely wet climates, such as this one (on the average, the state receives an impressive 40 inches of rain a year). As the raised bog develops, it becomes ombrotrophic, or, as some ecologists would prefer, ombrogenous. These terms refer to the fact that precipitation is the only form of water available to the bog; there is no flowing, oxygen-rich stream recharging the habitat's nutrient supply. In this stagnant regime, humic acid derived from the breakdown of dead plants builds up, and the pH level often drops to 4.0 or less. That's one thousand times as acidic as normal pH 7.0 tap water. While this raised bog form is a special case, peatlands in general occur all over the globe—on Caribbean islands and in such disparate places as Israel, South America, and Southeast Asia.

At 11.9 miles north of Alton Bog, a beautiful stand of paper birch enhances the median strip. Few North American trees produce the kind of instantaneous response I call, for lack of a better term, the Ah Factor. California has two of these, the giant sequoia and that bristling militaristic monocot, the Joshua tree. Closer to home, in more southern parts of New England, there are the elephantine sycamores of the old Connecticut River towns (see Tours 12 and 25). The birches here have no such grandeur, and it's even difficult to explain why their whiteness works as effectively as it does in the snowless season. Still, no other monotypic community takes one's breath away quite so dramatically.

Approximately 6.0 miles beyond the birch stand, you cross the Piscataquis River just before it meets the Penobscot. To the southeast lies the Enfield Horseback, one of the most famous eskers, or glacially formed ridges, in the United States. (For more on how these sinuous landforms came into being, see Tour 45.) In addition, here at river's edge the ever adaptable red maple puts forth its shallow roots in the soggy, unoxygenated soil of the banks. According to Pehr Kalm, the great eighteenth-century Swedish explorer of eastern North America, red maple wood was put to use by the colonists for a number of domestic needs: plates, spinning wheels, spools, furniture feet, and other sorts of turnery. Also, he related, the bark yielded a dark blue dye suitable for linen or worsted.

In the next 10 miles or so, a grab bag of different tree species and woodland regimes will delight the particularizer and frustrate the formulator of generalities. For example, taiga red spruces and balsam firs blend freely with white pine, a less cold-hardy conifer. While they certainly have not swept flowering plants from the scene, these softwoods are important members of the mixed community. As a group, their history extends back to the early part of the Age of Dinosaurs, and even beyond that to the Paleozoic era, when their ancestors, the so-called progymnosperms, first appeared. The progymnosperms were the descendants of an ancient vascular plant lineage. The fossil record, as incomplete as it is, suggests that the conifers and their fellow gymnosperms achieved a worldwide dominance in the Jurassic period, that important notch in geologic time when New England and the rest of North America began their separate identity in the sundering of a supercontinent. (For a hint or two on why the conifers still dominate the colder stretches of the Northern Hemisphere, see the discussion of their resistance to frost drought in Tour 49.)

Some 24.5 miles north of the Piscataquis, the interstate crosses the Penobscot. To your left the river splits into its East and West Branches. Just a few miles beyond the fork is Millinocket. This major lumber center is a company town in the most unequivocal sense of the word, because it was originally constructed, part and parcel, by the Great Northern Paper Company. Two other giants of the industry, the International Paper Company and the S. D. Warren Company, also have major mill facilities in the Pine Tree State and, accordingly, loom large on its economic landscape.

In many ways, the emphasis on lumber and the lingering frontier mentality are more reminiscent of the Far West than they are of the rest of New England. There is a solid historical basis for this. Previously under the jurisdiction of Massachusetts, Maine did not enter the Union as an independent state until 1820, as a result of the infamous compromise that also brought Missouri into the nation as a slave state. Though the seacoast and selected inland sites were settled by whites as early as the 1600s, Maine's great interior was opened up for large-scale settlement long after the rest of the region had been tamed. Consequently, this was the only New England state to experience a boom at the same time as the Midwest and West. Ironically, while the upland farm towns of New Hampshire and Vermont were losing half or more of their populations to the rich lands beyond the Erie Canal, Maine was in the midst of a belated expansion. To this day, the Pine Tree State retains its separateness, which manifests itself in everything from its own distinct dialects to its arresting place names. Many of the latter are Indian, French, or just downright quirky. Few are the Old English and biblical clone names that replicate themselves over and over across the rest of Yankee territory.

After leaving the Penobscot behind, you cross the small Salmon Stream, with wetland vegetation on the right. Just 0.7 mile beyond, Maine's highest peak,

Mount Katahdin, comes into view on your left. Its stark granite summit, 5,267 feet above sea level, is the first place in the United States to receive the rays of the morning sun. In an additional 4.5 miles, the viewpoint across Salmon Stream Pond is even better, and the road builders have obligingly put in a scenic turnout to the right. In 1846, long before this region had anything resembling paved roads, let alone scenic turnouts, Henry David Thoreau climbed the great mountain during a travel break in his sojourn at Walden Pond. He was awed by what he found. Katahdin revealed a face of nature that had no regard for mere human concerns:

> It was vast, Titanic, and such as man never inhabits. Some part of the beholder, even some vital part, seems to escape through the loose grating of his ribs as he ascends. He is more lone than you can imagine. . . . Vast, Titanic Nature has got him at a disadvantage, caught him alone, and pilfers him of some of his divine faculty. She does not smile on him as in the plains. She seems to say sternly, Why came ye here before your time. This ground is not prepared for you. It is not enough that I smile in the valleys? I have never made this soil for thy feet, this air for thy breathing, these rocks for thy neighbors. . . . Shouldst thou freeze or starve, or shudder thy life away, here is no shrine, nor altar, nor any access to my ear.

It is difficult, perhaps, to focus on the foreground in this vicinity, but if you do, you will see paper birch and another pale-barked tree that closely resembles it. This is the mature form of poplar, or quaking aspen. The life-styles of the two species are similar: both are pioneer trees that sprout on open ground, grow quickly, and have, by the standards of an oak or pine, a short life-span. Even their dangling flower clusters, or catkins, suggest they're related. In fact, the poplar belongs to the willow family, and the other to the quite separate birch family. This is an example of what biologists call homoplastic morphology, a resemblance between two organisms that isn't based on any close genetic connection.

Another term for this parallel development of unrelated look-alikes is convergent evolution. It's often the result of two different plants responding to the same environment by developing the same kind of trait. It can be the bane of paleobotanists, who often must speculate about the family relationships of long-extinct species with just a few bits of fragmentary evidence. For many years, for instance, an ancient group of plants known as pteridosperms were thought to be ferns; after all, the plants had highly compound fronds that were indistinguishable from the true item. Subsequently, the pteridosperms turned out to be something thoroughly unfernlike; they were an early form of the seed-bearing gymnosperms—in the same household, so to speak, with the conifers, the cycads, and the ginkgo. Once again, convergent evolution had tweaked the nose of a perfectly plausible preconception.

———————— ☙ ————————

In another 11.9 miles, you reach Exit 58 to Sherman. Do not take this turn; rather, use it to mark the beginning of the great Crystal, or Thousand Acre, Bog. Soon after, on the left, the slender silhouettes of black spruce signal the onset

Husband-and-wife trees. Two Norway spruces stand guard by an Aroostook County farmhouse.

of peatland conditions. At 2.4 miles above the exit, you'll have a nice view of the bogscape, with Katahdin standing behind it. If this isn't the chief glory of the North Country, what is? The vast complex, actually more than 1,500 acres, is a composite of various peatland types, including raised sections and less ombrotrophic fen areas. Note the road overpass a little farther up. You'll be going across this bridge shortly, if you choose to investigate the place more closely.

Bear off the interstate at Exit 59. Follow Route 159 East into Island Falls, to the Route 2 junction. Go south on 2. This is good agricultural land, part of the great Aroostook County potato belt described in Tour 52. Note the windbreak trees, set in rows at the edge of fields to lessen the wind's erosion of the topsoil. Notice the other trees planted on farm property. Besides the classic rows of sugar maples, you may spot a matched pair of husband-and-wife trees here and there. These were frequently set out by hopeful young couples when they first built on the land.

There will be another, more accessible example of a peatland later in this tour; but if you elect to examine part of Crystal Bog firsthand, and it is not mud season or a particularly wet spell otherwise, turn right onto the gravel road approximately 6.9 miles south of your turn onto Route 2. Head across the interstate overpass, being mindful of potholes and inquisitive farm dogs along the way. Soon, you'll come to the ghostly settlement of Crystal, and the line of the Bangor and Aroostook Railroad. Beyond that, you are definitely on the right side of the tracks, but do use caution. If the surface seems soggy, stop and check it before proceeding. After crossing the alder-fringed one-lane bridge, continue an additional 0.5 mile to the vicinity of the borrow pit. In my own experience, the road past this point is rockier and more deeply eroded; but you may find conditions considerably different if it has been graded recently.

The edge of a peatland section at Crystal Bog. The predominance of black spruce suggests the existence of highly acidic, nutrient-starved conditions.

The borrow pit is a great place to get out and observe an interesting part of the peatland, situated to the right of the road as you head in. The presence of Labrador tea, leatherleaf, small cranberry, and black spruce, not to mention certain species of sphagnum moss, suggest that it is part of the ombrotrophic bog proper. This fragile National Natural Landmark is one of the most ecologically significant examples of Maine's 700,000 acres of peatland. It may look like solid, easily explored ground, but bear in mind that the term *bog* derives from the Celtic *bocc*, meaning soft. Stay put on the sidelines unless you're an accomplished bog-trotter and know how to negotiate the undulating sphagnum mat without harming the precious plants or your own precious self.

———————— ∾ ————————

When you're ready to forge on, return to Route 2 the way you came in and turn right onto it, heading south. At the intersection of Route 158, bear right again onto 158 West. This takes you through Sherman Mills. Once again, be reminded that gas stations will be a scarce commodity in the miles to come. You may wish to fill up your tank here or in Patten, the next town up. Just beyond the interstate, you reach the Route 11 junction. Take careful note of your odometer reading here, and go right on 11 North. A few miles farther along, the road climbs Ash Hill. Katahdin is visible to the left, and you get one last splendid view of Crystal Bog on the right.

North of Patten the sights include MOOSE CROSSING signs, and lichens of the genus *Usnea* dangling from the tree branches. The composition of the forest varies from wetland species such as tamarack to pines and broad-leaved types characteristic of dry soils. Members of the northern hardwoods regime are present all the way to the Canadian border, but there is such a combination of taiga conifers and other species that it is fruitless to do more than admire the rapidly shifting scene and note its myriad details. (Speaking of tamarack, its species' largest known American specimen is located here in Aroostook County. It stands 92 feet tall and has a trunk circumference of 143 inches.)

As you approach the point 31.0 miles north of the Route 11–Route 158 junction back in Sherman Mills, keep a lookout on the right for an especially beautiful peatland that abuts the roadside. If you'd like, carefully pull onto the wide gravel shoulder and take a good look.

By now, you may have noticed that much of the northern Maine outback is not organized into named towns. Rather, it is subdivided into large squares designated either plantations (abbreviated "PLT" on maps) or numbered township sections. The second method conforms to the Bureau of Land Management cadastral surveying system, also used extensively across the Midwest. This locates a particular place by east-west township lines and north-south range lines. In this area, for instance, you may see a boundary marker for T8 R5 WELS; this stands for Township 8, Range 5, measured West of the Easterly Line of the State. It might sound confusing at first; but to the geologist or surveyor, this regularly spaced grid is preferable to the crazy-quilt mayhem of irregular boundaries found in other parts of New England.

On reaching the confluence of St. Croix Stream and the Aroostook River, you come to the settlement of Masardis, with its big sawmill and the piles of logs awaiting their rendezvous with the blade. After they've been cut down (or "harvested," as the timber-industry P.R. types euphemistically say), the best white

Sawtimber awaiting the blade at a mill in Masardis.

pine and hardwood trunks are used for sawtimber. The less desirable ones, crooked or otherwise flawed, are turned into fuel cordwood. The major operations, such as those at Millinocket, are the pulp mills, where spruce, fir, and some hardwood types are transformed into paper products. Trees selected for paper are usually 5 to 12 inches in caliper (trunk diameter measured at chest height). After felling, they are cut into 4- to 8-foot lengths called sticks, and then pulped. The ultimate result, of course, is this book and other indispensable accouterments of civilization. Unfortunately, the old conservative dictum about there being no free lunch is absolutely correct. The industry produces some particularly nasty chemical by-products, as well as huge clear-cut tracts that are often hidden from the casual eye by "beauty strips"—narrow lanes of surviving trees. What seems undisturbed forest to the motorist or boater is nothing less than a wasteland to the aviator overhead, who sees the full extent of the desolation beyond the cosmetic barriers. The price of paper is higher than we think.

At a point 6.9 miles north of Masardis, Route 163 meets Route 11 at Ashland. If you'd like a chance to stretch your legs and take a short, brisk climb to a beautiful vantage point, turn right onto Route 163 East. After 9.4 miles, pull off the road at the crest of the hill across from Haystack Mountain. The steep but brief trail to the nearby summit is marked with red blazes. It begins on the other side of the pavement. The mountain is composed of rhyolite—solidified lava that has the same chemical composition as granite, but a much finer-grained texture. As you proceed up the path, kick a piece or two of the rhyolite. It makes a distinctive *clink,* almost like broken crockery.

Most of Haystack's slope is covered with hardwoods festooned with foliose and fruticose lichens, notoriously slow-growing organisms that may take a decade to increase by half an inch. One special wildflower along the way is pipsissewa. This little white- or pink-flowered character is one of several wintergreen family representatives that grace the forest floors of New England. In colonial times, it was valued as a home remedy that supposedly cured heart disease and kidney problems, among others. At the summit, red spruce is very much in charge. The view in all directions is superb. To the east lies the broad, cultivated Aroostook plain; to the south, the V-shaped Squa Pan Lake; and away to the west, the forest, however thoroughly "harvested," still stretches far and wide.

When you've returned to your car, retrace your track to the junction of Routes 163 and 11 in Ashland. Turn right onto 11 North. In an additional 10.6 miles, you're greeted by Portage Lake, a lovely body of water that is sometimes still partly frozen in May. Beyond it, the taiga conifers put in a good showing. Though firs and spruces are indicative of colder conditions, they provide a more moderate habitat for wildlife in the winter. The northern hardwoods lose their foliage in the fall and expose associated animal life to the full blast of winter, but the evergreen taiga species create relatively windless microenvironments where temperature extremes are not so harsh.

Nevertheless, the spruce-fir community is almost a botanical desert. A pronounced lack of plant diversity can have catastrophic consequences. When a serious disease or pest appears on the scene, it may do sweeping damage. A vivid demonstration of this is the case of the dreaded spruce budworm, a native insect that in its larval stage feeds on taiga conifer buds and foliage. Its awesome destructive potential is illustrated by what it did between 1909 and 1918, when it killed 70 percent of all firs and spruces in Maine. Despite strenuous human efforts to combat the threat, outbreaks still occur.

One of the best parts of the route begins roughly 12 miles above Portage, where the roadway climbs into highlands formed from another extrusive igneous rock, basalt. About 4.3 miles farther on, a massive road-cut exposes the formation, on the heights just south of Winterville. If you can do so safely, pull over to the side and examine this half-billion-year-old lava flow. In places, the dark-hued basalt forms globular or egg-shaped lumps known as pillow lava. This indicates that the rock was extruded in a molten state underwater, in an eruption at the bottom of a long-vanished ocean or pond. What a perfect place this is to note how nature delights in gross contrasts. In the fullness of time, this sea floor or lake bed has been reborn as a hilltop. Once, marine organisms of the Ordovician period may have flitted past; today, terrestrial wildflowers like early saxifrage send their roots into its protective crevices. The calcium-containing basalt releases a good supply of plant nutrients as it breaks down.

The next town of any size is Eagle Lake, and 7.4 miles north of it is a scenic pullout to the right with an excellent view of the Fish River and its surrounding woodlands. In a few minutes, you've arrived "up there," on the bank of the fast-moving St. John River, at Fort Kent. This is the gateway to Canada. Turn right onto Route 1 at the intersection that marks its origin. The big river is visible on the left; see if you can make out the terraces it has formed. These shelflike surfaces above the modern banks make wonderful farmland and, sure enough, they are still largely under cultivation. In the vicinity of Frenchville, the channel islands are heavily wooded with bottomland trees, especially red maple. This broadly distributed species is a familiar sight, not only to the French speakers of this region, but also to their far-flung Cajun relatives in the heart of the Louisiana bayous. In that sultry southern locale, the tree blooms in February; here, it must wait until May. It is a memorable sight to see its branches put forth scarlet flowers as chunks of broken river ice go scraping by beneath them in the brown, foaming water.

The tour ends in Madawaska, New England's northernmost population center. On the way, you'll see good examples of floodplain vegetation, especially thickets of quick-spreading red osier dogwood, beaked willow, and speckled alder. Once you arrive in town, however, the emphasis shifts to the nature of the human environment. There are almost as many New Brunswick license plates to be seen as those from Maine. In the coffee shops and convenience stores, the sound of North American French blends with North American English. This free flow of cultures over an unarmed border enriches the place, and makes a good model for the rest of the planet.

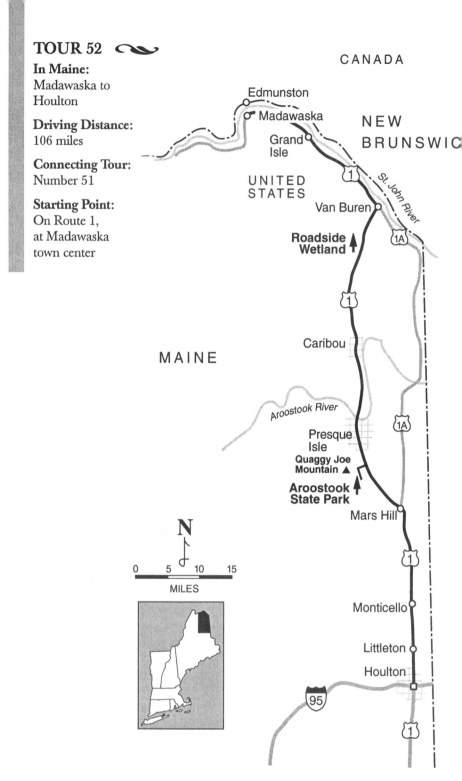

TOUR 52 ∾

In Maine:
Madawaska to
Houlton

Driving Distance:
106 miles

Connecting Tour:
Number 51

Starting Point:
On Route 1,
at Madawaska
town center

CANADA

Edmunston

Madawaska

Grand
Isle

NEW
BRUNSWIC

UNITED
STATES

Van Buren

1

St. John River

1A

**Roadside
Wetland**

1

MAINE

Caribou

1A

Aroostook River

Presque
Isle

**Quaggy Joe
Mountain ▲**

**Aroostook
State Park**

Mars Hill

N

0   5   10   15
MILES

1

Monticello

Littleton

Houlton

95

1

Tour 52
A World Apart: Aroostook

*I*f you had to pick a single part of New England that is most separate, most surprising, and most utterly unlike the rest of the region, it would have to be Maine's eastern Aroostook County, the subject of this tour. Anyone who equates all of the North Country with slumbering lakes and trackless forests is in for a jolt. At this farthest reach of Yankeedom, fir-clad hillsides give way to broad, flat land still heavily farmed for a special crop by a people who pronounce their *r*'s and build barns in the ground.

From downtown Madawaska, proceed eastward, more or less, on Route 1 South. In this area, in town and out, see if you can spot the tall, narrow trees that are often planted as windbreaks. Horticulturists describe the columnar habit of such plants as fastigiate. These are examples of an old cultivar, Lombardy poplar. As its name indicates, it was originally developed in northern Italy. The plant is closely related to our native poplar, or quaking aspen, which in this area is also very much in evidence. The Lombardy poplar has been used extensively in the United States since the late 1700s, both as an ornamental and as a fast-growing windbreak tree for farms. Unfortunately, it is especially prone to a fatal canker disease, and it's unusual to see a group of Lombardy poplars wholly unaffected by the malady. And a row of fastigiate trees with one or two dead snags mixed among the living is every bit as appealing as a missing-toothed sneer.

As noted at the conclusion of the previous tour, the town of Madawaska is a good place to see the beneficial effects of the world's longest undefended national border. To an outsider, the mighty St. John River seems to mark a political boundary about as profound as the one between North and South Dakota. Had you been visiting here in 1839, however, the situation would have been vastly different; a bloody showdown between Canada and the United States seemed just about inevitable.

The problem was that Maine's northern boundary was still not agreed upon when statehood was achieved in 1820. At one point, both sides agreed to let the king of the Netherlands mediate. And so he did, by formulating a compromise that was resoundingly rejected by the fledgling government of the Pine Tree State. In the years that followed, the governors of Maine and New Brunswick became pen pals, in a hellish sort of way. The chief executives of most states have to restrict their thundering oratory to internal subject matter, but the governor of Maine for once had the added perk of sending fuming letters full of dire threats to a foreign power. His New Brunswick counterpart, aware of his responsibility to defend the honor of the British Empire, responded admirably, with seething counter-invective. Unfortunately, backwoodsmen from both sides more accustomed to action than to ritual display became involved, and border incidents began to occur. Little blood was spilled, but tempers certainly flared.

When the crisis peaked, in 1839, forces of dubious military value concentrated in this area. The U.S. government sent one of its most famous generals, Winfield Scott, to assess the situation in Maine. As usual, Scott's luck held. He convinced both sides to simmer down and defuse what was being called, a trifle grandiosely, the Aroostook War. Then, in 1842, the Webster-Ashburton Treaty was signed, and at long last the boundary was fixed.

Route 1 follows the course of the St. John downstream. The river is large enough to have a substantial floodplain, and such wetland plants as tamarack, red osier dogwood, and speckled alder take advantage of the dampest locations. The alder is one of the species that have roots beneficially infected by nitrogen-fixing bacteria called actinomycetes. Northeastern Indians used the plant as a dye source; in addition, its stems provided them with arrow shafts. Thoreau noted that the plant's wood was used for powder cask hoops.

Alder thickets are especially prevalent between Grand Isle and Van Buren, some 25 miles from the start. If you do a survey of the different botanical and horticultural guidebooks that list the wild alders, you'll quickly learn that their genus, *Alnus,* is taxonomically muddled. Part of the problem is that the plants hybridize and present intermediate forms. Their common names are especially maddening; if you try to distinguish smooth alder from speckled alder, you'll find these titles sometimes refer to one species, sometimes to two, interchangeably. And in this case, the scientific names are also confused. My own snooping suggests that most examples of this genus in New England's low, wet habitats are either *Alnus rugosa* or hybrids linked to it. Throughout the book, I refer to these as speckled alder, though other writers have used different tags. Feel free to call them smooth alder, hazel alder, or, if you prefer, the Pan American Friendship alder. The plant itself has no such identity crisis, and in early May every year, it puts forth dangling flower tassels, which bob and sway in the cool wind.

———————————— ∾ ————————————

When you reach Van Buren, bear right with Route 1 as it angles southward and away from the river. Just south of town, American elm is a common sight on farm lots. It would be nice to think that the horrible Dutch elm disease that ravages trees throughout much of the rest of the country doesn't come this far north, but, sadly, it does. In most places, it is transmitted by an introduced elm bark beetle, but here and elsewhere close to the Canadian border, its main vector, or transmitting organism, is a native bark beetle.

On the right, about 7.3 miles south of Van Buren center, the presence of wetland terrain in the greater taiga regime is indicated by groups of arbor vitae and tamarack. This is a broad, low area that includes Orchard Bog. The tree community within sight of the road suggests that the nearest portion, at least, is a fen, a type of peatland in which there is some surface water flow. Topographic maps show a basic drainage pattern in this area, so it's plausible. Also in this vicinity, you'll find an excellent example of a marsh, which is characterized by open water and the absence of woody plants.

In the miles beyond, the Aroostook landscape is wide, level, and thoroughly tilled. Its rich soil, unlike much of New England's, is underlain with sweetening limestone that neutralizes harmful acidity. One almost expects to see a large sign reading WELCOME TO ILLINOIS, LAND OF LINCOLN. What sets this section of Maine

An Aroostook County pit barn. Note the raised, sodded flanks around the structure.

apart from its Midwestern analog is its main crop—potatoes, not corn and soybeans. Later in this tour, we'll take a closer look at the noble spud, its history, and the industry that has been built up around it.

One of the most eye-catching features of Aroostook farms is the pit barn style, seen over and over again. Barns here are built partly into the ground, or are flanked with thick sod slopes, for the same reason that root cellars were dug by Yankee farmers in generations past. This is definitely a fertile region, but it gets very cold in winter, with temperatures dipping to minus 30 or even 40 degrees Fahrenheit. Once harvested, the potatoes and other crops need to be protected from such bitter conditions.

Another interesting aspect of the regional building practices has to do with something that isn't here. Can you figure out what's missing? Look along the boundaries of the fields. There are no stone fences. That particular property-marking custom never really caught on here. And as you pass through the towns, you'll notice the general lack of prim Congregationalist meetinghouses, where all but the simplest ornament was considered distracting and downright Babylonian. Due to the influx of French Canadian people and their culture, Roman Catholic churches, with their distinctly different approach to ornament, often dominate the local skyline. In contrast, the section of New Brunswick just to the east was largely settled in the Revolutionary era by English-speaking loyalists from Connecticut, New York, and New Jersey. Even Benedict Arnold lived there awhile, after his defection to the British.

Between Caribou and Presque Isle the road parallels, then crosses, the stream that gives this county its name. In the Abnaki tongue, *aroostook* means "shining river." When you get to the heart of Presque Isle, note your mileage. After an additional 4.7 miles, take a right onto the paved access road to Aroostook State Park and follow the signs to the park's entrance and parking lot. Below you, just to the east, is Echo Lake; rearing up behind you, to the west, is the twin-peaked Quaggy Joe Mountain. If you're in the mood for a nice hour's hike, start at the Quaquajo Nature Trail sign at the parking lot, then follow the blue-blazed trail upward. If perchance you're doing this in mud season, wear ooze shoes. The path is corduroyed in places, but it's still likely to be a quagmire.

On the way upslope, the woods have a northern hardwoods overstory, with plenty of sugar maple and yellow birch, and a significant red spruce component for contrast. The role of sugar maple in syrup and candy production, described in Tour 45, is well known; the tree has also supplied wood for light pieces of furniture. Its diminutive relative, the striped maple, grows hereabouts, too. It and the shadbush, another Quaggy Joe resident, are two of the small trees adapted to living their entire life in the dense shade of the broad-leaved forest. Beneath them is an assortment of pteridophytes and flowering plants, ranging from evergreen Christmas fern to shinleaf, pipsissewa, trillium species, and bunchberry.

If you climbed nearby Haystack Mountain in Tour 51, you'll probably recognize that Quaggy Joe Mountain is made of the same igneous rock, once part of lava flows, named rhyolite. On reaching the saddle between the two peaks, you can proceed to either the northern or southern summit. In the saddle, you

506

An impressive foliose lichen on a tree at Aroostook State Park. This lichen species, Lobaria pulmonaria, *is also native to Europe, where in accordance with the Doctrine of Signatures, it was once used to treat pulmonary complaints, because of its superficial resemblance to human lung tissue.*

enter a miniature wonderland where reindeer moss, really a lichen, grows in large colonies along the stony ground. This important composite organism ranges northward to the Canadian tundra, where it serves as a main food supply for migrating caribou. Explorer-botanist Pehr Kalm observed that French Canadian traders used it as emergency rations: if their provisions ran out, they would boil the lichen and drink its broth.

Returning to the plant kingdom, you can also find some of its more primitive extant members here, the club mosses. They, too, are not really mosses, but vascular plants that occupy the free-sporing pteridophyte group along with the horsetails and ferns. Two alternative common names for the club mosses are ground pine and ground cedar. Indeed, they do resemble tiny conifers to some extent, but the similarity is only skin deep. They have been used herbally in dried and powdered form to treat urinary problems; additionally, pharmacists once used their dustlike spores for baby powder and a nonstick coating for pills.

———————— ∾ ————————

When you've soaked up the view from the high point of your choice, return to your car, drive back to Route 1, and resume your southward trek. The final leg, to Mars Hill and the interstate link at Houlton, contains pleasant farmland vistas, and it's a good locale to consider the most lucrative plant in sight. The potato is one species of the large genus *Solanum,* which also contains the eggplant and the legendary deadly nightshade. Other relatives in nearby genera include the tomato, the red pepper, the petunia, tobacco, and a pair of poisonous plants, jimsonweed and the sacred datura of the desert Southwest. The potato's genetic ties with the tomato have resulted in grafted hybrids with predictably awful names: topato, pomato, and so on.

The business end of the potato is its tubers. These underground starch-storage organs are modified rhizomes; they are stem tissue, not roots. If allowed to, potato plants will flower and then develop small fruit resembling tomatoes. Not surprisingly, given the family affiliations, the fruit, actually a berry, is toxic. This fact leads to a delightful account about Sir Walter Raleigh's interest in the species. When he introduced the potato to England, so the story goes, he decided to plant it on his own estate. It duly thrived. Eventually, the berries formed, and Sir Walter, ever the adventurer, picked and ate them. When he recovered, he angrily ordered the offending plants to be yanked up, and so his gardener discovered the much more digestible tubers.

The potato, a South American native, is one of several New World wonder plants that have transformed world agriculture. Among its attributes is its ability to provide twice as many calories per acre as wheat. The Spanish carried the potato from Peru to Europe in the 1500s, and it soon became a staple. In the early eighteenth century, Scots-Irish settlers established their own potato-centered farming practices in New England, and production reached the exporting stage by about 1745. The plant has not been without its problems. Massachusetts was hit with a severe potato disease before the Revolution, and there was the so-called late blight the following century. Still, the potato has ultimately had more staying power than any other crop except corn. The reason such northern places as the Aroostook region (and, dare I say it, Idaho) are ideal for potato raising is that the species needs cool summer temperatures and abundant rainfall. The long daylight hours of summertime are also important to growers, who need to have the plants flower fully for breeding purposes.

The plants are usually raised from sections of tubers called seed pieces. Harvesting is done, as you might imagine, in the fall. In recent years, potato consumption in the United States has declined; consequently, local farmers have experimented, not always successfully, with sugar beets as an alternative crop. If nothing else, it serves as a reminder that the area is not inexorably linked to a single type of plant. After all, a good part of Aroostook's reputation in the nineteenth century was based on growing wheat. For the time being, Maine remains a potato center, even though 90 percent of world production comes from the nations of northeastern Europe.

The tour ends at the crossroads town of Houlton, from which you can catch the interstate and other main roads to points south and west. Before you reach the terminus, though, take one last look at the bountiful soil stretching away from you on all sides. It has been estimated that a mere gram of this incredible life-sustaining medium contains 30,000 protozoa, 50,000 algae, and 400,000 fungi, not to mention an almost unbelievable 2.5 billion bacteria. The farmer's field, in common with the trackless forest, harbors hidden worlds.

508

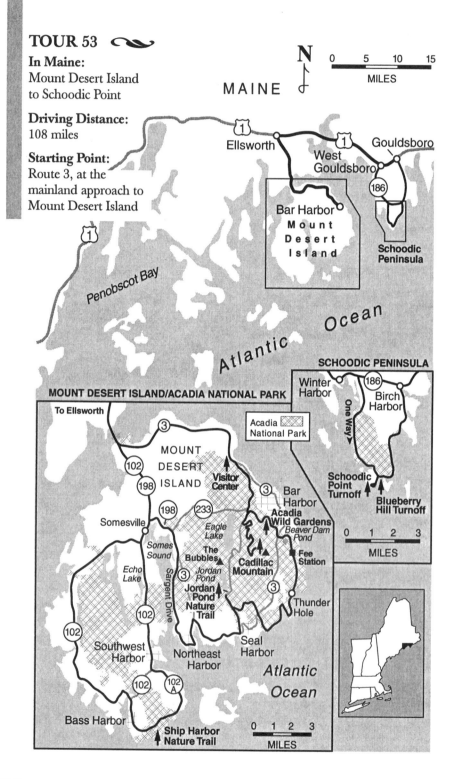

TOUR 53 ॐ

In Maine:
Mount Desert Island
to Schoodic Point

Driving Distance:
108 miles

Starting Point:
Route 3, at the
mainland approach to
Mount Desert Island

N

0   5   10   15
MILES

MAINE

MAINE

Ellsworth

West
Gouldsboro

Gouldsboro

186

Bar Harbor
**M o u n t
D e s e r t
I s l a n d**

**Schoodic
Peninsula**

1

Penobscot Bay

Atlantic Ocean

SCHOODIC PENINSULA

MOUNT DESERT ISLAND/ACADIA NATIONAL PARK

Winter
Harbor

186

Birch
Harbor

One Way

To Ellsworth

3

Acadia
National Park

**Schoodic
Point
Turnoff**

**Blueberry
Hill Turnoff**

**MOUNT
DESERT
ISLAND**

102

198

**Visitor
Center**

3

**Bar
Harbor**

0   1   2   3
MILES

198

233

*Eagle
Lake*

**Acadia
Wild Gardens**

*Beaver Dam
Pond*

Somesville

*Somes
Sound*

**The
Bubbles**

*Jordan
Pond*

**Cadillac
Mountain**

**Fee
Station**

*Echo
Lake*

3

**Jordan
Pond
Nature
Trail**

3

Sargent Drive

102

**Thunder
Hole**

102

**Southwest
Harbor**

**Northeast
Harbor**

*Seal
Harbor*

*Atlantic
Ocean*

102

102
A

Bass Harbor

**Ship Harbor
Nature Trail**

0   1   2   3
MILES

Tour 53
Acadia National Park

HIGHLIGHTS

Making the Scene at MDI ∾ The Essential Visitor Center ∾ Park Loop
Road ∾ A Local Extinction Event ∾ The Wild Gardens of Acadia ∾ A Hairy
and Tree-Chewing Fish, Hypnotic Snakes, and Malarial Watermelons ∾ Jordan
Pond Nature Trail and a View of the Bubbles ∾ Somes Sound, the Red Pine,
and Fjord Scenery ∾ Ship Harbor Nature Trail ∾ Avian Workaholics and the
Elusive White Spruce ∾ Sea Lavender and Silverweed ∾ Ellsworth's Hyde to
Acadia's Jekyll ∾ Schoodic's Shakespearean Setting and Sulking Seagulls ∾ The
Jack Pine ∾ Wildflowers of the Rocky Shore ∾ Sea Rocket and Friends ∾
A New and Ancient Kingdom ∾ A Seaweed Sanctuary

*S*ixteen years before the people of the *Mayflower* established their colony
at Plymouth, the far-ranging French explorer Samuel de Champlain and
his crew found themselves taken, quite literally, by the Down East coast
of Maine. In short, they'd run aground. Fortunately, the incident did not put
an end to Champlain's travels, nor did it diminish his powers of observation.
He accurately described the large and rugged island that lay before him, and
named it l'Isle des Monts-déserts. Hence, Mount Desert Island, our modern name
for one of the most beautiful places on the North American seacoast.

Today, Acadia National Park occupies a large portion of Mount Desert Island,
as well as part of the smaller Isle au Haute to the southwest, and Schoodic Point,
to the east. This tour will begin with a circuit of MDI, as the locals call it, and
will then return to the mainland for a sampling of Schoodic's charms. Before
you get under way, keep in mind that the driving tour and selected stops on
this itinerary are primarily of a botanical nature. It isn't a complete guide to
Acadia's scenic wonders. If you wish to incorporate a visit to Thunder Hole,
for example, plan ahead and make sure you get the additional information you
need at our first stop, the park's visitor center.

Take Route 3 from the starting point across the bridge spanning Mount
Desert Narrows. At the far end, follow 3 East by bearing left toward Bar
Harbor. This is largely a residential and resort area, so lawn trees intergrade
with wild arbor vitae, poplar, tamarack, black cherry, black locust, and
American elm.

Between Hulls Cove and Bar Harbor, 8.2 miles from the start, bear right at the national park entrance. Follow the signs to the visitor center and stop in for an overview of the park's attractions. The center also offers brochures, nature guides, and other informative literature. For this tour, I suggest you get the following: the official park map, the free Checklist of Common Flowering Plants (which includes conifers), and the "self-guiding" fliers for the Jordan Pond Nature Trail and the Ship Harbor Nature Trail. You may also find them available at the sites they describe, but it's best to play it safe and buy them here. For some reason, both pamphlets insist on referring to lichens as plants; other than that, they're helpful.

When you depart the visitor center complex, go straight on beautiful Park Loop Road. The tree community along the side contains everything from red oak to mountain ash, with paper birch, striped maple, bigtooth poplar, and black cherry as well. About 0.8 mile past the visitor center, you'll come to one of several scenic overlooks along the way. Here you get a sweeping view of Frenchman Bay, where warships manned by Champlain's compatriots used to seclude themselves to escape detection by their English adversaries. At a point 2.3 miles farther down the hill, take the Cadillac Mountain summit road. You'll soon see dramatic road-cuts into a Devonian period granite that has a distinctly pinkish cast. One characteristic feature of granite is that it forms joints, or fractures, where no significant displacement occurs. One of the trees that has gained a foothold on this stony surface is the conifer arbor vitae. At first, its presence is surprising, since it normally prefers low, damp soil. Try to figure out how this moisture-lover manages to persist on an exposed, rocky slope. Is there any reliable source of moisture nearby? What role do the atmosphere and local climate play in its survival?

At the mountain's terminus parking lot, you are just a quick walk from the summit. You'll notice balsam fir around you, in the stunted, windblasted form known as krummholz. The view of the island and its watery environs is magnificent. As you head back, look for other characteristic tree species on the lower slope: paper birch, pitch pine, and Canadian hemlock. When you return to the main roadway, resume your prior heading down the island. Along this stretch, you may marvel at the prevalence of the poplar, or quaking aspen, tree. This species is a pioneer, which does not persist in large numbers once a mature forest community has become well established. It rapidly—but temporarily—takes over environmental vacuums where competition is at a minimum. Take a moment to think on these things; what could have produced such a large vacuum habitat here? The evidence before you suggests that there was a local "extinction event" on Mount Desert Island; but what caused it? There has been no record of wholesale logging, volcanic eruptions, or killer asteroids in the recent past.

The answer is conflagration. In October 1947, the eastern side of the island suffered a devastating forest fire, with almost 9,000 acres burned. It must have been a shocking and heartbreaking sight for residents and nature lovers, but the

The predominance of poplars (quaking aspens) and birches along the Acadia roadway demonstrates their skill in filling environmental vacuums.

catastrophe was by no means unique in the annals of New England. In the year 1761, for example, a fire began in the vicinity of Lebanon, New Hampshire, on the western edge of that state. In the space of one month, it traveled across chains of hills and major rivers all the way to Maine's Atlantic shore. And in the centuries before that, native peoples routinely burned large tracts of woodland to open up land for agriculture and limit the growth of thickets and vermin. Indeed, fire is an excellent way to recycle nutrients. As the recent Yellowstone debate has pointed out, it may be in an ecosystem's long-term interest to suffer the flames every so often.

At 2.9 miles past the Cadillac Mountain road junction, you come upon the Sieur de Monts Nature Area. By all means turn right and visit the Wild Gardens of Acadia. If park attractions were rated the way restaurants and movies are, this small, superb facility would get five stars—or, better yet, five pine cones. As is the case at the more expansive Garden in the Woods in Framingham, Massachusetts (see Tour 19), an attentive visit here is worth a university-level course in botany.

The purpose of the Wild Gardens is to acquaint park visitors with the diversity of the island's plants and habitats, and it succeeds in doing so. On a fine summer day, the place is swarming with T-shirted, deck-shoed outlanders videotaping the bog flowers, the swaying pine branches, the airplane contrails, and each other. The plant kingdom rarely gets Standing Room Only audiences, so it's nice to see the sweet press of success here. Still, the most serious student will be delighted with the Wild Gardens' layout. Each Acadia plant environment, from roadside to beach to wetland, has its say in well-laid-out sections. My own obsession for copious, accurate plant tags is fully satisfied. It's the best opportunity you'll have to get acquainted with the species you'll see during the rest of this trip. The Wild Gardens of Acadia Committee members who volunteer their efforts deserve a warm round of applause for this masterpiece of design and public nature education.

———————————— ∽ ————————————

About 0.7 mile farther down the road, you pass Beaver Dam Pond. Its namesake, a common mammal here at Acadia, was the subject of one of Pehr Kalm's most interesting observations about French Canadian culture. Kalm, himself an Enlightenment-style Lutheran pastor, discovered that his French-speaking hosts, devout adherents to the Catholic faith, were allowed to eat beaver meat even on Fridays. The reason for this, he learned, was that His Holiness the Pope had declared the beaver a fish, based on the animal's aquatic tendencies. This gaffe might seem out of keeping with the Roman Church's numerous contributions to scientific learning, but in fact, the Pope was just following the lead of eminent zoologists of his day. In all fairness, Kalm, our Protestant source, believed that North American snakes could hypnotize squirrels and make them run down tree trunks into their open mouths. He also theorized that malaria was caused by eating watermelons.

———————————— ∽ ————————————

At 0.7 mile past the pond, you'll have a good view of the vegetated cliffs and the famous pinkish Cadillac Mountain granite. The Precipice pullout is a little farther down; this is a nesting area for peregrine falcons. Sometimes, as in 1991, it's closed to human use, to protect the birds. The next stop down the road is not for your benefit, but for Uncle Sam's. This is the "fee station," where you must settle your account with a smiling and courteous ranger. When you've done your part to defray the cost of park maintenance, think what it must be like to be relentlessly polite to thousands of people each day. The mind boggles.

Several miles south, in the Hunters Head area, spruce trees are prevalent. On this tour's next two stops, you'll have the chance to examine the subtle differences in the island's spruce community. As you approach the Seal Harbor area, make sure you continue straight on the Park Loop Road, and 0.7 mile beyond Jordan Pond House, turn left into the parking lot for the Jordan Pond Nature Trail.

Red spruce is a familiar sight around the parking area. At inland locations such as this, it is the dominant representative of its genus. The trail features fine lichen specimens, paper birches, and one of the park's most famous views, up the pond to the twin hills named the Bubbles. This is a classic glacially formed landscape. The prevailing north-south trend of this and other Mount Desert Island lakes reflects the ice sheet's direction of movement.

On leaving, go back down the road past Jordan Pond House, then bear right onto the linking road to Seal Harbor. Approximately 1.3 miles later, you'll be at the cove that goes by the same name; go right at the T-intersection. You are now on Route 3. The beach here is a worthy stop for marine-algae enthusiasts, but we'll reserve this tour's discussion of seaweeds until the final stop, in the Schoodic area. At 3.0 miles west of Seal Harbor, turn left onto Route 198 South; take it through the town of Northeast Harbor. Follow the signs for Sargent Drive, and at 0.9 mile past the last turn, bear right at the Y-intersection. You are now on Sargent Drive heading northward. Before long, you see why this unnumbered road was chosen: it parallels Somes Sound, often described as the only real fjord on the east coast of the continent. This is true only if one disqualifies the Hudson River valley in the vicinity of Bear Mountain, some miles upstream from the Big Apple.

Arbor vitae is predominant in the first couple of miles, before the view opens up; then paper birch and red pine take over on the stony flanks of road and water. The sound is another of the troughs, U-shaped in cross section, that owe their configuration to the grinding force of the glaciers. What sets this locale apart from its companions is that it has been flooded by the sea, and the island is cut almost in twain. It is good to see red pine growing wild here, and left to its own devices. Generally, it's encountered in the monocultural setting of the Civilian

White spruce thriving on the harsh winds and salt spray of the Mount Desert Island coastline. Note the rampart of jointed granite and the small cobble beach.

Conservation Corps plantations that dot the length and breadth of New England (see, for instance, Tour 38). When it is truly wild, however, it is an excellent indicator of soil that is too poor even for the white pine to endure.

Sargent Drive meets Route 198 in the upper reaches of the sound. Turn left there on 198 North. Near the following junction with Route 233, note the concentration of balsam poplar trees. This species, a familiar sight in many places near the Canadian border, is a cold-climate equivalent of the cottonwood. Stay on Route 198 until it joins with Route 102. Turn left onto 102 South. Past Somesville, the road runs along the eastern shore of Echo Lake, which, sure enough, is a north-south affair as well. Proceed resolutely through the Southwest Harbor landscape of boutiques and buoy-draped restaurants, and bear left onto Route 102A when it splits away from 102. In another 2.8 miles, you've reentered the national park at the lovely Seawall area. In an additional 1.9 miles, you will see the left-hand turnoff for Ship Harbor Nature Trail. Park here.

This is a very special site, featuring both a coastal forest and a maritime herbaceous-plant community. At the beginning of the trail, red spruce is the number-one conifer, but as you approach the water, its next-of-kin, the extremely cold-hardy white spruce, tends to replace it. The existence of white spruce here and in other locations along the Down East shore is particularly fascinating, for its main population is restricted to the Canadian taiga, far above the red spruce's limit of tolerance. Time was, though, when white spruce covered much of the New England interior. That was about ten thousand years ago, when the most recent continental ice sheet was staging its retreat northward.

Should your field guides leave you in any doubt, rest assured that white and red spruces are not the easiest trees to tell apart. As a basic rule, the red spruce needles point forward more and are dark green, while the white spruce needles are spread a bit more. They also tend to be glaucous—that is, whitish or pale blue. White spruce was a particular favorite of Native American canoe makers, who used the tree's roots to sew sections of their craft together. In northern European countries, spruces in general were important components in the ship-building trade. To quote John Josselyn: "Spruce is a goodly Tree, of which they make Masts for Ships, and Sail Yards: It is generally conceived by those that have skill in Building of Ships, that here is absolutely the best Trees in the World, many of them being three Fathom about, and of great length." In contrast, another conifer in the area, the arbor vitae, is a snap to identify. Its foliage is deployed in flattened sprays, and its habit is chubbier than either of the spruces. It is a member of the cypress family; its spruce companions belong to the pine family.

There are all sorts of things to see along the path, from interrupted fern, Schreber's aster and meadowsweet to lowbush blueberry and withe rod viburnum. Less obvious to most hikers is a bryophyte of the genus *Bazzania*. This is a ground-hugging, clump-forming liverwort, which lacks true stems and roots. Closer to the water, you may find it easy to strike up a meaningful relationship with the chickadees, who are afraid of everything except human beings. It certainly makes for a novel survival strategy. Still, nothing is more heart-warming than these little black-and-white workaholics darting about in the middle

The trunks of conifers along the Ship Harbor Nature Trail sport fantastic shapes.

of a soul-chilling ice storm. Their preferred habitat is a mixed stand of hard-woods and softwoods, where they utter their high-pitched profundities and raise their young.

At water's edge, only the toughest plants thrive. Mountain cranberry does well in this sea-level equivalent of mountaintop climatic harshness; sweet fern, a hardy nitrogen-fixing colonizer of poor soils, forms pygmy thickets close by. In summer, if you look in the right place, you'll also find herbaceous coastal species in abundance, among them orache, sea blite, seaside plantain, sea lavender, and silverweed. Orache and sea blite belong to the widely distributed goosefoot family, a group that in other cases has adapted admirably to alkaline desert environments. Separately, the seaside plantain is not one of its genus' weedy, imported species, but a certified native with atypical, narrow leaves. If you catch it when it is in blossom, examine the inflorescences. They are spikes composed of many tiny, four-petaled flowers. Even more striking is the sea lavender. It, too, has inflorescences that rise high above the foliage, but they are much more elaborate, and adorned with pink or rose-purple flowers. The silverweed, also found on the sandy shores of Lake Michigan, is a rose-family plant esteemed by herbalists for its astringent and tonic properties.

517

If in your waterfront explorations you spot a much-branched, spongelike organism growing in mats on bare rock surfaces, you've found a colony of reindeer moss lichen, or one of its closely allied species. Sadly, it is sometimes badly trampled here by thoughtless visitors. More extensive, healthier lichen communities can be found at our next destination, Schoodic Peninsula.

Depart Ship Harbor via Route 102A West. Continue on it through Bass Harbor. When it rejoins 102, take the latter past an excellent example of a tidal marsh. This route has two north-running branches; either one will do. Follow the signs for Route 102 all the way north to the Route 3 junction, near Mount Desert Island's connection with the mainland. Recross the Narrows bridge and head toward Ellsworth. This stretch, one of New England's worst commercial blightscapes, presents a nightmarish contrast to the untrammeled beauty of the national park. Both Ellsworth and Conway, New Hampshire (on the edge of White Mountain National Forest), demonstrate the Jekyll-and-Hyde nature of America's protected areas. The price of conserving our natural wonders seems to be that their surroundings become the paved-over trash heaps of tourism. The older, downtown section of Ellsworth, however, is a lovely place indeed. May the shop owners and merchants who've stuck to this nobler setting fare well.

Luckily, our track avoids the worst of the road rash. When Route 3 meets Route 1 North, take a right onto the latter and proceed east 17.9 miles to West Gouldsboro and the junction of Route 186. There, bear right onto 186 South. You should begin spotting signs for Schoodic Peninsula. This area is a pleasant contrast to the more popular Mount Desert Island. Even in midsummer, it evokes an isolated, uncrowded feeling. One almost expects to see Prospero, Miranda, and Caliban come stumbling out of the mist-shrouded woods. It would make the perfect setting for a Down East version of *The Tempest*.

After 7.5 miles on Route 186, turn right again at the big "Acadia National Park–Schoodic Point Section" sign. Carpets of reindeer moss and other lichens occupy every inch of bare ledge they possibly can. Arbor vitae, highly prized for its virtually rot-proof wood, is the predominant conifer. You arrive at the Frazer Point entrance to Acadia soon, and beyond, you'll notice outcrops of the same basic type of pinkish hornblende granite seen in many places on MDI. This Devonian period igneous rock apparently owes its origin to the collision of ancestral North America with Avalonia, an exotic terrane that was either its own separate minicontinent or the leading edge of something bigger. The collision resulted in one of New England's long-vanished mountain ranges, known, aptly, as the Acadian.

Just 3.9 miles past the entrance, you reach Schoodic Point; pull into the parking lot. Seagulls roughly the size of Cessnas slouch about the pavement, sullenly obstructing parking spaces and waiting for handouts. This Hitchcockian setting is not a good place to walk a lapdog; and if you have the bad sense to pull a peanut butter and jelly sandwich out of the cooler, you might just lose it, and possibly other things, in a flutter of wings.

A white spruce close by water's edge, at Schoodic Point. It has been contorted into a low, shrubby habit.

Before you head down toward the surf, pause to look at the woody plants by the parking lot. There, along with clumps of bayberry, you'll find two less common species. The first, black crowberry, is a prostrate shrub with tiny leaves and fruit in the form of black, berrylike drupes. It is found in alpine conditions and on the colder coastlines on both sides of the North Atlantic. This plant is so hardy that it grows within 10 degrees of the North Pole. Much more obvious in the parking lot area are the windblasted forms of white spruce, which could easily be mistaken for pines. To their lee, a trifle inland, real pines can be found, often in a shrublike form. This is jack pine, the most cold-hardy of its genus. Like the red pine seen along Somes Sound, this species has needles arranged in bundles of two. What sets it apart is the shortness of its leaves—they're only about an inch long—and its cones, which are narrower and often somewhat curved. While you have the chance, take a good look at it. The jack pine is a more common sight in Canada and on the shores of the western Great Lakes than in New England.

The rocky shore hardly seems the kind of place for a good wildflower display, but in middle to late summer, surprising things happen. First and foremost, consider the rock itself. The granite is dramatically transected with dark bands of diabase, a close relative of basalt. It was intruded in a molten state into fractures in the granite, and thereby formed the kind of stripy features known as dikes. A dike is always younger than its host rock, which makes perfect sense when you think about it. In this location, it's even possible to find older dikes that have been crossed by later ones.

To a person interested in both botany and geology, no sight could be more transcendent than that of seaside goldenrod growing and blooming here in the cracks of the stony surface. This stout, salt-tolerant herbaceous plant is joined by two other members of the composite family, the New York aster and the flat-topped white aster. I have also found the native angelica *(Angelica atropurpurea)* surviving in a sheltered spot in the rocks. Often indicative of swampy ground but sometimes seen at other seacoast sites such as this, its species has been used herbally much the same way its more famous European equivalent has, as a cure-all for everything from fevers to poisoning.

The final stop lies just a moment or two farther down the road, at the next scenic viewpoint, the Blueberry Hill turnoff. A different community of maritime plants has taken root here. Among the rugosa rose and lowbush blueberry shrubs grow all sorts of interesting herbaceous species, such as beach pea, the legume most likely to be seen at the shore. It's been suggested that coastal Indians used beach pea as a food supplement. Along with it are two members of the mustard family: sea rocket, a native; and wild radish, an alien. Both produce cross-shaped flowers of four petals apiece. Kalm reported that his French Canadian friends used sea rocket as a bread substitute. They would pound it and then mix it with flour. One of the interesting things about the mustard family as a whole is that its members are masters of chemical defense. They produce glycosides and associated enzymes pungent to the human nose and tastebuds, and often overtly repulsive to insects.

The last word on this tour has been reserved not for plants but for the marine algae that occupy the intertidal zone, just a few steps away. If you visit close to low tide, a separate way of life stretches out in front of you. Especially numerous are the brown algae of the wrack and rockweed genera. (For more on them, see Tour 41.) Traditionally, these chlorophyll-bearing organisms were for years grouped with the plants. In 1969, however, a new taxonomic system was proposed by R. H. Whittaker, and it has been widely accepted by the scientific community. In this modern scheme of things, there are five kingdoms instead of two; the algae belong to the one called the Protista. As you stand here, at the ever rising and falling edge of the ocean, it's not difficult to grant the seaweeds their own domain in the vastness of life.

(On leaving Schoodic Peninsula, you may wish to take the east branch of Route 186 back to Route 1. By doing so, you'll see Prospect Harbor and other traditional Down East towns that have not succumbed to strip-mall mania.)

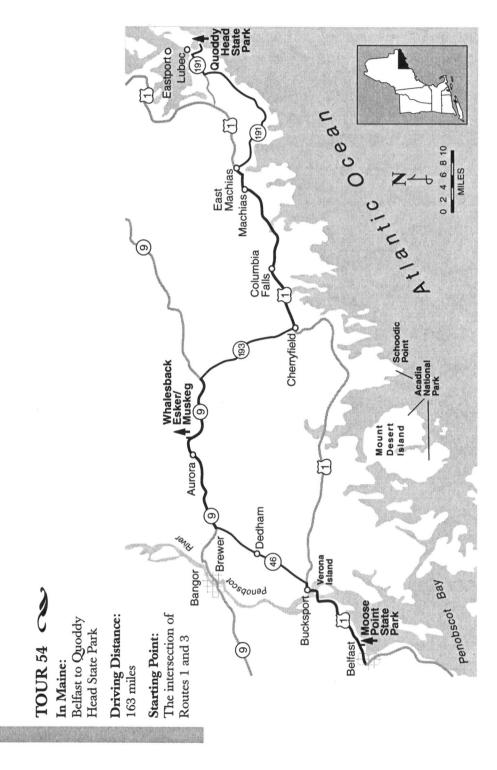

TOUR 54

In Maine:
Belfast to Quoddy
Head State Park

Driving Distance:
163 miles

Starting Point:
The intersection of
Routes 1 and 3

Tour 54
Down East to Quoddy Head

A Recommended Fall Foliage Tour

HIGHLIGHTS

The Pretty Port of Belfast ∾ Moose Point State Park ∾ The Apple Tree's
Other Uses ∾ Wild Lettuce ∾ Monster Plant Families ∾ Firing the Bracken ∾
A Cure for Visibility ∾ Racing Loads of Logs ∾ The Whalesback ∾ Remaining
Faithful to Fair Avalonia ∾ Blueberries! ∾ The Supreme Court Blows It,
Botanically Speaking ∾ A Grievous Peatland Pun ∾ A Taiga Tree Gets Gout ∾
Males Need Not Apply ∾ Machias versus the British Empire ∾ A Land of
Giant Tides ∾ Quoddy Head State Park ∾ Sea Cliffs and Surf ∾ The West
Quoddy Head Bog ∾ An Exhortation to Memory ∾ *Finis Est et Origo*

We live in a world of gridwork patterns that reflect our biases. Consider the seemingly innocent images of our maps: the Americas sit enthroned in the middle of the flattened globe, with the five remaining continents flung to the sides or corners. The top of the world is the North Pole; the bottom is the South Pole. When was the last time you thought of going "up" to Argentina?

A traditional trait of the New England mind is to reject prevailing trends, even to the point of turning the commonly accepted version of the world on its head. Such is the case with Yankee mariners who describe a cruise from New Hampshire to New Brunswick as going "down east." Their bias is not the chart or the compass rose, but the prevailing winds, which sweep from southwest to northeast along the coast; "down" means *downwind*.

In this tour, we will follow the New England tendency toward counterconvention. From Penobscot Bay, near the center of the Maine coast, we'll go down to the eastern tip, into a world that resists change more successfully than most places. The towns in this area, while never immune to the modern tourist trade, reflect the less cluttered ways of the nineteenth century. The forests, never immune to modern chain saws or imported insects, still reflect the ancient resin-scented times before the rise of the flowering plants. We're going Down East, and in doing so, we're traveling back along the jagged coastline of history.

Begin the tour by taking Route 1 North from the Belfast starting point. From the road, you get a striking view of the town's substantial port and bay-within-a-bay. Almost at once, you cross the Passagassawakeag River. At 3.8 miles from the starting point, you enter the town of Searsport; another 0.7 mile takes you to your first stop, Moose Point State Park. Turn right to enter and proceed down to the waterfront parking lot, located well past the gate.

The main attraction is obvious—a glorious view of greater Penobscot Bay, an exceptionally beautiful body of water. Plenty of interesting plants grow here, too, beginning with the speckled alders and the naturalized apple tree located to the right, along the drive as you approach the parking area. John Josselyn noted that the local Abnaki Indians used alder bark to soothe wounds and bruises.

The role of the apple tree in New England's orchard trade is described in Tour 31, but here we'll briefly consider a few other uses that this abundantly helpful plant has been put to. In earlier times, when an apple tree was felled after its fruit-producing days were over, its compact and well-knit wood was utilized for a number of things. In the domestic realm, it was preferred for spoons, mallets, and firewood; in the construction of old-time machinery, it was formed into cogs, wheels, and shuttles. And in common with the ash tree, it had a role in the sporting life, primarily as a prized material for golf clubs. Keep in mind that the apple genus is not native to northern New England. All the "wild" trees you see were once set out by a hopeful farmer or homeowner, and eventually abandoned to the encroaching forest or thicket.

The waterfront's botanical attractions are many. Indigenous wildflowers, including jewelweed, goldenrods, wild lettuce, and flat-topped white aster, mix with such aliens as Queen Anne's lace and yellow bedstraw, an example of the madder family. You may find yellow the bedstraw on the seaward edge of the lawn. If you do, note its whorled leaf arrangement. Most bedstraws are white-flowered, but as its name indicates, this one blooms a pleasant yellow. Also, if you spot wild lettuce, take a good look at it, too. This lofty herbaceous plant is, in botanical slang, a "comp"—a member of the world's largest plant family, the composites. Believe it or not, it is a real lettuce; genetically speaking, it is right next door to the cultivated types. This species is one of the most familiar sights along New England roadsides in late summer. (Since the biggest plant family has just been mentioned, can you venture any guesses about the next largest one? It's the orchid family. But if size is based not on the number of species but on the number of individual plants, the grand winner is the grass family.)

If time and tide permit, you should be able to locate copious seaweed specimens, mostly representing the brown algae. Two species are most common—knotted rockweed, of the genus *Ascophyllum,* and bladder wrack, of the genus *Fucus.* In Britain, bladder wrack has been employed in what sounds like the worst of many nose-crinkling herbal remedies, seaweed wine. This is a joint product of grapes and *Fucus* stems that supposedly cures hip diseases and other joint problems in children. In addition, wrack has been made into tablets to counteract obesity. In Scotland, it reaches its highest honor, where it is the badge emblem for Clan MacNeill.

A shoreline trail leads westward from the open area. It's a nice stroll, featuring sea breezes, bayberry, white spruce, red maple, Canadian hemlock, black cherry, bush honeysuckle, and much more. Before leaving the park, take a brief inspection tour of the trees near the parking lot. Among the various specimens is the Pine Tree State's namesake. Examine the needle foliage—it is arranged in five-leaf bundles, a sure sign that this is white pine, the tallest-growing plant in the Northeast.

—————————— ❧ ——————————

Leave the park by turning right onto Route 1 North. The roadside teems with plant life: bracken fern grabs the disturbed ground while behind it white ash and red maple contend with a full complement of conifers: tamarack, Canadian hemlock, spruces, balsam fir, and arbor vitae. As described in Tour 27, bracken is an aggressive invader with distinctly unpleasant characteristics. Nevertheless, it has had its uses, imaginary or otherwise. In rural parts of England, the fern was burned in the belief that rain would follow; this was known as "firing the bracken." People also believed that smoke from the fire repelled insects, which is not unreasonable. However, my favorite application has to do with the plant's spores, which were said to render a person invisible. If so, the bracken could be a distinct asset for many people during tax audits and extended family visits.

At 12.6 miles beyond Moose Point, the road bends to reveal an impressive view of the bridge connecting the mainland with Verona Island in the Penobscot River. About 0.7 mile farther along, you come to an overlook on the right. The rock exposed in this area is late Cambrian or early Ordovician pelite, a meta-morphosed mudstone. It is one of the main components of the Avalonian terrane that underlies the New England region. For more on the exotic origin of Avalonia, see Tour 53.

The road passes over the bridge, abruptly swings northward, and crosses a second span, over the Penobscot's Eastern Channel to Bucksport. Stay on Route 1 North by bearing right at the junction with Route 15, and note your mileage. In another 1.1 miles, turn left onto Route 46 North. This road heads inland east of Bangor and Brewer, across a rolling terrain of pastures, residential development, and wetlands. At the point 7.9 miles from the previous turn, Long Pond appears on the left. Farther north and beyond the Route 1A junction, balsam poplar, a north woods regular, is very much in evidence. Soon thereafter, you reach the T-intersection of Route 9. Turn right onto 9 East.

This scenic road is known as the Airline, though it probably should be renamed the Beeline, since it gives logging-truck drivers a chance to go supersonic on a hilly, frost-heaved, two-lane surface. If you dare sneak past one of their heavily loaded rigs as it inches up a steep grade, watch out. Hell hath no fury like a log jockey passed. On the next downhill, don't be surprised if you suddenly behold in your rearview mirror the spine-tingling sight of the same truck's chrome nose keeping station one-eighth of a millimeter behind your rear bumper.

This road parallels coastal Route 1. You are still south of the great Norumbega Fault, which marks the boundary between the Avalonian and interior terranes. (To repeat a point of semantics made elsewhere in the book, *terrane* is not a misspelling, but a geology term referring to an area's overall three-dimensional composition. We are not talking about just the surficial *terrain*, but also the grand structure of the underlying rock units. I wouldn't use these two potentially confusing words if they didn't each embody important and distinct concepts.)

About 8.3 miles from the turn onto Route 9, the woodland presents an ecologically significant tableau—low hemlock saplings snuggling under an overstory of mature spruces, birches, and maples. Hemlocks are a patient, shade-tolerant breed, willing to wait years for one of the big trees to topple and let some sunlight onto the forest floor. When it does, they shoot up and become part of the overstory themselves. Thus, the woodland community eventually changes its identity. This is not to suggest that conditions here or elsewhere must inexorably lead to hemlock dominance or any other kind of climax state. Every environment bears the potential for change, sudden or gradual, which will replace one apparent destiny with another.

When you reach the Route 179 junction in Aurora, continue straight but note your mileage. In another 4.4 miles, pull off at the left-hand roadside viewpoint. (There are other vantage points nearby, but this is the best.) The scene before you is the very essence of inland Maine. This is a seemingly endless expanse of muskeg, or peatland, punctuated by the impressive isolated hills of the Down East Range. Peatlands account for a minimum of 1 percent of our planet's land surface, the equivalent of half the United States. And the high ground where you are now is no less fascinating. It is known as the Whalesback, one of Maine's many eskers. These steep-sloped ridges are Ice Age features, formed from sediments lodged in the meltwater channels of glaciers. As a rule, eskers are aligned in the same direction in which their parent glaciers moved during their expansion phase. Nearby, you'll see much evidence of quarrying. Eskers are one of the best sources of construction-grade sand and gravel.

A sweeping view of Maine muskeg, from the top of the Whalesback esker.

When you resume your eastward trek on Route 9, notice the wet area below the esker, 1.2 miles beyond the scenic overlook described above. This also is part of the muskeg zone. Along with other conifers stand the narrow and often spindly forms of that quintessential bog tree, black spruce. You'll see the species again at the tour's end. Another 9.3 miles farther along, take the right-hand turn onto Route 193 South. In the townships of Beddington, Deblois, and Cherryfield, you are confronted with a landscape different from anything else described in this book—thousands of acres of commercial blueberry fields. This is Tom Thumb agriculture. The red-stemmed plants are the lowbush type, only a few inches high—one almost expects to see 6-inch-tall farmers driving pygmy tractors up and down the rows.

In actuality, the harvest of this heavenly fruit is back-breaking work. One fine mid-August afternoon I had the good luck to be here and watch the picking in progress. Hand-lettered signs were up, advertising instant employment at $4.00 an hour. From the looks of it, many local kids had signed on. Although I've never before felt any twinges of kleptomania, the sight of countless unguarded crates, stacked along the roadside and brimming over with mouthwatering berries, gave me my closest call with committing petit larceny. The pickers are usually called "rakers" after the tined, metal scoops used to painstakingly remove the berries from the branches.

The plants you see here are ericads of the genus *Vaccinium*, which includes cranberries, huckleberries, and bilberries. Lowbush blueberries have not gone through the same extensive breeding and hybridization that many crop plants have; but some growers, notably Elizabeth White of New Jersey, have come up with superior strains.

With such a luscious product passing by so close at hand, this is a fitting place to consider what fruit really is. By definition, a fruit is a fully developed flower ovary. Many people would agree with the United States Supreme Court 1893 ruling that the tomato is not a fruit but a vegetable; in fact, they're dead wrong. As any home gardener knows, tomatoes develop from flowers. In contrast, good examples of true vegetables are potatoes (tubers), carrots (roots), lettuce (leaf clusters), and globe artichokes (flower heads, but undeveloped ones). An easy test to separate the two types of produce is simple: does the mystery object contain seeds? If it does, it's a fruit. This test fails only when you come upon a specially bred seedless fruit variety; can you think of any examples? Here's a hint: what is it that cartoon characters are always slipping on?

For a while, Route 193 seems to promise to be blueberry fields forever, but eventually the woods return; the roadside becomes one of the most tamaracky stretches in New England. Also making an appearance are arbor vitae, red pine, poplar, and the herbaceous fireweed. This striking wildflower is a classic *r* environment species, as a rapid colonizer of burned land and other ecological vacuums. (For more on plants selected for *r* environments, not to mention *K* environments, see Tour 49.) In the United States, the plant has been put to herbal uses, primarily as an intestinal astringent. Pollen analysts have discovered that fireweed was present during the region's biggest ecological vacuum of all, the bare tundra that emerged from the melting Ice Age glaciers.

At the town of Cherryfield, you return to the trip's original road, Route 1. Turn left onto 1 North. Just a short distance away is Harrington. To its north, about 5 miles along, is the Great Heath, a gigantic peatland more than 4,000 acres in size. If you wish to see it, get directions locally; you'll have better access to an-other such place in a little while. The area surrounding this leg of the tour is a sphagnum lover's paradise. Other outstanding sites are in the vicinity of Jonesport, but were I to describe them all, this account would get hopelessly bogged down.

The road crosses the mouth of the Chandler River at Jonesboro. This is a good example of a tidal estuary and salt-marsh habitat. Salt marshes may not look particularly rampant with life, but they are one of the most productive ecosystems on earth, both in terms of total mass of aquatic vegetation and also in regard to the birds, insects, finfish, and shellfish they shelter. Beyond town, wild lowbush blueberry, virtually indistinguishable from its cultivated relatives, clambers over the sandy, nutrient-starved soil along the road shoulders. Would you gather, therefore, that it's a particularly tough customer? (You'd better.) The blueberry may be doing just fine, but one of the predominant taiga conifers, the balsam fir, is having problems with a nasty insect assailant.

The offender is the balsam woolly adelgid (ah-*del*-jid), a close relative of the pest that threatens Canadian hemlocks in southern New England. It was intro-duced from the Old World about a century ago. Amazingly, only female adelgids are present in the United States. These reproduce by egg-laying accomplished quite satisfactorily without the presence of males. This method is bound to be offensive to proponents of traditional family values, but it isn't a rare phenomenon in the insect world. One female green peach aphid, for instance, can bear large numbers of live, fully formed young. The aphid's reproductive shortcut is known as parthenogenesis.

Perhaps you have already spotted the effects of the balsam woolly adelgid. From a distance, infested trees look either flat-topped or notably sparse in their upper portions. Often, a plant's spindly leader stands virtually alone, with just a few surrounding branches. The insects spread from tree to tree mostly by being blown in the wind. Once they find a suitable spot, they settle in for life and sink mouth parts longer than their bodies deep into the tree tissue. One side effect of their activity is a telltale gouting, or swelling, of the tree's twigs and larger stems.

The port of Machias is situated 7.0 miles past Jonesboro. It is the handsomest of all Down East towns, with a consistently simple and functional architectural style. Folks here point to their illustrious Revolutionary forebears, who on June 12, 1775, cocked a snook at the Royal Navy by capturing its armed schooner *Margaretta* after a spirited sea fight. As a consequence, the British commander, Vice Admiral Samuel Graves, decided Machias and its populace deserved a complementary dose of sustained naval gunfire. Luckily, in the press of larger events this punishing bombardment never materialized.

Another picturesque tidal estuary can be seen on the way to the heart of East Machias. In the downtown area, be careful not to miss the intersection of Route 191 South. Take a right onto it. (It is marked by a sign for Naval Communications Unit, Cutler.) Now the loveliest portion of your tour has commenced. The coastal road leads you through an almost unpopulated land suffused with maritime light. No other part of New England, not even Cape Cod, is so given over to the North Atlantic. Near Holmes Bay, you'll see speckled alders, tamaracks, and a blueberry field here and there. Beyond lies Little Machias Bay. At low tide, broad mud flats, strewn with glacial erratics, are uncovered. Then, after the chaste fishing town of Cutler, where the rise and fall of the tide approaches 30 feet, the forests resume. By late summer, the roadside is tinted with goldenrods, pearly everlasting, and the red fruit of American mountain ash. By early fall, red maples are flaring in their wetland stations.

This road need never end, but eventually it does, at Route 189 in West Lubec. Turn right on 189 East, and after 4.3 miles, bear right again onto South Lubec Road. You should see a sign for your final destination, Quoddy Head State Park. Follow the succeeding signs the remaining 5.1 miles to the entrance, and park in the lot down the right-hand drive.

The complete name is West Quoddy Head, an ironic tag for the easternmost point of land in the United States. Take as much time as you possibly can. No single site in this book has more to offer the intellect or the soul. Head up the sea cliff trail. Flowering plants are by no means totally absent, but the taiga conifers rule—white and red spruce, and especially balsam fir. Nearest the cliff, they are dwarfed by the blast from the sea, and in places, you can see that their gouted stems have been attacked by the adelgid. In the final weeks of summer, the rayed heads of whorled and Lowrie's asters shine underneath, completing the cycle begun by the mat-forming mountain cranberry and other early blooming plants.

As the trail leads on toward Green Point, there are many spots where true mosses grow, and in one place a particular species of sphagnum is well settled in. Not far from it the careful eye will also discern lichens—not just *Usnea* tufts dangling from conifer branches but also the weird spire forms of *Cladonia amaurocrea* rising from the ground. All these organisms thrive in a cool, damp sea climate. Some of the cliffs are 200 feet above the surf, yet fog and spray reach this high-perched outpost of the land easily.

Before walking out all the way to the point, take the well-marked side trail to the right and brace yourself for a thrill. As the path proceeds, the sphagnum community beside it becomes more extensive. Then a boardwalk begins, and suddenly you're heading into the heart of West Quoddy Head Bog, a 7-acre gem of a peatland within earshot of the pounding surf. In accessibility, it matches the Charles Ward Reservation in Massachusetts (Tour 22) and the Black Spruce Bog in the Litchfield Hills of Connecticut (Tour 10); in instructive signage and overall dramatic impact, it exceeds both. Originally a pond, the depression was soon colonized by a sphagnum community, and over the course of eight thousand years the bog has developed into the complex habitat you see here. The classic Hydrarch Succession theory holds that the sphagnum mat is an intermediate step before trees take over. Here, however, the sphagnum is not constrained by theory–it is actually outcompeting the trees, and may thoroughly trounce them one day.

As you make the circuit of the boardwalk, you'll see species after species of rare and highly adapted plants. On the fringes stands that alley cat of conifers, the ever scraggly, ever amazing black spruce. See if you can catch one of its kind in the act of layering–that is, asexually propagating by sprouting new plants where its branches touch the ground. One specimen here, as the sign indicates, is eighty years old, but it is not even as tall as most human beings. Other woody plants include our friend the lowbush blueberry, sweet gale, bog rosemary, small cranberry, black crowberry, sheep laurel, leatherleaf, and Labrador tea. Labrador tea played a role in the American Revolution, being used as a substitute for real tea. As with many of the plants here, it is circumpolar in its native distribution. In northern Scandinavia, Laplanders place its branches in their grain storage areas to repel vermin.

The herbaceous plants of the bog are no less intriguing. Starting with everyone's favorites, the carnivorous kinds, both the round-leaved and spatulate-leaved sundews grow here, as well as the strangely beautiful pitcher plant. It has been suggested that these species originally came from a southern environment and subsequently adapted to New England's harsher conditions. (See Tour 42 for a description of their hunting methods.) Less acquisitive but just as striking are the bog goldenrod and cotton grass sedge. Still, one member of *Rubus,* the raspberry genus, is perhaps the star of the show. Known as baked apple berry or cloudberry, it is restricted to the arctic and other northern regions. Unlike its various kin, which are usually aggressive thicket shrubs, this species stays close to the ground, bears its white flowers singly, and doesn't even develop woody stems. Last but definitely most important are the predominant sphagnum species, the green or brownish *Sphagnum fuscum* and the striking red-tinted *Sphagnum capillifolium* var. *tenellum,* labeled here as *S. rubellum.* Both prefer the higher, drier parts of the mat, with the latter often growing around and just below the former.

Once you're back on the main trail, the final objective, Green Point, is not far away. The scenery is spectacular. Out in Lubec Channel, small fishing boats chug along the headland, and in the foreground, the black and greenish rocks–metamorphic pelite on the northern side of the promontory and coarse, intrusive gabbro on the southern–stand like the gunwale of a great iron ship embedded in a reef of narrow, cobbled beach.

The forest finally gives way to open ground at the point. One of the few woody plants abiding here, in the full force of the ocean winds, is the not so obvious but remarkable creeping juniper. Flat as a rug, the species is found in its wild state only in selected northern places, most of them quite challenging. Nonetheless, its prostrate habit and tolerance for the poorest soils has ensured its widespread horticultural use throughout the land. The list of cultivars developed from it is lengthy. If you are a gardener familiar with these cultivated varieties, you may be interested to see these–the original item, so to speak.

The herbaceous plants of Green Point are not quite so humble in their habit. The little, rock-loving harebell, representing the bluebell family, is equally at home

A sight not seen by most travelers: Quoddy Head State Park, the easternmost tip of America, in the snows of winter.

here as on the gentle heights of the lower Connecticut River valley. You may fine groundsel, orache, angelica, and seaside plantain, too. In the most seaward section of the rocks, other forms of life take over, as is the custom. Look for the handsome orange shore lichen; it is sometimes also found adorning old tombstones in Down East graveyards. In the tidal pools, you will find the first messengers from the kingdom Protista, the green *Enteromorpha* algae. Not far away, sprawled like fallen battalions awaiting the resurrection of the next rising tide, are the darker forms of brown algae. These include at least two wrack species, as well as knotted rockweed.

Whether your home is a few miles away or several thousand, this is an important place to remember, and not just with the camera's eye. The pervading sense of the primitive seems to set this rampart of the continent far from any human thought or action. Geologists say that just offshore runs the Fundy Fault, a main breach in the Old World island, Pangaea. When 200 million years ago North America regained its separateness and this new eastern ocean was being born, the conifers and other gymnosperms stood at the apex of the plant kingdom. Those were very different, subtropical times; but you can imagine the ancestor of this evergreen forest standing here then. Now, if conditions are right, you will see the cool sea mist drifting in through the stunted firs and hear it distort the antiphonal cries of the horn buoys. This is a fitting end for America, or, better yet, a fitting beginning.

531

*C*ommon and *T*axonomic *N*ames of *P*lants, *L*ichens, and *A*lgae

Ageratum (cultivated annual): *Ageratum houstonianum*

Ageratum (cultivated perennial): *Eupatorium coelestinum*

Agrimony: *Agrimonia gryposepala* and other *Agrimonia* species

African daisy: *Dimorphotheca sinuata*

Ailanthus: *Ailanthus altissima*

Alfalfa: *Medicago sativa*

All-heal: *Valeriana officinalis*

Alternate-leaved dogwood: *Cornus alternifolia*

Aluminum plant: *Pilea cadierei*

American beech: *Fagus grandifolia*

American bladdernut: *Staphylea trifolia*

American chestnut: *Castanea dentata*

American cranberry bush: *Viburnum trilobum*

American elm: *Ulmus americana*

American fly honeysuckle: *Lonicera canadensis*

American hazel: *Corylus americana*

American holly: *Ilex opaca*

American hornbeam: *Carpinus caroliniana*

American mountain ash: *Sorbus americana*

American prickly ash: *Zanthoxylum americanum*

American white hellebore: *Veratrum viride*

Amur cork tree: *Phellodendron amurense*

Amur maple: *Acer ginnala*

Angelica (wild): *Angelica atropurpurea*

Apple (cultivated): *Malus* cultivars

Arbor vitae: *Thuja occidentalis*

Arrow arum: *Peltandra virginica*

Arrowhead: *Sagittaria latifolia*

Arrow-leaved tearthumb: *Polygonum sagittatum*

Arrowwood: *Viburnum dentatum*

Arum family: Araceae

Artillery plant: *Pilea microphylla*

Astilbe: *Astilbe* x *arendsii*

Atlantic white cedar: *Chamaecyparis thyoides*

Austrian pine: *Pinus nigra*

Autumn elaeagnus: *Elaeagnus umbellata*

Baked apple berry: *Rubus chamaemorus*

Bald cypress: *Taxodium distichum*

Balm of Gilead: *Populus balsamifera* var. *candicans*

Balsam fir: *Abies balsamea*

Balsam poplar: *Populus balsamifera* or *P. balsamifera* var. *subcordata*

Bamboo: *Bambusa* and related genera

Baneberry: *Actaea* species
Basswood: *Tilia americana*
Bayberry: *Myrica pensylvanica*
Bayberry family: Myricaceae
Beach grass: *Ammophila breviligulata*
Beach heather: *Hudsonia tomentosa*
Beach pea: *Lathyrus japonicus*
Beach plum: *Prunus maritima*
Beaked hazel: *Corylus cornuta*
Beaked willow: *Salix bebbiana*
Bear oak: *Quercus ilicifolia*
Bearberry: *Arctostaphylos uva-ursi*
Bearberry willow: *Salix uva-ursi*
Beardtongue penstemon: *Penstemon digitalis*
Bebb willow: *Salix bebbiana*
Beech family: Fagaceae
Bellwort: *Uvularia* species
Benjamin fig: *Ficus benjamina*
Bigelow's sedge: *Carex bigelowii*
Bigtooth poplar: *Populus grandidentata*
Birch family: Betulaceae
Bird's-foot trefoil: *Lotus corniculatus*
Bitternut hickory: *Carya cordiformis*
Bittersweet family: Celastraceae
Bittersweet: *Celastrus* species
Black alder: *Ilex verticillata*
Black ash: *Fraxinus nigra*
Black birch: *Betula lenta*
Black cherry: *Prunus serotina*
Black chokeberry: *Aronia melanocarpa*
Black crowberry: *Empetrum nigrum*
Black-eyed Susan: *Rudbeckia hirta*
Black gum: *Nyssa sylvatica*
Black haw: *Viburnum prunifolium*
Black highbush blueberry: *Vaccinium atrococcum*
Black huckleberry: *Gaylussacia baccata*
Black locust: *Robinia pseudoacacia*
Black oak: *Quercus velutina*
Black snakeroot: *Sanicula marilandica*
Black spruce: *Picea mariana*
Black walnut: *Juglans nigra*
Black willow: *Salix nigra*
Bladder campion: *Silene vulgaris*
Bladder wrack: *Fucus vesiculosus*
Bleeding heart: *Dicentra spectabilis*
Bloodroot: *Sanguinaria canadensis*
Blue cohosh: *Caulophyllum thalictroides*
Blue highbush blueberry: *Vaccinium corymbosum*
Blue vervain: *Verbena hastata*

Bluebell family: Campanulaceae
Blue-eyed grass: *Sisyrinchium* species
Bluets: *Hedyotis* species
Bog bilberry: *Vaccinium uliginosum*
Bog goldenrod: *Solidago uliginosa*
Bog laurel: *Kalmia polifolia*
Bog rosemary: *Andromeda glaucophylla*
Boneset: *Eupatorium perfoliatum*
Borage family: Boraginaceae
Boston ivy: *Parthenocissus tricuspidata*
Bottlebrush grass: *Hystrix patula*
Bouncing Bet: *Saponaria officinalis*
Box elder: *Acer negundo*
Bracken: *Pteridium aquilinum* var. *latiusculum*
Bradford pear: *Pyrus calleryana* 'Bradford'
Braun's holly fern: *Polystichum braunii*
Bread tree: *Encephalartos altensteinii*
Bristly locust: *Robinia hispida* or *R. fertilis*
Bristly sarsaparilla: *Aralia hispida*
British soldiers: *Cladonia cristatella*
Broom crowberry: *Corema conradii*
Broomcorn: *Sorghum vulgare*
Buckbean: *Menyanthes trifoliata*
Buckwheat family: Polygonaceae
Bull briar: *Smilax* species
Bullhead lily: *Nuphar variegatum*
Bulrush: *Scirpus* species
Bunchberry: *Cornus canadensis*
Burning bush: *Euonymus alata*
Bush honeysuckle: *Diervilla lonicera*
Buttercup family: Ranunculaceae
Butternut: *Juglans cinerea*
Buttonbush: *Cephalanthus occidentalis*

Calla lily: *Calla palustris*
Callery pear: *Pyrus calleryana*
Camellia: *Camellia japonica*
Canada Mayflower: *Maianthemum canadense*
Canadian hemlock: *Tsuga canadensis*
Canadian yew: *Taxus canadensis*
Candlewood: *Pinus rigida*
Cape marigold: *Dimorphotheca sinuata*
Caraway: *Carum carvi*
Cardinal flower: *Lobelia cardinalis*
Carolina spring beauty: *Claytonia caroliniana*
Carrot (cultivated): *Daucus carota* var. *sativus*
Carrot family: Umbelliferae
Cashew family: Anacardiaceae
Catalpa: *Catalpa speciosa* or *C. bignonioides*

Cedar of Lebanon: *Cedrus libani*
Century plant: *Agave americana*
Checkerberry: *Gaultheria procumbens*
Cherry elaeagnus: *Elaeagnus multiflora*
Chestnut oak: *Quercus prinus*
Chestnut rose: *Rosa roxburghii*
Chicory: *Cichorium intybus*
Chinese cork tree: *Phellodendron chinense*
Chinquapin oak: *Quercus prinoides*
Chokecherry: *Prunus virginiana*
Christmas fern: *Polystichum acrostichoides*
Chufa: *Cyperus esculentus* var. *sativus*
Cinnamon fern: *Osmunda cinnamomea*
Clearweed: *Pilea pumila*
Clethra: *Clethra alnifolia*
Clinton's fern: *Dryopteris clintoniana*
Cloudberry: *Rubus chamaemorus*
Club moss: *Lycopodium* species
Colorado blue spruce: *Picea pungens* cultivars
Colorado spruce: *Picea pungens*
Coltsfoot: *Tussilago farfara*
Columbine: *Aquilegia canadensis*
Comfrey: *Symphytum officinale*
Common barberry: *Berberis vulgaris*
Common buckthorn: *Rhamnus cathartica*
Common cattail: *Typha latifolia*
Common elder: *Sambucus canadensis*
Common evening primrose: *Oenothera biennis*
Common fig: *Ficus carica*
Common juniper: *Juniperus communis*
Common lilac: *Syringa vulgaris*
Common milkweed: *Asclepias syriaca*
Common mullein: *Verbascum thapsus*
Common plantain: *Plantago major*
Common St. Johnswort: *Hypericum perforatum*
Common tansy: *Tanacetum vulgare*
Common walnut: *Juglans regia*
Composite family: Compositae
Concord grape: *Vitis labrusca* 'Concord'
Connecticut mahogany: *Prunus serotina*
Copper-leaved European beech: *Fagus sylvatica* 'Atropunicea'
Corn: see Maize
Corn lily: *Clintonia borealis*
Cotton grass: *Eriophorum virginicum* and related species
Cottonwood: *Populus deltoides*
Crab apple: *Malus* species and cultivars
Creeping juniper: *Juniperus horizontalis*

Creeping snowberry: *Gaultheria hispidula*
Creosote bush: *Larrea tridentata*
Crimson king maple: *Acer platanoides* 'Crimson King'
Crowberry: *Empetrum nigrum*
Cure-all: *Geum rivale*
Curled lichen: *Cetraria cucullata*
Cut-leaved toothwort: *Dentaria laciniata*
Cut-leaved water horehound: *Lycopus americanus*
Cycad: *Cycas* and related genera
Cypress family: Cupressaceae

Damask rose: *Rosa damascena*
Dandelion: *Taraxacum officinale*
Daphne: *Daphne mezereum*
Dark green bulrush: *Scirpus atrovirens*
Dawn redwood: *Metasequoia glyptostroboides*
Day lily (cultivated garden perennial): *Hemerocallis* cultivars
Day lily (escaped): *Hemerocallis fulva*
Dead nettle: *Lamiastrum galeobdolan* var. *variegatum*
Deadly nightshade: *Solanum dulcamara*
Devil's guts: *Cuscuta gronovii* and related species
Devil's paintbrush: *Hieracium aurantiacum*
Dewdrop: *Dalibarda repens*
Diapensia: *Diapensia lapponica*
Dill: *Anethum graveolens*
Dodder: *Cuscuta gronovii* and related species
Dog lichen: *Peltigera canina*
Dogbane family: Apocynaceae
Dogwood: *Cornus* species
Douglas fir: *Pseudotsuga menziesii*
Dove tree: *Davidia involucrata*
Downy chess: *Bromus tectorum*
Downy rattlesnake plantain: *Goodyera pubescens*
Duckweed: *Lemna minor* and related species
Dusty miller (wild): *Artemisia stellerana*
Dutchman's breeches: *Dicentra cucullaria*
Dwarf birch: *Betula glandulosa*
Dwarf ginseng: *Panax trifolius*
Dwarf huckleberry: *Gaylussacia dumosa*
Dwarf sumac: *Rhus copallina*

Early saxifrage: *Saxifraga virginiensis*
Elaeagnus: *Elaeagnus* species

Elecampane: *Inula helenium*
Elm family: Ulmaceae
Enchanter's nightshade: *Circaea quadrisulcata*
English oak: *Quercus robur*
English yew: *Taxus baccata*
Englishman's foot: *Plantago major*
Eurasian milfoil: *Myriophyllum spicatum*
European alder: *Alnus glutinosa*
European barberry: *Berberis vulgaris*
European beech: *Fagus sylvatica*
European larch: *Larix decidua*
European littleleaf linden: *Tilia cordata*
European mountain ash: *Sorbus aucuparia*
European snowball viburnum: *Viburnum opulus* 'Roseum'
Evening primrose family: Onagraceae
Evergreen pear: *Pyrus kawakamii*
Everlasting pea: *Lathyrus latifolius*

False Solomon's seal: *Smilacina racemosa*
Fern-leaved false foxglove: *Aureolaria pedicularia*
Few-fruited sedge: *Carex oligosperma*
Fir club moss: *Lycopodium selago*
Fireweed: *Epilobium angustifolium*
Flameleaf sumac: *Rhus copallina*
Flat-topped white aster: *Aster umbellatus*
Flax (cultivated): *Linum usitatissimum*
Fleabane: *Erigeron* species
Flowering dogwood: *Cornus florida*
Fly honeysuckle: *Lonicera canadensis*
Foam flower: *Tiarella cordifolia*
Forget-me-not: *Myosotis* species
Forsythia: *Forsythia* species and cultivars
Fox grape: *Vitis labrusca*
Fragrant bedstraw: *Galium triflorum*
Fragrant sumac: *Rhus aromatica*
Fragrant water lily: *Nymphaea odorata*
Franklin tree: *Franklinia alatamaha*
Franklinia: *Franklinia alatamaha*
Freshwater cord grass: *Spartina pectinata*
Fringed loosestrife: *Lysimachia ciliata*
Fringed sedge: *Carex crinita*
Fumitory family: Fumariaceae

Gagroot: *Lobelia inflata*
Gall-of-the-earth: *Prenanthes trifoliata*
Garden peony: *Paeonia lactiflora* and associated cultivars

Giant fir: *Abies grandis*
Giant reed: *Phragmites australis*
Ginkgo: *Ginkgo biloba*
Glasswort: *Salicornia* species
Globe artichoke: *Cynara scolymus*
Golden alexanders: *Zizia aurea*
Golden rain tree: *Koelreuteria paniculata*
Golden saxifrage: *Chrysosplenium americanum*
Goldenrod: *Solidago* species
Goldthread: *Coptis groenlandica*
Goosefoot family: Chenopodiaceae
Grass family: Gramineae
Gray birch: *Betula populifolia*
Gray dogwood: *Cornus racemosa*
Great blue lobelia: *Lobelia siphilitica*
Great Solomon's seal: *Polygonatum canaliculatum*
Green ash: *Fraxinus pennsylvanica*
Ground nut: *Apios americana*
Groundsel: *Senecio* species

Hackberry: *Celtis occidentalis*
Haircap moss: *Polytrichum juniperinum*
Hardhack: *Spiraea tomentosa*
Harebell: *Campanula rotundifolia*
Hartford fern: *Lygodium palmatum*
Harvest lice: *Agrimony gryposepala*
Hawthorn: *Crataegus* species
Hay-scented fern: *Dennstaedtia punctilobula*
Heath: *Erica* species
Heather: *Calluna vulgaris*
Heath family: Ericaceae
Hellweed: *Cuscuta gronovii* and related species
Hemp nettle: *Galeopsis tetrahit*
Herb Robert: *Geranium robertianum*
Hickory: *Carya* species
Highbush blueberry: *Vaccinium corymbosum* or *V. atrococcum*
Highland rush: *Juncus trifidus*
Himalayan pine: *Pinus wallichiana*
Hobblebush: *Viburnum alnifolium*
Hollyhock: *Alcea rosea*
Honesty: *Lunaria annua*
Honey locust: *Gleditsia triacanthos*
Hop hornbeam: *Ostrya virginiana*
Horned bladderwort: *Utricularia cornuta*
Horse balm: *Collinsonia canadensis*
Horse chestnut: *Aesculus hippocastanum*
Hosta: *Hosta* species and cultivars

Iceland moss: *Cetraria islandica*
Indian blanket: *Gaillardia pulchella*
Indian cucumber root: *Medeola viginiana*
Indian fig: *Opuntia ficus-indica*
Indian hemp: *Apocynum cannabinum*
Indian pipe: *Monotropa uniflora*
Indian tobacco: *Lobelia inflata*
Inkberry: *Ilex glabra*
Interrupted fern: *Osmunda claytoniana*
Irish moss: *Chondrus crispus*
Ironwood: *Carpinus caroliniana*

Jack-in-the-pulpit: *Arisaema triphyllum*
Jack pine: *Pinus banksiana*
Japanese andromeda: *Pieris japonica*
Japanese barberry: *Berberis thunbergii*
Japanese cedar: *Cryptomeria japonica*
Japanese climbing fern: *Lygodium japonicum*
Japanese honeysuckle: *Lonicera japonica*
Japanese knotweed: *Polygonum cuspidatum*
Japanese stewartia: *Stewartia pseudocamellia*
Japanese umbrella pine: *Sciadopitys verticillata*
Japanese yew: *Taxus cuspidata*
Japanese zelkova: *Zelkova serrata*
Jasmine: *Jasminum* species
Jewelweed (orange flowers): *Impatiens capensis*
Jewelweed (yellow flowers): *Impatiens pallida*
Jimsonweed: *Datura stramonium*
Joe-pye-weed: *Eupatorium maculatum* or *E. purpureum*
Joshua tree: *Yucca brevifolia*
Juneberry: *Amelanchier* species

Kaffir bread: *Encephalartos altensteinii*
Katsura family: Cercidiphyllaceae
Katsura tree: *Cercidiphyllum japonicum*
Kelp: *Laminaria* and related genera
Kentucky bluegrass: *Poa pratensis*
Knotted rockweed: *Ascophyllum nodosum*
Kudzu: *Pueraria lobata*
Kwanzan cherry: *Prunus serrulata* 'Kwanzan'

Labrador tea: *Ledum groenlandicum*
Lady Fern: *Athyrium filix-femina* and varieties
Lance-leaved coreopsis: *Coreopsis lanceolata*
Lance-leaved goldenrod: *Solidago graminifolia*
Large cranberry: *Vaccinium macrocarpon*
Large-leaved avens: *Geum macrophyllum*
Laurel: *Laurus nobilis*

Laurel family: Lauraceae
Leatherleaf: *Chamaedaphne calyculata*
Leatherwood: *Dirca palustris*
Legume order: Leguminales
Lettuce (cultivated): *Lactuca sativa*
Lily family: Liliaceae
Littleleaf linden: *Tilia cordata*
London plane tree: *Platanus* x *acerifolia*
Long beech fern: *Thelypteris phegopteris*
Lotus: *Nelumbo lutea* or *N. nucifera*
Lowbush blueberry: *Vaccinium angustifolium* or *V. vacillans*
Lowrie's aster: *Aster lowrieanus*
Lucerne: *Medicago sativa*
Lungwort: *Pulmonaria officinalis*

Madagascar periwinkle: *Catharanthus roseus*
Madder family: Rubiaceae
Magnolia family: Magnoliaceae
Maidenhair fern: *Adiantum pedatum*
Maidenhair spleenwort: *Asplenium trichomanes*
Maidenhair tree: *Ginkgo biloba*
Maize: *Zea mays*
Maleberry: *Lyonia ligustrina*
Manzanita: *Arctostaphylos* species
Maple family: Aceraceae
Maple-leaved viburnum: *Viburnum acerifolium*
Map lichen: *Rhizocarpon geographicum*
Marginal woodfern: *Dryopteris marginalis*
Marigold: *Tagetes* species and cultivars
Marsh elder: *Iva frutescens* var. *oraria*
Marsh mallow: *Hibiscus palustris*
Marsh marigold: *Caltha palustris*
Marsh St. Johnswort: *Hypericum virginicum*
Maté: *Ilex paraguariensis*
May apple: *Podophyllum peltatum*
Meadow beauty family: Melastomataceae
Meadowsweet: *Spiraea latifolia*
Meetinghouses: *Aquilegia canadensis*
Melilot: *Melilotus alba*
Metasequoia: *Metasequoia glyptostroboides*
Mexican fan palm: *Washingtonia robusta*
Mint family: Labiatae
Mockernut hickory: *Carya tomentosa*
Moneywort: *Lysimachia nummularia*
Morrow honeysuckle: *Lonicera morrowii*
Motherwort: *Leonurus cardiaca*
Mountain alder: *Alnus crispa*
Mountain ash: *Sorbus* species

Mountain cranberry: *Vaccinium vitis-idaea* var. *minus*
Mountain holly: *Nemopanthus mucronatus*
Mountain laurel: *Kalmia latifolia*
Mountain maple: *Acer spicatum*
Mountain sandwort: *Arenaria groenlandica*
Mountain woodfern: *Dryopteris campyloptera*
Mulberry family: Moraceae
Mulberry: *Morus* species
Mustard family: Cruciferae
Myrtle family: Myrtaceae

Nannyberry: *Viburnum lentago*
Narrow-leaved gentian: *Gentiana linearis*
New England aster: *Aster novae-angliae*
New Jersey tea: *Ceanothus americanus*
New York aster: *Aster novae-belgii*
New York fern: *Thelypteris noveboracensis*
Ninebark: *Physocarpus opulifolius*
Norfolk Island pine: *Araucaria heterophylla*
Northern white cedar: *Thuja occidentalis*
Norway maple: *Acer platanoides*
Norway pine: *Pinus resinosa*
Norway spruce: *Picea abies*
Nut sedge: *Cyperus esculentus*

Oconee bells: *Shortia galicifolia*
Opuntia (the species native to New England): *Opuntia humifusa*
Orache: *Atriplex patula*
Orchard grass: *Dactylis glomerata*
Orchid family: Orchidaceae
Organ-pipe cactus: *Lemaireocereus thurberi*
Oriental bittersweet: *Celastrus orbiculatus*
Osage orange: *Maclura pomifera*
Osier: *Salix viminalis*
Osmunda family: Osmundaceae
Ostrich fern: *Matteuccia struthiopteris* var. *pensylvanica*
Ox-eye daisy: *Chrysanthemum leucanthemum*
Oyster plant: *Tragopogon* species

Painted trillium: *Trillium undulatum*
Panicle hydrangea: *Hydrangea paniculata*
Panicled tick trefoil: *Desmodium paniculatum*
Paper birch: *Betula papyrifera*
Paper birch (montane variety): *Betula papyrifera* var. *cordifolia*
Partridgeberry: *Mitchella repens*

Pearly everlasting: *Anaphalis margaritacea*
Peony: *Paeonia* species
Peppermint: *Mentha* x *piperita*
Perennial salt-marsh aster: *Aster tenuifolius*
Persian walnut: *Juglans regia*
Persimmon: *Diospyros virginiana*
Petunia: *Petunia* x *hybrida*
Phragmites: *Phragmites australis*
Pickerel weed: *Pontederia cordata*
Pignut hickory: *Carya glabra*
Pin cherry: *Prunus pensylvanica*
Pin oak: *Quercus palustris*
Pine family: Pinaceae
Pink lady's slipper: *Cypripedium acaule*
Pipsissewa: *Chimaphila umbellata*
Pitch pine: *Pinus rigida*
Pitcher plant: *Sarracenia purpurea*
Plane tree family: Platanaceae
Plantain-leaved sedge: *Carex plantaginea*
Pointed-leaved tick trefoil: *Desmodium glutinosum*
Poison hemlock: *Conium maculatum*
Poison ivy: *Rhus radicans*
Poison sumac: *Rhus vernix*
Pokeweed: *Phytolacca americana*
Polypody: *Polypodium virginianum*
Pond cypress: *Taxodium distichum* var. *nutans*
Ponderosa pine: *Pinus ponderosa*
Poplar: *Populus tremuloides*
Popple: *Populus tremuloides*
Poppy family: Papaveraceae
Potato: *Solanum tuberosum*
Poverty birch: *Betula populifolia*
Prickly ash: *Zanthoxylum americanum*
Prickly cycad: *Encephalartos altensteinii*
Prickly pear (the species native to New England): *Opuntia humifusa*
Protea family: Proteaceae
Pukeweed: *Lobelia inflata*
Purple avens: *Geum rivale*
Purple cliff brake: *Pellaea* x *atropurpurea*
Purple-flowering raspberry: *Rubus odoratus*
Purple loosestrife: *Lythrum salicaria*
Purple love grass: *Eragrostis spectabilis*
Pussy willow: *Salix discolor*
Pussytoes: *Antennaria plantaginifolia* or *A. neglecta*

Quaking aspen: *Populus tremuloides*
Queen Anne's lace: *Daucus carota*

Rabbit's-foot clover: *Trifolium arvense*
Ragwort: *Senecio* species
Red cedar: *Juniperus virginiana*
Red clover: *Trifolium pratense*
Red maple: *Acer rubrum*
Red oak: *Quercus rubra*
Red osier dogwood: *Cornus sericea*
Red pine: *Pinus resinosa*
Red spruce: *Picea rubens*
Red top: *Agrostis gigantea*
Red trillium: *Trillium erectum*
Redwood family: Taxodiaceae
Reindeer moss: *Cladina rangiferina, C. subtenuis,*
  and other *Cladina* species
Rhodora: *Rhododendron canadense*
Riga fir: *Pinus sylvestris*
River birch: *Betula nigra*
Rock tripe: umbilicate lichens of *Umbilicaria,
  Lasallia,* and other genera
Rocky Mountain columbine: *Aquilegia
  caerulea*
Rose azalea: *Rhododendron prinophyllum*
Rose family: Rosaceae
Rosebay rhododendron: *Rhododendron maximum*
Rough hawkweed: *Hieracium scabrum*
Round-headed bush clover: *Lespedeza capitata*
Round-leaved sundew: *Drosera rotundifolia*
Round-leaved yellow violet: *Viola rotundifolia*
Round-lobed hepatica: *Hepatica americana*
Royal fern: *Osmunda regalis* var. *spectabilis*
Rue family: *Rutaceae*
Rugosa rose: *Rosa rugosa*
Rye (cultivated): *Secale cereale*

Sacred datura: *Datura wrightii*
Sago palm: *Cycas revoluta*
Saguaro: *Carnegiea gigantea*
Sakhalin knotweed: *Polygonum sacchalinense*
Salt-marsh aster: *Aster tenuifolius* or *A. subulatus*
Salt-meadow cord grass: *Spartina patens*
Saltwater cord grass: *Spartina alterniflora*
Sargent weeping hemlock: *Tsuga canadensis*
  'Pendula'
Sassafras: *Sassafras albidum*
Saxifrage family: Saxifragaceae

Scarlet elder: *Sambucus pubens*
Scarlet oak: *Quercus coccinea*
Schreber's aster: *Aster schreberi*
Schwedler maple: *Acer platanoides* 'Schwedleri'
Scotch fir: *Pinus sylvestris*
Scots broom: *Cytisus scoparius*
Scots heather: *Calluna vulgaris*
Scots pine: *Pinus sylvestris*
Scrub oak: *Quercus ilicifolia*
Sea blite: *Suaeda maritima*
Sea lavender: *Limonium carolinianum*
Sea lettuce: *Ulva lactuca*
Sea rocket: *Cakile edentula*
Seaside goldenrod: *Solidago sempervirens*
Seaside plantain: *Platago maritima*
Sedge family: Cyperaceae
Self-heal: *Prunella vulgaris*
Sensitive fern: *Onoclea sensibilis*
Sequoia: *Sequoiadendron giganteum*
Serviceberry: *Amelanchier* species
Shadblow: *Amelanchier canadensis*
Shadbush: *Amelanchier* species
Shagbark hickory: *Carya ovata*
Sharp-lobed hepatica: *Hepatica acutiloba*
Sharp-winged monkey flower: *Mimulus alatus*
Sheep laurel: *Kalmia angustifolia*
Shepherd's purse: *Capsella bursa-pastoris*
Shining club moss: *Lycopodium lucidulum*
Shining sumac: *Rhus copallina*
Shinleaf pyrola: *Pyrola elliptica*
Shore lichen: *Xanthoria parietina*
Showy orchis: *Orchis spectabilis*
Silk oak: *Grevillea robusta*
Silk tree: *Albizia julibrissin*
Silver maple: *Acer saccharinum*
Silverweed: *Potentilla anserina*
Silvery spleenwort: *Athyrium thelypteroides*
Skullcap: *Scutellaria* species
Skunk cabbage: *Symplocarpus foetidus*
Slender glasswort: *Salicornia europaea*
Slender nettle: *Urtica gracilis*
Small cranberry: *Vaccinium oxycoccos*
Smaller enchanter's nightshade: *Circaea alpina*
Smokebush: *Cotinus coggygria*
Smooth alder: *Alnus rugosa*
Snapdragon family: Scrophulariaceae
Snow lichen: *Cetraria nivalis*
Soapberry family: Sapindaceae
Soft maple: *Acer saccharinum*

Solomon's seal: *Polygonatum biflorum*
Sow thistle: *Sonchus* species
Spanish moss: *Tillandsia usneoides*
Spatterdock: *Nuphar advena*
Spatulate-leaved sundew: *Drosera intermedia*
Spearmint: *Mentha spicata*
Speckled alder: *Alnus rugosa*
Sphagnum moss: *Sphagnum* species
Spicebush: *Lindera benzoin*
Spikenard: *Aralia racemosa*
Spotted joe-pye-weed: *Eupatorium maculatum*
Spotted knapweed: *Centaurea maculosa*
Spotted wintergreen: *Chimaphila maculata*
Spring beauty: *Claytonia virginica* or *C. caroliniana*
Staghorn fern: *Platycerium bifurcatum*
Staghorn sumac: *Rhus typhina*
Starflower: *Trientalis borealis*
Stargrass: *Hypoxis hirsuta*
Starry false Solomon's seal: *Smilacina stellata*
Steeplebush: *Spiraea tomentosa*
Stiff club moss: *Lycopodium annotinum*
Stinging nettle: *Urtica dioica* or *Laportea canadensis*
Stout goldenrod: *Solidago squarrosa*
Striped maple: *Acer pensylvanicum*
Sugar maple: *Acer saccharum*
Sugarberry: *Celtis laevigata*
Summer grape: *Vistis aestivalis*
Sundew: *Drosera* species
Swamp azalea: *Rhododendron viscosum*
Swamp dewberry: *Rubus hispidus*
Swamp dogwood: *Cornus amomum*
Swamp horsetail: *Equisetum fluviatile*
Swamp loosestrife: *Decodon verticillatus*
Swamp rose: *Rosa palustris*
Swamp smartweed: *Polygonum coccineum*
Swamp white oak: *Quercus bicolor*
Sweet fern: *Comptonia peregrina*
Sweet gale: *Myrica gale*
Sweet gum: *Liquidambar styraciflua*
Sweet joe-pye-weed: *Eupatorium purpureum*
Sycamore: *Platanus occidentalis*
Sycamore maple: *Acer pseudoplatanus*

Tacamahac: *Populus balsamifera*
Tall lettuce: *Prenanthes altissima*
Tall meadow rue: *Thalictrum polygamum*
Tamarack: *Larix laricina*

Tatarian honeysuckle: *Lonicera tatarica*
Teosinte: *Zea mexicana*
Thornless honey locust: *Gleditsia triacanthos* var. *inermis*
Three-square bulrush: *Scirpus americanus*
Three-toothed cinquefoil: *Potentilla tridentata*
Tick trefoil: *Desmodium* species
Tickseed: *Coreopsis lanceolata*
Tiger lily: *Lilium lancifolium*
Timothy: *Phleum pratense*
Tobacco (cultivated for smoking): *Nicotiana tabacum*
Tomato: *Lycopersicon lycopersicum*
Trailing arbutus: *Epigaea repens*
Tree of heaven: *Ailanthus altissima*
Tree peony: *Paeonia suffruticosa*
Trillium: *Trillium* species
Trout lily: *Erythronium americanum*
Trumpet creeper: *Campsis radicans*
Tulip tree: *Liriodendron tulipifera*
Tumbleweed: *Salsola kali*
Tupelo: *Nyssa sylvatica*
Tupelo family: Nyssaceae
Turkey oak: *Quercus cerris*
Turtlehead: *Chelone glabra*
Tussock sedge: *Carex stricta*
Tyrol knapweed: *Centaurea vochinensis*

Umbrella magnolia: *Magnolia tripetala*

Valerian: *Valeriana officinalis*
Venus fly trap: *Dionaea muscipula*
Violet: *Viola* species
Viper's bugloss: *Echium vulgare*
Virginia creeper: *Parthenocissus quinquefolia*
Virginia knotweed: *Tovara virginiana*
Virginia meadow beauty: *Rhexia virginica*
Virginia rose: *Rosa virginiana*
Virgin's bower: *Clematis virginiana*
Vomitwort: *Lobelia inflata*

Walking fern: *Asplenium rhizophyllum*
Wall rue: *Asplenium ruta-muraria*
Water carpet: *Chrysosplenium americanum*
Water hemlock: *Cicuta maculata*
Water parsnip: *Sium suave*
Water plantain: *Alisma plantago-aquatica* var. *americana*
Water smartweed: *Polygonum amphibium*

Weak-stalked bulrush: *Scirpus debilis*
Weeping fig: *Ficus benjamina*
Weeping willow: cultivars of *Salix babylonica,*
   S. alba
Wheat: *Triticum aestivum*
White ash: *Fraxinus americana*
White baneberry: *Actaea pachypoda*
White clover: *Trifolium repens*
White-fringed orchis: *Habenaria blephariglottis*
White mulberry: *Morus alba*
White mulberry (cultivated for silkworms):
   Morus alba var. *multicaulis*
White oak: *Quercus alba*
White pine: *Pinus strobus*
White poplar: *Populus alba*
White snakeroot: *Eupatorium rugosum*
White spruce: *Picea glauca*
White sweet clover: *Melilotus alba*
White trillium: *Trillium grandiflorum*
Whorled aster: *Aster acuminatus*
Whorled loosestrife: *Lysimachia quadrifolia*
Wild bergamot: *Monarda fistulosa*
Wild geranium: *Geranium maculatum*
Wild ginger: *Asarum canadense*
Wild grape: *Vitis* species
Wild indigo: *Baptisia tinctoria*
Wild leek: *Allium tricoccum*
Wild lettuce: *Lactuca canadensis*
Wild mint: *Mentha arvensis*
Wild parsnip: *Pastinaca sativa*
Wild radish: *Raphanus raphanistrum*

Wild sarsaparilla: *Aralia nudicaulis*
Wild strawberry: *Fragaria virginiana* or *F. vesca*
Willow family: Salicaceae
Wineberry: *Rubus phoenicolasius*
Winged euonymus: *Euonymus alata*
Winterberry: *Ilex verticillata*
Wintergreen: *Gaultheria procumbens*
Wisteria: *Wisteria floribunda* or *W. sinensis*
Witch hazel: *Hamamelis virginiana*
Witch hazel family: Hamamelidaceae
Withe rod: *Viburnum cassinoides*
Wood anemone: *Anemone quinquefolia*
Wood aster: *Aster divaricatus*
Wood rush: *Luzula multiflora*
Wood sorrel: *Oxalis montana*
Woodland horsetail: *Equisetum sylvaticum*
Woodland sunflower: *Helianthus divaricatus*
Woodsia: *Woodsia* species
Woolly hudsonia: *Hudsonia tomentosa*
Wrack: *Fucus* species

Yarrow: *Achillea millefolium*
Yellow bedstraw: *Galium verum*
Yellow birch: *Betula alleghaniensis*
Yellow false foxglove: *Aureolaria flava*
Yellow hawkweed: *Hieracium* species
Yellow jewelweed: *Impatiens pallida*
Yellow-eyed grass: *Xyris caroliniana* and related
   species
Yew family: Taxaceae

Glossary

Italicized terms in the descriptions have their own entries in this glossary.

Abscise. In botany, to detach or cut away from.

Acadian mountains. The ancient mountains that rose in New England in the *Devonian period*.

Acid rain. Precipitation rendered acidic by its interaction with human-produced atmospheric contaminants.

Actinomycete. A form of bacterium especially abundant in soil.

Adelgid. A type of plant-feeding insect.

Aerenchyma. The tissue found in some plants that contains air chambers. These air chambers apparently help aquatic plants overcome the problems of gas exchange and buoyancy.

Agaric. A form of fungus that has been used in herbal medicine.

Age of Dinosaurs. See *Age of Reptiles*.

Age of Mammals. A popular name for the *Cenozoic era*.

Age of Reptiles. A popular name for the *Mesozoic era*.

Algin. A substance derived from kelp and other brown algae. It is used as an ingredient in a variety of food, beverage, and industrial products.

Alien. In botany, a non-native plant species that reproduces successfully in the wild without human assistance.

Alleghenian mountains. The ancient mountains that rose on the eastern side of North America near the close of the *Paleozoic era*.

Allochthonous. Referring to a large unit of rock that was formed in one place, and was subsequently moved to another location. See *klippe*.

Alternate. In botany, referring to buds, leaves, or branches that are arranged along a stem singly, in a staggered, unpaired arrangement.

Angiosperm. A plant that is a member of the division Magnoliophyta, the flowering plants.

Annual. In botany, a plant that completes its entire individual life cycle in a single growing season.

Anorthosite. A type of *igneous rock* common on the moon but relatively rare on earth. It is, however, an important constituent of the Adirondack Mountains of northeastern New York State.

Anthocyanin pigments. The plant-cell pigments that are most often responsible for the red tones in autumn foliage coloration.

Apically dominant. Referring to a growth pattern in plants in which the *leader* is favored over the lower, side branches.

Apothecium. A cup-shaped reproductive structure found on many lichens.

Apple pine. See *pumpkin pine*.

Arboretum. An institution where living tree specimens, and usually shrubs and other plant types as well, are displayed and studied.

Aril. A fleshy protective jacket that encloses a seed.

Arkose. A sandstone containing a substantial amount of feldspar grains in addition to the usual quartz grains.

Asexual propagation. Any type of natural or human-induced reproduction of plants that does not involve the sexual cycle in which two gametes fuse.

Autochthonous. Referring to a large unit of rock that is still situated in its original location.

Avalonia. The *suspect terrane* that many geologists believe joined with ancestral New England in the *Devonian period*.

Backshore. The part of a beach that lies between the high-tide mark and sand dunes or other inland landforms. The backshore usually contains one or more *berms*.

Banner tree. Synonymous with *flag tree*.

Basalt. An important and prevalent *extrusive, igneous rock* that is fine-grained and dark-colored (though it may weather to a light reddish brown).

Bast. Fibers derived from the *phloem* tissue of linden, basswood (originally "bast wood"), and other plants. Bast was most often used for cordage and cloth.

Beauty strip. A forestry term for a narrow boundary of surviving vegetation that prevents boaters and motorists from seeing the extent of *clear-cutting* done by logging companies.

Bedrock. A rock unit that is a fixed part of the earth's crust (distinct, for example, from a detached boulder). See *ledge*.

Berm. A terracelike or benchlike surface found on the *backshore* portion of a beach.

Berry. In precise botanical usage, the type of fleshy or pulpy *fruit* that has one or more seeds.

Biennial. In botany, a plant that completes its entire individual growth cycle in two growing seasons. Characteristically, a biennial flowers and sets fruit only during the final growing season.

Bisexual plant. Synonymous with a *perfect plant*.

Blaze. A trail marker in the form of a painted or carved symbol on a tree or post.

Board-and-batten. A traditional style of construction in which the joints between vertical wooden boards on a building's exterior are covered with narrow protective or decorative lumber strips.

Bog. In general usage, any wet and inaccessible ground. In more precise ecological usage, the type of *peatland* characterized by high water acidity, little or no water circulation, and a paucity of available nutrients.

Botany. The branch of biology that deals with four of the five kingdoms of life on earth: plants, fungi, Protists (algae, etc.), and bacteria.

Bottomland. Low, level land along a river or smaller stream. Bottomland is usually considered a prime agricultural zone because of the high fertility of its soil.

Bract. A modified leaf that serves a specialized function: for example, the bright red bracts subtending the otherwise small and visually insignificant flowers of the poinsettia plant.

Broad-leaved tree. An *angiosperm* tree.

Brookie. An angler's nickname for brook trout.

Brownstone. A reddish brown *Mesozoic era* sandstone, such as the Portland Arkose, which has been extensively quarried in the lower Connecticut River Valley for use in building construction.

Bryophyte. A *nonvascular plant.* Mosses, liverworts, and hornworts are bryophytes.

Cabbage pine. A white pine tree with a low, rounded *habit* reminiscent of a head of cabbage.

Caliper. A tree's trunk diameter, as measured at chest height.

Cambium. The zone of growth in plant tissue where new cells—for instance, cells for new *phloem* and *xylem*—are created.

Cambrian period. The first period of the *Paleozoic era.* The Cambrian lasted approximately from 570 million to 505 million years ago.

Capitulum. The top of a sphagnum moss plant, composed of a cluster of short branches. The term is Latin for "little head."

Carbohydrate. An important type of organic compound. Many crucial components of plant tissue—including sugars, *cellulose,* and starch—are carbohydrates.

Carboniferous period. The period of the *Paleozoic era* that lasted approximately from 360 million to 285 million years ago. In the United States, geologists usually break up the Carboniferous into two separate periods, the *Mississippian* and the *Pennsylvanian.*

Carotenoid pigments. Leaf-tissue pigments that are often responsible for the yellow and orange tones in autumn foliage coloration.

Carr. The type of *peatland* characterized by an extensive stand of shrubs (e.g., alders or willows) growing on a sphagnum moss mat.

Catkin. An elongated spike or cluster of flowers, all of which are the same sex.

Cellulose. A *carbohydrate* composed of glucose that is the main constituent of plant cell walls.

Cenozoic era. The current era of the geologic time scale. It began approximately 65 million years ago.

Central Lowland. The area of southern New England that is underlain by Mesozoic era sedimentary and igneous rocks, from the northern Massachusetts border to New Haven, Connecticut.

Character. In botany, any identification trait that helps to distinguish one plant type from another.

Chlorophyll. The green pigment, essential to *photosynthesis,* that is present in plants, algae and some bacteria.

Chlorosis. A condition of unhealthy plants in which their foliage turns yellowish. Chlorosis is often caused by the deficiency of available iron or other nutrients.

Circinate vernation. The method by which ferns and some other plants unroll and deploy their growing leaves from their original coiled *fiddlehead* position.

Circumglobal species. A plant species that is native to most of the earth's continents.

Circumpolar species. A plant species that is native to the northern regions of both Eurasia and North America.

Cirque. Pronounced "serk." An amphitheater-shaped hollow on a mountainside. Cirques are carved by alpine (mountain-formed) glaciers.

Clapboard. Pronounced "*dab*-erd." A traditional style of construction in which the wooden boards on a building's exterior are mounted horizontally. Often these boards are thicker on one side.

Clear-cutting. The timber industry practice of cutting down all trees in a particular tract, rather than selectively choosing the trees that best fit their needs.

Clonescape. A term used by the author to describe the look of modern American shopping districts—characterized by gas stations, franchise food restaurants, and the like—in which the same building styles and layouts are repeated from one end of the country to the other, regardless of differing climates, cultural factors, or other geographical considerations.

Club moss. A *pteridophyte* of ancient pedigree that belongs to genus *Lycopodium.* See *lycopod.*

Cohesion-Adhesion-Tension Theory. A theory that seeks to explain how plants, including very tall trees, transport water and nutrients from the roots to the uppermost branches and leaves. The theory cites a combination of three important factors: the cohesiveness of the rising water molecules, which permits them to stick to each other; the adhesiveness of the water molecules, which allows them to get a good grip on the sides of the plant cells they rise through; and the tension or upward pull on the water caused by the leaves' *transpiration.* The effectiveness of these factors rests on the fact that the channels provided by the cells are very narrow.

Collenchyma. A type of plant tissue that provides a stem with extra support for structural strength.

Composite. A plant that is a member of the composite family (Compositae).

Compound. Referring to a leaf that is composed of two or more distinct *leaflets.*

Conifer. A cone-bearing plant. Botanists usually restrict the term to members of the order Coniferales (also known as the Pinales), which includes the pine, cypress, redwood, podocarp, araucaria, and plum yew (but not true yew) families.

Connecticut mahogany. The black cherry tree, or wood therefrom.

Continental glaciation. The formation of a massive, lobed ice sheet that spreads over a large portion of a continent.

Contractile roots. Roots that tend to pull a plant's exposed stems or foliage downward, to maintain a prostrate *habit.*

Controlled burning. A practice first developed on this continent by American Indians,

546

and now widely used in parkland maintenance, to control the growth of unwanted vegetation by deliberately setting fire to selected tracts of ground.

Convergent evolution. The development of similar traits in unrelated forms of life.

Cordaite. A type of early *gymnosperm* common in the *Pennsylvanian period* "coal swamp" wetlands of North America and other continents. Cordaites apparently became extinct by the end of the *Paleozoic era.*

Cordate. Heart-shaped. Botanists usually restrict the term to the base of a leaf that has the rounded shapes of the top of a stylized heart.

Cortex. A term with more than one meaning in botany. In this book it is used to indicate the outermost part of a lichen's interior tissues.

Cradle. In forestry usage, a hollow in the woodland floor formed by the toppling of a tree and the removal of soil by the wrenching action of its roots.

Cretaceous period. The third and final period of the Mesozoic era. The Cretaceous lasted approximately from 144 million to 65 million years ago.

Cross-pollination. The type of *pollination* involving the interaction of two or more plants. Cross-pollination offers the advantage of keeping a community of plants genetically diverse.

Crown. In botany, the system of leaf-bearing branches of a tree—as distinct from the trunk below it. It also sometimes is used to refer to the base of a tree's trunk, where the roots begin.

Crozier. In botany, synonymous with *fiddlehead.*

Cretaceous period. The third and final period of the Mesozoic era. The Cretaceous lasted approximately from 144 to 65 million years ago.

Crustose lichen. A lichen with an extremely flat form, often resembling a coat of paint or a layer of powder.

Culm. The stem of a grass plant.

Cultivar. A contraction of "cultivated variety"; that is, a variety of a plant species that has been developed by human beings for a particular horticultural or agricultural use.

Cusp. In botany, the sharp, short point found on the tips of some leaves.

Cuticle. The outermost, protective layer of leaf tissue. It is composed of *cutin* and wax.

Cutin. A fatty substance that is a main constituent of a leaf's *cuticle* layer.

Cycad. An unusual type of *gymnosperm* that appears to resemble a small, stout-trunked palm tree. Cycads were an important part of the world's plant community in the *Mesozoic era,* but they are now restricted to just a few species and habitats.

Cycadeoid. A type of *gymnosperm* common in the *Mesozoic era,* but now extinct. As its name implies, it resembled the *cycads,* especially in its foliage; but other aspects of this enigmatic group of plants set it distinctly apart.

Day use area. A key term in the vocabulary of Parkspeak (the trade language of parks officials and other ecobureaucrats). A day use area is simply a part of a park set aside for visitors who, unlike campers, do not plan to stay overnight.

Deal. In forestry, the timber of the Norway spruce tree.

Deciduous. Referring to plants that shed all or most of their foliage at a particular time; for example, in autumn or during a season of drought.

DED. An acronym for *Dutch elm disease.*

Dendrology. The branch of botany that deals specifically with the study of trees.

Depositional coastline. A coastline with beaches that are growing upward and outward, due to the arrival of new sediment deposits.

Desire line. A term used by park officials to denote a footworn trail made by people taking a shortcut between established paths.

Development. A widely used euphemism for building projects, however tacky, redundant, or environmentally irresponsible.

Devonian period. The fourth period of the *Paleozoic era*. The Devonian lasted approximately from 408 million to 360 million years ago.

Diabase. An *igneous rock* similar in composition to *basalt*, but differing from the latter by being *intrusive*, rather than *extrusive*.

Dicot. A plant belonging to class Magnoliopsida, one of the two main groups of *angiosperms*. Dicots characteristically have seeds that can be separated into two "halves"—as can easily be demonstrated with a bean seed. In addition, a dicot usually has flower parts arranged in fours or fives, as well as leaves that are not parallel-veined.

Dike. In geology, a relatively narrow band of rock that cuts through other, older rock units.

Dimorphic. Capable of having two distinct forms. For example, white mulberry leaves are dimorphic because they can be either lobed or unlobed.

Dioecious. Having flowers of only one sex on a particular plant. The term also refers to the nonflowering plants (e.g., the ginkgo) that are unisexual, too.

Diorama. A closed-in display that depicts a particular environment, using models in the foreground and a painted backdrop.

Dolerite. Synonymous with *diabase*.

Double fertilization. The mode of reproduction characteristic of *angiosperms*, in which the two male gametes contained in a pollen grain are both involved in the process of fertilization. In this way both the seed's embryo and its associated *endosperm* tissue are formed.

Down East. A Maine term that indicates both direction (northeastward—and hence, generally downwind—along the seacoast, toward New Brunswick) and location (the northeasternmost half of the state's seacoast area).

Drumlin. A geologic term denoting a fairly low, elongated, and rather humpbacked hill formed when the margin of a glacier molds and streamlines a *till* deposit.

Drupe. The type of fruit with a fleshy exterior enclosing a hard-coated seed. Cherries and plums are good examples of drupes.

Duff. The layer of fallen conifer needles that helps to create *podzol* soil conditions.

Dune. A hill or ridge of wind-blown sand.

Dutch elm disease. A deadly plant disease affecting various elm species. In this country the most significant victim is the American elm. Dutch elm disease is caused by the fungus *Ophiostoma ulmi*, which in turn is transmitted by both the European and North American elm bark beetles. The disease is actually composed of various strains, which vary in aggressiveness and virulence.

Ecotone. The transition zone between two distinct habitats.

Ecotourism. The industry that caters to the public demand for travel emphasizing natural—or naturalistic—settings.

Emergent. In wetland ecology, an aquatic plant that rises above the water's surface.

End moraine. A ridgelike deposit of *till* formed at the front of a glacier.

Endosperm. The nourishing tissue that serves as a food source for the embryo of an *angiosperm* seed.

Ensilage. Chopped-up *fodder* that is stored in dairy farm *silos*.

Epiphyte. A nonparasitic plant that grows on another plant.

Ericad. A plant that is a member of the heath family (Ericaceae).

Erosional coastline. A coastline with dwindling beaches or rocky shores that are being gradually eaten away by the effects of storms, wave action, currents, or other similar effects.

Erratic. In geology, a detached rock that has been carried from its point of origin by a moving glacier.

Escape. A good example of the American penchant for turning verbs into nouns. In plant ecology, an escape is an *alien* species, intended only for cultivation or for the garden, which nevertheless manages to establish itself in the wild.

Esker. A long, snakelike ridge that was formed by running water depositing sand and gravel in a tunnel inside a glacier.

Essence. In agriculture, an oil or other extract derived from aromatic plants.

Eukaryotic. Referring to an organism that has cells with distinct, highly specialized parts enclosed in membranes. Four of the five kingdoms of life—the *Protists,* fungi, plants, and animals—are eukaryotic.

Evergreen. A plant that keeps its green foliage for more than one growing season. Actually, most individual evergreen leaves are shed eventually and replaced by new growth, but the shedding process is gradual.

Exfoliate. To peel off or fall away from.

Exotic terrane. A *terrane* that was formed in a place different than its present location.

Extinction event. A point in earth history that marks the disappearance of one or more forms of life.

Extrusive rock. *Igneous rock* that was still in its molten state when it reached the earth's surface.

Family. In biology, a group of closely related *genera.*

Fan. In architecture, a semi-elliptical, Federal-style ornament composed of radiating, fan-like slats.

Fanlight. A type of Federal-style *fan,* in which panes of glass take the place of the wooden slats.

Fascicle. A bundle of leaves or other plant parts. Applied to pine trees, it refers to a set of needles joined together at their common base. Applied to sphagnum moss, it refers to a cluster of branches attached at one point on the plant.

Fastigiate. Referring to a tree with closely spaced, upward-pointing branches that produce a tall, columnar *habit.*

Fen. The type of *peatland* characterized by moderate water acidity, some water circulation, and at least some available nutrients.

Fiddlehead. In botany, a young fern leaf in its initial coiled position.

First-growth forest. A mature woodland that has never been logged or cleared for agriculture.

Fixed dune. A *dune* that is held in place by a mantle of vegetation.

Fjord. A deep, narrow, flooded valley that was originally a glacial trough.

Flagging. The rapid dieback of tree branches that is especially characteristic of Dutch elm disease.

Flag tree. A tree in an exposed location that has its branches permanently twisted to leeward by the force of the prevailing wind. Compare with *krummholz*.

Flatlander. A distinctly pejorative term used by northern New Englanders to describe human beings from anywhere else—including their close neighbors in the rather unflat country of the Massachusetts and Connecticut Berkshires. When flatlanders decide to settle in rural Vermont and New Hampshire, they can be counted on to refurbish dilapidated barns, interfere at town meetings, and insist on road repairs several decades before they're absolutely necessary.

Flora. The plant community of a particular area.

Flower. The reproductive structure of *angiosperms*.

Fodder. Food for such domesticated animals as cattle and horses.

Foliage. The leaves of one or more plants.

Foliose lichen. A lichen of relatively flat habit with an upper surface different from the lower.

Foreshore. The part of a beach that is alternately covered and exposed by the action of the tides.

Frond. A fern leaf.

Frost drought. A wintertime affliction of plants, in which their *xylem* tissue is frozen, with the result that their upper stems and leaves are deprived of water.

Fruit. A ripened ovary or ovaries that develop from an *angiosperm* flower.

Fruticose lichen. A branched lichen with no readily discernible difference between its upper and lower surfaces. Fruticose lichens may be either upright and stemlike, or dangling and hairlike.

Fucoxanthin. An olive green pigment found in brown algae.

Gabbro. A dark-colored *igneous rock* that is the *intrusive* equivalent of the *extrusive* rock *basalt*.

Gall. A swelling on a plant stem or leaf analogous to a tumor. Galls are caused by the presence of such foreign organisms as fungi, mites, and wasps. They are rarely if ever divided into three parts.

Genus. A group of closely related *species*. Plural, *genera*.

Ginkgophyte. Any *gymnosperm* of the ancient order of plants now represented only by the ginkgo tree species.

Girdle. To kill a tree, however slowly, by making a ringlike incision all the way around its trunk. In the process the continuity of the *phloem* tissue is destroyed, with the result that the tree cannot supply its lower portions with the food manufactured in the leaves.

Glaciomarine deposit. A deposit of coastal sediments carried into seawater by the outwash streams of a melting glacier.

Glaucous. Referring to a plant part covered with a substance that imparts a white or pale blue tint.

Gneiss. A *metamorphic rock* derived from *granite* or similar *igneous rocks*. Gneiss is distinguished by its alternating bands of light and dark minerals.

Gradient. In geology, the steepness of a slope.

Granite. An *intrusive, igneous,* quartz-bearing rock that has mineral crystals visible to the naked eye.

Granodiorite. A type of coarse-grained *intrusive, igneous rock.*

Gymnosperm. Any member of the class of plants that produce "naked" seeds not enclosed in true fruit. Gymnosperms past and present include *cordaites, conifers, cycads, cycadeoids, yews,* and *ginkgophytes.*

Habit. A plant's overall shape or silhouette, as seen from a little distance.

Hardpan. Compacted, impermeable soil.

Hard-rock geology. An insider's term for the branch of earth science that deals with *igneous* and *metamorphic rocks,* and with the *terranes* formed by them.

Hardwood. In forestry, an *angiosperm* tree.

Haustorium. A rootlike structure found in fungi and parasitic plants. It penetrates a host organism's tissue and robs it of food, water, or nutrients.

Heat island. An area along a city street where temperatures soar far above normal.

Herb. In general botanical usage, any plant that does not develop distinctly woody tissue: for example, many wildflowers and ferns. In a more restricted sense, a herb is a plant that is either grown or sought out in the wild for its purported medicinal, culinary, or domestic uses.

Herbaceous. Not *woody.*

Herbalist. A person skilled in the selection, preparation, and use of *herbs.*

High-energy environment. In geology, an exposed seacoast where strong wind action and vigorous surf conditions prevail.

Hip. In botany, the type of fruit produced by rose plants.

Hobby farm. A farm operated by a person who earns the bulk of his or her living elsewhere.

Holdfast. In botany, the lower part of a marine alga, which anchors it to a solid surface.

Homoplastic morphology. The manifestation of *convergent evolution,* in which two unrelated organisms independently develop a common form.

Horsetail. A *pteridophyte* of ancient lineage, belonging to the genus *Equisetum.* Some horsetail species are known as scouring rush.

Horticulture. The science and art of growing and maintaining ornamental plants and fruit trees.

Husband-and-wife trees. A pair of matching trees planted in front of a farmhouse.

Hybrid. A "cross," or offspring, of two dissimilar parents. In horticulture, hybrids between different *species* are common; there are even a few examples of crosses between *genera.*

Iapetos (or Iapetus) Ocean. An ocean extant in the *Paleozoic era,* roughly analogous to the modern North Atlantic.

Igneous rock. A rock formed directly from the solidification of *magma.*

Indicator species. An organism that is a good benchmark of a particular condition, habitat, or environment.

Inflorescence. A group or cluster of flowers and its immediate supporting structure.

Intervale. A northern New England term for a tillable valley floor situated between hills or mountains.

Intrusive rock. *Igneous rock* that solidified beneath the earth's surface.

Involucre. An arrangement of *bracts* or other leaflike structures attached at the base of some *inflorescences* or fruit.

Isidium. A small cylindrical or fingerlike reproductive structure found on some lichens.

Joint. In geology, a fracture in granite, or other rock types, where no significant displacement has occurred.

Juglone. A toxic substance secreted by walnut tree roots. It discourages the growth of nearby competing plants.

Jurassic period. The second period of the *Mesozoic era.* The Jurassic lasted approximately from 208 million to 144 million years ago.

Kame. A moundlike feature composed of sediments originally deposited at the edge of, or in a hole in, a glacier.

Kame moraine. A *moraine* that contains *kame* structures.

Kame terrace. A terracelike structure found in some river valleys that owes its origin to the accumulation of sediments between a glacier and the exposed valley wall.

***K* environment.** An ecology term denoting a mature, stable, and densely populated environment, where many different species are present, but only a few individuals of each species. See *r environment.*

Kettle. In geology, a pond or other depression created by the gradual melting of a detached and at least partly buried block of glacial ice.

Kitchen garden. A domestic garden, especially common in colonial times, which was devoted to small food crops and cooking herbs.

Klippe. A rock unit isolated by erosion after being moved along a fault from its point of origin. In New England, klippes are often composed of older rocks that have come to rest atop younger rocks—which is a reversal of the normal geologic sequence.

Knee. In botany, an exposed bump or prominence on the roots of a bald cypress tree. Originally, it was thought that knees help the tree's roots absorb needed oxygen; however, recent research suggests that this may not be true.

Krummholz. A stand of trees stunted and deformed by high winds and other harsh conditions at higher elevations. From the German words for "crooked wood."

Laminated arch. A covered-bridge arch composed of separate, long sheets of wood set one atop another and strapped together.

Landform. Any feature on the earth's surface that has been produced by natural forces: for example, mountains, lakes, valleys, *moraines,* and even anthills.

Layering. A horticultural technique of *asexual propagation* in which a stem is induced to form roots while it is still connected to its parent plant.

Leader. In horticulture, the uppermost continuation of a tree's trunk. In the ideal, well-shaped tree, the leader is the tallest central and vertical extension of the *crown.*

Leaflet. A separate subdivision of a compound leaf. Leaflets are often confused with the leaves of which they form only a part; for instance, the cloying jingle about poison ivy ("leaves of three, let it be") should really be "*leaflets* of three. . . ."

Leaf-peeper. Another classic Yankee exclusionist term, but a rather gentle one. Leaf-peepers are tourists who cruise the region in the fall, in search of "peak foliage"—the best autumn leaf coloration. Leaf-peepers are usually so dazzled by the chromatic display that they drive as slowly as the natives always do.

Ledge. A New England term, synonymous with *bedrock*.

Legume. A plant that is a member of the order Leguminales. This order is composed of the bean, mimosa, and caesalpinia families, all of which have characteristic fruit in the form of pods. Common legumes include peas, peanuts, clovers, and locust trees.

Lenticel. A wartlike or ridgelike bump of raised tissue found on the stems of some woody plants.

Lignin. The wood-forming substance in plant tissue.

Lignite. A low-grade brownish coal. In contrast to bituminous coal and anthracite, lignite is a potential source of fossils, because it contains recognizable wood tissue and other plant parts.

Living fossil. An extant organism that has also left its trace in the fossil record. The ginkgo and the dawn redwood are two classic examples of living fossils.

Low-energy environment. A geology term denoting an environment (such as a protected tidal marsh) where wave and surf action is minimal.

Lycopod. A plant of the *pteridophyte* genus *Lycopodium*. Better known as the club mosses or ground pines, the lycopods are of ancient ancestry.

Lycopsid. A more general term than *lycopod*, denoting any member of the class Lycopsida. This class includes the lycopods and such extinct, tree-sized plants as *Lepidodendron*.

Magma. Molten rock.

Marine alga. Any type of algal *Protist* that lives in salt water. Plantlike, multicellular marine algae are often called *seaweeds*.

Marsh. A *wetland* characterized by a general lack of trees and shrubs. Instead, *herbaceous* plants (for example, cattail or tussock sedge) dominate.

Meander. In geology, a pronounced bend or loop in the course of a river.

Medulla. As used in this book, the internal tissue of a lichen.

Megabotany. An ad hoc term, used by the author in this book to denote the study of large-scale patterns in plant communities.

Megaflora. An ad hoc term denoting a series of large-scale plant communities and their ecological settings.

Megasporangiate cone. The female reproductive structure of the *conifers*. It bears the ovules and, after fertilization, the seeds. The megasporangiate cones of pines are the "pine cones" of common parlance.

Meristem. A zone of tissue where new plant cells are formed.

Mesozoic era. The penultimate era of the geologic time scale. The Mesozoic lasted approximately from 245 million to 65 million years ago. It is often nicknamed the Age of Reptiles or the Age of Dinosaurs.

Metamorphic rock. Rock that has been altered from its original form or composition without being reduced to a molten state.

Microsporangiate cone. The male, pollen-producing reproductive structure of *conifers*. Microsporangiate cones are usually smaller than their female *megasporangiate* counterparts, and thus are usually not noticed by the casual observer.

Mississippian period. The fifth period of the *Paleozoic era*. The Pennsylvanian lasted approximately from 360 million to 320 million years ago. See *Carboniferous period*.

Monadnock. A freestanding mountain or hill that rises prominently above an area of otherwise low relief.

Monkeytail. See *fiddlehead*.

Monocot. A plant belonging to the class Liliopsida, one of the two main groups of *angiosperms*. Monocots characteristically have seeds that cannot be separated into two "halves." Usually, monocot flower parts are arranged in threes or multiples thereof; the leaves are often parallel-veined.

Monoculture. A community of cultivated plants (e.g., farm crops or street trees) composed of only one plant type.

Monoecious. Having both male and female flowers that are borne separately on one plant.

Monotypic stand. A more general term than the preceding. It denotes any plant community, wild or cultivated, composed of only one plant type.

Moraine. A ridge or other high area composed of *till* and other debris left behind by retreating glaciers.

Mucro. A small, sharp tip, as found, for example, on the ends of some leaves.

Mulch. A protective material applied to the soil surface. It discourages weed growth and moderates soil temperatures. It may be either organic (e.g., wood chips) or inorganic (e.g., plastic sheeting).

Muskeg. A synonym for *peatland*, most often used in Maine and Canada.

Mutualism. The interaction of two different types of organisms in which both types benefit.

Mycorrhiza. The symbiotic association of a beneficial fungus and its host plant root.

Neogene period. The second and current period of the *Cenozoic era*. It began approximately 24 million years ago.

Nonvascular plant. A plant lacking well-defined tissues that conduct water, nutrients, and food throughout its interior. Synonymous with *bryophyte*.

Nucleated community. A geographer's term denoting a traditional agricultural town, in which the houses and other buildings are grouped together in a central and sometimes fortified area, with outlying communal farm fields.

Ombotrophic. Referring to a *peatland* that receives its moisture from precipitation rather than from surface water flow.

Ombrogenous. Synonymous with *ombotrophic*.

Opposite. In botany, referring to leaves, buds, or branches arranged in pairs along the stem.

Orbicular. Circular in shape. The synonym *orbiculate* is sometimes used instead.

Ordovician period. The second period of the *Paleozoic era*. The Ordovician lasted approximately from 505 million to 438 million years ago.

Outcrop. An exposed section of *bedrock*.

Outwash plain. A plain composed of sediments deposited in front of a melting glacier.

Overburden. Any structure, surface, or material that separates a geologist from *bedrock*.

Oxbow. In geology, either a tightly curved river *meander* or the narrow, arc-shaped lake that is formed when such a meander is cut off at its neck.

Paleobotany. The study of ancient plant life and its evolution, as revealed in the fossil record.

Paleogene period. The first period of the *Cenozoic era*. The Paleogene lasted approximately from 65 million to 24 million years ago.

Paleozoic era. The era of the geologic time scale that lasted from 570 million to 245 million years ago.

Palladian window. A common design element in Federal (and also Georgian) architecture. A Palladian window has a large, arched center window that is flanked by two smaller, rectangular ones.

Palmate. Referring to a leaf that is fan-shaped, with lobes or leaflets radiating from a central point.

Palmately compound. A *compound* leaf with leaflets that radiate in a circular or semicircular pattern from a single point of attachment.

Paludification. The flooding of a previously dry tract of land, resulting in the creation of a *wetland*.

Pangaea. The supercontinent that existed from the late *Paleozoic era* to the early *Mesozoic era*. It contained all the earth's major land masses and stretched virtually from pole to pole.

Panicle. A type of *inflorescence* that has flower-bearing branches growing out from a vertical axis.

Parthenogenesis. A process in which organisms produce their young without the normal sexual process of fertilization of an egg.

Peatland. A type of *wetland* in which the soil contains mostly organic matter.

Pelite. A type of fine-grained *sedimentary rock,* formed from mud deposits.

Peneplain. A type of *landform* that has been reduced by the forces of erosion to a low, gently rolling surface.

Pennsylvanian period. The sixth period of the *Paleozoic era*. The Pennsylvanian lasted approximately from 320 million to 285 million years ago. See *Carboniferous period.*

Perennial. In botany, a *herbaceous* plant that completes its individual growth cycle in three or more growing seasons.

Perfect plant. A plant that bears flowers each of which contains both male and female reproductive parts.

Perigynous. Referring to a stamen or other flower part that is attached around a flower's ovary, rather than above or below it.

Permian period. The seventh and final period of the *Paleozoic era*. The Permian lasted approximately from 285 million to 245 million years ago.

Petal. A modified leaf that is part of a flower's corolla. It is often brightly colored, and it often serves as an attraction device for pollinating animals.

Petiole. The leaf stem that connects the leaf blade to its branch.

Petroglyph. An image or character carved into a rock surface.

Phenocryst. A large, distinctly visible mineral crystal found in certain types of *igneous rock.*

Phloem. The zone of internal plant tissue that takes food manufactured in the leaves to other parts of the plant.

Photosynthesis. The biochemical process in which light is absorbed by chlorophyll and other pigments to turn water and carbon dioxide into the *carbohydrates* the plant uses for food.

Phycobilin. A red pigment found in the red algae.

Phyllite. A *metamorphic rock* containing minerals that give it a silvery sheen.

Pillow. In forestry usage, a high, rounded area in the woodland floor situated between *cradles*.

Pillow lava. A lava that flowed and cooled underwater. It has a characteristic globular or pillowlike shape.

Pit barn. A barn that is either partially sunken into the ground or flanked with sod slopes, to protect the crops stored within from very low temperatures.

Plant material. A rather unfeeling term much used by landscape architects to denote the living plants they use in their designs.

Plant regime. The community of plants living in a particular environment.

Plate tectonics. A widely accepted concept in modern geology that describes how large pieces of the earth's crust, known as plates, move relative to one another.

Pleistocene epoch. The subdivision of the *Neogene period* that lasted from approximately 2 million to about 10,000 years ago. The Pleistocene was the Ice Age epoch, when continental glaciers covered large portions of North America and Eurasia.

Plume. In geology, a source of heat originating deep in the earth's interior that causes molten rock to rise upward toward the surface in one particular area.

Pluton. A large body of subterranean *igneous rock*.

Pneumatophore. A part of the root system of some wetland trees that sticks upward into the air, and which possibly helps the plant absorb oxygen.

Podzol. A type of soil characteristic of the taiga and other areas where *conifers* are predominant. Podzol is generally covered with a layer of fallen conifer needles, and is acidic and nutrient poor.

Pollination. The act of bringing pollen from an anther (a male flower part) to a stigma (a female flower part).

Polyembryony. The existence of more than one embryo in a seed.

Precambrian time. The first and longest major division of the geologic time scale. It lasted from the formation of the earth, some 4.6 billion years ago, to approximately 570 million years ago.

Prickle. Bark tissue that has been modified into a sharp and pointed structure.

Primary growth. Plant tissue growth that manifests itself in the lengthening, rather than the thickening, of the stem.

Progradational. Referring to a shoreline that is growing outward.

Prokaryotic. Referring to an organism that has cells without distinct, highly specialized parts enclosed in membranes. Only one of the five kingdoms of life—Monera (the bacteria)—is prokaryotic.

Prothallus. A small vegetative body that is a fern's gamete-producing generation.

Protist. A member of kingdom Protista, which includes the algae, the slime molds, and the diatoms.

Pteridophyte. A member of the most ancient group of *vascular plants*. Pteridophytes reproduce by free spores rather than by seeds. They include the ferns, the horsetails, the club mosses, and many extinct types.

Pteridosperm. A seed fern; that is, an extinct form of plant that superficially resembled a true fern, though in fact it reproduced by seeds rather than by free spores.

Pulldown. A tree felled by a storm or other natural forces.

Pulp. In forestry, the soft substance (mostly *cellulose*) derived from wood and used to make paper.

Pumpkin pine. A term used by some carpenters and antiques experts to refer to well-aged pine wood that has seasoned to a rich, orange-brown tone. Also called *apple pine*.

Raised bed. A garden or landscaping bed built up above ground level.

Raised bog. A bog that is dome-shaped, rather than level, in cross section. Raised bogs are restricted to very wet environments.

Raker. A Down East Maine term for a worker who harvests blueberries.

Red bed. A unit of sandstone or other similar *sedimentary rock* that has a reddish cast.

Regression. In geology, the retreat of an ocean or lake and the corresponding emergence of land.

r environment. A term denoting an ecological vacuum—for instance, a newly flooded field or a recently mowed roadside. This environment tends to be recolonized by one or just a few species especially adept at reproducing rapidly and spreading over a large area. See *K environment*.

Resin. A chemical compound composed of turpentine and *rosin*. Resin is secreted by the wood tissue of *conifers* as a chemical defense against fungal diseases and some animal pests.

Rhizine. A small, rootlike projection on the undersides of some lichens.

Rhizome. A horizontal underground stem.

Rhyolite. A fine-textured *extrusive, igneous rock*.

Ring dike complex. A ring-shaped pattern of *igneous rock* that cuts upward through preexisting rock units.

Road-cut. An *outcrop* exposed by a road excavation.

Rosaceous. Referring to a rose family plant.

Rosin. A waxy substance that is one of the primary constituents of *resin*.

Safrole. A toxic chemical compound found in sassafras plants.

Salt hay. Salt-marsh grasses harvested for *fodder*.

Salt marsh. A seacoast *wetland* that lies within the influence of the daily tides.

Salt pan. An area above a salt marsh that is usually flooded only at the *spring tide*. Standing seawater left there after the retreat of the spring tide evaporates, leaving salt deposits.

Samara. A type of fruit with winglike projections.

Sawtimber. Logs destined for the sawmill, where their wood will be processed for lumber. This in contrast to trees cut down for paper pulp.

Scarp. A linear cliff face.

Schist. A common *metamorphic rock* that has its minerals arranged in a layered pattern.

Scree. Synonymous with *talus*.

Scorpioid. A botany term referring an *inflorescence* that has a coiled shape resembling a scorpion's tail.

Seaside heath. A seaside plant community dominated by dwarfed shrubs and salt-tolerant wildflowers.

Seaweed. A popular term for a marine multicellular alga.

Secondary growth. Plant tissue growth that manifests itself in the thickening, rather than the lengthening, of the stem.

Second-growth forest. A woodland that has grown back after having been logged or cleared for agriculture.

Sedimentary rock. Rock formed by the accumulation of rock particles or the remains of organisms, or by chemical precipitation.

Seed. The fertilized ovule of a *gymnosperm* or *angiosperm*. The seed contains the plant embryo.

Seed fern. See *pteridosperm*.

Seed piece. A small section of a potato *tuber* sown to produce a complete potato plant.

Serotinous. Referring to a plant structure, such as a cone or flower, that delays its opening.

Sessile. Referring to a leaf or other structure attached directly to the stem without a *petiole*.

Set out. A common New England term, synonymous with "to plant outdoors."

Shrub. A woody plant, usually less than 15 feet tall at maturity, which has more than one main stem.

Sidelight. A narrow, vertical window placed beside a main doorway in Federal-style buildings.

Silo. A tall dairy farm structure used to store chopped-up fodder.

Silurian period. The third period of the *Paleozoic era*. The Silurian lasted approximately from 438 million to 408 million years ago.

Simple. In botany, denoting a leaf that is not subdivided into distinct *leaflets*.

Snag. In forestry, a dead tree that is still standing.

Soft-rock geology. An insider's term for the branch of earth science that deals with *sedimentary rocks* and the *terranes* formed by them.

Softwood. A forestry term, synonymous with *conifer*.

Soredium. A powdery reproductive structure on the surface of a lichen.

Spadix. A cylindrical or club-shaped cluster of flowers, as found in arum family plants.

Spathe. The sheathlike modified leaf that surrounds the *spadix*.

Species. The taxonomic rank below *genus* that is composed of organisms of one identifiable type. One general criterion for a species is that its members can interbreed only with each other; but as anyone who has seen a mule or hybrid petunia knows, it just isn't that simple.

Sphenopsid. A *pteridophyte* that is a member of the extant horsetail family, or of other related but extinct families.

Spine. A leaf or leaf part that has been modified into a sharp, pointed projection.

Split-rail fence. A straight fence made of wooden posts and split logs.

Spore. In botany, a reproductive cell (or group of cells) that can develop into a plant's gamete-producing generation.

Spring ephemeral. A forest wildflower that accomplishes its growth and flowering cycle before the onset of summer.

Spring tide. The larger-than-normal high tide that occurs near the time of the full moon.

Squaretail. Synonymous with *brookie.*

Staddle. A wooden post erected in a tidal marsh as a support to dry *salt hay.*

Stick. In forestry, a log section that has been cut to a standard length for pulping.

Stipe. A fern frond's *petiole;* also, the stemlike structure of kelps and other *marine algae.*

Stipule. A structure found at the base of a leaf in some plants. A stipule can take the form of a winglike or wispy appendage (as in roses), of a small leaf (as in some willows), or of a sharp spine.

Stock. A wonderful Anglo-Saxon word with a multitude of meanings. In horticulture, it refers both to the lower, root-bearing component of a grafted plant, and, quite separately, to a mustard family plant. In geology, it denotes a type of underground body of *igneous rock.*

Stolon. A horizontal stem that grows aboveground, along the soil surface.

Stoma. A tiny pore on a leaf or stem where the plant exchanges gases with the atmosphere. The plural form is "stomata."

Street tree. An ornamental tree planted along a thoroughfare.

Striation. A groove, scratch, or other linear mark on a rock surface, caused by the motion of an overriding glacier.

Stromatolite. A dome- or mushroom-shaped structure formed by layers of cyanobacteria ("blue-green algae") growing on calcium-rich sediments.

Subshrub. A rather equivocal botany term for a small plant that has shrublike tendencies, so to speak, without being wholly convincing.

Succulent. A plant with fleshy stems or leaves that can store large quantities of water in its tissues. Cacti are one familiar form of succulents; but not all succulents are cacti.

Sucker. An upward-growing shoot arising from a root or from the base of a plant's trunk.

Sugarbush. A stand of sugar maple trees that are tapped for sugar production.

Sun scald. A condition in which a plant's leaf tips turn brown when there is not enough water reaching the foliage from the roots.

Supercooling. A process used by some cold-hardy plants to prevent the formation of destructive ice crystals in their buds and foliage.

Suspect terrane. A *terrane* that appears to have originated in another location.

Swamp. A *wetland* dominated by trees or shrubs.

Syenite. A type of *intrusive, igneous rock* distinguished from granite by its lack of quartz.

Symbiosis. A sustained interaction—beneficial or not—between two different types of organism.

Taconian mountains. The ancient mountains that rose in the area of the current Green and Berkshire Mountains during the *Ordovician period.*

Talus. A body of rock rubble lying at the foot of a cliff or hillside.

Tannin. A brown-tinted chemical compound found in the oaks and other plants.

Taproot. The primary, downward-pointing root that emerges from a sprouting seed. The taproot of some plants persists into maturity; but in many others it does not.

Target specific. Referring to a pesticide or herbicide that harms only the intended species, without affecting other organisms.

Terrain. The combination of *landforms* that make up a region's surface.

Terrane. A large-scale, three-dimensional assemblage of rock units that forms both a region's surface and its subsurface.

Thallus. The tissue forming the body of a lichen or of a primitive plant that is not differentiated into stems and leaves.

Thorn. A branch that has been modified into a sharp, pointed projection.

Tidal inlet. A seacoast channel through which salt water moves landward and then seaward, in accordance with the tides.

Tidal marsh. Synonymous with *salt marsh.*

Till. Mixed and unlayered rock debris and finer sediments deposited by a glacier.

Tombolo. An exposed sandbar or other strip of sediments that runs between an island and the mainland.

Topiary. The art of training or pruning woody plants into ornamental shapes.

Transgression. In geology, the encroachment of an ocean or lake on what had been dry land.

Transpiration. The process by which water in a plant is lost to the atmosphere. Most transpiration occurs at the *stomata.*

Traprock. A term for dark-hued, fine-textured *igneous rocks,* especially *basalt* and *diabase.*

Tree. A woody plant, usually 15 feet or taller at maturity, that has one main trunk or a compact group of several trunks.

Triassic period. The first period of the *Mesozoic era.* The Triassic lasted approximately from 245 million to 208 million years ago.

Tuber. An enlarged underground stem—for instance, a potato—that contains food reserves for its plant.

Tulipomania. A bizarre episode in the history of seventeenth-century Holland, in which tulip bulbs were the basis of wild financial speculation.

Turgor. The state of being filled or swollen with water or sap.

Turpentine. The solvent that is one of the primary chemical constituents of *resin.*

Umbilicate lichen. A lichen that is attached to its substrate by a central strand or umbilicus.

Understory. The lower layer of small trees, shrubs, and other plants that grows under the canopy of the larger forest trees.

Urushiol. The oil in poison ivy and poison sumac that causes an allergic rash.

Variegated. Referring to a leaf that is marked with different colors or shadings.

Vascular plant. A plant with well-defined tissues that conduct water, nutrients, and food throughout its interior. The vascular plants are the *pteridophytes,* the *gymnosperms,* and the *angiosperms.*

Vector. In horticulture and in medicine, an organism that transmits a disease to another organism.

Vulnerary. A herbal cure used to heal wounds.

Wainscoting. Pronounced "*wain*-scutting." Ornamental wood paneling on walls indoors. Wainscoting is often placed on the lower portion of the wall.

Water gap. A notch or gap in a high ridge, through which a river flows.

Water table. The upper boundary of a zone saturated with groundwater.

Water table pond. A pond that forms when the *water table* rises above the pond's bottom surface.

Weatherboard. A vertical board used on the exterior of building. The edges of weatherboards are set flush against one another.

Weed. Any plant that is in the wrong place—by human standards.

Wetland. A habitat where water is at or near the ground's surface throughout the year.

Whaleback. Synonymous with *drumlin*.

White wood. Synonymous with *deal*.

Wildflower. A popular term for any herbaceous plant that grows and reproduces successfully without sustained human assistance.

Windfall. In its original, colonial-era meaning, a timber tree that could be legally taken and sold because it had fallen to the ground without human intervention.

Windthrow. A tree that has been toppled by the wind.

Wolf tree. A tall woodland tree with a spreading crown that blocks sunlight from other plants in its vicinity.

Woody. Referring to plants with tissue containing the strengthening substance *lignin*.

Worm fence. A wooden fence set in a zigzag pattern.

Xylem. The zone of internal plant tissue that carries the water and nutrients absorbed by the roots upward to the leaves.

Zygomorphic. Referring to an irregularly shaped flower—one that is symmetrical only along the vertical axis.

Annotated Bibliography and Recommended Reading

The following texts were consulted in the making of this book. Works I especially recommend are marked as follows:

I   Introductory material for adults

A   Advanced material for the more serious student of the subject

C   Children's books—usually of interest to adults, too

PLANT SCIENCES

Allen, Oliver E., and the Editors of Time-Life Books. 1978. *Pruning and Grafting*. From the Time-Life Encyclopedia of Gardening. Alexandria, Va.: Time-Life.

A.M.C. Guide to Mountain Flowers of New England. 1974. Boston: Appalachian Mountain Club. A sturdy and well-done little book, with the best information-to-mass ratio in the whole genre.

Arnberger, Leslie P. 1962. *Flowers of the Southwest Mountains*. 3rd ed. Globe, Ariz.: Southwestern Monuments Association.

Arnold, Augusta Foote. 1901. *The Sea-beach at Ebb-tide*. New York: Dover.

Baumgardt, John Philip. 1982. *How to Identify Flowering Plant Families*. Portland, Ore.: Timber Books.

Beattie, Mollie, Charles Thompson, and Lynne Levine. 1983. *Working with Your Woodland*. Hanover, N.H.: University Press of New England. An excellent guide for both the amateur woodlot manager and the professional forester.

Benson, Lyman, and Robert A. Darrow. 1981. *Trees and Shrubs of the Southwestern Deserts.* 3rd ed. Tucson: University of Arizona Press.

Blanchard, Robert O., and Terry A. Tattar. 1981. *Field and Laboratory Guide to Tree Pathology.* New York: Academic.

Bothfeld, Diane. 1991. How much can a cow eat? *University of Vermont Extension News* 8:5.

Britton, Nathaniel Lord, and Addison Brown. 1913. *An Illustrated Flora of the Northeastern United States and Canada.* 2nd ed. 3 vols. New York: Dover.

I Brockman, C. Frank. 1986. *Trees of North America.* Rev. ed. New York: Golden. Still the one best pocket tree guide in the known universe. Even exalted academic botanists, who usually eschew any keyless guide, use it—when no one's looking.

Brown, Clair A. 1972. *Wildflowers of Louisiana and Adjoining States.* Baton Rouge: Louisiana State University Press.

I Brown, Lauren. 1979. *Grasses: An Identification Guide.* Boston: Houghton Mifflin. Brown and her publisher are to be heartily congratulated for producing a general guide to a vast but often overlooked group of plants. Note, however, that a significant portion of the book is devoted not to grasses but to other grasslike plants.

Burk, John. 1981. *The Gardens and Botanical Facilities of Smith College.* Northampton, Mass.: Smith College.

Chase, A. R. 1987. *Compendium of Ornamental Foliage Plant Diseases.* St. Paul, Minn.: APS Press.

Cobb, Boughton. 1963. *A Field Guide to the Ferns and Their Related Families of Northeastern and Central North America.* Boston: Houghton Mifflin.

I *Common Weeds of the United States.* 1971. Agricultural Research Service of the U.S. Department of Agriculture. New York: Dover. A bit dated, but includes good illustrations.

I Conard, Henry S., and Paul L. Redfearn, Jr. 1979. *How to Know the Mosses and Liverworts.* 2nd ed. Dubuque, Iowa: William C. Brown. A commendable attempt to present the fascinating but overlooked world of nonvascular plants to the nonspecialist. Some readers find the long, involved keys difficult to use.

Connecticut's Coastal Marshes. 1961. Bulletin No. 12. New London: Connecticut Arboretum.

A Coombes, Allen J. 1985. *Dictionary of Plant Names.* Portland, Ore.: Timber Press.

I ——. 1992. *Trees.* New York: Dorling Kindersley. In the splendid Eyewitness Handbooks series. An intelligently and gorgeously designed handbook to the trees of the world, featuring the admirable "lexivisual" format. It includes many New England native and ornamental species.

Craul, Philip J. 1991. Urban Soil: Problems and Promise. *Arnoldia* 51(1): 23–32.

A Cronquist, Arthur. 1988. *The Evolution and Classification of Flowering Plants.* 2nd ed. New York: New York Botanical Garden. One of the most important botany texts of modern times. A comprehensive exposition of angiosperm evolution, presented by one of the towering figures of American plant taxonomy. Not for the comic book set.

A Crum, Howard, and Lewis E. Anderson. 1981. *Mosses of Eastern North America.* 2 vols. New York: Columbia University Press. The standard comprehensive reference for this region's nonvascular plants.

Del Tredici, Peter. 1984. Propagating leatherwood: A lesson in humility. Reprinted in *Arnoldia* 51(4): 63–66.

——. 1990. The trees of Tian Mu Shan: A photo essay. *Arnoldia* 50(4): 16–22.

I ——. 1991. Ginkgos and people—a thousand years of interaction. *Arnoldia* 51(2): 2–15. A good introduction to the ancient and unique ginkgo.

DiGregorio, Mario, and Jeff Wallner. 1989. *A Vanishing Heritage: Wildflowers of Cape Cod.* Missoula, Mont.: Mountain Press. Excellent photos and ecological perspective.

◣ Dirr, Michael A. 1990. *Manual of Woody Landscape Plants.* 4th ed. Champaign, Ill.: Stipes. The horticulture student's bible. An ocean of information, with tart wisecracks about certain plant types.

——. 1991. Sweet pepperbush: A summer sensation. *Arnoldia* 51(3): 18–21.

Edland, H. 1969. *The Pocket Encyclopedia of Roses.* 3rd ed. New York: Macmillan.

Egler, Frank E., and William A. Niering. 1965. *Yale Nature Preserve, New Haven.* Vegetation of Connecticut Natural Areas, No. 1. Hartford: Connecticut Geological and Natural History Survey.

——. 1967. *The Natural Areas of the McLean Game Refuge.* Vegetation of Connecticut Natural Areas, No. 3. Hartford: Connecticut Geological and Natural History Survey.

——. 1976. *The Natural Areas of the White Memorial Foundation.* Vegetation of Connecticut Natural Areas. Litchfield, Conn.: Friends of the Litchfield Nature Center and Museum.

Elbert, Virginie F., and George A. Elbert. 1989. *Foliage Plants for Decorating Indoors.* Portland, Ore.: Timber Press.

Ferguson, Barbara, ed. 1987. *All About Growing Fruits, Berries and Nuts.* San Francisco: Ortho.

Garden Lore of Ancient Athens. 1963. American School of Classical Studies at Athens. Princeton, N.J.: Institute for Advanced Study.

◣ Gifford, Ernest, and Adriance S. Foster. 1989. *Morphology and Evolution of Vascular Plants.* 3rd ed. New York: W. H. Freeman. An in-depth look at plant structures and their evolution. Beautiful writing, even though it is necessarily couched in rigorous scientific language.

Glattstein, Judy, ed. 1991. *Gardener's World of Bulbs.* Plants and Gardens Series 47(2). New York: Brooklyn Botanic Garden.

Gleason, Henry A. 1952. *The New Britton and Brown Illustrated Flora of Northeastern United States and Adjacent Canada.* 3 vols. New York: Hafner.

◣ Gleason, Henry A., and Arthur Cronquist. 1991. *Manual of Vascular Plants of the Northeastern United States and Adjacent Canada.* 2nd ed. New York: New York Botanical Garden. The ultimate wild-plant key to this region. Not a picture guidebook.

◣ Gould, Stephen Jay. 1977. *Ontogeny and Phylogeny.* Cambridge, Mass.: Belknap. This already classic work on the development of organisms—taken as individuals and as species—presents ideas central to the understanding of modern evolution theory. Rigorous scientific language.

——. 1991. More light on leaves. *Natural History,* 2:16–23. In Gould's "This View of Life" series. A look at Goethe and the perils of academic overspecialization. Worth hunting for; it deserves thoughtful reading. It is a good example of Gould's popular essay style, which does not insult the intelligent reader by refraining from the elaboration of ideas.

Graf, Alfred Byrd. 1974. *Exotic Plant Manual.* 4th ed. East Rutherford, N.J.: Roehrs. A condensation of Graf's *Exotica,* the standard houseplant identification manual.

Gray, Asa. 1878. *School and Field Book of Botany.* Rev. ed. New York: American Book.

Grieve, M. 1931. *A Modern Herbal.* 2 vols. New York: Dover. The botanical taxonomy is long out-of-date, and the system of plant entries by common names is often aggravating, but the anecdotes and the wealth of information make it all worthwhile.

A Guide to the Sedgwick Gardens at Long Hill. n.d. Milton, Mass.: The Trustees of Reservations.

I Hale, Mason E. 1979. *How to Know the Lichens*. 2nd ed. Dubuque: William C. Brown. A good first look at these rugged mutualistic organisms. Identification by relatively straightforward botanical keys.

The Harvard Forest Models. 1975. 2nd ed. Cambridge, Mass.: Harvard University Printing Office.

Hendry, George. 1990. Making, breaking, and remaking chlorophyll. *Natural History* 5:36–41.

Hill, Lewis. 1989. *Christmas Trees*. Pownal, Vt.: Garden Way.

A Hitchcock, A. S. 1950. *Manual of the Grasses of the United States*. 2 vols. 2nd ed., revised by Agnes Chase. New York: Dover. The taxonomy and distribution maps are now a bit out-of-date, but this is still a standard reference work for the grasses. Each species is illustrated.

A *Hortus Third*. 1976. Liberty Hyde Bailey Hortorium. New York: Macmillan. A massive one-volume plant "dictionary"—more like an encyclopedia—which is the horticulturist's sine qua non.

Hsueh Chi-ju. 1985. Reminiscences of collecting the type specimens of *Metasequoia glyptostroboides*. Reprinted in *Arnoldia* 51(4): 17–21.

Hugo, Nancy Ross. 1990. Challenging the biggest champ. *American Forests* 96(1-2): 8–10.

Huxley, A. J. 1971. *Garden Terms Simplified*. New York: Winchester.

I Huxley, Anthony. 1974. *Plant and Planet*. New York: Penguin. A noted British botanical writer details the plant kingdom's contributions to humankind and to the whole system of life.

Huxley, Anthony, ed. 1973. *Evergreen Garden Trees and Shrubs*. London: Blandford.

C Jaspersohn, William. 1991. *Cranberries*. Boston: Houghton Mifflin. A ravishingly designed first look at growing and harvesting a classic Yankee crop. The book looks almost edible.

Jenkins, Charles F. 1946. Asa Gray and his quest for *Shortia galacifolia*. Reprinted in *Arnoldia* 51(4): 4–11.

I Johnson, Charles W. 1985. *Bogs of the Northeast*. Hanover, N.H.: University Press of New England. A well-written, well-illustrated exposition of the peatland environment for nonspecialists. Very highly recommended.

A Johnson, Warren T., and Howard H. Lyon. 1988. *Insects That Feed on Trees and Shrubs*. 2nd ed. Ithaca, N.Y.: Cornell University Press. A massive photographic survey of insects—and arachnids—that attack woody plants. The illustrations are gruesome, and irresistible.

Koller, Gary L., and Michael A. Dirr. 1979. *Street Trees for Home and Municipal Landscapes*. Jamaica Plain, Mass.: Arnold Arboretum.

I Lansky, Mitch. 1992. *Beyond the Beauty Strip*. Gardiner, Me.: Tilbury House. An environmentalist's voice, crying in the clear-cut wilderness, in response to the tidal wave of sophisticated forest-industry P.R. Focused on the politics, economics, and ecosystem of Maine; but it bears a message of even greater scope.

Lee, Cara. 1985. *West Rock to the Barndoor Hills*. Vegetation of Connecticut Natural Areas, No. 4. Hartford: Connecticut Geological and Natural History Survey.

I Lellinger, David B. 1985. *A Field Manual of the Ferns and Fern-allies of the United States and Canada*. Washington, D. C.: Smithsonian Institution Press. The one best and most up-to-date guide to pteridophytes. Identification by keys and crisp color photographs.

I Lincoff, Gary H. 1981. *The Audubon Society Field Guide to North American Mushrooms*. New York: Alfred A. Knopf.

Livingston, Philip A., and Franklin H. West, eds. 1978. *Hybrids and Hybridizers: Rhododendrons and Azaleas for Eastern North America.* Newtown Square, Pa.: Harrowood.

Lund, Bruce. n.d. *Massachusetts Field Guide to Inland Wetland Plants.* Lincoln: Massachusetts Audubon Society.

Mayes, Vernon O., and Barbara Bayless Lacy. 1989. *Nanisé: A Navajo Herbal.* Tsaile, Ariz.: Navajo Community College Press.

McQueen, Cyrus B. 1990. *Field Guide to the Peat Mosses of Boreal North America.* Hanover, N.H.: University Press of New England. A handy and fairly nontechnical guide to the sphagnum mosses.

Merrill, E. D. 1948. *Metasequoia,* another "living fossil." Reprinted in *Arnoldia* 51(4): 8–16.

Mickel, John T. 1979. *How to Know the Ferns and Fern Allies.* Dubuque, Iowa: William C. Brown. A good general-interest guide to the pteridophytes. It covers all of North America. Note that its taxonomic treatment sometimes differs from Lellinger, above.

"The national register of champion trees." 1990. American Forestry Association. *American Forests* 96 (1-2): 11–39.

New England Wild Flower Society. *Wild Flower Notes 1987* 2(3). Framingham, Mass.

New England Wild Flower Society. *Wild Flower Notes 1989: Wetlands* 4(1). Framingham, Mass.

New England Wild Flower Society. *Wild Flower Notes 1990: Meadows and Meadow Gardening* 5(1). Framingham, Mass.

New England Wild Flower Society Newsletter. Fall 1991. Framingham, Mass.

New Pronouncing Dictionary of Plant Names. 1984. Chicago: American Florists. A brochure-sized publication worth its weight in gold.

Niehaus, Theodore F., Charles L. Ripper, and Virginia Savage. 1984. *A Field Guide to Southwestern and Texas Wildflowers.* Boston: Houghton Mifflin.

Niering, William A., and Frank E. Egler. 1966. *The Natural Area of the Audubon Center of Greenwich.* Vegetation of Connecticut Natural Areas, No. 2. Hartford: Connecticut Geological and Natural History Survey.

Niering, William A., and Richard H. Goodwin. 1973. *Inland Wetland Plants of Connecticut.* Bulletin No. 19. New London: Connecticut Arboretum. Applicable to most of southern and central New England.

Niering, William A., and R. Scott Warren. 1980. *Salt Marsh Plants of Connecticut.* Bulletin No. 25. New London: Connecticut Arboretum. Applicable to most of southern and central New England.

Peattie, Donald Culross. 1953. *A Natural History of Western Trees.* Boston: Houghton Mifflin.

——. 1964. *A Natural History of Trees of Eastern and Central North America.* Boston: Houghton Mifflin. Peattie, in addition to being a botanist in his own right, was one of this century's best nature writers. This hefty book is a wonderful compendium of tree lore—a sort of dendrological *Golden Bough.*

Perrin, Noel. 1980. *Making Maple Syrup.* Bulletin A-51. Pownal, Vt.: Garden Way.

Perry, T. O. 1989. Tree roots: Facts and fallacies. *Arnoldia* 49(4): 2–21.

Peterson, Lee Allen. 1977. *Edible Wild Plants.* Boston: Houghton Mifflin.

Peterson, Roger Tory, and Margaret McKenny. 1968. *A Field Guide to Wildflowers of Northeastern and North-central North America.* Boston: Houghton Mifflin. After years of trying out this book's various competitors, I must say this is still the best.

Petrides, George. 1973. *A Field Guide to Trees and Shrubs.* Boston: Houghton Mifflin.

Petrides, George, and Olivia Petrides. 1992. *Western Trees.* Boston: Houghton Mifflin.

Phillips, Roger. 1977. *Wild Flowers of Britain.* New York: Quick Fox.

I ——. 1978. *Trees of North America and Europe.* New York: Random House. A lovely photo book with rather spotty coverage. A good place to find European species now commonly planted in the United States.

I Phillips, Roger, and Martyn Rix. 1988. *Roses.* New York: Random House. The one best photographic survey of cultivated roses currently available.

I ——. 1989. *Shrubs.* New York: Random House. The emphasis of this visually stunning book is on cultivated shrub species and varieties, with rhododendrons and azaleas by the bucketfuls.

I ——. 1991. *Perennials.* Vols. 1 and 2. New York: Random House. The Phillips-Rix team does it yet again, with an extensive photographic survey of cultivated perennial herbaceous plants from around the world.

Pirone, Pascal P. 1978. *Diseases and Pests of Ornamental Plants.* 5th ed. New York: John Wiley and Sons.

I *Poison Ivy, Sumac and Oak.* 1993. American Academy of Dermatology. Schaumberg, Ill. This brochure gives good tips on how to avoid and treat rashes caused by these plants.

Polunin, Oleg. 1977. *Trees and Bushes of Britain and Europe.* St. Albans, Herts., Eng.: Paladin.

Powell, Charles C., Jr. 1985. *Disease Control in the Landscape.* Ohio Cooperative Extension Service Bulletin 614. Columbus: Ohio State University.

A Raven, Peter H., Ray F. Evert, and Susan E. Eichhorn. 1986. *Biology of Plants.* 4th ed. New York: Worth. A superb and handsome text for the undergrad majoring in botany or another life science.

Read, Christopher. 1991. Review of *A Reunion of Trees,* by Stephen A. Spongberg. *Horticulture* 69(6): 74–76.

Rehder, Alfred. 1946. On the history of the introduction of woody plants into North America. Reprinted in *Arnoldia* 51(4): 22–29.

I Russell, Howard S. 1982. *A Long, Deep Furrow: Three Centuries of Farming in New England.* Abridged ed. Hanover, N.H.: University Press of New England.

I Sawyers, Claire E., and Barbara B. Pesch, eds. 1986. *American Gardens: A Traveler's Guide.* Plants and Gardens Series 42(3). New York: Brooklyn Botanic Garden.

I Schiffer, Nancy, and Herbert Schiffer. 1977. *Woods We Live With.* Exeter, Pa.: Schiffer.

Schultz, Jack C. 1991. The multimillion-dollar gypsy moth question. *Natural History* 6:40–45.

Simonds, Roberta L., and Henrietta H. Tweedie. 1978. *Wildflowers of the Great Lakes Region.* Chicago: Chicago Review Press.

A Sinclair, Wayne A., Howard H. Lyon, and Warren T. Johnson. 1987. *Diseases of Trees and Shrubs.* Ithaca, N.Y.: Cornell University Press. The worthy companion to Johnson and Lyon's sumptuously illustrated book on insect foes. Just as luridly fascinating and informative.

I Sloane, Eric. 1965. *A Reverence for Wood.* New York: Ballantine. A small but highly recommended paperback by a gifted illustrator and student of traditional wood crafts.

Sokolov, Raymond. 1991. Grasping the nettle. *Natural History* 8:72–75.

Spongberg, Stephen A. 1979. Notes on persimmons, kakis, date plums, and chapotes. Reprinted in *Arnoldia* 51(4): 47–54.

———. 1990. The first Japanese plants for New England. *Arnoldia* 50(3): 2–11.

I   Steele, Frederic L. 1971. *A Beginner's Guide: Trees and Shrubs of Northern New England.* Concord, N.H.: Society for the Protection of New Hampshire Forests.

I   Stern, Kingsley R. 1991. *Introductory Plant Biology.* 5th ed. Dubuque, Iowa: William C. Brown. A good introductory botany text for those not majoring in the life sciences.

A   Still, Steven M. 1988. *Manual of Herbaceous Ornamental Plants.* Champaign, Ill.: Stipes.

Stipes, R. Jay, and Richard J. Campana, eds. 1981. *Compendium of Elm Diseases.* St. Paul, Minn.: APS Press.

I   Symonds, George W. D. 1958. *The Tree Identification Book.* New York: William Morrow. This wonderful old warhorse, with its black-and-white photos and simple yet clever identification system, is enshrined in the hearts of naturalists. It covers wild trees of the Northeast, and a few ornamentals.

I   ———. 1963. *The Shrub Identification Book.* New York: William Morrow. The best layman's book on an oft-neglected subject. Format and coverage similar to his tree identification book.

I   Taylor, Sally L., and Martine Villalard. 1985. *Seaweeds of the Connecticut Shore: A Wader's Guide.* Bulletin No. 18. New London: Connecticut Arboretum.

A   Taylor, William Randolph. 1962. *Marine Algae of the Northeastern Coast of North America.* 2nd ed. Ann Arbor: University of Michigan Press.

Tessman, Jim. 1991. Christmas trees: Better quality from better care. *University of Vermont Extension News* 8:7.

I   Thomas, Joseph D., ed. 1990. *Cranberry Harvest.* New Bedford, Mass.: Spinner. The history of a unique form of agriculture. A lovely book.

A   Thompson, D'Arcy W. 1992. *On Growth and Form.* New York: Cambridge University Press.

Tweed, William. 1988. *The General Sherman Tree.* Three Rivers, Calif.: Sequoia Natural History Association.

I   Viertel, Arthur T. 1970. *Trees, Shrubs and Vines.* Syracuse, N.Y.: Syracuse University Press. This not very well-known manual for horticulturists features well-rendered leaf illustrations, which are the main method of identification.

Visser, J. M., trans. 1986. *Spring-flowering Bulbs.* Venissieux, France: Horticolor S.A.R.L.

I   Watts, Mary Theilgaard, and Tom Watts. 1970. *Winter Tree Finder.* Berkeley, Calif.: Nature Study Guild. One of the Guild's handy booklets. A good introduction to the field of wintertime botany.

Western Garden Book. 1988. Menlo Park, Calif.: Sunset Publishing.

Whatley, Michael E. 1988. *Common Trailside Plants of Cape Cod National Seashore.* Eastham, Mass.: Eastern National Park and Monument Association.

Wiggers, Friedrich Heinrich. 1780. *Primitiae Florae Holsaticae.* Kiel: Letteris Mich. Frider.

C   Wiggers, Raymond. 1991. *Picture Guide to Tree Leaves.* New York: Franklin Watts. Intended for grade-schoolers, but also useful for adults. It covers many North American tree types.

Wilson, B. F. 1970. *The Growing Tree.* Amherst: University of Massachusetts Press.

Woodward, F. I. 1987. *Climate and Plant Distribution.* New York: Cambridge University Press.

Wyman, Donald. 1961. The *Forsythia* story. Reprinted in *Arnoldia* 51(4): 34–37.

New England Field Guides, Natural History Surveys, and Atlases

Baldwin, Henry I. 1980. *Monadnock Guide.* Rev. 3rd ed. Concord, N.H.: Society for the Protection of New Hampshire Forests.

I Brady, John, and Brian White. 1983. *Fifty Hikes in Massachusetts.* Woodstock, Vt.: Backcountry.

I Butcher, Russell D. 1977. *Field Guide to Acadia National Park, Maine.* New York: Reader's Digest.

I Johnson, Charles W. 1980. *The Nature of Vermont.* Hanover, N.H.: University Press of New England. If only every state had this kind of nontechnical summary of its natural history.

I Jorgensen, Neil. 1977. *A Guide to New England's Landscape.* Chester, Conn.: Globe Pequot. Another must for the New England naturalist. The geology information is dated, but the descriptions of New England plant regimes make a very good introduction to the region's plant ecology.

I *Maine Atlas and Gazetteer.* 1989. 14th ed. Freeport, Me.: DeLorme. All atlases in this series are extremely helpful to the plant explorer; they are much more detailed than conventional road maps, but are nonetheless easy to read.

I *New Hampshire Atlas and Gazetteer.* 1985. Freeport, Me.: DeLorme.

I Pope, Eleanor. 1981. *The Wilds of Cape Ann.* Gloucester, Mass.: Resources for Cape Ann.

I Robbins, Sarah Fraser, and Clarice M. Yentsch. 1973. *The Sea Is All About Us.* Salem, Mass.: Peabody Museum of Salem.

I Sammartino, Claudia F. 1981. *The Northfield Mountain Interpreter.* Berlin, Conn.: Northeast Utilities. An eclectic compendium of natural and cultural history.

Taffe, William J., and John E. Ross. 1977. *A Guide to the Physical Environment of New Hampshire.* Plymouth, N.H.: Plymouth State College Environmental Studies Center.

I *Vermont Atlas and Gazetteer.* 1988. 8th ed. Freeport, Me.: DeLorme.

Geology and Paleobotany

I Adams, George F., and Jerome Wyckoff. 1971. *Landforms.* New York: Golden. Another terrific Golden Guide.

Andrews, Henry A. 1980. *The Fossil Hunters: In Search of Ancient Plants.* Ithaca, N.Y.: Cornell University Press.

Bates, Robert L., and Julia A. Jackson, eds. 1984. *Dictionary of Geological Terms.* 3rd ed. New York: Anchor.

A Behrensmeyer, Anna K., et al. 1992. *Terrestrial Ecosystems Through Time.* Chicago: University of Chicago Press. An excellent survey of the history of life on earth, from the perspective of overall plant and animal communities and their habitats.

I Bell, Michael. 1985. *The Face of Connecticut.* Bulletin 110. Hartford: Connecticut Geological and Natural History Survey. Connecticut is renowned, or should be, for the excellence and quantity of its state natural history guides. This, simply put, is the best of the lot. It skillfully places Connecticut's geologic legacy in a human context. Don't miss the account of James Gates Percival.

A Billings, Marland P. 1955. *Geologic Map of New Hampshire.* Washington, D.C.: U.S. Geological Survey.

——. 1956. *The Geology of New Hampshire. Part II: Bedrock Geology.* Concord, N.H.: New Hampshire Planning and Development Commission.

——. 1972. *Structural Geology.* 3rd ed. Englewood Cliffs, N.J.: Prentice Hall.

I Billings, Marland P., et al. 1979. *The Geology of the Mount Washington Quadrangle, New Hampshire.* Concord: New Hampshire Planning and Development Commission.

Bloom, Arthur L. 1959. *The Geology of Sebago Lake State Park.* State Park Geologic Series, No.1. Augusta: Maine Department of Economic Development.

Colbert, Edwin H. 1970. *Fossils of the Connecticut Valley: The Age of Dinosaurs Begins.* Bulletin No. 96. Hartford: Connecticut Geological and Natural History Survey.

A Davis, William Morris. 1909. *Geographical Essays.* New York: Dover. Writings by one of the founders of geomorphology. It includes references to the Connecticut River valley.

A Doll, Charles G. 1961. *Centennial Geologic Map of Vermont.* Montpelier, Vt.: Vermont Geological Survey. Showing the state's bedrock geology.

A ——. 1970. *Surficial Geologic Map of Vermont.* Montpelier, Vt.: Vermont Geological Survey.

I Francis, Jane E. 1991. Arctic Eden. *Natural History* 1:56–63. A well-written, beautifully illustrated account of the ancient dawn redwood forests that once grew near the North Pole.

I Gilman, Richard A., et al. 1988. *The Geology of Mount Desert Island.* Augusta, Me.: Maine Geological Survey. A particularly good introductory geology guidebook. Currently, it is up-to-date—many of its companions are not.

Goldthwait, James Walter, Lawrence Goldthwait, and Richard Parker Goldthwait. 1947. *The Geology of New Hampshire. Part I: Surficial Geology.* Concord, N.H.: New Hampshire Planning and Development Commission.

Heald, Milton T. 1950. *The Geology of the Lovewell Mountain Quadrangle, New Hampshire.* Concord, N.H.: New Hampshire Planning and Development Commission.

Hitchcock, C. H. 1878. *The Geology of New Hampshire.* Parts III–V. Concord, N.H.: Edward A. Jenks.

Jackson, Kern C. 1970. *Textbook of Lithology.* New York: McGraw-Hill.

Kaye, Clifford A. 1960. *Surficial Geology of the Kingston Quadrangle, Rhode Island.* Map. Washington, D.C.: U.S. Government Printing Office.

——. 1960. *Surficial Geology of the Kingston Quadrangle, Rhode Island.* Geological Survey Bulletin 1071-I. Washington, D.C.: U.S. Government Printing Office.

I Kendall, David L. 1987. *Glaciers and Granite.* Camden, Me: Down East. A first-rate geology guide to Maine, specifically for the layman. Very highly recommended. It includes road tours and an appreciation of plate tectonics. Beautifully illustrated.

Kruger, Frederick C. 1946. *The Geology of the Bellows Falls Quadrangle.* Concord, N.H.: New Hampshire Planning and Development Commission.

I Little, Richard D. 1986. *Dinosaurs, Dunes, and Drifting Continents: The Geohistory of the Connecticut Valley.* Greenfield, Mass.: Valley Geology Publications. A delightful excursion into the big valley's dinosaur-ridden past. Good for kids and adults alike. Filled with unpretentious, comprehensible illustrations and text.

I ——. 1989. *Exploring Franklin County: A Geology Guide.* Greenfield, Mass.: Valley Geology Publications. Recommended for geologically curious residents of—and visitors to—north-central and northwestern Massachusetts. Easy-to-follow road tours with mileage logs.

Lull, Richard Swann. 1953. *Triassic Life of the Connecticut Valley.* Bulletin No. 81. Rev. ed. Hartford, Conn.: Connecticut Geological and Natural History Survey.

Lyons, John B., et al. 1986. *Interim Geologic Map of New Hampshire*. Reston, Va.: U.S. Geologic Survey.

I   McPhee, John. 1980. *Basin and Range*. New York: Farrar Straus Giroux. This and the next three titles are beautifully written overviews of modern geology and geologists. Superb nonfiction.

I   ——. 1982. *In Suspect Terrain*. New York: Farrar Straus Giroux.

I   ——. 1986. *Rising from the Plains*. New York: Farrar Straus Giroux.

I   ——. 1993. *Assembling California*. New York: Farrar Straus Giroux.

Meyers, T. R. 1947. *Foundry Sands of New Hampshire*. Concord, N.H.: New Hampshire Department of Resources and Economic Development.

Moore, George E., Jr. 1949. *The Geology of the Keene-Brattleboro Quadrangle, New Hampshire and Vermont*. Concord, N.H.: New Hampshire Division of Resources and Economic Development.

——. 1964. *Bedrock Geology of the Kingston Quadrangle, Rhode Island*. Geological Survey Bulletin 1158-E. Washington, D.C.: U.S. Government Printing Office.

Novotny, Robert F. 1969. *The Geology of the Seacoast Region, New Hampshire*. Concord, N.H.: New Hampshire Department of Resources and Economic Development.

Orville, Philip M. 1968. *Guidebook for Fieldtrips in Connecticut*. Guidebook No. 2. Hartford, Conn.: Connecticut Geological and Natural History Survey.

Pankiwyskyj, Kost A. 1965. *The Geology of Mount Blue State Park*. State Park Geologic Series, No.3. Augusta, Me.: Maine Department of Economic Development.

Petersen, Morris S., J. Keith Rigby, and Lehi F. Hintze. 1973. *Historical Geology of North America*. Dubuque, Iowa: William C. Brown.

Pollack, Samuel J. 1964. *Bedrock Geology of the Tiverton Quadrangle, Rhode Island–Massachusetts*. Geological Survey Bulletin 1158-D. Washington, D.C.: U.S. Government Printing Office.

Raup, David M. 1991. *Extinction: Bad Genes or Bad Luck?* New York: W. W. Norton.

I   Raymo, Chet, and Maureen E. Raymo. 1989. *Written in Stone: A Geological History of the Northeastern United States*. Chester, Conn.: Globe Pequot. A good introduction written in a thoroughly jargon-free style.

Reineck, H.-E., and I. B. Singh. 1980. *Depositional Sedimentary Environments*. 2nd ed. Berlin: Springer-Verlag.

I   Rodgers, John. 1985. *Bedrock Geological Map of Connecticut*. Washington, D.C.: U.S. Geological Survey. A beautiful, up-to-date, and highly detailed map.

A   Schafer, J. P. 1965. *Surficial Geologic Map of the Watch Hill Quadrangle, Rhode Island–Connecticut*. Washington, D.C.: U.S. Geological Survey.

I   Scully, Vincent, et al. 1990. *The Great Dinosaur Mural at Yale: The Age of Reptiles*. New York: Harry N. Abrams. One of the co-authors, Leo J. Hickey, a paleobotanist, gives an excellent introduction to the evolution of ancient plants, as shown in the great Peabody Museum painting.

Stanley, Steven M. 1987. *Extinction*. New York: Scientific American.

A   Stewart, Wilson N., and Gar W. Rothwell. 1992. *Paleobotany and the Evolution of Plants*. 2nd ed. New York: Cambridge University Press.

I   Strahler, Arthur N. 1966. *A Geologist's View of Cape Cod*. Garden City, N.Y.: Natural History Press.

A   Taylor, Thomas N., and Edith L. Taylor. 1993. *The Biology and Evolution of Fossil Plants*. Englewood Cliffs, N.J.: Prentice Hall.

Thornbury, William D. 1954. *Principles of Geomorphology.* New York: John Wiley and Sons.

A ——. 1965. *Regional Geomorphology of the United States.* New York: John Wiley and Sons. Including an overview of New England's landforms.

Van Diver, Bradford B. 1985. *Roadside Geology of New York.* Missoula, Mont.: Mountain Press.

I ——. 1987. *Roadside Geology of Vermont and New Hampshire.* Missoula, Mont.: Mountain Press. Fascinating road tours for the layman interested in the region's geology.

White, Robert S. 1991. Ancient floods of fire. *Natural History* 4:50–61.

I Wiggers, Raymond. 1993. *The Amateur Geologist: Explorations and Investigations.* New York: Franklin Watts. Intended for grades 7 and up. Includes general information on mineral and fossil sites in the Northeast and other regions.

Wyckoff, Jerome. 1971. *Rock Scenery of the Hudson Highlands and Palisades.* Glens Falls, N.Y.: Adirondack Mountain Club.

A Zen, E-an, ed. 1983. *Bedrock Geologic Map of Massachusetts.* Washington, D.C.: U.S. Geological Survey.

History, Anthropology, Geography, and Architecture

I Allen, Richard Sanders. 1983. *Covered Bridges of the Northeast.* Rev. ed. Lexington, Mass.: Stephen Greene.

Andrews, Edward Deming. 1940. *The Gift to Be Simple: Songs, Dances and Rituals of the American Shakers.* New York: Dover.

I ——. 1953. *The People Called Shakers: A Search for the Perfect Society.* New York: Dover.

Attenborough, David. 1987. *The First Eden: The Mediterranean World and Man.* Boston: Little, Brown.

Benes, Peter, ed. 1985. *American Speech: 1600 to Present.* Boston: Boston University.

I Blumenson, John J.-G. 1981. *Identifying American Architecture.* 2nd ed. Nashville, Tenn.: AASLH Press. An excellent handbook, with cogent descriptions of American architectural styles from 1600–1945.

Brown, Richard D. 1978. *Massachusetts: A History.* New York: W. W. Norton.

Bruce, Nona B., and Barbara Bullock Jones. 1990. *The Fort at No. 4, 1740–1760.* Charlestown, N.H.: Old Fort No. 4 Associates.

I Burns, Deborah E., and Lauren R. Stevens. 1988. *Most Excellent Majesty: A History of Mount Greylock.* Pittsfield, Mass.: Berkshire Natural Resources Council. A lovely book of greater significance than its specific subject might suggest, since it chronicles the cause of conservation and the public interest against the forces of private greed and "development."

A Carroll, Charles F. 1973. *The Timber Economy of Puritan New England.* Providence, R.I.: Brown University Press.

Clark, Charles E. 1990. *Maine: A History.* Hanover, N.H.: University Press of New England.

The Connecticut Arboretum: Its First Fifty Years. 1981. New London, Conn.: Connecticut Arboretum.

I Cronon, William. 1983. *Changes in the Land.* New York: Hill and Wang. The story of the human exploitation of New England's landscape and natural resources. Engaging, provocative, and highly recommended.

I Delaney, Edmund. 1983. *The Connecticut River: New England's Historic Waterway.* Chester, Conn.: Globe Pequot.

Fuller, Mary Williams. 1930. *The Story of Deerfield*. Deerfield, Mass.: Pocumtuck Valley Memorial Association.

Goodwin, Richard H. 1991. *The Connecticut College Arboretum: Its Sixth Decade and a Detailed History of the Land*. Bulletin No. 32. New London, Conn.: Connecticut College Arboretum.

I Heffernan, Nancy Coffey, and Ann Page Stecker. 1986. *New Hampshire—Crosscurrents in Its Development*. Grantham, N.H.: Thompson and Rutter.

Irwin, R. Stephen. 1984. *Hunters of the Eastern Forest*. Blaine, Wash.: Hancock House.

I Jennison, Peter S. 1989. *Roadside History of Vermont*. Missoula, Mont.: Mountain Press. Excellent descriptions and anecdotes, arranged town by town.

I Leach, Douglas Edward. 1958. *Flintlock and Tomahawk*. New York: W. W. Norton. The classic account of King Philip's War. Very highly recommended.

Lewis, Thomas R., and John E. Harmon. 1986. *Connecticut: A Geography*. Boulder, Colo.: Westview.

A Malone, Joseph J. 1979. *Pine Trees and Politics*. New York: Arno.

McCrum, Robert, William Cran, and Robert MacNeil. 1986. *The Story of English*. New York: Penguin.

Meeks, Harold A. 1986. *Time and Change in Vermont*. Chester, Conn.: Globe Pequot.

Morrison, Hugh. 1952. *Early American Architecture*. New York: Dover.

Morrissey, Charles T. 1981. *Vermont, a History*. New York: W. W. Norton.

Pendle, George. 1963. *A History of Latin America*. New York: Pelican.

I Randall, Willard Sterne. 1990. *Benedict Arnold: Patriot and Traitor*. New York: Quill.

Richardson, Robert D., Jr. 1986. *Henry Thoreau: A Life of the Mind*. Berkeley, Calif.: University of California Press.

I Russell, Howard S. 1980. *Indian New England before the Mayflower*. Hanover, N.H.: University Press of New England.

I Sewall, Richard B. 1974. *The Life of Emily Dickinson*. New York: Farrar Straus Giroux.

I Sheldon, Asa. 1862. *Yankee Drover*. Hanover, N.H.: University Press of New England. A window onto the life and mind of a nineteenth-century New Englander.

Simmons, William S. 1986. *Spirit of the New England Tribes*. 1986. Hanover, N.H.: University Press of New England.

I Sloane, Eric. 1955. *Our Vanishing Landscape*. New York: Ballantine.

Stravinsky, Vera, and Robert Craft. 1978. *Stravinsky in Pictures and Documents*. New York: Simon and Schuster.

I Tolles, Bryant F., Jr. 1979. *New Hampshire Architecture*. Hanover, N.H.: University Press of New England.

Walker, William H. C., and Willard Brewer Walker. 1984. *A History of World's End*. 2nd ed. Milton, Mass.: The Trustees of Reservations.

I Wikoff, Jerold. 1985. *The Upper Valley*. Chelsea, Vt.: Chelsea Green.

C Wilbur, C. Keith. 1978. *The New England Indians*. Chester, Conn.: Globe Pequot. A splendid picture book. Very informative.

Wissler, Clark. 1966. *Indians of the United States*. Garden City, N.Y.: Anchor.

ORIGINAL SOURCES AND OTHER WORKS

Basho, Matsuo. 1966. *The Narrow Road to the Deep North and Other Travel Sketches.* Translated by Nobuyuki Yuasa. London: Penguin.

Benson, Robert. 1989. *The Connecticut River.* Boston: Bulfinch.

I Beston, Henry. 1928. *The Outermost House.* New York: Viking.

A Bradford, William. 1952. *Of Plymouth Plantation.* New York: Alfred A. Knopf.

Daniel, Glenda. 1984. *Dune Country.* Rev. ed. Athens, Ohio: Swallow Press.

I Emerson, Ralph Waldo. 1968. *Essays and Journals.* Garden City, N.Y.: Doubleday. Including an illuminating character sketch of Thoreau. Shame on any American who doesn't strike up an acquaintance with Emerson!

A Johnson, Susanna, and John C. Chamberlain. 1814. *A Narrative of the Captivity of Mrs. Johnson.* Bowie, Md.: Heritage.

A Josselyn, John. 1672. *New-Englands Rarities Discovered.* Boston: Massachusetts Historical Society. A reprint of the original text, with the original plant illustrations. New England's first botany field guide.

A ——. 1988. *John Josselyn, Colonial Traveler: A Critical Edition of Two Voyages to New England.* Edited by Paul J. Lindholdt. Hanover, N.H.: University Press of New England. If for no other reason, read it for the hyperbolic Restoration prose: in turn pithy, paranoid, and pedantic. The best section contains Josselyn's railing against his critics, real or imagined.

A Kalm, Pehr. 1770. *English Version of Peter Kalm's Travels in North America.* New York: Dover.

A Marsh, George Perkins. 1864. *Man and Nature.* Cambridge, Mass.: Belknap. A benchmark work, but its florid Victorian circumlocutions may be offensive to some.

Pliny the Elder (Gaius Plinius Secundus). 1991. *Natural History, a Selection.* Translated by John Healy. New York: Penguin. The world as it appeared to a very curious and very credulous first-century Roman.

Raleigh, Sir Walter. 1984. *Selected Writings.* New York: Penguin.

Richardson, Wyman. 1955. *The House on Nauset Marsh.* New York: W. W. Norton.

Smith, Captain John. 1988. *Captain John Smith: A Select Edition of His Writings.* Chapel Hill: University of North Carolina Press.

Stevens, Phineas. 1747. *Three Day Siege: Captain Phineas Stevens' Dispatch to Colonel William Williams.* Charlestown, N.H.: Fort No. 4 Associates.

I Thoreau, Henry David. 1864. *The Maine Woods.* New York: Perennial.

A ——. 1962. *The Journal of Henry D. Thoreau.* 2 vols. New York: Dover.

I ——. 1984. *Walden and Other Writings.* New York: Bantam.

I ——. 1984. *Cape Cod.* Orleans, Mass.: Parnassus Imprints.

Waldron, Larry. 1983. *The Indiana Dunes.* New York: Eastern Acorn.

*I*ndex

New Canaan Nature Center, Connecticut, 38–41, *40*
New England Aster, 354
New England Wild Flower Society, 19, 181
New Hampshire, history of, 253, 285, 293–94, 295, 301, 302, 362, 368, 395, 429, 482–84
New Haven, Connecticut, 49, 53, 57, 65, 95
New Jersey Tea, 417
New London, Connecticut, 73
New York Aster, 520
New York Fern, 123
Newark, Vermont, 426
Newburyport, Massachusetts, 197
Newfane, Vermont, 278
Newport, Rhode Island, 131–33
Niantic, Connecticut, 70
Ninebark, 84
Ninigret National Wildlife Refuge, Rhode Island, 83–84, *85*
Noank, Connecticut, 79
Nonvascular Plants, 104, 189, 203, 314, 412. *See also* Bryophytes
Norfolk Island Pine, 91
North Adams, Massachusetts, 269
North Common Meadow Reservation, Massachusetts, 224
North Conway, New Hampshire, 445
North Hero Island, Vermont, 335–37
North Hero State Park, Vermont, 335–37, *336*
North River, Massachusetts, 163
North Stratford, New Hampshire, 427
North Sugarloaf Mountain, Massachusetts, 232
North Walpole, New Hampshire, 292
Northampton, Massachusetts, 173, 245, 249–50, 484
Northeast Kingdom of Vermont, 93, 347–49, 421, 423–27
Northern Hardwoods Regime, 22–23, 37, 116, 117, 125, 161, 260, 265, 269, 281, 313, 316, 346, 354, 363, 382, 383, 411, 413, 419, 420, 427, 430, 433, 439, 440, 450, 466, 470, 497, 499, 505
Northern White Cedar. *See* Arbor Vitae
Northfield, Massachusetts, 238
Northfield Mountain, Massachusetts, 238–39; Recreation and Environmental Center, 238
Northford, Connecticut, 53, 173
Norumbega Fault Zone, Maine, 482, 525
Norwalk, Connecticut, 100
Norway, Maine, 233
Norway Maple, 60, 81, 99, 141, 196, 230, 276, 324, 339, 458, 484
Norway Pine. *See* Red Pine
Norway Spruce, *16,* 81, 103, 163, 195, 196, 262–63, 269, 312, 318, 322, *495*
Norwell, Massachusetts, 161
Notchview Reservation, Massachusetts, *ii,* 263–65

Nulhegan River, Vermont, 427
Nut Sedge, 336–37

Oaks, 13, 52, 59, 63, 81, 83, 96, 114, 116, 124, 139, 154, 187, 212, 292, 316, 458, 482, 487
  See also individual species' names
Oats, 322
Ocean Spray (cranberry cooperative), 144
Oconee Bells, 181–82
Odiorne Point State Park, New Hampshire, 392–94, *393*
Ogunquit, Maine, 482
Old Man of the Mountains, New Hampshire, 419–20
Old South Church, 305
Old Sturbridge Village, Massachusetts, 97, 177–80, *178, 179*
Old Town Hill Reservation, Massachusetts, 195–97, *196*
Olive Family, 392
Olmsted, Frederick Law, 164, 165, 169, 250
Onions (cultivated), 95
Opuntia, 53, 183. *See also* Prickly Pear
Orache, 26, 67, 163, 165, 517, 531
Orange, New Hampshire, 411
Orange-flowered Jewelweed, 260–61, 347
Orchard Bog, Maine, 505
Orchard Grass, 262, 421
Orchards, 27, 95, 179, 224, 289, 318
Orchid Family, 206, 243, 251, 524
Orchids, 26, 202, 251. *See also* individual species' names.
Ordovician Period, 8, 88–89, 433, 499, 525
Orford, New Hampshire, 437–38, *438*
Organ Pipe Cactus National Monument, Arizona, 269
Oriental Bittersweet, 84, 189
Oriental Plane Tree, 132
Orleans, Vermont, 347
Orono, Maine, 492
Orthodox Congregational Church, Petersam, Massachusetts, 224
Osage Orange, 173–74
Osier, 287
Osmunda Family, 101, 204
Ossipee Mountains, New Hampshire, 458, 459–62
Ossipee Pine Barrens, 25, 463
Ossipee River, New Hampshire and Maine, 465
Ostrich Fern, 270, 279, 365–66, 475
Ottauquechee River, Vermont, 306, 307, 308
Ox-eye Daisy, 281, 352, 354, *425,* 426, 439
Oyster Plant, 282
Ozone Layer, development of, 8

Pachaug State Forest, Connecticut, 125–26, *126*
Painted Trillium, *30, 363*

About the Author

Raymond Wiggers' passion for botany led him to thoroughly explore New England's plant life. After receiving a bachelor's degree in geology from Purdue University and subsequent graduate training in botany and horticulture, he served as editor and press officer for the New York City Parks Department. He later owned and operated a horticulture and plant care business in New Hampshire and worked as a botanist and naturalist for the National Park Service. His articles on horticulture, botany, and forestry have been published in various New England newspapers. Raymond Wiggers lives in Springfield, Illinois, where he is museum editor and curator of publications for the Illinois State Museum.

Also by Raymond Wiggers:

PICTURE GUIDE TO TREE LEAVES (FRANKLIN WATTS, 1991)
THE AMATEUR GEOLOGIST (FRANKLIN WATTS, 1993)